Gram-Negative Bacterial Infections and Mode of Endotoxin Actions

Pathophysiological, Immunological, and Clinical Aspects

Edited by
Bernhard Urbaschek, Renate Urbaschek,
and Erwin Neter

Springer-Verlag
Wien New York

Gram-Negative Bacterial Infections
and
Mode of Endotoxin Actions
Pathophysiological, Immunological,
and Clinical Aspects

Immuno-Symposium, Vienna, September 1973
Sponsored by Immuno AG, Vienna

With 155 partly colored figures

Prof. Dr. Bernhard Urbaschek
Dr. Renate Urbaschek
Abteilung für Immunologie und Serologie
am Institut für Hygiene und Medizinische Mikrobiologie
der Fakultät für Klinische Medizin Mannheim
der Universität Heidelberg
Mannheim, Federal Republic of Germany

Prof. Dr. h. c. Erwin Neter, M. D.
Departments of Microbiology and Pediatrics
State University of New York at Buffalo
Buffalo, N.Y., U.S.A.

Library of Congress Cataloging in Publication Data

Urbaschek, B   1922-
   Gram-negative bacterial infections,and mode of
endotoxin actions.
   Includes index.
   1. Bacterial diseases. 2. Gram negative bacteria.
3. Endotoxin. I. Urbaschek, R., 1936-   joint author
II. Neter, Erwin, 1909-   joint author. III. Title.
   DNLM: 1. Bacterial infections---Immunology. 2. En-
dotoxins---Pharmacodynamics. 3. Gram negative bacteria---
Pathogenicity. WG200 U721
RC115.U7   616.9'2   74-34099

ISBN-13:978-3-7091-8398-4   e-ISBN-13:978-3-7091-8396-0
DOI: 10.1007/978-3-7091-8396-0

Preface

The relevance of international symposia and congresses has been questioned with increasing frequency. All too often, presentations at such meetings are meaningful only for very few members of the audience; fundamental scientific principles and relationships between disciplines are not always taken into account.

Under present conditions rapid sharing of new information is highly desirable. However, many journals do not lend themselves to speedy exchange of information, nor do many provide the opportunity for the publication of discussion by various specialists of problems arising from different disciplines. Symposia may lend themselves, perhaps more readily, to the identification of current knowledge and the exchange of thoughts between scientists from various fields. It was particularly opportune to have the present symposium held in Vienna, where as early as 1365 the University was founded and where the medical sciences have played a dominant role in the investigation of the nature of diseases. It was the aim of the organizer to provide in this city the opportunity for the exchange of information and of thoughts between scientists from different disciplines and different countries.

One of the many current, important problems relates to the fundamental basis of the pathogenesis of gram-negative infections. Information on immune mechanisms against these infections is inadequate; the relative role of non-specific resistance and specific immunity has not been fully elucidated. Problems regarding active immunization and the effectiveness of passively administered immunoglobulins still exist even today. The part played by endotoxins in the pathogenesis of these infections probably was overestimated in the past, and yet bacterial break-down products and endotoxins play a significant role. With these considerations in mind, the organizer of this symposium included sessions on endotoxins and on the pathophysiology of gram-negative infections and invited representatives of several disciplines, such as pathology, pharmacology, medical microbiology, and biochemistry, in addition to clinicians. Also, it seemed to him to be particularly important to invite additional representatives of these disciplines to participate in the discussions and to acquaint young scientists and physicians with these proceedings.

To make this volume accessible to a larger audience, all contributions are being published in English. This policy, obviously, necessitated the translation of several contributions and numerous discussion remarks. For reasons beyond the control of the authors, the chairmen of the sessions, the editors, and the publishers, the publication of the volume was delayed. The index of this book was prepared to include all major topics without overloading it unduly.

Mannheim-Heidelberg, December 1974

Bernhard Urbaschek

Acknowledgements

The program was developed in a relatively short period of time. This was made possible through the cooperation of the authors and the chairmen of the sessions. The organization was aided particularly by Drs. Bloch, Braude, Cottier, Formal, Landy, Lowbury, Lundsgaard-Hansen, Marget, Mergenhagen, Neter, Shubin, Smith, Springer, and Wundt. With the untimely passing of Professor Bloch I lost a close friend and effective counselor, I miss him very much.

Sincere appreciation is expressed to the chairmen of the sessions, Drs. Neter, Braude, Landy, Springer, Lundsgaard-Hansen, Giertz, Lüscher, Shubin, and Lowbury for their excellent cooperation and help before and during the Symposium and especially for their support in the publication of this book. For the translation into English of several manuscripts and the time-consuming and painstaking redaction of the contributions I am particularly indebted to my wife, Dr. Renate Urbaschek. For his important and valued help and consultations I especially thank Dr. Erwin Neter. I express sincere appreciation also to my secretary, Mrs. Udeberga Schücker, who contributed so much through her dedicated efforts. Particular thanks are due also to the publishers, Springer-Verlag, for numerous suggestions and for help in the preparation of this volume.

The days of the Symposium were filled with presentations and discussions from early morning until evening. The Symposium was made possible only because of the generous support given by Immuno AG., Vienna, who contributed so much to the success of the scientific sessions and to the enjoyment of the days in Vienna.

Bernhard Urbaschek

Contents

Chapter III.
Nonspecific Resistance and Endotoxin Tolerance

Chapter IV. Pathophysiology

IX

Chapter VI. Pharmacology

Chapter VII. Blood Coagulation

Chapter VIII. Clinical Aspects Including Shock

Chapter IX. Burn Disease

Contributors

ALEXANDER, J. W.
Transplantation Division
Department of Surgery
University of Cincinnati
Cincinnati, OH 45229
U.S.A.

BARTL, R.
Abteilung für Knochenmarks-
diagnostik der Universität München
D-8000 München 2
Federal Republic of Germany

BELLER, F. K.
Universitäts-Frauenklinik
D-4400 Münster
Federal Republic of Germany

BERRY, L. J.
Department of Microbiology
University of Texas
Austin, TX 78712
U.S.A.

BITTER-SUERMANN, D.
Institut für Medizinische Mikro-
biologie der Universität Mainz
D-6500 Mainz
Federal Republic of Germany

BLEYL, U.
Pathologisches Institut
der Universität Heidelberg
D-6900 Heidelberg
Federal Republic of Germany

BONA, C.
Institut Pasteur
Immunothérapie Expérimentale
F-75015 Paris
France

BRAUDE, A. I.
Department of Medicine
University Hospital
San Diego, CA 92103
U.S.A.

BROBMANN, G. F.
Chirurgische Universitätsklinik
D-7800 Freiburg
Federal Republic of Germany

CHEDID, L.
Institut Pasteur
Immunothérapie Experiméntale
F-75015 Paris
France

CLOWES, G. H. A., JR.
Department of Surgery
Harvard Medical School
Boston City Hospital
Boston, MA 02118
U.S.A.

COTTIER, H.
Pathologisches Institut
der Universität Bern
CH-3000 Bern
Switzerland

EIBL, H.
Immuno AG
A-1220 Wien
Austria

FEY, H.
Veterinär-Bakteriologisches
und Parasitologisches Institut
der Universität Bern
CH-3000 Bern
Switzerland

FORMAL, S. B.
Department of Applied Immunology
Walter Reed Army Institute of
Research
Walter Reed Army Medical Center
Washington, DC 20012
U.S.A.

FRITSCHE, C. D.
Institut für Medizinische
Mikrobiologie und Serologie der
Städtischen Krankenanstalten
D-6700 Ludwigshafen
Federal Republic of Germany

GANS, H.
Department of Surgery
Cornell-Medical Center
New York, NY 10021
U.S.A.

GIERTZ, H.
Medizinisch-Biologische Forschungs-
laboratorien der BASF AG
D-6700 Ludwigshafen
Federal Republic of Germany

GLASER, E.
Medizinische Kliniken und Poli-
kliniken der Universität Gießen
D-6300 Gießen
Federal Republic of Germany

GRAEFF, H.
I. Frauenklinik der
Universität München
D-8000 München 2
Federal Republic of Germany

GRAEVENITZ, A. VON
Department of Laboratory Medicine
Yale University
Yale-New Haven Hospital
New Haven, CT 06504
U.S.A.

GREISMAN, S. E.
University of Maryland
Department of Medicine
Baltimore, MD 21201
U.S.A.

GRÖSCHEL, D.
Section of Microbiology
University of Texas
Houston, TX 77025
U.S.A.

HADDING, U.
Institut für Medizinische Mikro-
biologie der Universität Mainz
D-6500 Mainz
Federal Republic of Germany

HEENE, D. L.
Zentrum für Innere Medizin
am Klinikum
der Universität Gießen
D-6300 Gießen
Federal Republic of Germany

HENNEMANN, H.-H.
III. Medizinische Klinik Mannheim
der Universität Heidelberg
D-6800 Mannheim 31
Federal Republic of Germany

HESS, M. W.
Pathologisches Institut
der Universität Bern
CH-3000 Bern
Switzerland

HEY, D.
Innere Abteilung II
Städtisches Krankenhaus
D-7120 Bietigheim
Federal Republic of Germany

HUGO, R. VON
I. Universitäts-Frauenklinik
D-8000 München 2
Federal Republic of Germany

HUSER, H.-J.
Pathologisches Institut
der Universität Bern
CH-3000 Bern
Switzerland

HUTH, E. F.
Kinderklinik Mannheim
der Universität Heidelberg
D-6800 Mannheim
Federal Republic of Germany

HUTH, K.
Innere Abteilung des
Diakonissen-Krankenhauses
D-6000 Frankfurt 1
Federal Republic of Germany

IRWIN, J. W.
Microcirculatory Laboratory
Massachusetts
General Hospital
Boston, MA 02114
U.S.A.

JACHERTZ, D.
Institut für Hygiene und
Medizinische Mikrobiologie
der Universität Bern
CH-3000 Bern
Switzerland

JOHNS, M. A.
Division of Infectious Diseases
Boston University Medical Center
Boston, MA 02118
U.S.A.

KOSLOWSKI, L.
Chirurgische Klinik und Poliklinik
der Universität Tübingen
D-7400 Tübingen
Federal Republic of Germany

KUHN, M.
Abteilung für Experimentelle
Chirurgie
der Universität Bern
CH-3000 Bern
Switzerland

LANDY, M.
Schweizerisches Forschungsinstitut
Medizinische Abteilung
CH-7270 Davos-Platz
Switzerland

LASCH, H. G.
Medizinische Kliniken und Poli-
kliniken der Universität Gießen
D-6300 Gießen
Federal Republic of Germany

LOWBURY, E. J. L.
Medical Research Council
Industrial Injuries and Burns Unit
Birmingham Accident Hospital
Birmingham B 15 1 NA
England

LÜSCHER, E. F.
Theodor-Kocher-Institut
der Universität Bern
CH-3000 Bern 9
Switzerland

LUNDSGAARD-HANSEN, P.
Abteilung für Experimentelle
Chirurgie
der Universität Bern
CH-3000 Bern
Switzerland

MARGET, W.
Kinderklinik der
Universität München
D-8000 München 2
Federal Republic of Germany

McCABE, W. R.
Division of Infectious Diseases
Boston University Medical Center
Boston, MA 02118
U.S.A.

McCALLUM, R. E.
Department of Microbiology and
Immunology
University of Oklahoma
Oklahoma City, OK 73190
U.S.A.

MELA, L.
University of Pennsylvania
Harrison Department of Surgical
Research
Philadelphia, PA 19104
U.S.A.

MERGENHAGEN, S. E.
Laboratory of Microbiology and
Immunology
National Institutes of Health
Bethesda, MD 20014
U.S.A.

MÜLLER-BERGHAUS, G.
Zentrum für Innere Medizin am
Klinikum der Universität Gießen
D-6300 Gießen
Federal Republic of Germany

NEKARDA, K.
Zentrum für Innere Medizin am
Klinikum der Universität Gießen
D-6300 Gießen
Federal Republic of Germany

NETER, E.
Departments of Microbiology
and Pediatrics
State University of New York
at Buffalo
Children's Hospital
Buffalo, NY 14222
U.S.A.

NEUHOF, H.
Zentrum für Innere Medizin am
Klinikum der Universität Gießen
D-6300 Gießen
Federal Republic of Germany

NOWOTNY, A.
Department of Microbiology
and Immunology
Temple University
Philadelphia, PA 19140
U.S.A.

OEHLER, G.
Zentrum für Klinische Chemie,
Immunologie und Humangenetik
der Universität Gießen
D-6300 Gießen
Federal Republic of Germany

PUSZTAI-MARKOS, Zs.
Abteilung für Medizinische Mikro-
biologie der Rheinisch-Westfälischen
Technischen Hochschule Aachen
D-5100 Aachen
Federal Republic of Germany

RAYNAUD, M.
Institut Pasteur
F-92380 Garches
France

RENOUX, G.
Faculté de Médicine
Université de Tours
Laboratoire d'Immunologie
F-37 Tours
France

SANDBERG, A. L.
Humoral Immunity
Laboratory of Microbiology
and Immunology
National Institutes of Health
Bethesda, MD 20014
U.S.A.

SCHAAL, K. P.
Hygiene-Institut
der Universität Köln
D-5000 Köln 41
Federal Republic of Germany

SCHAUER, A.
Pathologisches Institut
der Universität Göttingen
D-3400 Göttingen
Federal Republic of Germany

SCHMAHL, F. W.
Zentrum für Innere Medizin
der Universität Gießen
D-6300 Gießen
Federal Republic of Germany

SCHOENENBERGER, G. A.
Forschungsabteilung
Department für Chirurgie
Universitätskliniken
CH-4004 Basel
Switzerland

SCHUMER, W.
Department of Surgery
University of Illinois
West Side Administration
Chicago, IL 60612
U.S.A.

SHUBIN, H.
University of Southern California
Center for the Critically Ill and
the Shock Research Unit
Los Angeles, CA 90027
U.S.A.

SMITH, H.
Department of Microbiology
University of Birmingham
Birmingham B 15 2 TT
England

SPAULDING, E. H.
Department of Microbiology
and Immunology
Temple University
Philadelphia, PA 19140
U.S.A.

SPRINGER, G. F.
Department of Immunochemistry
Research
Evanston Hospital
Northwestern University
Evanston, IL 60201
U.S.A.

URBASCHEK, B.
Abteilung für Immunologie und
Serologie
Institut für Hygiene und Medi-
zinische Mikrobiologie Mannheim
der Universität Heidelberg
D-6800 Mannheim
Federal Republic of Germany

URBASCHEK, R.
Abteilung für Immunologie und
Serologie
Institut für Hygiene und Medi-
zinische Mikrobiologie Mannheim
der Universität Heidelberg
D-6800 Mannheim
Federal Republic of Germany

WIEDEMANN, B.
Hygiene-Institut der
Johann-Wolfgang-Goethe-
Universität
D-6000 Frankfurt 70
Federal Republic of Germany

WISSLER, J. H.
Schweizerisches Forschungsinstitut
Medizinische Abteilung
CH-7270 Davos-Platz
Switzerland

WUNDT, W.
Institut für Hygiene und Medi-
zinische Mikrobiologie Mannheim
der Universität Heidelberg
D-6800 Mannheim
Federal Republic of Germany

ZELLNER, P. R.
Abteilung für Verbrennungen,
plastische und Handchirurgie
der Berufsgenossenschaftlichen
Unfallklinik
D-6700 Ludwigshafen-Oggersheim
Federal Republic of Germany

Chapter I

Microbiology and Epidemiology

The Nature of Specific and Nonspecific Gram-Negative Infections Including Enteropathogenic E. coli Enteritis

E. Neter

It is a great privilege for me to open the first session of this symposium, and I would like to take this opportunity to express our appreciation to Professor Urbaschek and Professor Lasch for their leadership in its organization. This session is devoted to microbiologic and epidemiologic aspects of gram-negative infections.

Infections due to gram-negative bacteria have assumed ever-increasing importance during the past several years. This is due, to a large extent, to the facts that many members of these groups cause disease in subjects with low resistance and such opportunistic infections occur with increased frequency because of the very triumphs of medicine. Thus, prematures, patients with cystic fibrosis, leukemia or other malignancies, or with certain immune deficiency diseases are kept alive for longer periods of time than heretofore. Corticoids and immunosuppressant agents, which may interfere with host resistance, are used more widely. Microorganisms of low virulence, which fail to cause disease in otherwise healthy subjects, may be responsible for serious and even fatal infection in these highly susceptible subjects. In fact, a volume could be written on the pathogenicity of nonpathogenic microorganisms! Many of these infections develop in hospitalized subjects, and the topic of nosocomial infections will be discussed on the basis of extensive personal experience by Dr. von Graevenitz.

The second reason why gram-negative infections have become evermore important in recent times is the recognition of the frequency and long-range impact of certain infections, such as those of the urinary tract. It is estimated that asymptomatic bacteriuria may be present in 1% of seemingly healthy females, and this percentage increases to 5, 6, or 7% in pregnant women. In addition to asymptomatic bacteriuria, overt disease is no longer considered to be an unimportant event. Needless to say, early diagnosis and appropriate management becomes increasingly important, as will be discussed by Dr.Marget, who has made significant contributions to this field.

The third reason why infections by gram-negative bacteria are important today is due to the fact that salmonellosis, shigellosis, and enteropathogenic E. coli enteritis are still around, even in economically advanced countries although typhoid fever has been largely eliminated there. In a recent report from the National Research Council of the U. S. A. it was estimated that some two million Salmonella infections occur in this country per year. This figure is derived from the fact that some 20,000 isolates are reported to the Center for Disease Control (CDC) and that perhaps only 1% of bacteriologically proved infections come to the attention of public health authorities. Similarly, the recent experience in Central America has clearly shown that even Shigella dysenteriae or Shiga infection can assume epidemic proportions at the present time.

On the basis of the above considerations it is evident that a few members of the family Enterobacteriaceae usually cause enteric disease, and it is these microorganisms which can be characterized by enteropathogenicity. Many other members of the group also have pathogenic potential, but the disease they produce is usually extraintestinal in nature. The former diseases, then, may be considered specific and the others nonspecific infections.

In addition to the unresolved problems of pathogenicity and virulence unanswered questions remain regarding epidemiologic features of various infections. There can be no doubt that enteropathogenic E. coli (EEC) or Dyspepsiecoli has become less virulent than it was two to three decades ago. More and more patients are now seen in Out-Patient Departments rather than having to be admitted as in-patients for intensive care. Furthermore, throughout the years E. coli O26 has been less contagious than E. coli O111 or O127 for reasons that are to be elucidated in the future. Another poorly understood aspect of enteric infections relates to the infecting dose of microorganisms and the incisive conditions allowing adequate multiplication within the intestinal tract.

To me, at least, it has been a perplexing problem why Shigella sonnei should be significantly more common in certain European countries and the U.S.A., whereas Shigella flexneri infections are more prevalent in Japan. If either of these infections were essentially absent, these epidemiologic features could be understood. This, however, is not the case.

More comprehensive information on the epidemiology of Salmonella gastroenteritis, shigellosis, and enteropathogenic E. coli enteritis is obtained by combining bacteriologic and immunologic approaches to the diagnosis of these infections. The documentation of the immune response is particularly useful when bacteriologic examination yields negative results, either because of technical reasons or because the pathogen may no longer be present. Numerous studies were carried out in our laboratory on siblings and parents of children with clinical shigellosis or salmonellosis. Evidence of infection was obtained by the documentation of a specific immune response on family members without clinical illness and even in the absence of the pathogen in the feces. Illustrative examples are shown in Tables I and II.

Much important new information on the pathogenesis of enteric infection has emerged during the last few years, a topic which will be discussed by Dr. Formal, who has contributed significantly to this field.

Of particular interest, too, is the unanswered question why antibiotic treatment usually prolongs the excretion by patients with Salmonella gastroenteritis, in spite of in vitro susceptibility to the drugs, whereas similar treatment of patients with enteropathogenic E. coli infection or shigellosis is effective, both clinically and bacteriologically.

Factors of virulence and host resistance will be discussed by Dr. Smith. I would like to restrict myself, therefore, to a very few comments. In my judgment, neither pathogenicity nor virulence can be defined solely in terms of the microbe but must always be related to host factors. An intriguing question relates to the problem why, with few exceptions, salmonellae are pathogenic for numerous animal species under natural conditions in contrast to shigellae which have a far more restricted host range. To the best of my knowledge it is also unknown why a few salmonellae, such as S. typhi, affect only a few species and often invade the blood stream, while most other salmonellae usually remain localized within the gastrointestinal tract. In all likelihood, it is host factors rather than characteristics of the microbe that relate to invasiveness. In our own studies on patients with malignant diseases bacteremia due to S. typhimurium and S. derby occurred with unusual frequency.

Numerous other members of the family Enterobacteriaceae are attracting attention because they contribute to clinical illness to a significant extent. Such infections may well be described as nonspecific to indicate that these gram-negative bacteria, including members of the family Enterobacteriaceae, Pseudomonas and others, are not responsible for enteritis or colitis or a clinical entity such as typhoid fever. Rather, they can cause infection in various parts of the body, including infection of the urinary and respiratory systems, sepsis, wound infections, and meningitis.

During the past few years considerable improvement has been made in the identification of certain members of the family Enterobacteriaceae, as will be discussed by Dr. Fritsche. Many diagnostic laboratories now use important biochemical tests for characterization rather than inspecting a MacConkey agar plate alone. In addition, changes in nomenclature have been made. For these reasons, it has become increasingly difficult to compare the occurrence of certain infections at various times, and many facilities storing records do not have available information as to the precise methodologies and nomenclatures used during different years. Particularly in this computer age such information should be transmitted and stored for more accurate retrieval of information.

Usually the diagnosis of opportunistic infection is based on bacteriologic examination, including cultures of sputum, urine, wound, etc. Although this approach to diagnosis is indispensable, the following shortcomings must be kept in mind:

1. For technological and other reasons the pathogen may not be recovered from a given specimen.

2. The presence of a suspected pathogen does not necessarily indicate its pathogenic role, because it may be either a contaminant or the patient may be a carrier of this microorganism and the disease is due to another one.

For these reasons, we have studied in considerable detail the immune response to suspected pathogens to gain a better understanding of their etiologic role. I would like to illustrate the usefulness of this approach with the following examples.

Antigenic analysis of various pathogens has clearly shown relationships between certain serotypes or serogroups and various infections. It is now generally recognized that certain serotypes of E.coli are responsible for enteritis of infants. Certain other serotypes are found with remarkable frequency in urinary tract infection, raising the question of the relationship between antigenic structure and uropathogenic potential. Extensive studies in our laboratories have revealed, however, that these serotypes are encountered with almost identical frequency also in infections other than those of the urinary tract, namely in peritonitis, meningitis, septicemia, and even in patients with cystic fibrosis (Table III).

The question that hopefully will be taken up during the discussion period relates to the problem of "persistence". There can be little question that certain enteric pathogens may remain viable within the inner milieu for many months or even years in the absence of clinical disease. For example, typhoid bacilli have been known to survive under these conditions even for a few decades. With the increased knowledge on L forms, cell wall-free bacteria, an attractive hypothesis for the survival has become available. Production of the cell wall results in the emergence of vegetative forms and may lead to a relapse of the clinical disease. I hope that Dr. Braude, who has particular expertise in this area, will comment on this problem. One word of caution may be offered at this time to clinicians who, as Dr. Marget can testify, frequently see relapses of urinary tract infection to E. coli. The fact is that, more often than not, these relapses are truly new infections caused by different serotypes of this species. Serogrouping of the organism and the antibody response of such patients clearly lead to this conclusion. Illustrative examples are shown in Tables IV and V. It is evident that in one of these patients a specific antibody response occurred to four isolates at different times and in the other subject to six different isolates of E. coli.

Chemotherapy and chemoprophylaxis of gram-negative infections have posed increasingly serious problems, in part because of the emergence of drug-resistant strains. As will be discussed by Dr. Schaal, strains resistant to several antibiotics have been encountered with increasing frequency. Of great theoretical interest is the analysis of this phenomenon, indicating that transfer of DNA from one strain to another and even from one species to another may code for antibiotic resistance. These R factors are examples of similar genetic materials which may have profound effects on virulence and other characteristics of gram-negative bacteria as will be discussed by Dr. Wiedemann. One conclusion is evident. Antibiotics, like other drugs, should be used when indicated, and physicians must remember that no microorganism has yet become resistant to aseptic technique.

I am certain that the presentations to follow will be highly informative and will stimulate discussion in depth.

6

References

Neter, E., Drislane, A.M., Harris, A.H., Jansen, G.T.: New Engl. J. Med. 261: 1162 (1959).
Neter, E.: Amer. J. Public Health 52: 61 (1962).
Neter, E.: Ped. Clinics of North America 11: 517 (1964).
Neter, E., Steinhart, J., Calcagno, P.L., Rubin, M.I. In: Kass, E.: Progress in Pyelonephritis, p. 129 (F.A. Davis Co., Philadelphia 1965).
Elsea, W.R., Partridge, R.A., Neter, E.: Public Health Reports 82: 347 (1967).
Han, T., Sokal, J.E., Neter, E.: New Engl. J. Med. 276: 1045 (1967).
Diaz, F., Neter, E.: Amer. J. med. Sci. 259: 340 (1970).
Diaz, F., Mosovich, L.L., Neter, E.: J. infect. Dis. 121: 269 (1970).
Neter, E., Oberkircher, O.R., Rubin, M.I., Steinhart, J.M., Krzeska, I.: Pediat. Res. 4: 500 (1970).
Neter, E.: Yale J. Biol. Med. 44: 241 (1971).
Neter, E.: Antibody response of patients with infection of the urinary tract: Aid to understanding of pathogenesis, and to diagnosis and prognosis. In: Losse, H., Kienitz, M.: Pyelonephritis, p. 101 (Georg Thieme Verlag, Stuttgart 1972).
Neter, E.: Opportunistic pathogens: Immunological aspects. In: Prier, J.E., Friedman, H.: Opportunistic Pathogens (University Park Press, Baltimore 1973).
National Academy of Sciences, Committee on Salmonella: An evaluation of the Salmonella problem. Nat. Acad. Sci. (Washington, D.C. 1969).

Table I.

Immunologic evidence of subclinical shigellosis

Patients	Diarrhea	Shigella in feces	Shigella antibody titers
Index case	+	+	640
Sibling 1	-	-	160
Sibling 2	-	-	640
Sibling 3	-	-	320

Table II.

Immunologic evidence of subclinical salmonellosis

Patients	Diarrhea	Salmonella in feces	Salmonella antibody titers
Index case	+	+	2560
Sibling 1	-	+	5120
Sibling 2	-	+	2560
Sibling 3	-	-	1280

Table III.

Serogroups of E. coli isolated from children with urinary and other infections

Number of strains

Serogroups	Urinary isolates		All others	
O1	133 ⎤		8 ⎤	
O2	12		1	
O4	170		60	
O6	320		28	
O7	115	> 982 = 61%	1	> 130 = 51%
O11	52		-	
O15	13		3	
O62	7		8	
O75	160 ⎦		21 ⎦	
All others	623		126	
Total	1,605		256	

Table IV.

Differentiation between chronic and repeated E. coli infection of the urinary tract

Dates of blood specimens	O60	E. coli O91	O38	NT
		Antibody titers		
March	80	80	40	< 20
May	1280	40	40	< 20
December	320	1280	160	< 20
July	160	160	1280	< 20
September	80	160	640	640

Table V.

Differentiation between chronic and repeated E. coli infection of the urinary tract

Dates of blood specimens	O60	O7	E. coli O91	O38	NT	O1
			Antibody titers			
March	80	< 20	80	40	< 20	< 20
May	1280	20	40	40	< 20	< 20
December		80	1280	160	< 20	< 20
July				1280	< 20	< 20
September					640	< 20
October						160

Gram-Negative Bacteria: The Determinants of Pathogenicity

H. Smith

In contrast to the success of public health measures in controlling the worst effects of infectious disease, the mechanisms of most microbial infections remain obscure (48). Nevertheless with increasing realisation that many troublesome aspects of infectious disease have not been removed by improved public health or by the antibiotic era, interest in the determinants of microbial pathogenicity is re-awakening with a view to designing fresh control measures. Much of the renewed activity has involved the gram-negative organisms and I have been asked to survey current knowledge of their pathogenicity. In the time available I can only summarise the main points under headings describing the cardinal requirements for pathogenicity: 1. ability to enter the host by surviving on and penetrating mucous surfaces; 2. ability to multiply in the nutritional conditions of the host tissues; 3. ability to inhibit host defence mechanisms; and 4. ability to damage the host. The general methods for investigating the determinants of microbial pathogenicity and the difficulties encountered have been described (48) and will not be re-iterated here.

Entry: Survival on and Penetration of Surface Membranes

Although some gram-negative pathogens can gain direct access to tissues such as Pseudomonas aeruginosa in burns most infections occur on or over the mucosal surfaces of the respiratory, alimentary and urogenital tracts. Defence of these mucous surfaces against microbial attack relies on: 1. competition with and interference by surface commensals; 2. presence of bactericidal or bacteriostatic materials in mucous secretions; and 3. mechanical flushing action of moving mucus or lumen contents (45). These defences are overcome in infectious disease but how is this accomplished? Electron and light microscopy (45, 57) indicate at least three types of early attack on mucosal surfaces. First bacteria can attach to the mucosa, multiply on it but not penetrate; this occurs in infections with some strains of Escherichia coli and possibly with Vibrio cholerae, although some authors (57) believe the latter multiplies in the lumen without attachment. Second there can be attachment and subsequent phagocytosis by the mucosal cells with consequent surface damage there as for Shigella and some E. coli infections. Finally there can be attachment and passage into the underlying tissues either through the mucosal cells or between them, as occurs in salmonellosis. Some light on the nature of commensal competition with pathogens has come from mixed culture experiments in vitro and in germ free animals, and some of the bacteriostatic materials have been recognised. For example the fatty acids from intestinal fusiforms are, in a reducing environment, inhibitory for salmonellae and shigellae. However we still do not know how these defences are overcome in the early stages of dysentery or typhoid fever (45). Similarly mechanisms whereby invading bacteria interfere with inhibitory materials in mucous secretions are largely unknown but presumable they are similar to those discussed below under inhibition of humoral defences. We are beginning to know a little of the nature of the materials which stick some gram-negative bacteria to mucous surfaces and even of those which might promote their phagocytosis by the surface cells. These microbial products render the invading organisms safe from the flushing action of mucus or lumen contents and help the pathogen to compete with commensals for room on the mucosa. Genetic transfer and other experiments (2, 3, 26, 53) with enteropathogenic strains of E. coli have indicated the importance of the K88 protein antigen in attaching this pathogen to the brush border in the upper small intestine. This view is supported by the fact that mucosal immunity to E. coli enteric infections may rely on antibodies to the K88 antigen (44, 51). Observations on mutants or hybrids of E. coli and Shigella flexneri indicated that the structure of the polysaccharide side chain of the somatic 0 antigen may be one factor controlling attachment and penetration of the organisms into mucosal cells (20).

Multiplication In Vivo

Virulent bacteria must multiply on mucous surfaces or within the host tissues to produce their disease syndrome. Two qualities are needed for multiplication. First, an inherent ability to multiply in the nutritional conditions of the host tissue and second, an ability to inactivate or not to stimulate host defence mechanisms which would otherwise kill or remove the bacteria. The effects of these two qualities in vivo are not easy to separate and often it is difficult to assess their relative importance. Ability to multiply is discussed here.

Avirulence can arise from inability to grow and divide in the environment in vivo. Thus, nutritionally deficient mutants of Salmonella typhi and Pasteurella pestis were avirulent unless injected with their required nutrients (6). However, for most bacteria, the tissues and body fluids probably contain sufficient nutrients to support some growth. Few naturally occurring strains will be avirulent due solely to inability to grow in the host. Nutritional considerations will, however, affect rate of growth in vivo. The more rapid it is, the greater the chance of establishing the infection against activity of the host defence mechanisms. We can measure bacterial multiplication rates in vitro but what about those in vivo? The number of viable bacteria in the tissues of an infected host can be counted at any time after inoculation. These numbers are however the resultants of multiplication and destruction or removal. True bacterial division rates in vivo have been measured by the following method. Pathogens, genetically labelled with biochemical markers retained by a known proportion of the progeny at each division were used. Division rates were determined by measuring the proportion of organisms with the marker at various times after inoculation into animals. Remarkable results were obtained. In the spleens of mice some time after inoculation Salmonella typhimurium divided at only 5-10 % of the maximum rate in vitro, and a similar slow rate of multiplication of E. coli was observed during the critical primary lodgement period of infection (41).

Although the general nature (amino acids, sugars, purines) of the nutrients that support microbial growth in the tissues is known the precise nutritional conditions which determine the growth rate or population size in any particular infection are not (47) except in a few instances involving gram-negative bacteria. Proteus mirabilis persists in and causes severe damage to the kidney because a powerful urease (4) enables the bacteria to use the urea at this site for rapid growth and for production of ammonia which damages the tissue. Brucellosis in many animals (e. g. humans, rats, guinea pigs and rabbits) is a relatively mild and chronic disease; the causative organisms do not grow prolifically and have no marked affinity for particular tissues. However, in pregnant cows, sheep, goats and sows there is an enormous growth of brucellae in the placentae, the foetal fluids and the chorions, leading to the characteristic climax of the disease - abortion. The presence of erythritol, a growth stimulant for brucellae, in only the susceptible tissues of the susceptible species explains this heavy localised growth (46). The rapid growth of Pseudomonas aeruginosa in burns may be determined by a similar localisation of a particular nutrient but as far as I am aware this has not been investigated.

Inhibition of Host Defence Mechanisms

To increase within the host tissues metabolic ability to multiply in the nutritional environment is not enough. Pathogenic bacteria must also produce compounds - aggressins - to inhibit host defence mechanisms which otherwise would destroy them.

Aggressins act in the decisive, primary lodgement period of infection (36), that is during the first few hours when the few invading bacteria are most vulnerable to the protective reaction of the host. At this early stage, aggressins must inhibit non-specific bactericidal mechanisms; not only those already existing in or on the tissues but also those agencies, especially phagocytic cells that are mobilised by inflammatory processes soon after the tissues are irritated. If some bacteria survive the primary lodgement and grow, spread of infection is opposed by the fixed phagocytes of the reticuloendothelial system (lymph nodes, spleen, liver); and again to make headway bacteria need aggressins, possibly different from those operating during the early lodgement phase. To break through the protection of previously immunised animals or of animals several days after infection bacteria must either be numerous or well endowed with aggressins, since the host defence mechanisms are of increased efficiency being supplemented by antibodies capable of direct neutralisation of microbial products and by cellular immune processes. The clinical outcome of the disease depends on the interplay of these defensive reactions of the bacteria and of the host, and varies from complete subjugation of the host to complete destruction of the bacteria. It also includes near stalemate in chronic infections and carrier states where the factors contributing to persistence of the stalemate are still obscure.

Various types of aggressins are described below.

Inhibitors of blood and tissue bactericidins

Body fluids (blood, saliva, mucus) contain a variety of bactericidal factors such as basic polypeptides, lysozyme, complement (acting with 'natural' antibody) and possibly a system involving the iron-binding protein transferrin. Resistance to these bactericidins has been associated with virulence in strains of many gram-negative species such as Escherichia, Vibrio, Salmonella, Proteus, Shigella and Brucella (5, 22). Clearly then, virulent strains of these species produce aggressins to counteract humoral defences. However only rarely have they been identified, for examples a cell-wall component containing protein, carbohydrate, formyl residues and 35 - 42 % lipid for Br. abortus (13) and the acidic polysaccharide, K antigens of E. coli (22).

Inhibitors of the action of phagocytes

Phagocytes vary in origin, morphology, constituents and bactericidal function (24, 42). Bacteria can interfere with their action by preventing one or more of the following processes, phagocytic mobilisation (inflammation), contact ingestion and intracellular killing (49). Several aggressins that interfere with the last two processes have been identified for gram-negative bacteria.

Inhibitors of ingestion

Once engulfed in phagosomes many gram-negative bacteria are usually destroyed and digested (11). Resistance to ingestion, thus avoiding intracellular destruction is the main aggressive mechanism of these bacteria. Examples of these aggressins are the O somatic antigens of some Enterobacteriacae (17, 33, 34, 38, 39, 59), the Vi antigen (poly-N-acetyl-D-galactosaminouronic acid) of S. typhi (8, 11), the acid polysaccharide K antigens of E. coli (22), the protein carbohydrate envelope substance of P. pestis (11) and a surface slime from Ps. aeruginosa (29).

Knowledge of the chemical structure of the O antigens of mutants of the Enterobacteriacae is now being used to investigate the relation between structure and antiphagocytic activity. Resistance of E. coli to phagocytosis by mouse polymorphonuclear leucocytes appears to depend on a complete saccharide component of the cell-wall lipopolysaccharide; a mutant lacking only colitose in its side chain was significantly more phagocytosis susceptible(and less virulent) than the wild type and a mutant lacking galactose, glucose, N-acetyl glucosamine and colitose was even more so (34). Similarly, work with mutants and with media inducing phenotypic change has indicated that not only a complete sugar sequence in the core but also a complete 0-specific polysaccharide side chain in the 0 antigens is necessary for full phagocytosis resistance and virulence of S. typhimurium for mice; the tetra-saccharide sequences abesquosyl-mannosyl-rhamnosyl-galactose have been suggested as the determinant group, acetyl and glucosyl groups being less important (17, 21, 38, 39, 55, 59).

While we are beginning to learn something of the connection between structure and aggressive activity, the mode of action of these aggressins is still not clear. They may interfere with ingestion by purely mechanical means, by inhibiting adsorption of serum opsonin and possibly by rendering the bacterial surface less foreign to the host.

Inhibition of digestion

Some bacteria resist the phagocytic bactericidins which destroy others and thus they can survive or grow intracellularly. Within the cells these bacteria are protected from natural and injected antibacterial agents and hence they produce diseases which are often chronic and beset with the complications of hypersensitivity. Amongst the gram-negative organisms the brucellae are typical 'intracellular' pathogens and ability to grow within phagocytic and other cells is probably the most important aspect of their pathogenicity. It is seen in infected animals and in cell maintenance culture in vitro (16). In addition, bacteria whose virulence is determined in part by resistance to phagocytic ingestion can, when ingested under certain circumstances, survive and grow intracellularly. Examples are salmonellae (1, 43), shigellae (37, 62) and plague bacilli (7, 25).

Although much has been learned of the many and different bactericidal mechanisms of polymorphonuclear and monocytic phagocytes (24, 42) practically nothing is known about what happens to the bacteria when they are killed by these mechanisms. And studies of how intracellular pathogens resist phagocytic bactericidins have only just begun.

Virulent strains of Br. abortus survived and grew in the mixed phagocyte population of bovine 'buffy coat' more than avirulent strains (59). This was neither due to a greater ability of the virulent strains to use the nutritional conditions within the phagocytes nor to their higher catalase content which may have afforded a greater protection against the bactericidal action of phagocytic hydrogen peroxide (46). The superior ability of virulent strains to survive intracellularly appears to be due to the production under the growth conditions occurring in vivo, and simulant ones in vitro, of a cell-wall substance which interferes with the bactericidal mechanisms of the phagocytes. Virulent brucellae obtained from infected bovine placental tissue (50), or from cultures in laboratory media supplemented by bovine placental extracts of foetal fluids (16), had an increased ability to survive intracellularly compared with the same strain grown in laboratory media. Cell-wall preparations of organisms from infected bovine placenta and from supplemented media inhibited intracellular destruction of an avirulent strain of Br. abortus (16, 50). Finally, the antigen which appeared responsible for the inhibition of the phagocytic bactericidins was removed from Br. abortus grown in supplemented medium by washing with an ether water mixture (19). This material, which prevented intracellular destruction of Br. abortus, was serologically different from the other cell-wall material described previously which interfered with the humoral bactericidins of bovine serum (19). The two aggressins may prove to be similar structurally but differ, as the 0 antigens of the Enterobacteriaceae, in the detailed structure of their side chains.

Inhibition of the immune responses

Surprisingly, more work has been done on immunosuppression by viruses than

by bacteria. Yet there is enough information to indicate that this might be an important aggressive effect of gram-negative bacteria. E. coli contain cytoplasmic factors which inhibited antibody response to sheep red blood cells (28). L-asparaginase
of E. coli prevented the appearance of plaque-forming cells making antibody to
sheep blood cells (18). Cell mediated immunity seemed to be depressed in burned
patients infected with Ps. aeruginosa (57). The bacterial compounds responsible for
these effects are unknown but endotoxin may be involved since in vivo it can decrease
antibody formation (15).

Damage to the Host

Gram-negative bacteria can cause harm by a number of mechanisms which vary
in importance with the bacterial species or even the strain involved. Liberation of
endotoxin, production of an exotoxin, tissue invasion and evocation of hypersensitivity appear to be the main host-damaging processes that can occur.

Liberation of endotoxin

When endotoxins are extracted from cell-walls of many different gram-negative
bacteria by fairly drastic means (treatment with trichloracetic acid or warm aqueous
phenol) and injected into animals they all produce similar toxic manifestations -
pyrexia, diarrhoea, prostration and death. In some infections, there is little doubt
that endotoxins are liberated from the cell-wall of the invading bacteria and are
responsible for pathological effects, such as pyrexia, leucopenia, shock and death
in typhoid fever (61), pyrexia and shock in brucellosis of man (54) and abortion in
brucellosis of domestic animals (58, 60). On the other hand, in many other gram-
negative infections it appears that endotoxin may not contribute significantly to the
pathology. First, endotoxins have the same biological properties no matter from
which bacterial species they are obtained, yet the pathological effects of gram-negative infections vary enormously. Second, avirulent strains of gram-negative species including the E. coli of the normal gut contain much endotoxin yet when growing enterically they do not liberate enough to harm the host. Third, an old experiment often forgotten, mice bred resistance to endotoxin succumbed to oral infection
with Salmonella typhimurium (23). Endotoxin participation in host damage is more
likely when there is large scale bacterial invasion of blood and tissues where perhaps bacterial lysis and liberation of endotoxin can occur more readily than on body
surfaces. This deeper invasion happens in typhoid fever and in brucellosis but not
in cholera and dysentery and two factors already mentioned can contribute to the invasion. First there are aggressins, such as the Vi and O antigens of salmonellae and
the cell-wall products from brucellae, which inhibit humoral and cellular bactericidins. Second nutritional requirements can determine bacterial localisation in certain tissues where endotoxin may be released. As mentioned previously the massive
growth of brucellae in placentae and other foetal tissues, which leads to abortion in
certain domestic animals, is determined by the presence in the susceptible tissues
of a growth stimulant for brucellae, erythritol.

Production of exotoxins

In the past decade there have been spectacular advances in our understanding of
the role of exotoxins in the acute diarrhoeal diseases of man and animals, the so-
called 'enterotoxic enteropathies'. This has been due to discarding mouse toxicity
tests for biological and animal tests in which bacteria and their products were put
into the gut lumen and their effects in this site observed. Investigations on cholera
formed the template for those on other diseases and the voluminous research has
been summarised recently (9). Only the main points are given here.

An enterotoxin from V. cholerae, responsible for the gross fatal fluid loss from
the intestine which occurs in cholera, was recognised in the first instance by using
two tests. First a ligated segment of small intestine in a living rabbit would fill

with fluid following intraluminal injection of V. cholerae and its products. Second V. cholerae and its products caused fluid accumulation and diarrhoea in suckling rabbits when introduced into the gut lumen by a gastric tube. The extracellular enterotoxin has been purified; it is a protein and is different from the cell-wall endotoxin. It acts by increasing the normal secretion of the small intestine, probably by activating adenyl cyclase present in the intestinal epithelial membrane, thereby raising intracellular CAMP levels which in turn would affect electrolyte transport.

Using similar 'gut reaction' tests, enterotoxins have also been demonstrated for the strains of E. coli that produce diarrhoea in domestic animals and human infants; the enterotoxins are different form the endotoxins, protein in nature and plasmid transmitted (9, 52). Furthermore they act like cholera toxin in activating adenyl cyclase (9, 14). There seems little doubt that these enterotoxins cause the diarrhoea in many E. coli infections.

Enterotoxin production has also been demonstrated for Vibrio parahaemolyticus (9) which causes food poisoning in Japan and for Shigella dysenteriae (27). In these cases the role of enterotoxin in the pathogenesis of the natural diseases is not as clear as for the two previous examples. This is particularly so for the dysentery bacillus in view of the experiments summarised below.

Ps. aeruginosa appears to produce several extracellular products which may play significant roles in the damage following burn infections. Liu (32) demonstrated in burn-like lesions of rabbit skin and in filtrates of vigorously shaken cultures in rabbit serum or broth, a toxin different from endotoxin. It killed mice in shock and lowered the blood pressure of rabbits. A recently investigated vascular permeability factor (30) is probably the same toxin. Necrotising and lethal elastase (35) and collagenase (10) preparations have also been described. They may be responsible for the vasculities which sometimes follow Ps. aeruginosa infection in burns or in leukemia patients. The collagenase preparation induced petechiae and haemorrhages in mice but its relation to Liu's toxin of the permeability factor is not clear. Thus, even if the endotoxin of Ps. aeruginosa is not set free in infection, the organism seems to be able to produce extracellularly a number of other harmful products whose activities fit well with the pathology of infection.

Tissue invasion

Investigation in animal models and in volunteers by Formal and his colleagues (12, 31) using different strains of E. coli and Shigella dysenteriae have shown that invasion of intestinal mucosal surfaces is the damaging process in some E. coli infections and in dysentery. Two strains of E. coli isolated from American soldiers in Vietnam produced enterotoxin but no mucosal damage in the rabbit ligated loop model; in volunteers they caused diarrhoea resembling cholera. Two other strains were not enterotoxigenic but they penetrated into Hela cells in vitro, into ileal and conjunctival cells of guinea pigs and into ileal cells of rabbits. These cell-penetrating strains produced a dysentery-like illness in volunteers with tenesmus, urgency, hyperpyrexia and hypotension. Obviously E. coli can cause disease by two mechanisms, one cholera-like and the other dysentery-like. In similar experiments ten organisms of two strains of Shigella dysenteriae which were both cell penetrating and enterotoxigenic produced typical dysentery in volunteers. A strain which was cell penetrating and not enterotoxigenic produced dysentery in volunteers but large inocula were needed. A strain which was enterotoxigenic but not cell-penetrating failed to produce dysentery even when taken in large numbers. Hence mucosal invasion is important in dysentery and the role of the enterotoxin has yet to be defined.

The mechanism of mucosal cell-penetration is not yet clear nor the microbial products responsible for damage but endotoxins are one possibility and proteolytic enzymes another.

Evocation of hypersensitivity mechanisms

The classical work on tuberculosis showed that severe and continuing damage can result from primary sensitisation of host tissues to non-toxic microbial products and evocation of hypersensitivity phenomena by subsequent or continuing microbial attack. In many bacterial diseases the host becomes hypersensitive to bacterial products. It is easy to speculate but hard to prove that such hypersensitivity plays an important role in a disease process (40, 46). The view that hypersensitivity to the endotoxin of E. coli caused scouring in piglets has been dispelled by recognition of E. coli enterotoxins (40). Nevertheless damage from hypersensitivity seems to occur in some gram-negative infections especially in chronic states such as that in brucellosis. Also Arthus type, antigen-antibody complex, reactions may contribute to basal membrane damage in renal infections with Proteus mirabilis and E. coli.

This survey of the pathogenic activities of the gram-negative bacteria shows that these activities are as diverse and as interesting as their serology but that they are less well understood.

References

1. Baskerville, A., Dow, C., Curran, W.L., Hanna, J.: Brit. J. exp. Path. 53: 641 (1972).
2. Bertschinger, H.U., Moon, H.W., Whipp, S.C.: Infect. Immun.5: 595 (1972).
3. Bertschinger, H.U., Moon, H.W., Whipp, S.C.: Infect. Immun.5: 606 (1972).
4. Braude, A.I., Siemienski, J.: J. Bacteriol. 80: 171 (1960).
5. Braun, W., Siva Sankar, D.V.: Ann. N.Y. Acad.Sci. 88: 1021 (1960).
6. Burrows, T.W.: Ann. N.Y. Acad.Sci. 88: 1125 (1960).
7. Cavanaugh, D.C., Randall, R.: J. Immunol. 83: 348 (1959).
8. Clarke, W.R., McLaughlin, J., Webster, M.E.: J. biol. Chem. 230: 81 (1958).
9. Craig, J.P.: Symp.Soc.gen.Microbiol. 22: 129 (1972).
10. Diener, B., Carrick, L., Benk, R.S.: Infect. Immun. 7: 212 (1973).
11. Dubos, R. J., Hirsch, J. G.: Bacterial and mycotic infections of man, 4th ed. (Lippincott, Philadelphia, 1965).
12. Dupont, H.L., Formal, S.B., Hornick, R.B., Snyder, M.J., Libonati, J.P., Sheahan, D.G., LaBrec, E.H., Kalas, J.P.: New Engl. J. Med. 285: 1 (1971).
13. Ellwood, D.C., Keppie, J., Smith, H.: Brit. J. exp. Path. 48: 28 (1967).
14. Evans, D.J., Chen, L.C., Curlin, G.T., Evans, D.G.: Nature, Lond. 231: 137 (1972).
15. Finger, H., Fresenius, H., Angerer, M.: Experientia 27: 456 (1971).
16. Fitzgeorge, R.B., Smith, H.: Brit. J. exp. Path. 47: 558 (1966).
17. Friedberg, D., Shilo, M.: Infect. Immun. 2: 279 (1970).
18. Friedman, H., Chakrabarty, A. K.: Transplant. Proc. 3: 826 (1971).
19. Frost, A.J., Smith, H., Witt, K., Keppie, J.: Brit. J. exp. Path. 53: 587 (1972).
20. Gemski, P. Jr., Sheahan, D.G., Washington, O., Formal, S.B.: Infect. Immun. 6: 104 (1972).
21. Germanier, R., Fürer, E.: Infect. Immun. 4: 663 (1971).
22. Glynn, A.A.: Symp. Soc. gen. Microbiol. 22: 75 (1972).
23. Hill, A.B., Hatswell, J.M., Topley, W.W.C.: J. Hyg. 40: 538 (1940).
24. Hirsch, J.G.: Symp. Soc. gen. Microbiol. 22: 59 (1972).
25. Jansen, W.A., Surgalla, M.J.: Science, 163: 950 (1969).
26. Jones, G.W., Rutter, J.M.: Infect. Immun. 6: 918 (1972).
27. Keusch, G.T., Mata, L.J., Grady, G.F.: Clin. Res. 18: 442 (1970).
28. Kirpatovski, I.D., Stanislavski, E.S.: Transplant. Proc. 3: 831 (1971).
29. Kobayashi, F.: Jap. J. Microbiol. 15 (4): 301 (1971).

30. Kusama, H., Sus, R.H.: Infect. Immun. 5: 363 (1972).
31. Levine, M.M., DuPont, H.L., Formal, S.B., Hornick, R.B., Takeuchi, A., Gangarosa, E.J., Snyder, M.J., Libonati, J.P.: J. infect. Dis. 127: 261 (1973).
32. Liu, P.U.: J. infect. Dis. 116: 481 (1966).
33. Lüderitz, O., Staub, A.M., Westphal, O.: Bact. Rev. 30: 192 (1966).
34. Medearis, D.N. Jr., Camitta, B.M., Heath, E.C.: J. exp. Med. 128:399 (1968).
35. Meinke, G.J., Barum, J., Rosenberg, B., Benk, R.S.: Infect. Immun. 2: 583 (1970).
36. Miles, A.A., Miles, E.M., Burke, J.: Brit. J. exp. Path. 38: 79 (1957).
37. Nakamura, M., Jackson, K.E., Cross, W.R.: Infect. Immun. 2: 570 (1970).
38. Nakano, M., Saito, K.: Jap. J. Microbiol. 12: 471 (1968).
39. Nakano, M., Saito, K.: Nature, Lond. 222: 1085 (1969).
40. Parish, W.E.: Symp. Soc. gen. Microbiol. 22: 157 (1972).
41. Polk, H.C., Miles, A.A.: Brit. J. exp. Path. 54: 99 (1973).
42. Rebuck, J.W.: Symp. J. reticuloendothel. Soc. 12, No. 2 (1972).
43. Roantree, R.J.: Ann. Rev. Microbiol. 21: 443 (1967).
44. Rutter, J.M., Jones, G.W.: Nature, Lond. 242: 531 (1973).
45. Savage, D.C.: Symp. Soc. gen. Microbiol. 22: 25 (1972).
46. Smith, H.: Bact. Rev. 32: 164 (1968).
47. Smith, H.: Host factors influencing microbial proliferation in vivo. In: Dunlop, R.H., Moon, H.W., Resistance to infectious disease, p. 141 (Modern Press, Saskatoon 1970).
48. Smith, H.: Symp. Soc. gen. Microbiol. 22: 1 (1972).
49. Smith, H.: Microbial interference with host defence mechanisms.In: Prophylaxis of infectious and other diseases by means of vaccination and the use of immunoglobulins. In press. Karger, Basel 1973.
50. Smith, H., Fitzgeorge, R.B.: Brit. J. exp. Path. 45: 174 (1964).
51. Smith, H.W.: J. med. Microbiol. 5: 345 (1972).
52. Smith, H.W., Linggood, M.A.: J. med. Microbiol. 4: 301 (1971).
53. Smith, H.W., Linggood, M.A.: J. med. Microbiol. 4: 467 (1971).
54. Spink, W.W.: Brucellosis as a model for metabolic studies on bacterial shock and inflammation. In: Stoner, H.B., Threlfall, C.J., The biochemical response to injury, p. 361 (Blackwell Scientific Publications, Oxford 1960).
55. Stendall, O., Edebo, L.: Acta path. microbiol. scand. 80b: 481 (1972).
56. Stone, H.H., Given, K.S., Martin, J.D.: Surg. Gynec. Obstet. 124: 1067 (1967).
57. Takeuchi, A.: Ergebn. Pathol. 54: 1 (1971).
58. Urbaschek, B.: Nature, Lond. 202:883 (1964).
59. Valtonen, V.V., Makela, P.H.: J. gen. Microbiol. 69: 107 (1971).
60. Williams, A.E., Keppie, J., Smith, H.: Brit. J. exp. Path. 43: 530 (1962).
61. Wilson, G.S., Miles, A.A.: Topley and Wilson's Principles of Bacteriology and Immunity. 5th ed. (Edward Arnold, London 1964).
62. Yee, R.B., Buffenmeyer, C.L.: Infect. Immun. 1: 459 (1970).

Studies on Shigellosis and Salmonellosis

S. B. Formal, P. Gemski, Jr., R. A. Giannella, and W. R. Rout

Introduction

Diarrheal disease of bacterial etiology is a world-wide problem and no population is spared. In geographical areas where malnutrition, congested living conditions and poor sanitation exist, enteric infections represent not only a striking cause of morbidity but also of mortality.

At present, two general mechanisms are recognized by which enteric pathogens cause intestinal disease and acute diarrhea. In the first, in which cholera may be considered a prototype, the organism multiplies in the lumen or on the surface of the epithelium of the small intestine and produces a toxin which causes the small intestinal epithelial cell to secrete electrolytes and water (1). Mucosal invasion does not occur and intestinal morphology remains normal. In the second type, as exemplified by bacillary dysentery, bacterial invasion of the intestinal mucosa is a necessity for the initiation of the disease process (2). In contrast to the "toxigenic diarrheas", diarrheal disease due to invasive bacteria result in an acute colitis or enterocolitis. With the exception of Shigella dysenteriae 1 enterotoxins have not been identified in invasive strains. In addition to dysentery bacilli, other etiologic agents of the "invasive diarrheas" include those salmonellae which produce gastroenteritis and certain Escherichia coli serotypes (3, 4). In this communication we shall discuss some aspects of the pathogenesis of the disease syndromes caused by invasive organisms.

Bacillary Dysentery

Most knowledge concerning the role of mucosal invasion as a requisite for virulence stems from studies carried out with Shigella flexneri. These investigations demonstrated unequivocally that penetration of the mucosal epithelial cell by the pathogen followed by bacterial multiplication are essential steps in the pathogenesis of shigellosis (2). Tests to determine invasiveness include: 1. direct visualization of bacilli in the mucosa of experimental animals; 2. the ability of an organism to produce keratoconjunctivitis in rabbits or guinea pigs (5); and 3. the ability to penetrate monolayers of HeLa cells (2). Mutant strains which failed to invade epithelial cells of the intestine uniformly failed to cause disease. Thus it is apparent that knowledge of the mechanism of epithelial cell penetration is of paramount importance in understanding the pathogenesis of bacillary dysentery.

In spite of its importance in pathogenesis, information concerning the process of mucosal invasion is at the moment limited. There is no doubt that several attributes of the virulent bacterial cell are responsible for its ability to invade. One property of the cell which may be involved in penetration is the nature of its somatic antigen. We have recently addressed ourselves to the question of O-antigen specificity and its effect on the invasiveness of pathogens (6).

Shigella flexneri hybrids with the somatic antigens of Escherichia coli. The similarity of chromosomal position for somatic antigen genes of E. coli and Sh. flexneri (near the histidine operon, his), has enabled us to construct by intergeneric hybridization techniques, Sh. flexneri derivatives which express either the E. coli O-25 or O-8 somatic antigen rather than their native parental serotype. Table I is a summary of the serological properties of Sh. flexneri hybrids obtained from matings with the O-25 E. coli Hfr W3703 and O-8 E. coli Hfr 59.

With hybrids derived from matings with the O-25 E. coli W3703 donor, three agglutinin classes were found. Class C, in which cells agglutinated only in Sh.flexneri antisera (type 2 and Y) represents his+ hybrids which did not inherit E. coli determinants. Among those hybrids which inherited antigen 25, two serological patterns were distinguished. Class A hybrids agglutinated strongly in factor 25 serum and in the Sh.flexneri type 2 serum but were unreactive in the group factor Sh. flexneri Y serum. It thus appears that these hybrids had replaced their native group antigens with the E. coli O-25 in the process of genetic recombination. The type specific 2 antigen, whose gene(s) map near the pro locus (distal to the his region) remains unchanged and is still expressed in such hybrids. With Class B hybrids, agglutination was seen in all three sera, although it was weak in the Sh.flexneri group factor serum. These results suggest that Class B hybrids may be diploid for the his chromosomal region, conserving their native Sh.flexneri group factors in addition to gaining antigen 25 from E. coli. With his+ hybrids derived from the antigen 8 donor, E. coli Hfr 59, only two distinct agglutinin classes were evident (Table I, bottom). Class A hybrids agglutinated only in factor 8 serum, being unreactive in any of the Sh.flexneri antisera. Class B hybrids behaved serologically as typical Sh.flexneri 2a.

Virulence of Sh.flexneri hybrids with E. coli antigenic characteristics. The Sereny test for keratoconjunctivitis (5, 2) was employed in preliminary screenings to determine the virulence of Sh.flexneri hybrids expressing E. coli antigens. A provocation of keratoconjunctivitis by Sh.flexneri reflects the ability of the organism to invade intestinal epithelium (2). Both E. coli Hfr 59 and Hfr W3703 failed to evoke keratoconjunctivitis and hence were considered avirulent. When the 88 his+, antigen 8 positive hybrids derived from a mating with Hfr 59 (Table I, class A, bottom) were so tested, none caused keratoconjunctivitis. Subsequent testing of these hybrids for sensitivity to "rough-specific" bacteriophages revealed that 77 of 88 were lysed by at least one of them. Thus, the avirulence of O-8 hybrids could be due to their rough state. Nevertheless, the remaining 11 hybrids, which were insensitive to the "rough-specific" phages and agglutinated strongly in O-8 antisera, still were avirulent.

Sereny tests on Sh.flexneri hybrids which inherited E. coli antigen-25 revealed that some of these were able to penetrate epithelial cells. Screening tests were limited to class A hybrids (Table I, W3703 donor) since this type of hybrid had replaced Sh.flexneri group factors with the O-25 donor antigen. Six of 44 such hybrids provoked a positive Sereny test and when tested with the rough-specific phages, scored as smooth. As in the case of the O-8 Sh.flexneri strains, many, but not all, of the avirulent O-25 hybrids appeared to be rough, showing sensitivity to at least one of the rough-specific phages employed.

Upon reisolation of organisms from diseased eyes, the virulent cells were serologically indistinguishable from those employed in the initial challenge, being agglutinated only by E. coli O-25 and Sh.flexneri type 2 antisera. Because previous studies indicated that type 2 antigen is not essential for maintenance of virulence (11) we excluded this antigen from a virulent O-25 hybrid strain by hybridization with a E. coli K-12 donor. One such strain O-25 hybrid 547-1-7 which does not produce type 2 antigen was subjected to further study of virulent properties.

Both the parent Sh.flexneri 2a strain and the O-25 hybrid derivative 542-1-7 were fed to groups of starved-opiated guinea pigs at a dose of about 1×10^8 cells. Animals were sacrificed 24 hours later and their intestines histologically examined to assess the extent of bowel damage. Sections of the small and large intestine from guinea pigs fed either the parent his⁻ Sh.flexneri 2a strain or the O-25 hybrid 542-1-7 showed marked alterations of the mucosal architecture, and an acute inflammatory reaction. Control animals, fed either a known avirulent strain or broth alone, revealed no alterations in intestinal morphology.

Our finding that Sh.flexneri hybrids with E. coli O-8 antigen are avirulent while those with the chemically related E. coli O-25 factor can conserve virulence thus indicates that the O-repeat unit composition of surface lipopolysaccharides can be altered within limits without significant alteration to virulence. The uniform aviru- lence of all the O-8 Sh.flexneri hybrids, however, may indicate that the chemical composition and structure of the O-repeat unit is indeed one determining factor for epithelial penetration by Sh.flexneri. Studies of Simmons (7) have revealed the group antigenic determinants of Sh.flexneri 2a consist of a N-acetyl-glucosamine-rham- nose-rhamnose repeat unit and that the attachment of α-glucosyl secondary side chains to a rhamnose of this repeat unit confers type 2 specificity. Similar studies on the chemical composition of E. coli O-8 strains have revealed that the immuno- dominant sugar of the O-repeat unit is a D-mannose (8). This distinct difference in O-repeat unit composition of the strains employed is further illustrated by our find- ing that those Sh.flexneri which replace their group antigens with the O-8 repeat unit do not express their type-specific 2 antigens (see Table I). Since the genes controlling type 2 specificity are distal to the his chromosomal segment and were not altered in our hybridization procedure, it appears that the α-glucosyl side chains (type factor 2) can not be expressed by these hybrids. A similar situation was previously noted in studies dealing with the transfer of Sh.flexneri 2a antigens to E. coli K-12 recipients. Hybrids which contained the genes for type factor 2 did not ex- press this antigen unless the genes for Sh.flexneri group factor were present (9). The possibility that the chemical composition of the O-repeat unit is indeed a deter- mining factor for epithelial cell penetration by Sh.flexneri is supported, to some de- gree, by our finding that Sh.flexneri expressing antigen 25 can conserve their pene- trating ability and virulence. Although the chemical components of the O-25 lipopo- lysaccharide (LPS) layer have not been fully described, it has been established that rhamnose is present in its O-repeat unit (10). In addition, our observation that Sh. flexneri O-25 hybrids continue to express type-specific antigen 2 (see Table I) indi- cates that the α-glucosyl side chains conferring this serological specificity can be functionally linked to the O-25 repeat unit, presumably to rhamnose as in Sh.flex- neri 2a. Thus although being serologically distinct, the O-repeat units of the Sh. flexneri group factor and the E. coli O-25 antigen bear some similarity. This simi- larity may be reflected in the conservation of virulence by such O-25 hybrids.

A locus which controls the capacity of Sh.flexneri to provoke keratoconjunctivitis. The complexity of the penetration step in pathogenesis of bacillary dysentery is further illustrated by our identification of a genetic locus on the genome of Sh.flex- neri which controls its ability to evoke a positive Sereny test (11). The locus, which we have termed kcpA in reference to its involvement in provoking keratoconjuncti- vitis, has been mapped between the lac and gal chromosomal markers in close pro- ximity to the purE allele. This conclusion is based on both conjugational and trans- ductional analyses. We observed that Sh.flexneri 2a hybrids which received the lac- gal chromosomal segment from various E. coli K-12 Hfr derivatives lost the abili- ty to evoke a positive Sereny test. At the present time, the nature of the alteration to Sh.flexneri cells which results in a loss of penetrating ability and hence, virulence, remains obscure.

Diarrhea and Dysentery in Shigellosis

Following penetrating and multiplication of shigellae in the gastro-intestinal mu- cosa, events are set in motion which result in fever, alteration in intestinal motor function, inflammation und ulceration of the mucosa, and fluid and electrolyte loss into the intestinal lumen. With respect to fluid loss, patients with shigellosis may present in three different ways; classical dysentery characterized by multiple li- quid stools of small volume containing blood, mucus, and pus, with uncomplicated watery diarrhea alone, or with a combination of diarrhea and dysentery.

Neither the site nor the mechanism of intestinal fluid loss in shigellosis has been defined. In an attempt to examine these questions, small and large intestinal fluid and electrolyte transport were examined in the Sh.flexneri infected Rhesus monkey, which is a natural host of this disease (12). Intestinal perfusion studies in non-infected, control animals revealed net fluid and electrolyte absorption in the jejunum, ileum and colon.

In animals with shigellosis, the most consistent physiologic lesion was found in the colon, net fluid secretion being regularly observed. The response of the small intestine as contrasted to the colon, was more variable but abnormalities in fluid and electrolyte transport correlated with the presence of watery diarrhea.

In animals with dysentery only, transport in the small intestine was normal. In animals with diarrhea only, or with diarrhea and dysentery, jejunal fluid transport was abnormal in that net secretion was observed. Although an acute colitis was seen in all animals, this being a reflection of bacterial invasion of the mucosa, jejunal morphology remained normal and bacterial invasion of the jejunal mucosa was not observed.

The combination of fluid secretion by the jejunum in the absence of bacterial invasion or tissue damage suggests the possibility of enterotoxin elaboration by Sh. flexneri which acts in this region. Enterotoxin production by Sh.flexneri has not been demonstrated however. Indeed, Sh.dysenteriae 1 is the only dysentery serotype in which an enterotoxin has been identified (13). At present this enterotoxic activity has not been separated from the classical neurotoxic and cytotoxic activity, suggesting that a single toxin is responsible for all three activities. In view of the recently demonstrated enterotoxic activity of Shiga toxin, its role in pathogenesis was investigated (15).

Comparative studies on toxigenic Sh.dysenteriae 1 employing penetrating and non-penetrating derivatives. Initial experiments were conducted with Sh.dysenteriae 1 strain 3818. This strain behaves similarly to a strain of Sh.flexneri 2a previously studied, in that two colonial types are observed, an opaque (O) and a translucent (T) form. As with Sh.flexneri 2a, the T colonial form of strain 3818 was able to penetrate epithelial cells as evidenced by an ability to evoke keratoconjunctivitis in the guinea pig; the O form lacked this property. Comparative titrations for toxin production by the O and T forms of strain 3818 were performed to determine whether the O form produced toxin and if so, whether the amount elaborated was significantly lower than its T form parent. Dilutions of sterile culture filtrates of both the O or T form were tested for their cytotoxicity to HeLa cells. The preparation from a T-colony form of strain 3818 killed HeLa cells at a dilution of 1:800 while that from the O-colony form had a cytopathogenic effect at a dilution of 1:1600. The finding of comparable levels of toxin in both the O (non-penetrating) and T (penetrating) form of strain 3818 suggests no significant involvement of toxin in the initial steps of epithelial cell penetration.

Experiments with the T and O derivatives of 3818 were conducted in the rabbit ileal loop model. Sterile filtrates from overnight broth cultures of both colonial forms contained enterotoxic activity as evidenced by positive ileal loops. Likewise when bacterial cells of both the penetrating and non-penetrating derivatives were inoculated, positive loops resulted. However, when fluorescent antibody studies were performed on sections of these positive ileal loops, a dramatic difference in the distribution of organisms was observed. Organisms of the T-form (Sereny positive) strains were seen within epithelial cells of intestinal mucosa as well as in the bowel lumen, while the cells from the O-form (Sereny negative) derivatives were present only in the lumen. Thus, it is apparent that despite production of toxin by both the O and T form, only the T form had the ability to penetrate the intestinal epithelial cell.

The toxigenic O and T forms of strain 3818 were fed to groups of Rhesus monkeys in dosage levels of 5×10^{10} cells. The results of this study (summarized in Table II) were unequivocal, in that 9 of 15 animals fed the penetrating strain exhibited evidence of disease; 4 of these animals died with acute dysentery while the remaining

5 diseased monkeys suffered severe diarrhea. There was no evidence of paralysis in any of the animals. None of the monkeys which were fed the toxigenic but non-penetrating strain 3818-O showed any signs of illness. Three additional monkeys with classical dysentery caused by strain 3818-T were sacrificed for pathological examination. At necropsy, gross abnormalities of all three were confined to the colon. The small intestine appeared normal. Histologically the colonic mucosa was involved by acute colitis, which was indistinguishable from that caused by Shigella flexneri infections previously reported. From these findings it is evident that toxin production alone is not responsible for the capacity of Sh.dysenteriae to cause clinical disease in monkeys.

It is evident that shiga toxin does not act in a manner of classical enterotoxin such as cholera enterotoxin. Cholera toxin is elaborated in the lumen or on the surface epithelium of the small intestine and causes the epithelium to secrete fluid and electrolytes. Invasion of the mucosa by cholera vibrio does not occur.

Salmonella Enterocolitis

Salmonellosis, depending upon the host species and the particular Salmonella species, occurs in two general syndromes, namely enteric fever and gastroenteritis. For instance Salmonella typhimurium produces a typhoid-like (enteric fever) picture in mice, rats and guinea pigs while in human beings and monkeys, the disease is usually a gastroenteritis (16, 17, 18).

We have studied the reactions elicited by several strains of S. typhimurium in various laboratory models with the objective of obtaining some insight into the pathogenesis of gastroenteritis (15, 16, 17, 18, 19). As summarized in Table III, strains which invaded the rabbit ileal mucosa induced mucosal inflammation and fluid exsorption. Non-invasive strains produced neither inflammation nor fluid secretion. While all invasive strains elicited an acute inflammatory reaction, not all such strains evoked fluid secretion. Furthermore, there was no correlation in the ability of invasive strains to evoke fluid secretion and the intensity of mucosal inflammation. These observations indicate that, as is the case in shigellosis, mucosal invasion is an essential step in the pathogenesis of Salmonella enteritis. However, invasion and acute inflammation alone are not sufficient stimuli for fluid secretion. Presumably other bacterial factors, which function subsequent to mucosal invasion, are necessary for fluid secretion (4).

As in our studies with shigellosis, perfusion experiments were carried out in normal and S. typhimurium infected Rhesus monkeys (20). In control monkeys, net water absorption was observed in the jejunum, ileum and colon. In monkeys with either mild or severe diarrhea caused by Salmonella, net colonic secretion was observed. Moreover, in those animals with mild diarrhea, net jejunal water absorption was also impaired. In contrast, in monkeys with severe diarrhea, net water secretion was seen in the jejunum and ileum as well as the colon. In the ileum and colon, the severity of the morphologic damage, the degree of bacterial invasion, and the intraluminal Salmonella concentrations, correlated with the severity of the transport abnormalities and with the severity of the observed diarrhea. As was the case in shigellosis, despite the presence of a transport defect in the jejunum, there was no morphologic evidence of bacterial invasion and jejunal Salmonella concentrations were minimal.

Thus, Salmonella-induced diarrhea in the Rhesus monkey may be largely dependent upon abnormalities in colonic transport. In animals with mild diarrhea, the colonic transport abnormalities may be the prime determinant of the observed diarrhea. In monkeys with severe diarrhea, however, the severe diarrhea may be the result of fluid transport abnormalities in the jejunum and ileum superimposed on that in the colon. To what extent the present findings in monkeys can be extrapolated to human salmonellosis is uncertain, and must await human perfusion studies.

Conclusions

Both shigellosis and Salmonella gastroenteritis are infections of the gastrointestinal tract. On the basis of our experiments with both genera, invasion of the intestinal mucosa is essential for these organisms to induce disease. In Rhesus monkeys, both infections consistently affect the colon morphologically and physiologically and the ileum was involved in the Salmonella infection but not in shigellosis. In salmonellosis and in cases of shigellosis where watery diarrhea is a component of the disease, the jejunum is in a net secretory state. However, in spite of these physiological abnormalities in the jejunum in shigellosis and salmonellosis, invasion of the jejunum was not seen and significant histological alterations did not occur. The stimulus to jejunal secretion is not known.

Certainly much further work concerning the pathogenesis of these infections are required. The mechanism of the penetration of intestinal cells must be understood, for this property is one of the major determinants of pathogenicity of the infecting organism. The process of fluid loss into the bowel must be understood in diarrheal diseases in which mucosal invasion occurs and no enterotoxins have been implicated. In addition, studies must be initiated to understand how derangements in intestinal motility result following bacterial invasion, because these alterations are very likely responsible, at least in part, for the abdominal discomfort which occurs in these clinically important infections. Should these various problems be understood, new methods for control and treatment of these diseases can be devised.

References

1. Formal, S.B., DuPont, H.L., Hornick, R.B.: Annu. Rev. Med. (1973).
2. LaBrec, E.H., Schneider, H., Magnani, T.J., Formal, S.B.: J. Bact. 88: 1503 (1964).
3. DuPont, H.R., Formal, S.B., Hornick, R.B., Snyder, M.J., Libonati, J.P., Sheahan, D.G., LaBrec, E.H., Kalas, J.P.: New Engl. J. Med. 285: 1 (1971).
4. Giannella, R.A., Formal, S.B., Dammin, G.J., Collins, H.: J. clin. Invest. 52: 441 (1973).
5. Serény, B.: Acta Microbiol. Acad. Sci. Hung. 2: 293 (1955)
6. Gemski, P., Sheahan, D.G., Washington, O., Formal, S.B.: Infect. Immunity 6: 104 (1972).
7. Simmons, D.A.R.: Bact. Rev. 35: 117 (1971).
8. Schmidt, G., Fromm, I., Mayer, H.: Europ. J. Biochem. 14: 357(1970).
9. Formal, S.B., Gemski, P. Jr., Baron, L.S., LaBrec, E.H.: Infect. Immunity 1:279 (1970).
10. Lüderitz, O., Staub, A.M., Westphal, O.: Bact. Rev. 30: 192 (1966).
11. Formal, S.B., Gemski, P. Jr., Baron, L.S., LaBrec, E.H.: Infect. Immunity 3: 73 (1971).
12. Rout, W.R., Formal, S.B., Giannella, R.A., Dammin, G.J.: Gastroenterology, in press, 1974.
13. Keusch, G.T., Grady, G.F., Mata, L.J., McIver, J.: J. clin. Invest. 51: 1212 (1972).
14. Gemski, P. Jr., Takeuchi, A., Washington, O., Formal, S.B.: J. infect. Dis. 126: 523 (1972).
15. Kent, T.H., Formal, S.B., LaBrec, E.H.: Arch. Path. 81: 501 (1966).
16. Kent, T.H., Formal, S.B., LaBrec, E.H.: Arch. Path. 82: 272 (1966).
17. Maenza, R., Powell, D.W., Plotkin, G.R., Formal, S.B., Jervis, H.R., Sprinz, H.: J. infect. Dis. 121: 475 (1970).
18. Powell, D.W., Plotkin, G.R., Maenza, R.M., Solberg, L.I., Catlin, D.R., Formal, S.B.: Gastroenterology 60:1053 (1971).

19. Giannella, R. A., Washington, O., Gemski, P., Formal, S. B.: J. infect. Dis. 128: 69 (1973).
20. Rout, W. R., Giannella, R. A., Formal, S. B., Dammin, G. J.: Submitted for publication, 1973.

Table I.

Slide agglutination of His[+] Sh. flexneri hybrids recovered from matings with E. coli Hfr donors W3703 and Hfr 59

Donor strain	Agglutinin class	ANTISERA				No. in each class
		E. coli O-25	E. coli O-8	Sh. flexneri type 2	Sh. flexneri Y (3, 4)	
W3703 (O-25)	A	+	NT*	+	-	144
	B	+	NT	+	+	23
	C	-	NT	+	+	56
Hfr 59 (O-8)	A	NT	+	-	-	88
	B	NT	-	+	+	159

*Not tested

Table II.

The virulence of Sh. dysenteriae 1 strain 3818T and its mutant non-penetrating derivative

Strain	Toxin Production	Sereny* Test	Rabbit Ileal* Loop	Invasion of* Guinea Pig Intestine	Clinical Disease* in Monkeys
3818-T	+	+	6/6	+	9/15
3818-O	+	-	4/4	-	0/15

*No. positive
Total tested

Table III.

Behavior of various strains of Salmonella typhimurium in laboratory models

Strains	Sereny test	Tissue culture Invasion Cell death	Mouse LD$_{50}$ (Log 10)	Starved guinea pig (Log 10) LD$_{50}$	Diarrhea in monkeys	Fluid secretion in rabbit ileal loop	Mucosal invasion and inflammation in rabbit ileal loops
TML	0**	+/+	1.0	2.0	20/25*	+	+
M 206	0	+/0	8.5	>8.0	0/5	+	+
SL1027	0	+/+	5.0	5.0	1/13	0	+
LT-7	0	+/0	6.0	ND	ND	0	+
PG-41	0	+/+	7.4	ND	ND	0	+
9SR-2	0	0/0	7.5	ND	ND	0	0
THAX1	0	0/0	7.8	ND	ND	0	0

* Number positive / Number tested

** Negative

Extrachromosomal Characters as Determinants of Pathogenicity in Enterobacteriaceae

B. Wiedemann

There is not much known about the determinants of pathogenicity in Enterobacteriaceae, although the biological properties of the endotoxins are fairly well established; however, this is only one component of the pathogenicity. There is a great variety of organisms causing similar infections, but also a great variety of diseases like enteritis, typhoid, pyelonephritis, bronchitis, etc. The Enterobacteriaceae have various characteristics, which enable this group of organisms to cause infectious processes. The extrachromosomally determined ones are of special interest within this group.

In the cytoplasm of Enterobacteriaceae covalently closed circular DNA, which is physically independant of the chromosome, is found commonly. These DNA-particles determine different characters of the bacterial host. Some of these plasmid-borne determinants are closely related to the pathogenicity of the host bacteria. Some of them are effective pathogenic markers, others are indirectly linked with the pathogenicity of Enterobacteriaceae. The size of the plasmid DNA is about 100 times smaller than that of the chromosome. Plasmids are 0.5 to 40 /um in length and have a molecular weight of 10 to 100 million dalton. Some of the plasmids code for only a few genes, sufficient for one character, others for several markers, such as transferability or sex functions. The transferability is one of the most important determinants for the biologic potency of plasmids. This character enables a donor cell carrying such a plasmid to mediate conjugation with plasmidless bacteria. During this process genes carried on the same plasmid, on other plasmids or even on the chromosome can be transferred. Thus the recipient receives the donor functions (Fig. 1). In the following five different plasmids will be mentioned:
1. The enterotoxin plasmid, coding for the production of enterotoxin
2. The K88-plasmid, coding for the K88 protein antigen
3. The Hly-plasmid, coding for the production of β-haemolysin
4. The col factor, coding for the production of colicins, and
5. The R-factor, coding for the resistance against different drugs.

The Enterotoxin Factor (17)

Strains of Escherichia coli that are enteropathogenic for pigs have been shown to synthesize two forms of enterotoxin, described as heat-labile and heat-stable. The enterotoxins cause fluid accumulation when introduced into ligated intestinal loops prepared in pigs or rabbits (11, 17). The heat-stable and the heat-labile enterotoxins are both under plasmid control (15, 17). One class of plasmid codes for both enterotoxins, and another for the heat-stable toxin only. These plasmids are not always transferable by their own transfer systems, but can be mobilized by other transferable extrachromosomal elements. Recently, and E. coli strain implicated in infantile diarrhea and another incriminated in human adults diarrhea have both been shown to possess a plasmid, that codes for enterotoxin production (14, 18).

The K88-Plasmid

The K88-plasmid codes for the K88 surface antigen which is found on E. coli strains. These strains are almost exclusively found to be enteropathogenic for swine (12). The K88-antigen seems to be the only surface antigen which is trans-

ferable. The transfer is not mediated by its own transfer system, but is mobilizable
by other transferable agents. The K88-antigen seems to play an important part in
the colonisation of the bacteria in the upper small intestine of swine (18). Thus it
is an important factor in the pathogenicity of E. coli.

The Hly-Plasmid

The production of haemolysin in E. coli could be shown to be a plasmid-borne
character in many strains (16). Although the role of haemolysins in the pathogenicity of gram-negative bacteria is not yet defined, many workers have been able to
demonstrate that haemolysin can be a pathogenic determinant. Sojka (19) reported
that most strains of E. coli that cause diarrhea in pigs are haemolytic. Haemolytic
E. coli has also been implicated in enteritic and systemic infections in dogs and
cats. Even in experimental pyelonephritis haemolysin seems to be a pathogenic determinant (6). Plasmids coding for haemolysin production give a high rate of transfer in vitro.

The Colicin-Plasmid

A variety of related substances produced by E. coli strains are called colicins.
They are capable of killing E. coli but are ineffective to the producer strain. Colicin production in the majority of cases is determined by plasmids. Most of the col-
factors are self transmissible, others can be mobilized by transferable plasmids.
The carriage a colicin-plasmid is clearly a potential asset to an E. coli strain in
competition with others. However, some studies have failed to show that colicin
acts in the intestine to produce the effects that would be predicted on the basis of
in vitro activity (4, 7). Braude and Simienski (1) on the other hand have been able
to demonstrate that colicin production is an advantage to the organism in urinary
tract infection.

The Resistance-Plasmid (R-Factor)

The plasmid, which has caused the greatest interest in medical microbiology is
the R-factor. It determines resistance of the host cell to a great variety of drugs,
mostly by the production of enzymes, which inactivate the drugs. There are some
other mechanisms of drug resistance, but they are not as well examined as the enzyme production (5). Although there is no direct relationship between R-factors and
the pathogenicity of bacteria, they should be mentioned in this context. As stated below, these plasmids might play an important part in the evolution of pathogenic organisms. Furthermore R-factors endow their host cells with additional information,
whereas organisms with chromosomal resistance such as gentamycin-resistant
strains often lose their pathogenicity with the acquisition of resistance (21).

Plasmids are quite common in Enterobacteriaceae. In a single cell as many as
five different plasmids can be found, which can be distinguished from each other by
their size and different functions (2). These different species of plasmids interact
with each other and with the chromosome under certain circumstances. Small plasmids can aggregate to give larger ones, forming an unique DNA molecule which
codes for many different characters. Therefore, when there is a selective pressure
against only one of these characters, all the determinants will be selected. These
aggregated characters can be transferred to recipient strains en bloc. Alternatively large plasmids can fragment into small ones (Fig. 2), which if small enough will
replicate under relaxed control (3). Thus multiple copies of one plasmid arise in a
single cell. These result in a corresponding increase of gene products, for example in the production of more antibiotic inactivating enzymes, so that the population
becomes more resistant. This reaction can be enhanced by specific selection pressure (13). Furthermore the transfer factors enable other determinants not present
on the same DNA to be transferred by conjugation and thus to be spread easily in a
bacterial population.

Another phenomenon that can be of evolutionary value is that plasmids can integrate into the chromosome. During the process of excision from the chromosome, chromosomal genes may stick to the plasmid. Thus these genes can become mobile like R- or col-factors. This has not only been shown for the lac-gene but also for chromosomal resistance genes (8), (Fig. 3). Kontomichalou (10) could demonstrate that enzymes, naturally produced by chromosomal genes in Klebsiella strains, are similar to those produced by R-factor genes in E. coli. All these Klebsiella strains had been isolated before the use of the corresponding drug. Tschäpe (20) and others (8) demonstrated the pickup of chromosomal genes by plasmids. This is good evidence of the way in which R-factors could be developed in nature. These events of chromosomal pickup may be rare in nature, but they are frequent enough to react on a selection pressure and endow bacteria with evolutionary advantage, as the example of the development of R-factors has clearly demonstrated.

In epidemiologic studies, there has been no indication that with the use of antibiotics, plasmids which code for pathogenic characters are more prevalent. However, not many surveys on this subject exist. The evolution of multiresistant R-factors has demonstrated that with a selection pressure on a single gene many others can be selected simultaneously. In our studies we found that R-factor bearing Salmonella typhimurium strains carry colicinogenic factors more frequently than those which are sensitive to antibacterial drugs (Table I). The R-factor is commonly found to be genetically linked to the col-gene.

There is another relationship between the pathogenicity of bacteria and plasmids. The transferability of extrachromosomal elements depends on the surface of the recipient cell. K-antigens and O-antigens of bacteria, which form smooth colonies, inhibit the transfer of plasmids. Cells from which the capsule and the outer part of the O-antigen have been removed by mutation act as better recipients (Fig. 4) compared to the parent strain. Simultaneously with the degradation of the cell wall a decrease in the pathogenicity can be observed. One could conclude that the facts provide an anti-evolutionary mechanism. Thus only apathogenic rough strains can receive plasmids in a conjugational process. Jarolmen and Kemp (9) transferred R-factors in vivo, in the intestine of pigs, to Salmonella cholerasuis, with simultaneous application of antibacterial drugs. These authors found only R-factor bearing apathogenic Salmonella strains in the fecal flora of the pigs, while the original population of pathogenic strains vanished. However, as most of the R-factor bearing strains isolated from clinical specimen have intact K- and O-antigens, cell wall and capsule are only partly a barrier for the uptake of plasmids.

As an overall picture one might say that plasmid-borne characters may code for pathogenicity. These characters are especially involved in evolutionary processes, as different plasmids are in a special interrelation and also interact with chromosomal markers. A selection of resistant plasmids, which are linked with genes coding for pathogenicity, could enhance the spread of these characters.

References

1. Braude, A.I., Simienski, J.S.: J. clin. Invest. 47: 1763 (1969).
2. Christiansen, C., Christiansen, G., Leth Bak, A., Stenderup, A.: J. Bact. 114: 367 (1973).
3. Clowes, R.C.: Bact. Rev. 36: 361 (1972).
4. Craven, J.A., Minitas, O.P., Barnum, D.A.: Amer. J. vet. Res. 32: 1775 (1971).
5. Curtis, N.A.C., Richmond, M.H., Stanisich, V.: J.gen. Microbiol. 79: 163 (1973).
6. Fried, F.A., Vermeulen, C.W., Ginsburg, M.J., Cone, C.M.: J. Urol. 106: 251 (1971).

7. Ikari, N. S., Kenton, D. M., Young, V. M.: Proc. Soc. exp. Biol. Med. 130: 1280 (1969).
8. Kameda, M., Harada, K., Suzuki, M., Nakajema, T., Mitsuhashi, S.: Jap. J. Microbiol. 16: 205 (1972).
9. Jarolmen, H., Kemp, G.: J. Bact. 99: 487 (1969).
10. Kontomichalou, P., Papachristou, E., Kotsahi, S., Levis, G.: Chemotherapy, Athens, Greece 1973.
11. Moon, H. W., Sorensen, D. K., Sautter, J. H.: Amer. J. vet. Res. 27: 1317 (1966).
12. Ørskov, I., Ørskov, F.: J. Bact. 91: 69 (1966).
13. Rownd, R., Kasamatsu, H., Michel, S.: Ann. N. Y. Acad. Sci. 182: 187 (1971).
14. Skerman, F. J., Formal, S. B., Falkow, S.: Infection Immunity 5: 662 (1972).
15. Smith, H. W., Gyles, C. L.: J. med. Microbiol. 3: 387 (1970).
16. Smith, H. W., Halls, S.: J. gen. Microbiol. 47: 153 (1967).
17. Smith, H. W., Halls, S.: J. gen. Microbiol. 52: 319 (1968).
18. Smith, H. W., Linggood, M. A.: J. med. Microbiol. 4: 301 (1971).
19. Sojka, W. J.: Escherichia coli in domestic animals and poultry. Buchs, p.104 (England, Commonwealth Agricultural Bureaux 1965).
20. Tschäpe, H.: Z. allg. Microbiol. 13: 693 (1973).
21. Weinstein, M. J.: Acta path. Microbiol. Scand. Suppl. 241: 99 (1972).

Table I.

Incidence of Colicinogeny in Relation to Antibiotic Resistance Patterns in 654 Strains of S. typhimurium

	Sensitive	Resistant All Strains	Resistant Transferable Strains
No of strains	458	196	152
No of col$^+$ strains	23	19	16
% of col$^+$ strains	5.0	9.7	10.5

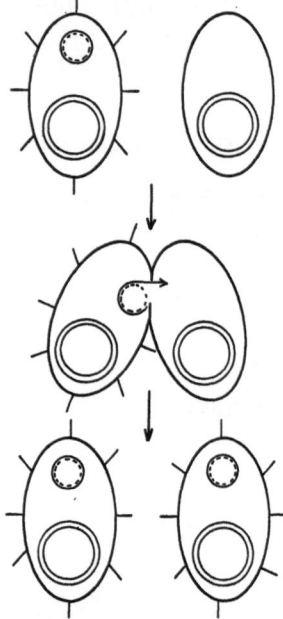

Fig. 1. Bacterial cells carrying plasmids with transferability form sexpili. In a process of conjugation plasmids can be transferred and thus endow the recipient with donor functions.

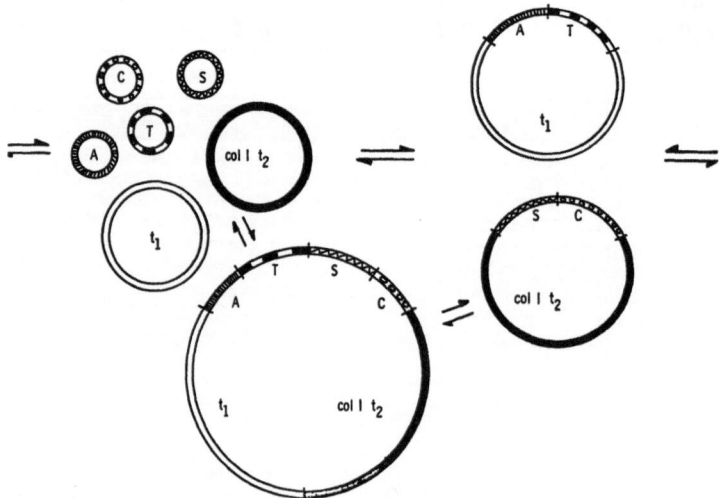

Fig. 2. Plasmids can aggregate to big ones and segregate to small ones. The arrows pointing outwards indicate that there are more possibilities than shown here.

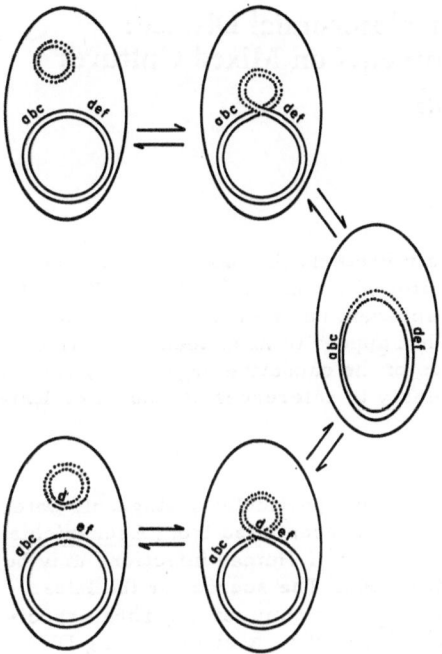

Fig. 3. Plasmids can integrate into the chromosome. During the process of excision the plasmid can pick up chromosomal genes.

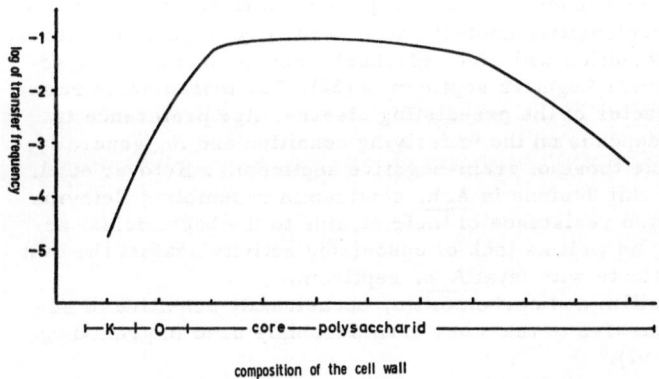

Fig. 4. The outer cell wall structure (K- and O-antigen) prohibits a conjugational transfer. The rough mutants without O- and K-antigen act as good recipient strains whereas a forthgoing destruction of the cell wall results in a worse transfer again.

Gram-Negative Rods as Agents of Nosocomial Disease: Some Recent Developments, with Comments on Mixed Cultures

A. von Graevenitz

This paper deals with infections due to certain bacteria that occur chiefly, but not exclusively, in hospitalized patients. We followed Brachman's (4) definition of nosocomial infections as "those which are not apparent on admission but develop afterwards and, when clinically diagnosed, do not appear to have been incubating at the time of admission". For the identification of the causative organisms, the reader is referred to the bibliography, particularly to references 16 and 61 d. Emphasis has been on recent publications.

1. Septicemia due to Aeromonas hydrophila (A. h.)

The natural habitats of A. h. are outdoor and indoor (particularly stagnant) water sources. In the clinical laboratory, A. h. has to be differentiated from Escherichia coli, Plesiomonas shigelloides, and Vibrio spp. (62, 66). Human infections may be exogenous (through contaminated water) or endogenous. The source for the latter seems to be the intestine: intestinal carriage of A. h. and summer diarrhea associated with A. h. have both been described (62, 66). A selective agar containing DNA, toluidine blue, and ampicillin serves to detect the organism from the feces (66). Cross infections have so far not been observed. In the first half of 1973, only five strains of A. h. were isolated from inpatients at Yale-New Haven Hospital.

A. h. septicemia, which may also be community-acquired, occurs predominantly in patients with certain preexisting conditions. Of the more than 40 cases reported in the literature (13, 26, 33, 34, 38, 48, 54, 62, 67, with further references), about 35 % occurred in leukemics, 30 % in patients with hepatobiliary disease (Laennec's cirrhosis, cholangitis, cholecystitis), and 16 % in patients with other malignancies. The association with liver cirrhosis may be part of the association of that disease with gram-negative septicemia (30). The male:female ratio is 2:1 regardless of the character of the preexisting disease. Age preference is not conspicuous. Prognosis depends on the underlying condition and is, generally, poor. The symptoms resemble those of gram-negative septicemia. Ketover et al. (33) have recently described skin lesions in A. h. septicemia resembling Ecthyma gangraenosum. They also noted resistance of their strains to the bactericidal action of normal human serum; as well as lack of opsonizing activity against the own strains in the sera of two patients with fatal A. h. septicemia.

A. h. is resistant to ampicillin and carbenicillin, occasionally sensitive to cephalotin, and still largely sensitive to the other antimicrobials used in gram-negative septicemia (33, 38, 62, 67).

2. Septicemia due to Herbicola-Lathyri (H.-L.) Bacteria (Erwinia sp., Enterobacter agglomerans)

These bacteria are ubiquitous in nature, and are by now well characterized biochemically (17). With 58 strains isolated at Yale-New Haven Hospital from inpatients in the first half of 1973, they ranked ninth among gram-negative rod isolates. Own experiments with selective media for culture of H.-L. bacteria from human stools have not met with success.

Sporadic blood isolates of H.-L. bacteria have only been reported five times since 1928 (61 b). Chills, tachycardia, fever, polymorphonuclear leucocytosis with a left shift, but no shock or fatal outcomes were recorded. Two patients had underlying disease.

In 1970 and 1971, nosocomial H.-L. septicemias were observed on a large scale in the United States. The bottle caps of various intravenous solutions were found contaminated with a variety of microbes, among which H.-L. bacteria and Enterobacter cloacae were most conspicuous (21, 39, 43 a, 43 b). In certain lots, 24 % to 47 % of the bottles were found contaminated (21). The fluids (5 % dextrose in half or normal strength saline, normal saline) rarely yielded growth immediately upon opening but were able to support growth of the bacteria upon contamination from the caps or from artificial sources. Even then, they remained optically clear (for a discussion of the mechanisms of contamination and growth, see Felts et al. (21)).

In 11 of the 125 H.-L. septicemias reported to the Center for Disease Control (43 a, 43 b), the following symptoms were observed (21): fever up to 41° C in all patients (which started with a mean interval of 4 1/2 days after i.v. placement and lasted for a mean of 3.7 days), and chills, phlebitis, hypotension in some. Abatement of symptoms was not associated with antimicrobial treatment (endotoxin effect?); defervescence most often occurred after stoppage of intravenous fluids or after change of the i.v. site (which also meant replacement of the fluid administration system). Patients with underlying disease were more often affected. But such patients tend to be more often on i.v. fluids; and H.-L. septicemia has also been observed in young patients without predisposing factors (43 a). Other data (39) support these observations. Death may occur if underlying disease is present (21, 39), or if subsequent renal tubular damage develops (58).

It could be shown that H.-L. bacteria produce endotoxin (21). Normal serum killed 99 % of the inoculum in one hour; phagocytosis also occurred normally, but both functions were curtailed if the serum was heated for 30 minutes to 56° C (21).

Three smaller outbreaks of H.-L. septicemia have been recorded recently. One, in the United States in 1973, was due to contamination of 5 % Dextrose in Lactated Ringer solutions with H.-L. bacteria, Citrobacter freundii, and E. cloacae (43 c, 43 d). A Center for Disease Control report also mentions a 1972 outbreak in Great Britain (43 c). A third outbreak, involving H.-L. bacteria and Pseudomonas fluorescens, apparently from an acid-citrate-dextrose solution used in a blood bank, was reported from Denmark in 1973 (20).

In the U.S. cases, several biotypes (17) of H.-L. bacteria were apparently involved (43). Antibiotic sensitivities (21, 39) confirmed earlier data asserting variable susceptibility to ampicillin, carbenicillin, and cephalotin, but rare resistance against other drugs used in gram-negative septicemia (61 c).

3. Septicemia due to Bacteroides fragilis (B.f.)

With the improvement of anaerobic techniques, B.f. has been isolated more frequently from clinical sources in recent years. Its human habitats are the large intestine and the female lower genital tract. Using neomycin blood agar in the GasPak system (BioQuest, Cockeysville, Md.) and a disc technique (23), we diagnosed 226 strains from inpatients in the first half of 1973 (6th rank of all gram-negative rods isolated).

B.f. septicemia accounted for 6.5 % of all gram-negative septicemias during that period. Certain pertinent features have been stressed (3, 19, 25, 31, 37, 46, 52, 70): About 75 % of the cases are nosocomial. At least two-thirds of the patients have chronic gastrointestinal disorders (mostly carcinoma of the colon) and/or a history of recent abdominal surgery; about 20 % have gynecological disorders, mostly accompanied by recent surgery. Sources for the bacterium are the aforementioned habitats and, occasionally, lung (46) or decubitus (52) infections. Diabetes mellitus is a predisposing condition. The median age for gastrointestinal B.f. septicemia is over 40 years. At the time of the first positive blood culture, over half of the patients have been treated with antimicrobials ineffective against B.f.

(see below). The clinical symptomatology is that of a gram-negative septicemia with a tendency towards high leucocytosis (46). The fatality rate is about 33 %; underlying gastrointestinal malignancy, age over 40 years, incorrect antibiotic treatment, shock, and development of metastatic abscesses worsen the outlook; but septicemia due to septic abortion has a better prognosis and runs a milder course (25).

Due to the slow generation time of B.f., the average positive blood culture is detected only five days after inoculation (as opposed to 1 - 2 days for Enterobacteriaceae in our own experience). Therefore, some strains will show up only in the routine anaerobic subculture. 12 to 30 % of all B.f. in blood cultures are mixed: four of the 20 cases observed in 1972 at Yale-New Haven Hospital were mixed (three with other gram-negative rods). Therefore, the use of a routine selective medium for anaerobes, such as neomycin blood agar, seemed advisable to us in all cases of gram-negative rod bacteremia (64).

B.f. is resistant to penicillin, ampicillin, the cephalosporins, the polymyxin, and the aminoglycoside antibiotics (preoperative bowel preparation!). At present, the best drugs for the treatment of B.f. septicemia seem to be clindamycin, chloramphenicol, and possibly carbenicillin (46, 52).

4. Urinary Tract Infection (UTI) due to Serratia marcescens (S.m.)

S.m. is by now a frequent inhabitant of chronic disease wards. In the first half of 1973, 176 strains were isolated from inpatients at Yale-New Haven Hospital, making S.m. the 7th ranking of all gram-negative rods. About half of the strains came from the UT.

The natural habitat of S.m. are soil and water sources. Fecal carriers in humans have occasionally been reported. A recent investigation using Mueller-Hinton agar with vancomycin, colistin, nystatin, methyl red, and sorbitol found no rectal colonization in patients carrying S.m. in the urinary tract (36): only two of 89 patients with no S.m. in the UT carried S.m. in the stool. But with the help of a different selective medium containing DNA, toluidine blue, colistin, and cephalotin (6), this laboratory isolated S.m. strains with identical sensitivities from feces and urines of 12 patients out of 20 with urine cultures positive for S.m..Most evidence, however, supports transmission of urinary S.m. from patient to patient through the hands of personnel (36), less often through nebulizer liquids (51). The correlation of S.m. UT findings to high patient density is significant (1, 36). Clinical spread has been verified by serotyping (16), phage typing (28), marcescin sensitivity (59), and marcescin production (18), the latter method being the most reliable one.

Our former (7) and recent experience with UTI due to S.m. corroborates that of other authors (1, 12, 22, 35, 68, 69). About two-thirds of the cases occur in patients over 50 years. This would be in line with UTI caused by most other organisms, but, in contrast to them (44), S.m. shows a predilection for males over 50 (male:female ratio about 2.5:1). In one series, only about two-thirds of the cases were nosocomial (68); in our experience, the percentage is higher (up to 90 %). The organism generally makes its appearance in the UT after the second week of hospitalization; often as a supercolonizing or superinfecting agent, and mixed with other potential pathogens in at least one-third of the cases. Almost all patients have been instrumented or catheterized, ci. two-thirds with an indwelling catheter. Over three-quarters have received multiple antimicrobials to which S.m. is resistant (ampicillin, penicillin, cephalotin, colistin, nitrofurantoin). Half of the patients have UT abnormalities or are postoperative, and up to one-third may be diabetic. In a recent study, it was found that patients with S.m. colonization differed significantly from non-colonized ones in: duration of hospitalization, mean number of antimicrobials received previously, incidence of supercolonization or superinfection, and Foley catheterization (incidence and duration) (36). There was no difference in the incidence of underlying chronic debilitating conditions, age, or sex.

S. m. strains are nowadays most often sensitive to gentamycin and nalidixic acid (63, 68, 69). In line with the results of other authors (36), we found 54 UT strains isolated in the first quarter of 1973 less sensitive than 28 respiratory strains to carbenicillin (7 % vs. 79 %), tetracycline (0 % vs. 25 %), kanamycin (35 % vs. 96 %), and chloramphenicol (15 % vs. 86 %). No difference was noted with gentamycin or nalidixic acid.

5. Urinary Tract Infection (UTI) due to Proteus rettgeri (P. r.)

P. r. is the least frequently isolated Proteus species in the United States (22, 63, 65); in Great Britain, P. morganii was reported to be the least common species in three series (8, 14, 40). UT strains comprise the majority of isolates. This laboratory recovered only eight strains from inpatients in the first half of 1973.

P. r. shows more often multiple antimicrobial resistance than do other Proteus species; nalidixic acid, gentamycin, and carbenicillin are the most effective drugs (63). There is a predilection for males (about 4:1) (47, 65), and, like in other UTIs, for the age group over 50 years (65). Most of the P. r. strains are nosocomial in origin, the patients giving a history of indwelling catheterization, previous hospitalization(s), and/or previous antimicrobial treatment (14, 47, 65). Random and cross infections as well as colonization have been observed. In many cases, P. r. disappeared upon catheter removal. The source of random isolates may be the stool (14). Mixed isolates are more frequent than pure ones (14, 22, 47, 65).

Recently, three outbreaks of P. r. UT infections or colonizations have been recorded in the United States, the strains being resistant to streptomycin, sulfonamides, ampicillin, cephalotin, carbenicillin, kanamycin, gentamycin, tetracycline, colistin, nitrofurantoin, chloramphenicol, nalidixic acid, and rifampin. The first outbreak involved 51 patients in a North Carolina hospital and was caused by a lactose-positive P. r. strain unable to transfer antibiotic markers to E. coli K 12 (60). A common source could not be traced down, although the utility room, but none of the personnel, was found contaminated. Four patients also had stool cultures positive for P. r..Identity of the strains was proven by proticine testing (9). Two-thirds of the patients were colonized only. In the second outbreak involving ten patients in an Illinois hospital (10), no source was found either. Spread was controlled by contact isolation of patients and through use of gloves by personnel. None of the four deaths was directly attributable to P. r. infection. The third outbreak, in a Missouri hospital, was due to an indole-negative strain different from P. mirabilis in ornithine and inositol reactions (57).

6. Urinary Tract Infection (UTI) due to Providencia stuartii (P. s.)

Providencia strains also belong to the rare gram-negative rods: only 25 were isolated from inpatients at Yale-New Haven Hospital in the first half of 1973, mostly from urines. Speciation is often omitted. The Center for Disease Control reported more strains of P. alcalifaciens (4 biogroups) than of P. stuartii (2 biogroups). However, 35 of 45 strains isolated in one series (55) from clinical specimens and representing at least nine different biotypes, belonged to P. stuartii.

Nosocomial infection, chiefly of the UT, is the usual type of Providencia disease (5, 14, 22, 40, 55). However, a recent report from the U.S. Army Institute of Surgical Research (11) stresses the rising incidence of P. s. pneumonia and septicemia in burn patients, surpassing similar infections due to Pseudomonas aeruginosa but showing a similar fatality rate. UTI in burn patients due to P. s. was even seen more frequently than UTI due to E. coli, Proteus spp., Enterobacter spp., and P. aeruginosa. Since many strains were found on admission as colonizers, the authors have suggested a widespread geographical distribution of the organism (11). Indwelling catheterization or other UT instrumentation and intensive antimicrobial treatment are found in the history of most patients with P. s. UTI. One group, patients with neurogenic bladder dysfunction, particularly with paraplegia, seems to be preferred. Most of them are male and younger than 40 (8, 22, 40, 55). The strains are fairly resistant to antimicrobials; the most effective ones being genta-mycin, carbenicillin, nalidixic acid (5, 55, 63) and, possibly, trimethoprim (8). A rising occurrence of gentamycin-resistant strains has been reported recently (11). Infection seems more frequent than colonization, and the majority of strains is found in mixed culture (22). Cross infections as well as endemic ones have been ob-served (5, 55). Transmission occurs probably through mechanisms similar to those operating in P. rettgeri. The source of endemic strains may be the stool; gastroin-testinal infections associated with Providencia have been reported (22).

7. Urinary Tract Infection (UTI) due to Pseudomonas cepacia (P.c.)

P. cepacia (P. multivorans, P. kingii, EO-1) (61 d) is the prototype of pseudo-monads that resist the action of quaternary ammonium compounds and of chlorhexi-dine (29) and even multiply in their presence. Its natural habitat are soil and water sources. It was a very rare isolate at Yale-New Haven Hospital in the first half of 1973 (2 strains).

Contaminated solutions of disinfectants, when used as catheter-irrigating fluids, have given rise to nosocomial UTI and colonization (29, 41). Such solutions appear optically clear. A recent report reviewed 26 patients with P.c. in the urine (15). All had significant underlying disease and had been instrumented (12 from proven contaminated catheter kits). Most had received antibiotics prior to the positive cul-ture. The mean duration of hospitalization at the time of the initial isolation of P.c. was 19 days. The organism disappeared from the urine independently of the anti-microbial regimen. P.c. is resistant to most antimicrobial agents except sulfon-amides, chloramphenicol, and nalidixic acid (29, 61 d).

8. Pseudomonas maltophilia in Nosocomial Disease

This bacterium is the third most frequent non-fermentative gram-negative rod isolated in clinical laboratories. At Yale-New Haven Hospital, 40 strains (vs. 403 of P. aeruginosa and 53 of Acinetobacter anitratus) were isolated during the first half of 1973 from inpatients. In spite of its requirement for methionine, it is a ubi-quitous organism (61 d). However, there are relatively few cases of infection re-ported in which P. maltophilia could be incriminated as the sole infectious agent. Most strains occur either in mixed culture and/or are found as supercolonizers in hospitalized patients (24, 27 a, 27 b, 49). Even blood isolates are significant only in a minority of cases (56, 61 a, and further own observations). Consistent sensi-tivity is nowadays observed only against chloramphenicol and nalidixic acid (61 d).

9. Some Comments on Mixed Cultures Involving Gram-Negative Rods

Quite a few strains of the aforementioned gram-negative rods will be found in mixed culture, the interpretation of which may present problems to bacteriologist and physician alike. A few remarks on this topic are in order.

Correct collection, transport, storage, preparation, selection of media, inocul-ation, and incubation are essential but cannot be discussed here. Particular empha-sis must be placed on the recovery of anaerobes from blood, body fluids, wounds, pus, tissues, and transtracheal aspirates.

A mixed culture involving gram-negative rods may represent:
1. resident normal flora (i.e., non-enteropathogenic E. coli in the large in-testine),
2. contaminants originating:
   a) from a non-patient source (i.e., air, equipment),
   b) from bacteriological media,
   c) from a site adjacent to the one to be cultured (i.e., throat in sputum cultures, skin in cultures of blood or body fluids),
3. colonizing organisms,

4. infecting organisms,
- or any combination of these.

The bacteriologist's knowledge of what constitutes normal flora is critical for the interpretation since the physician has no chance for a review of the culture. Repeat cultures for the elimination of possible sources of contamination are called for in cases 2 a and 2 b, and in case 2 c if blood or body fluid cultures seem likely to be contaminated with skin flora. Bacterial counts do not differentiate contaminants from infecting agents since both may occur in low density.

More difficult is exclusion of contaminants from sputa and urines. Washing of sputum or quantitation of its flora are of no value in eliminating or differentiating throat flora (2, 53). The presence of oropharyngeal epithelial cells would strongly suggest contamination; but in many instances, transtracheal aspiration remains the only solution (53), particularly if anaerobic infection is suspected. Quantitation of organisms is, on the other hand, of prime importance in urines (32). If a mixed culture with an overall or individual count between 10^4 and 10^5 per ml is found in repeat specimens, percutaneous suprapubic aspiration may be called for (42). Quantitative cultures of wounds may also be valuable since many infections, including gram-negative ones, (but not those with Clostridium tetani or beta-hemolytic streptococci), develop only if bacteria are present in quantities of over 10^5 per gram of tissue (50).

Serological tests (agglutination, precipitation, opsonization, complement fixation) can be used to measure the host response, but they may react to mere colonization as well. The test found to be of greatest value in gram-negative infections is the indirect hemagglutination test (45 a - c). In quantitative culture, colonizers may be the ones with low counts (e. g., $<10^5$ in wounds, see above), but they may show low or high counts in urines. But in contrast to infecting agents, they do not elicit a local tissue response, or a systemic cellular or clinical one.

Thus, the differentiation between colonizing and infecting agents must be made on clinical and/or microscopical grounds. The presence of polymorphonuclear leukocytes has long been used to differentiate colonization from infection in patients with intact cellular defenses - provided contamination has been excluded. The greatest difficulties arise, however, if colonizers and infecting agents are present at the same time. If the two have different antimicrobial sensitivities, the outcome of in vivo chemotherapy may, in retrospect, give a clue as to the infecting agent.

References

1. Allen, S. D., Conger, K. B.: J. Urol. 101: 621 (1969).
2. Bartlett, R. C., Melnick, A.: Conn. Med. 34: 347 (1970).
3. Bodner, S. J., Koenig, M. G., Goodman, J. S.: Ann. intern. Med. 73:537 (1970).
4. Brachman, P. S.. In: Williams, R. E. O., Shooter, R. A.: Infection in Hospitals, p. 329 (Davis, Philadelphia 1963).
5. Brühl, P., Müller, U.: Münch. med. Wschr. 108: 467 (1966).
6. Cate, J. C.: Proc. 7th Intern. Congr. Chemother., p. 763 (1971).
7. Clayton, E., von Graevenitz, A.: JAMA 197: 1059 (1966).
8. Colley, E. W., Frankel, H. L.: Paraplegia 2: 132 (1964).
9. Craddock, M. E., Traub, W. H.: Experientia 27: 980 (1971).
10. Cross, A., Landau, W., Lavin, S., Edwards, L. D.:Abstr. Intersci. Conf. Antimicrobiol. Agents chemother. 172 (1972).
11. Curreri, P. W., Bruck, H. M., Lindberg, R. B., Mason, A. D., Pruitt, B. A.: Ann. Surg. 177: 133 (1973).
12. Davis, J. T., Foltz, E., Blakemore, W. S.: JAMA 214: 2190 (1970).
13. DeFronzo, R. A., Murray, G. F., Maddrey, W. C.: Amer. J. digest. Dis. 18: 323 (1973).
14. Dutton, A. A. C., Ralston, M.: Lancet I: 115 (1957).

15. Ederer, G. M., Matsen, J. M.: J. infect. Dis. 125: 613 (1972).
16. Edwards, P. R., Ewing, W. H.: Identification of Enterobacteriaceae (Burgess, Minneapolis, Minn. 1972).
17. Ewing, W. H., Fife, M. A.: Enterobacter agglomerans - The Herbicola-Lathyri Bacteria. Center for Disease Control, Atlanta, Ga. (1971).
18. Farmer, J. J.: Appl. Microbiol. 23: 218 (1972).
19. Felner, J. M., Dowell, V. R.: Amer. J. Med. 50: 787 (1971).
20. Felsby, M., Munk-Andersen, G., Siboni, K.: J. med. Microbiol. 6:413 (1973).
21. Felts, S. K., Schaffner, W., Melly, M. A., Koenig, M. G.: Ann. intern. Med. 77: 881 (1972).
22. Fields, B. N., Uwaydah, M. M., Kunz, L. J., Swartz, M. N.: Amer. J. Med. 42: 89 (1967).
23. Finegold, S. M., Harada, N. E., Miller, L. G.: J. Bact. 94: 1443 (1967).
24. Gardner, P., Griffin, W. B., Swartz, M. N., Kunz, L. J.: Amer. J. Med. 48: 735 (1970).
25. Gelb, A. F., Seligman, S.: JAMA 212: 1038 (1970).
26. Gifford, R. R. M., Lambe, D. W., McElreath, S. D., Vogler, W. R.: Amer. J. med. Sci. 263: 157 (1972).
27. Gilardi, G.: a) Amer. J. clin. Pathol. 51: 58 (1969); b) Ann. intern. Med. 72: 211 (1972).
28. Hamilton, R. L., Brown, W. J.: Appl. Microbiol. 24: 899 (1972).
29. Hardy, P. C., Ederer, G. M., Matsen, J. M.: New Engl. J. Med. 282:33 (1970).
30. Jones, E. A., Crowley, N., Sherlock, S.: Postgrad. med. J., Suppl. 43: 7 (1967).
31. Kagnoff, M. F., Armstrong, D., Blevins, A.: Cancer 29: 245 (1971).
32. Kass, E. H.: Trans. Ass. amer. Physicians 69: 55 (1956).
33. Ketover, B. P., Young, L. S., Armstrong, D.: J. infect. Dis. 127: 284 (1973).
34. Kovarik, J. L., Sides, L. J., Becky, J. R.: Rocky Mtn. med. J. 70: 36 (1973).
35. Lancaster, L. J.: Arch. intern. Med. 109: 536 (1962).
36. Maki, D. G., Hennekens, C. H., Phillips, C. W., Shaw, W. V., Bennett, J. V.: J. infect. Dis. 128: 579 (1973).
37. Marcoux, J. A., Zabransky, R. J., Washington, J. A., Wellmann, W. E., Martin, W. J.: Minn. Med. 53: 1169 (1970).
38. McCracken, A. W., Barkley, R.: J. clin. Path. 25: 970 (1972).
39. Meyers, B. S., Bottone, E., Hirschman, S. Z., Schneierson, S. S.: Ann. intern. Med. 76: 9 (1972).
40. Milner, P. F.: J. clin. Path. 16: 39 (1963).
41. Mitchell, R. G., Hayward, A. C.: Lancet I: 793 (1966).
42. Monzon, O. T., Ory, E. M., Dobson, H. L., Carter, E., Yow, E. M.: New Engl. J. Med. 259: 764 (1958).
43. Morbidity and mortality weekly report. Center for Disease Control, Atlanta, Ga.: a) Spec. Suppl. 20/9 (1971); b) 20/11 (1971); c) 22/11 (1973); d) 22/13 (1973).
44. Mou, T. W., Siroty, R., Ventry, P.: J. amer. Geriat. Soc. 10: 170 (1962).
45. Neter, E.: a) Yale J. Biol. Med. 44: 241 (1971); b) In: Losse, H., Kienitz, M., Pyelonephritis, p. 101 (Thieme, Stuttgart 1972); c) In: Prier, J. E., Friedman, H., Opportunistic Pathogens (University Park Press, Baltimore, Md., in press).
46. Nobles, E. R.: Ann. Surg. 177: 601 (1973).
47. Omland, T.: Acta path. microbiol. scand. 48: 221 (1960).
48. Pearson, T. A., Mitchell, C. A., Hughes, W. T.: Amer. J. Dis. Child. 123: 579 (1972).
49. Pedersen, M. M., Marso, E., Pickett, M. J.: Amer. J. clin. Path. 54: 178 (1970).
50. Robson, M. C., Krizek, T. J., Heggers, J. P.: Curr. Probl. Surg. (1973).

51. Sanders, C. V., Luby, J. P., Johanson, W. G., Barnett, J. A., Sandford, J. P.: Ann. intern. Med. 73: 15 (1970).
52. Schoutens, E., Labbé, M., Yourassowsky, E.: Path. Biol. 21: 349 (1973).
53. Schreiner, A., Digranes, A., Myking, O., Solberg, C. O.: Infection 1: 137 (1973).
54. Slotnick, I. J.: Ann. N. Y. Acad. Sci. 174: 503 (1970).
55. Solberg, C. O., Matsen, J. M.: Amer. J. Med. 50: 341 (1971).
56. Sonnenwirth, A. C.: Ann. N. Y. Acad. Sci. 174: 488 (1970).
57. Sonnenwirth, A., Hermann, G. J.: Abstr. Annu. Meet. Amer. Soc. Microbiol. M 190 (1972).
58. Soule, T. R., Cunningham, G. R.: JAMA 223: 1265 (1973).
59. Traub, W. H., Raymond, E. A., Startsman, T. S.: Appl. Microbiol. 21: 837 (1971).
60. Traub, W. H., Craddock, M. E., Raymond, E. A., Fox, M., McCall, C. E.: Appl. Microbiol. 22: 278 (1971).
61. von Graevenitz, A.: a) Med. Welt 54: 177 (1965); b) JAMA 216: 1485 (1971); c) Zbl. Bakt. I. Abt. Ref. 233: 47 (1973); d) Prog. clin. Path. 5: 185 (1973).
62. von Graevenitz, A., Mensch, A.: New Engl. J. Med. 278: 245 (1968).
63. von Graevenitz, A., Nourbakhsh, M.: Med. Microbiol. Immunol. 157:142 (1972).
64. von Graevenitz, A., Sabella, W.: J. Med. 2: 185 (1971).
65. von Graevenitz, A., Spector, H.: Yale J. Biol. Med. 41: 434 (1969).
66. von Graevenitz, A., Zinterhofer, L.: Health Lab. Sci. 7: 124 (1970).
67. Washington, J. A.: Ann. intern. Med. 76: 611 (1972).
68. Wilfert, J. N., Barrett, F. F., Ewing, W. H., Finland, M., Kass, E. H.: Appl. Microbiol. 19: 345 (1970).
69. Wilkowske, C. J., Washington, J. A., Martin, W. J., Ritts, R. E.: JAMA 214: 2157 (1970).
70. Wilson, W. R., Martin, W. J., Wilkowske, C. J., Washington, J. A.: Mayo clin. Proc. 47: 639 (1972).

Anaerobic Bacteria and their Role in Human Infections

E. H. Spaulding

I appreciate the opportunity to extend some of the remarks Dr. von Graevenitz made about anaerobic bacteria and their role in human infections. I have three points to make.

My first point is that when proper collection and laboratory methods are used, anaerobic bacteria can be recovered from about 40 % of specimens from wounds, post-operative infections, abscesses, infections of the respiratory tract and particularly those of the female genital tract (which yield in our hands 85 to 90 % positives). But special methods are necessary in order to obtain such results. Specimens should be collected with needle and syringe, injected into a tube containing O_2-free gas and delivered to a laboratory prepared to process them with minimum exposure to air.

My second point is that about one-half of the anaerobic species we have found in human infections are obligately anaerobic gram-negative bacilli. Therefore I wish to direct the attention of this seminar to the potential significance of these previously neglected species.

The two principal genera of gram-negative anaerobic bacilli are Bacteroides and Fusobacterium. These bacteria contain endotoxin that is similar chemically to the endotoxin in the facultative pathogenic gram-negative bacilli. It was recently reported, for example, that the polysaccharides of species in both genera may contain ketodeoxyoctonate. The sera of patients with systemic Bacteroides infection may be positive for the Limulus test and show a rise in specific antibody titers when successive specimens are compared. Therefore, they may play more of a role in gram-negative shock than is presently suspected.

My third point is that anaerobes are typical opportunistic pathogens. Anaerobic infections usually stay localized in persons who are otherwise healthy; there is only a slight antibody response, and clinical recovery can be expected. But when the defense mechanisms of the host are lowered severely, fatality rates in some reports have been as high as 75 %. Consequently I believe that clinicians should consider the possibility of anaerobic infections more often than they have in the past.

I should like to present some data to support the statements I have just made. In a series of about 850 specimens from a variety of sources, including 100 blood cultures, anaerobes were recovered from 38 % (Table I). Note that the average number of different species per positive specimen was 2.3, a finding that reflects the multiple anaerobic flora so characteristic of anaerobic infections. Note also that about one-fourth (83) of the 322 anaerobe-positive specimens contained only anaerobes, i. e., no facultatives (aerobes), thus indicating that they were etiologic factors in the infectious process, at least at the times the specimens were collected.

I would like to call your attention to the high percentage of gram-negative bacilli among the 750 anaerobic isolates in this study (Table II). Note that 47 % are gram-negative bacilli, and that almost one-third (30 %) are anaerobic cocci. Gram-positive bacilli account for only 22 %, and the spore formers (Clostridium) for only 4 % of the total. Thus, in human anaerobic infections one is dealing primarily with two groups, gram-negative bacilli and cocci (most gram-positive).

The data in Table III serve to emphasize the frequent occurrence of one species, Bacteroides fragilis. Although 88 different species were identified among our 750 isolates, this one species accounted for 15 % of the total. B. fragilis is also the species whose pathogenicity has been documented most thoroughly, and in contrast to most anaerobes it is highly resistant to penicillin. Therefore we believe clinical microbiology laboratories which do anaerobic bacteriology should determine whether the anaerobic gram-negative bacilli they isolate are, or are not, <u>B. fragilis</u>. The March 1974 issue of Applied Microbiology will contain an article describing a practical method for doing this.

My final comment is made to emphasize a statement by Dr. von Graevenitz that there is now a need to determine the pathogenic significance of anaerobes in human infections. I will extend this need to include intensive studies on the pathogenesis of mixed anaerobic infections. To this end we reported recently preliminary results with a rabbit virulence test which is promising and is being further studied in our laboratory.

Table I.

Incidence of Anaerobes in Human Infections
  (Anaerobe Lab. - Temple University)

Facultative Species	Anaerobes	
-	-	235
+	-	298
+	+	239
-	+	83
		855

Incidence of anaerobes - 38 %
750 anaerobes isolated,
or ave. of 2. 3/pos. spec.

Table II.

Genus Identification of 750 Anaerobes from Human Infections
        (Anaerobe Lab. - Temple University)

Bacteroides	268	Eubacterium	69
Fusobacterium	68	Lactobacillus	24
Unident. G neg. bac.	14	Propionibacter	23
(47 %)	350	Actino.& Bifido.	11
		Unident. G pos. bac.	11
		(18 %)	138
Peptococcus	108		
Peptostrept.	82	Clostridium (4 %)	29
Veillonella	19	Misc. (1 %)	10
Unident. cocci	14		
(30 %)	223		

Table III.

Species Cultured Most Frequently from Human Infections
(Anaerobe Laboratory - Temple University)

Bact. fragilis	15. 1 %	Pc. magnus	3. 6 %
Bact. melaninogenicus	9. 5	Peptostreptococcus anaerobius	4. 8
Fusobact. nucleatum	2. 3	Psc. intermedius	4. 1
Fusobact. russii	1. 9	Eubact. lentum	4. 3
Peptococcus prevotii	6. 0	Cl. perfringens	1. 6
Pc. asaccharolyticus	3. 7		56. 9 %

Immunoresponse to Gram-Negative Bacteria in Infants: An Aspect of Pathogenesis of Urinary Tract Infections

W. Marget, M. Westenfelder, and C. Galanos

Hemagglutination Antibody Response in Infants

Gram-negative infections in young infants between 3 and 12 months lead to specific antibody responses (ABR). In all cases the ABR was found to be of short duration with maximum titres being reached within two weeks and then decreasing rapidly. In all cases effective chemotherapy was administered either immediately or after testing the sensitivity of the causative organism, i.e. not later than two days after manifestation of the infection.

The most common causative agents in infant infections were found to be Klebsiella, Pseudomonas, E. coli and Proteus. ABR to these four types of infections were measured by the indirect hemagglutination technique (3) using a mixture of antigens (polyvalent antigen) from a number of serologically distinct strains of each species (eight of the most common E. coli types described by Andersen; 16 common Klebsiella types, Proteus reference types; and 42 different Pseudomonas phage types isolated in our hospital).

The curves of Fig. 1 and 2 illustrate examples of the ABR during infections with these four agents. In Fig. 1 the first curve shows the ABR in post-operative Klebsiella septicaemia in a seven month old child. The second curve demonstrates the rapid high increase of the ABT within ten days in a new-born infant with Ps. aeruginosa. In Fig. 2 the first curve illustrates the response to a Pseudomonas infection and compares the titre obtained with the polyvalent antigen with that obtained with the homologous antigen. The second curve shows an E. coli septicaemia in a premature infant diagnosed as apparent respiratory distress syndrome. The last curve demonstrates an umbilical Proteus infection in a one week old infant. Here the titre obtained with the polyvalent agent is also compared to that obtained with the homologous antigen. From these curves it can be seen that the pattern of ABR is characterized by a sharp increase in antibody titres. Although the titres obtained with the polyvalent antigen are lower than those obtained with the homologous, the former are nevertheless high enough to be of significant value. Antibody titres obtained with the polyvalent antigen are low only in infections in very small prematurely born children.

The shaded area in each curve shows the level of normal antibody titres in healthy children at that particular age. In these investigations the results became more specific in younger children since the normal titre, in particular the E. coli titre, decreases in healthy, bacterially uninfected children from a mean of about 1:40 in the first few weeks after birth (Fig. 3) to almost zero between the second and twelfth week. It then rises to between 1:320 and 1:640, this adult titre being reached at about six years of age. With the other gram-negative bacteria the titre in healthy children is considerably lower, as shown in the first and second curves of Fig. 2.

The present results show that infants promptly respond to gram-negative infections with the production of specific antibodies. The titres lie well above those of the normal background and can be measured by the indirect hemagglutination method. Similar results were obtained with a large number of other cases and the reliability of the hemagglutination method as a diagnostic tool has been statistically verified.

Antibody Response in Urinary Tract Infection

These results prompted us to investigate whether the indirect HA method employed above would be of diagnostic value in urinary tract infection (UTI) or in cases of significant bacteriuria.

In a screening program involving over 1500 children between 3 and 6 months of age, from the healthy population, serum titres and bacterial counts in the urine of these children were investigated. The isolates obtained are shown in Table I. In 83 cases there was significant bacteriuria, 75 with monocultures. In 24 cases, significant bacteriuria on two or more occasions was observed over a period of 14 days. Table II shows some of the E. coli types found in these children. (The E. coli typing was carried out by Dr. Sietzen, Microbiological Institute, Frankfurt). The results of K-antigen typing are not yet available. It can be seen that 0:4 and 0:6 are mostly responsible for these early infections in the children investigated. The selection of these E. coli types may have its origin in the interrelation of blood group antigenicity. Table III shows that in the case of types 0:4 and 0:6, which are the most frequent types in the first significant bacteriuria, there is a serological cross-reaction with blood group substances especially those of A and B.

Fig. 4 shows a typical case of asymptomatic bacteriuria comparable with the other 82 cases. The ABR in this child was observed directly after isolation of the causative organism, which was E. coli in the first infection and Klebsiella in the second. With the exception of 14 cases with Proteus infection which showed a low response, all the rest of the 83 children exhibited antibody titres higher than 1:160. The difference with regard to the E. coli species between the healthy children and children with significant bacteriuria was statistically highly significant (Fig. 3). In our screening program the antibody titre of healthy children and infected children were compared at the same time and under the same conditions (Table IV).

Only 15 of the infected children showed leukocyturia at the commencement of the investigation. Despite the fact that only ten children were treated for a short time, the infection disappeared in 13 children. Two infants presented a bilateral pathological isotope nephrogram. On the other hand, eight children with obstructive UTI showed leukocyturia and other symptoms, such as fever and vomiting. A pathological isotope nephrogram was seen in four of these cases.

In 15 of the 83 children with bacteriuria we examined the urine which was collected by suprapubic bladder puncture or catheterisation. The results were positive in only eight children despite the increased ABR. Twenty-five children demonstrated a changing causative organism in repeated occurrences of significant bacteriuria. The majority of the children with pathological findings were either not treated or treated for a short time only. In most cases bacteriuria and leukocyturia disappeared without clinical manifestation.

It is of interest to note that only about 0.5 % of the cases investigated in this study demonstrated bacteriuria and that this manifested itself clinically in an even smaller percentage in the 3 to 16 month age group.

The immune responses measured so far represent antibodies which are specific for the O-polysaccharide chains of the lipopolysaccharide component of gram-negative bacteria. Recently it was shown that the lipid A which is a common constituent of the lipopolysaccharides of gram-negative bacteria is also immunogenic and that anti-lipid A antibodies cross-react with a large number of lipopolysaccharides which are otherwise distinct in their O-specificity. These antibodies were also found in the serum of healthy adults and in older children with recurrent UTI. Anti-lipid A antibodies were not detected in the serum of children up to 12 months.

Lipid A has a high affinity for tissues and membranes, where it can attach itself and thereby passively sensitize for the fixation of anti-lipid antibodies (2). Such cell or tissue-bound immune complexes may then lead to local damage through activation of cellular and humoral defence mechanisms. In experimental model Beinker and

Westenfelder could show that LPS administered in the renal pelvis of rats, distributes itself homogeneously in the parenchyma (1). In an experimental model using adult dogs and puppies we investigated the possible role of lipid A, and anti-lipid A was injected in the renal pelvis of both the young and adult animals under ligation of the ureter (4). Histological investigation showed the presence of lipid A in the parenchyma of both groups of animals. All adult dogs developed an acute abacterial interstitial nephritis. This was not observed in any of the puppies. The serological tests showed the presence of anti-lipid A antibodies only in the serum of the adult dogs. Anti-lipid A antibodies were not detected in any of the animals prior to lipid A injection. These results show that lipid A can induce the development of an interstitial nephritis (IN) in the absence of infecting bacteria. It is not yet clear why only adult dogs developed IN, although lipid A was also found in the kidney of the puppies. However, since the development of IN always coincided with the presence of anti-lipid A antibodies, the latter may possibly be involved in the renal damage observed. In all cases the histological changes disappeared after 6 weeks. These changes were the result of a single injection of lipid A.

The inability of puppies to respond to lipid A with an IN seems comparable to the inability of the most young infants to respond to UTI with symptoms, although the latter, unlike puppies, are capable of low ABR. This property of young infants and puppies may be explained by the phylogenetic relation of the UT to the intestine. The lower part of the UT is derived from the intestine and/or the cloaca (Fig. 5). Thus the property of tolerating and clearing effectively infecting micro-organisms which is shown by the intestine, may also be shared by the UT during the first year of life. Thus bacterial infections and possible infiltration of the kidney with LPS and Lipid A proceed without a pathological reaction. Only when extremely large amounts of living bacteria appear in the UT in the obstructive UTI is it possible to establish the existence of clinically relevant disease. However, even here it does not signify chronic UTI, since these children are often cured subsequent to surgery.

The observations of untreated self-resolving significant bacteriuria in young infants and the presence of lipid A in the kidney of puppies without a pathological reaction suggest that in the early stages of life bacterial colonization of the UT alone, may not be the cause of recurrent UTI.

The question remains as to what colonization of the urinary tract in the first year of life signifies in regard to UTI, since it leads to an ABR considerably lower than that in normal infections and the capacity of eliminating particular strains is also extremely high. Pyelonephritis in older children affects both kidneys at the onset of the illness as investigations by our deceased coworker Wichman (5) demonstrated, and this could be a precondition of the autoimmune mechanism. The initiating factor may be a reaction of lipid A and the tissue, which is probably mediated through ABR. One could speculate that this initial reaction leads to the exposure of otherwise cryptic antigens, thus priming the organism to its own antigenic determinants. Without question there remains much research to be accomplished on this subject.

References

1. Beinker, K. H.: Zur Rattenpyelonephritis: Injektion von Lipopolysaccharid (Endotoxin) aus Escherichia coli- und Salmonella-Keimen. Vergleichende histopathologische und immunfluoreszenzmikroskopische Untersuchungen. Dissertation, München 1973.
2. Galanos, C., Lüderitz, O., Westphal, O.: Eur. J. Biochem. 24: 116 (1971).
3. Neter, E., Steinhart, J., Calcagno, P. L., Rubin, M. I.: Urinary tract infection in children. I. Studies on antibody response. In: Kass, E.: Progress in Pyelonephritis, p.129 (F. A. Davis Company, Philadelphia 1965).
4. Westenfelder, M., Galanos, C., Madsen, P. O.: Experimental lipid A induced nephritis in the dog. In Press.

5. Wichmann, U., Marget, W.: Arch. Kinderheilk. 182: 240 (1971).

Table I.

ISOLATES

48 x E.COLI
14 x PROTEUS SP.
 5 x STAPH. AUREUS
 3 x ENTEROCOCCUS
 2 x KLEBSIELLA
 2 x PS. AERUGINOSA
 1 x STREPTOCOCCUS
────────
75 STRAINS

Table II.

E.COLI ISOLATES

 4 children 04
10 children 06
 2 children 075
 2 children 022
 1 child   023
 1 child   08
 1 child   02

Table III.

FREQUENT AND RARE E.COLI STRAINS IN UT RELATING TO BLOOD GROUPS
(BLOOD GROUP ANTISERA ABSORBED WITH UT ISOLATES)

name	type	AB *)tube #	B	A	homol. antigen	name	type	AB	B	A	homol. antigen
1.M.G.	04	7	5	4	Ø	14.M.D.	023	9	8	8	Ø
2.W.W.	"	7	5	5	Ø	15.M.K.	022	8	8	8	Ø
3.N.W.	"	7	6	4	Ø	16.Mi.K.	" "	7	7	9	Ø
4.C.Z.	"	9	7	–	3	16. " "	" "	9	7	8	Ø
4." "	"	7	7	6	1	17.R.M.	075	7	7	8	Ø
5.A.H.	06	8	5	5	Ø	18.A.W.	02	9	7	8	Ø
6.C.K.	"	7	5	5	Ø	19.K.W.	075	7	9	9	Ø
7.S.M.	"	8	5	4	Ø						
8.T.M.	"	7	5	4	Ø	AB erythrocytes agglutination					
9.S.O.	"	6	5	6	Ø						
9." "	"	8	7	6	Ø	CONTROLS:					
10.A.S.	"	7	4	4	Ø						
11.C.S.	"	7	6	6	Ø	anti AB s. vs. AB erythroc.	9				
11." "	"	7	6	4	Ø						
12.G.S.	"	7	6	4	Ø	anti B s. vs. AB erythroc.	8				
12." "	"	7	7	5	Ø						
13.K.Z.	"	7	6	5	Ø	anti A s. vs. AB erythroc.	9				

(FREQUENT — rows 1–13; RARE — rows 14–19)

*)Haemaggl.titer 1 : 10 = tube # 1

Table IV.

COMPARISON OF E.COLI ABR IN CHILDREN
WITH AND WITHOUT SIGNIFICANT E.COLI
BACTERIURIA IN OUR SCREENING PROGRAM
----

once positive:
    n = 22   titer : $\bar{x}$ = 1 : 165
controls (same period and investigation)
    n = 80   titer : $\bar{x}$ = 1 : 68.6
    t = 2.84 p $<$ 0.01
----

more than one E.coli signif.bacteriuria
    n = 10   titer : $\bar{x}$ = 1 : 208
controls (same period and investigation)
    n = 80   titer : $\bar{x}$ = 1 : 68.6
    t = 3.5   p $<$ 0.001

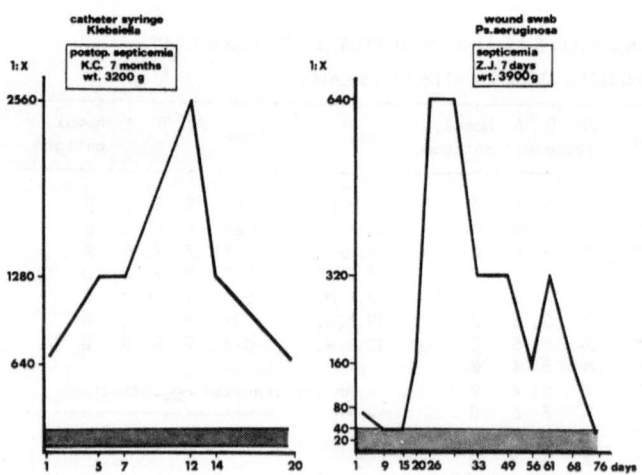

**Antibody titer of different bacterial infections in infants, determined by the
indirect Haemagglutination   [W.Fischer]**

    polyvalent antigen ———
    homologuos titer – – – –

Fig. 1.

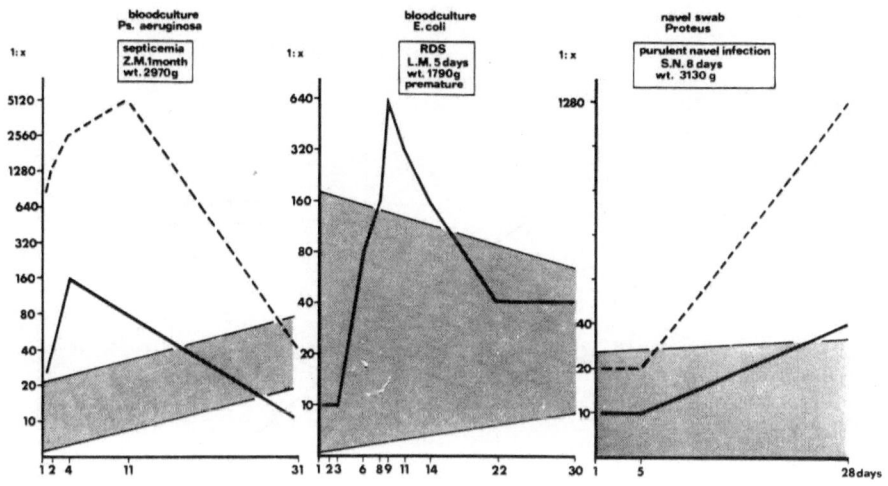

Antibody titer of different bacterial infections in infants, determined by the indirect Haemagglutination [W.Fischer]

polyvalent antigen ———
homologuos titer — — —

Fig. 2.

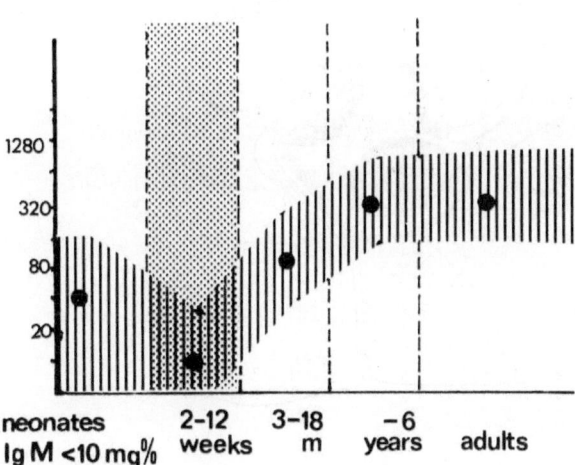

neonates IgM <10 mg%   2-12 weeks   3-18 m   −6 years   adults

Fig. 3.

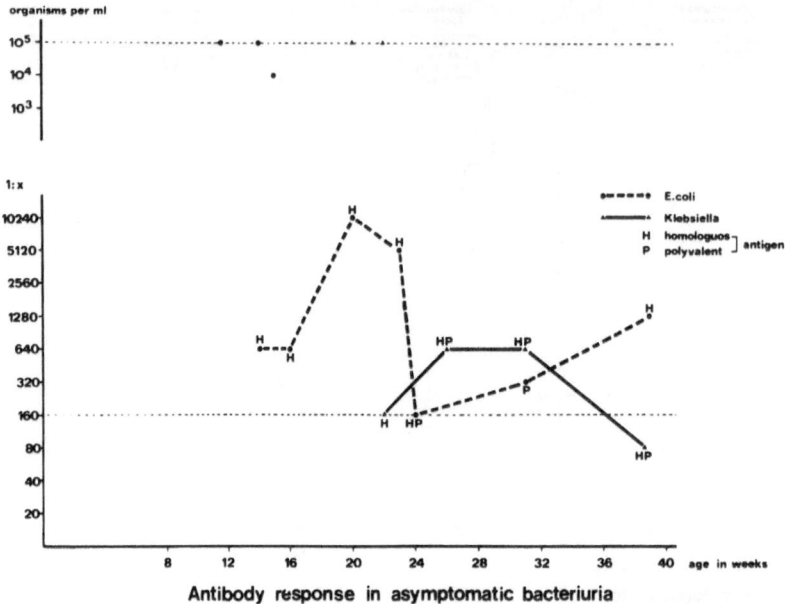

Antibody response in asymptomatic bacteriuria

Fig. 4.

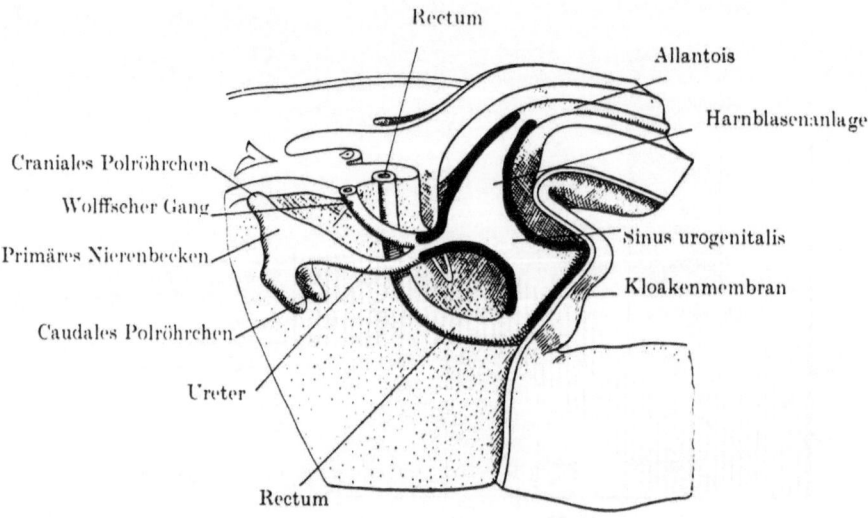

Fig. 5.

The Differentiation of the So-Called Coliform Bacteria in Medical Specimens

C. D. Fritsche and R. Lütticken

Coliform bacteria are characterized by growth on ordinary culture media within 48 hours, with colonies like those of E. coli. The families and genera of bacteria having these common characteristics, are represented in Table I.

With the aid of group characteristics, with the demonstration of the indophenol-oxidase, the aerobic and anaerobic growth and the estimation of motility, the coliforms can be divided into provisional groups. Table I begins with the aerobic genera and families, followed by groups of bacteria, which are able to grow under aerobic and anaerobic conditions: Vibrio, Aeromonas-Plesiomonas, the Enterobacteriaceae and the genus Pasteurella (remark: Pseudomonas aeruginosa is able to grow under anaerobic conditions in the presence of nitrite).

Further differentiation is shown in the Tables II, III and IV. In these we have summarized species and genera of bacteria with similar biochemical patterns.

Table II demonstrates bacteria with biochemical reactions like those of lactose non-splitting E. coli-strains.

Providencia and Edwardsiella tarda are diagnosed by estimation of motility, H_2S-production, and activity of the phenylalanine-deaminase.

The remaining bacteria are nonmotile, anaerogenic strains. Members of the genus Shigella must be defined by serological methods: Indol-producing strains are found only in the Sh. dysenteriae-, Sh-flexneri- and Sh.boydii-group. If the result of this examination is negative, the differentiation must be made between Alcalescens dispar and Pasteurella multocida: Alcalescens dispar is able to grow on MacConkey-agar, in contrast to Pasteurella multocida. On the KCN-medium this behaviour is the opposite.

In Table III bacteria with positive oxidase-reactions and facultative anaerobic growth are listed.

The most important problem at this time is the delimitation of Aeromonas-Plesiomonas and Vibrio. Both genera demonstrate motility, positive oxidase reaction, and facultative anaerobic growth; Vibrio often has the tendency of aerobiosis. They differ in the degradation of amino acids and in the sensitivity to novobiocin. Within the genus Vibrio it is necessary to separate Vibrio el tor from NAG-vibrios by serological methods.

The oxidase-positive Pasteurella multocida-group is characterized by absence of motility, no growth on MacConkey-agar, no degradation of gelatin and no gas production from dextrose.

Table IV summarizes certain Enterobacteriaceae and the Yersinia,until now called Pasteurella,with the common characteristics of a negative oxidase reaction, as well as facultative anaerobic growth.

It is very difficult to determine Citrobacter because of its inconstant biochemical reactions. The only constant pattern is its motility and absence of lysine decarboxylation. If it demonstrates the biochemical pattern of Salmonella, LDC-negative Salmonella must be excluded.

Salmonella paratyphi A is a deviating Salmonella; this must be remembered when isolating an aerogenic, LDC-negative "typhoid"-strain in blood cultures.

The Enterobacter-Serratia-group with the Hafnia is characterized by motility, predominantly no H_2S-production, negative indol-reaction, but citrate and ornithine degradation. Sometimes the delimitation of the species is very difficult. Enterobacter cloacae is unable to decarboxylate lysine, Enterobacter aerogenes, in contrast to other species, often splits lactose. The Hafnia strains are to be separated from Enterobacter aerogenes by inositol and sorbitol utilisation. Enterobacter liquefaciens and Serratia marcescens are related very closely. If they produce no gas from dextrose, they often can be separated only by the degradation of arabinose. Strains that do not split arabinose, are resistant to colistin, and possess a positive LDC-reaction, are diagnosed as Serratia.

The typical Klebsiella is nonmotile. Klebsiella ozaenae is marked by the production of a tenacious slime and inconstant biochemical behaviour. Typical strains of Klebsiella, however, promptly splits urea, citrate, and lactose with positive LDC-reaction, and the gas production from dextrose is usually very strong. There exists no common opinion about the separation of slime producing Klebsiellae as Kl. pneumoniae. The presence of indol producing Kl. oxytoca is confirmed by the change in colour of an iron containing gluconate medium (7).

Kl. rhinoscleromatis is very rare, one should only look for it when there are the typical clinical symptoms.

Near the Klebsiellae is listed the Yersinia-group of Pasteurellae, because these strains are very easily put into the Klebsiella-group. They are diagnosed by the following pattern: splitting of urea, no gas production, lactose and lysine degradation. These Pasteurellae demonstrate the typical motility, motile at 22° C, nonmotile at 37° C. Both species differ in the ability to decarboxylate ornithine.

Since the Yersinia strains often split urea in a delayed manner at first cultivation, it is necessary to examine the motility of strains with the described criteria at 22° C and 37° C. Y. enterocolitica sometimes produces indole and therefore can be confused with Proteus morganii or Alcalescens dispar.

Finally some comments on the Pseudomonas-Alkaligenes-group are presented.

Pseudomonas aeruginosa often demonstrates typical oxidase-positive colonies, a typical odour, a resistance pattern to antibiotics, and the production of pyocyanine. When the diagnosis is doubtful, growth at 42° C and fluorescein production on a gluconate medium in the absence of iron prove the presence of this species. Ps. fluorescens, sometimes found in specimens from the respiratory tract, produces fluorescein on iron-free media.

Pseudomonas maltophilia is often not diagnosed as Pseudomonas strain, because it seldom possesses indophenol-oxidase. It is an oxidase-negative, aerobic, motile rod, with a strong alcalization effect on peptone media, gelatin hydrolysis and production of a brown pigmentation of its colonies on sheep blood agar within 4 days at 37° C. It is methionine-dependent, it splits Na-citrate without methionine only in a small amount, but splits it promptly in the presence of 20 mg/l methionine.

The genus Alkaligenes, belonging to the Achromobacter-group, demonstrates the characteristics of the Pseudomonads, but it is unable to produce pigments and to degradate gluconate and gelatin. If diagnosis is doubtful, it is necessary to examine the inability of these strains to utilize carbohydrate on media containing dextrose as sole source of carbon.

The Acinetobacter-group is divided into carbohydrate splitting and not splitting species. In this group it is possible to identify only Acinetobacter anitratus. The delimitation of further species is under investigation.

This short lecture is made to demonstrate the possiblity of estimating bacteria found in medical specimens. The patterns summarized in the Tables and the text may not always allow a complete differentiation, and in these cases special papers and books must be studied (1-6).

In the future, however, it would be advised to perform a complete differentiation because each group of bacteria possesses different pathogenic properties and the determination of its importance in mixed infections is possible only after correct diagnosis of the species.

References

1. Bergey´s Manual of Determinative Bacteriology, 7th ed. (Tindall and Cox, Baillière 1957).
2. Cowan, S.T., Steel, K.J.: Manual for the Identification of Medical Bacteria (University Press, Cambridge 1970).
3. Edwards, P.R., Ewing, W.H.: Identification of Enterobacteriaceae, 3rd ed. (Burgess Publishing Company 1972).
4. Gilardi, G.L.: Amer. J. clin. Path. 51: 58 (1969).
5. Sonnenwirth, A.C.: Ann. N.Y. Acad. Sci. 174: 488 (1970).
6. Winkle, St.: Mikrobiologische und Serologische Diagnostik (Fischer, Stuttgart 1955).
7. Korth, H., Ørskov, I., Pulverer, G.: Zbl. Bakt. I. Orig. 211: 105 (1969).

Table I.

The Gram-negative Coliform Bacteria.

Bacteria / Reactions	Acineto bacter	Achromo bacter	Ps.malto philla	Pseudo monas	Alkaligenes	Moraxella	Vibrio	Aeromonas Plesio monas	P.multo cida	Entero bacteria	Yersinia (P.pseudo tuberculosis)
Oxydase	⊘	⊘	⊘ / (+)	+	+	+	+	+	d	⊘	⊘
growth — aerobic	+	+	+	+	+	+	+ +	+	+	+	+
growth — anaerobic							+	+	+	+	+
motility	⊘	+	+	+	+	⊘	+	+ (−)	⊘	d	22°C + d 37°C ⊘
carbohydrate utilisation	d		+	+	⊘	⊘	+	+	+	+	+
gas-production				⊘			⊘	d	⊘	d	⊘

Spheroids Gemella: Spheroids fak.anaerob Katalase ⊘

Pigment production, Methionine dependence

Pigment production

increase of growth by serum

Novobiocin + (10 γ)

Novobiocin ⊘ (10 γ)

Table II.

The Shigella, Alcaleszens dispar, Providencia, Edwardsiella, and Pasteurella multocida (Oxydase-negative).

	Shigella	Alcaleszens	Providencia.	Edwardsiella tarda	P.multocida
H$_2$S - production	-	-	-	+	d
Urease	-	-	-	-	-
Indol -production	d	+ (-)	+	+	+
Citrat - utilisation	-	-	+ often delayed	-	-
Dextrose (Gas)	- (d)	-	d	+	-
Lactose - degradation	-	d	-	-	-
Lysine - DC	-	d	-	+	-
Arginine - DH	-	d	-	-	-
Ornithine - DC	-	d	-	+	+
Methylred - R.	+	+		+	-
KCN	-	-	+	-	+
Mac Conkey	+	+			-
motility	-	-	+	+	-
	Aggl. in Sh.dys- enteriae - flexn- eri - und boydii - Sera	No Aggl. in Shigella Sera	Phenylalanin - Deaminase + Colistin-resist.	Phenylalanin - Deaminase negative	pathogenic for laboratory animals

Table III.

Aeromonas-Plesiomonas, Vibrio, and the Oxydase-positive Pasteurella multocida (with P. pneumotropica, P. haemolytica).

	Aeromonas	Vibrio	P. multocida	P. pneumotropica	P. haemolytica
Oxydase	+	+	+	+	+
Growth facultative anaerobic	+	+ / (+)	+	+	+
motility	+	+	-	-	-
Mac Conkey	+	+	-	-	- (d)
Gelatin hydrolysis	d	+	-	-	-
Dextrose (Gas)	d	-	-	-	-
Lactose -degradation	d	(+)	-	d	- (d)
Indol -production	+	+	+	+	-
Urease	-	-	-	+	d
Lysine - DC	d	+ (-)	-	-	-
Arginine -DH	+	-	-	-	-
Ornithine - DC	- (d)	+ (-)	+	d	- (d)
H$_2$S - production	- (d)	-	d	-	- (d)
haemolysis	d	d	-	-	+
	Novobiocin ∅ (10 γ)	Novobiocin + Vaiter: Polymy- xin B & Cholera- -Phage Ⅳ s. in Anti-Inaba or Ogawa +	pathogenic for laboratory animals.		

Table IV.

Some Enterobacteria and members of the Genus Yersinia.

	Citrobacter	S. para-typhi A	Hafnia	Enterobacter aerogenes	Enterobacter cloacae	Enterobacter liquefac.	Serratia	Klebsiella	Kl. ozaenae	Y. pseudo-tuberculos	Y. entero-colitica
Oxydase	−	−	−	−	−	−	−	−	−	−	−
Growth facultative anaerobic	+	+	+	+	+	+	+	+	+	+	+
motility	+	+	+	+	+	+ (−)	+	−	−	22°C + 37°C −	22°C + 37°C −
H_2S production	+ (−)	d	−/(+)	−	−	−	−	−	−	−	−
Urease	d	−	−	−	d	− (+)	− (+)	+	− (+)	+	+
Indol production	− (+)	−	−	−	−	−	−	− (+)	−	−	d
Citrate-utilisation	+ (−)	−	d	+	+	+ (−)	+	+	d	−	−
Dextrose (Gas)	+	+	+	+	+	+ (−)	d	+	d	−	−
Lactose-degradation	(+)/+	−	−/(+)	+ (−)	+ (−)	− (+)	−	+	d	−	−
Lysine-DC	−	−	+	+	−	+ (−)	+	+	d	−	−
Arginine-DH	d		−	−	+	−	−	−	− (+)	−	−
Ornithine-DC	− (+)		+	+	+	+	+	−	−	−	+
	Delimitation of LDC − negative Salmonella	O2 + H2 +	Inositol ø Sorbitol ø	Inositol + Sorbitol +		Arabinose + (−)	Arabinose −	Kl.oxytoca: Pigment on Gluconate with Iron V.R +(−) Malonate +(−)	Serotype 4,5,6. V.P. ø Malonate ø		Mac Conkey + pathogenic for guinea pigs: ø

Changes and Present Status of Resistance to Antimicrobial Drugs in Gram-Negative Bacteria

K. P. Schaal

Apart from its natural (non-acquired) resistance the actual antibiotic sensitivity pattern of a pathogenic species of bacteria results from several different factors. These factors comprise not only features of the antimicrobial agent like chemical structure or mode of action or the special abilities of an individual microorganism. They also include method of application and dosage of the antibiotic substance, composition of case material, and epidemiologic and hygienic peculiarities. Therefore antibiotic sensitivity patterns are primarily of strictly local character and cannot be generalized without further considerations. Even when an evaluation of sensitivity patterns has been performed in an area, it is not easy to deduce changes in sensitivity or therapeutic plans from a single investigation because the situation can deteriorate very quickly by changes of the usual therapeutic schemes, by regrouping of the local spectrum of pathogens, or by acquisition of resistance factors. This is especially true for many of the gram-negative bacteria which are well-known as microbes with a high potential for increasing resistance to antibiotics. In order to obtain results of common interest in spite of the above-mentioned considerations, we tried to delineate basic types of changes in bacterial sensitivity in the region of Cologne with which the current antibiotic sensitivity patterns can be correlated.

Materials and Methods

Results of sensitivity tests of 8144 strains of different species of gram-negative bacteria were evaluated. Strains were isolated from clinical specimens at the Institute of Hygiene in Cologne during the first half of each of the following years: 1965, 1967, 1971, 1972, and 1973. In order to minimize the influence of etiologically insignificant contaminants, we only included strains derived from suppurative processes or infections of the respiratory system. Specimens originated from the university clinics, various hospitals, and practicing physicians in the area of Cologne.

Sensitivity tests were performed according to the method routinely used at our institute: sparse inoculation of pure cultures on DST-agar plates (Oxoid, Code-No. CM 261), application of antibiotic containing disks (multodiscs, Oxoid), pre-diffusion at +4°C for about 3 hours, incubation at 37°C for 10 to 12 hours. For dosage of the antibiotics on the disks see Tables I - III. All clear inhibiting areolas were read as sensitive.

For this survey only the most frequently occurring gram-negative bacteria were considered: Escherichia coli, Klebsiella, Enterobacter species, Serratia marcescens, Proteus species including Providencia, and Pseudomonas aeruginosa. Biochemical criteria were used for their identification.

Results and Discussion

We were able to demonstrate three relatively well distinguishable types of changes in the antibiotic sensitivity patterns in the region of Cologne during the last 9 years. For these comparative studies we selected four groups of bacteria the taxonomical classification of which seemed to be sufficiently constant during the years evaluated, namely E. coli, Pseudomonas aeruginosa, Proteus group, and Enterobacter-Klebsiella-group.

The 1st type of temporal sensitivity pattern embraces

substances which showed a nearly complete activity against all gram-negative bacteria tested without regard to natural resistance. Included in this group are colistin, nitrofurane derivates, the combination of neomycin and bacitracin, and gentamycin as represented in Fig. 1. Although this group of antibiotic and chemotherapeutic agents has been used for many years, their efficacy decreased only slightly or not at all. The percentage of sensitive strains among most of the primarily sensitive species dropped only exceptionally below the 80 %-mark.

It should be mentioned that the apparent loss of sensitivity in 1972 is an artifact resulting from an obvious change in our case material, and is not due to a sudden change in bacterial resistance. This break-down of the average sensitivity occurred not only in gentamycin but practically in all other antibiotics and was nearly fully restored in 1973. In 1972 the number of specimens from the university clinics increased disproportionately in our case material. Therefore the average share of highly resistant hospital strains increased too and caused a depression of the average percentage of sensitivity (1, 2). By the end of 1972 we were able to reduce the number of specimens from the university clinics to a normal proportion and the average percentage of sensitive strains immediately returned to the values of 1971.

The 2nd t y p e o f t e m p o r a l s e n s i t i v i t y p a t t e r n was shown by all penicillin derivates and by the cephalosporines. This type, represented by ampicillin (Fig. 2), is characterized by varying levels of the average sensitivity according to the different bacterial species, but with essentially no change in sensitivity over a long time. In our opinion this is due to a sort of levelling effect which occurs when in a defined region the antimicrobial usage and epidemiological situation remain relatively stable for a longer time. The sensitivity levels of E. coli to ampicillin proved the best of the lot, in the range of 50 to 60 % sensitive strains. Against germs of the Enterobacter-Klebsiella-group ampicillin shows little efficacy. In the Proteus-group there is still a decreasing sensitivity which has not yet levelled off. That may depend on changes in the species composition within the Proteus-group. With regard to Pseudomonas aeruginosa which is generally resistant to the classical penicillin derivates, the sensitivity level to carbenicillin has settled down at 50 to 60 % too (5).

The 3rd t y p e o f t e m p o r a l s e n s i t i v i t y p a t t e r n is characterized by a surprising and for some species considerable increase of sensitivity during the last few years, after the average sensitivity had fallen to a minimum during the period from about 1965 to 1967 (3, 4). In this group of antibiotics and chemotherapeutics are the classical broad spectrum antibiotics chloramphenicol and tetracyclines, and also streptomycin, kanamycin, the sulpha drugs, and probably also the combination of sulphamethoxazole and trimethoprime. The group is represented here by tetracycline (Fig. 3). We suppose that this increase of sensitivity could be caused by a reduced usage of these drugs in the region of Cologne and by a repression of highly resistant hospital strains by hygienic measures.

All gram-negative bacteria examined show this recovery of sensitivity with the exception of E. coli. The latter species remained at a rather constant resistance level for several years. One could speculate that this common human commensal is not able to change its sensitivity patterns as quickly as other gram-negative bacteria which primarily belong to the free living saprophytes and invade the human body often as opportunists.

Having shown representatives of the three main patterns of sensitivity changes in the last 9 years, we will now present current antibiotic sensitivity patterns in the region of Cologne. Table I shows the average activity of penicillins and cephalosporines against gram-negative bacteria. Penicillin G is not the drug of choice in gram-negative infections although there are still some strains especially of E. coli and of the Proteus-group which are sensitive to higher doses of this classical antibiotic. The patterns of ampicillin are in general similar to those of penicillin G

(100 U). In spite of its reputation as a broad-spectrum penicillin derivative it would nearly always be of no use against infections caused by Klebsiella, Serratia marcescens, or Pseudomonas aeruginosa. Carbenicillin is tested in our laboratory in a high dosage as is necessary for treatment of Pseudomonas infections. Nevertheless only slightly more than 50 % of our Pseudomonas strains are proved to be sensitive. The inhibitory potency of carbenicillin for other microbes is known to be lower than that of ampicillin which, however, can be compensated for by a higher dosage.

The cephalosporines, as representative of which we tested cephaloridin, also show no basic differences from the sensitivity patterns of ampicillin with one exception: in our material Klebsiella has a distinctly higher sensitivity, in the range of about 50 % of resistant strains.

The sensitivity patterns of typical bacteriostatic drugs are much less uniform than those of penicillins and cephalosporines (Table II). The least efficacy is demonstrated by the sulphonamides. Only Enterobacter and Proteus species still retain a marked sensitivity to sulpha drugs. Concerning the other gram-negative bacteria it remains to be seen if the beginning restoration of their sensitivity to sulphonamides which was detectable in 1972 and 1973 will continue. There is no doubt that the combination of sulphonamides with trimethoprime has a higher antibacterial activity than sulphonamides alone, but an increase of resistance to this combination started also soon after its introduction, especially in Klebsiella and Pseudomonas strains.

For chloramphenicol and tetracycline Table II shows the already mentioned intensive restoration of the average sensitivity apart from the naturally high resistance of Pseudomonas aeruginosa. Bacteriostatic drugs of generally high activity remain nitrofurazone and also nitrofurantoin although these substances have been used for many years.

The sensitivity patterns of all antibacterial agents shown in Table III generally look rather favourable. Gentamycin, colistin and the combination of neomycin with bacitracin show the previously mentioned low tendency to induce increases of resistance. The nearly total ineffectiveness of colistin against Serratia and Proteus species is not due to an acquired, but to a natural resistance. Therefore gentamycin remains the drug of choice in emergency situations until the individual sensitivity tests have been performed. The usual combination with cephalotin broadens the spectrum of efficiency only slightly in gram-negative infections, but it protects much better in the case of mixed infections of gram-negative a n d gram-positive bacteria. Concerning the use of colistin it should be considered that this therapy can induce changes in the causative agents to naturally resistant Serratia or Proteus strains. In spite of statistically favourable sensitivity patterns of streptomycin and kanamycin the range of application of these substances is limited because of their pronounced toxicity.

Independent of the local situation in Cologne, it is possible to derive the following deductions of common interest from our data:
1. Increases or decreases of bacterial sensitivity to antibiotics may be influenced by preferences in antibiotic usage, dosage, or prevailing indications for antibacterial therapy as is already known for some years.
2. At present it does not seem possible to prevent the development of highly resistant strains especially in the area of hospitals and university clinics.
3. The spreading of such hospital strains within the hospital or among out-patients should be hindered by all possible means in order to keep the average sensitivity at acceptable levels.
4. Recommendations for antimicrobial therapy in emergeny cases are, apart from gentamycin, only of local value and have to be corrected periodically according to the changes in sensitivity patterns.
5. In order to prevent unnecessary decreases of efficacy the antibiotic therapy should be started after sensitivity tests have been performed whenever it is possible. Indis-

criminate usage of multiple combinations, routine non-specific antibiotic treatment, or unnecessarily prolonged therapy are dangerous because these factors appear to promote the development of resistance.

We are aware of the difficulties which may arise in practice from our above mentioned demands. But it seems to be the only way to keep the currently useful antibacterial agents effective for a longer time.

Summary

Results of sensitivity tests on 8144 strains of different species of gram-negative bacteria which had been examined at the Institute of Hygiene in Cologne during the last 9 years are reviewed. The presentation shows that the development of resistance to antibiotics depends on several conditions such as preference in antibiotic usage and dosage of the drugs, kinds of diseases, and special epidemiological situations. Therefore no general statement about the development of resistance is justified without the knowledge of the special local conditions. In the region of Cologne the decrease of bacterial sensitivity to most of the classical antimicrobial drugs stopped during the last 5 years and in 1973 a striking recovery of the effectiveness occurred. Some substances like gentamycin, colistin, or nitrofurazone showed only a very small reduction in their efficacy if at all, in spite of their use for several years. Administration of gentamycin possibly in combination with cephalothin remains the most effective therapy against all infections by gram-negative bacteria. In spite of these rather favourable results, as a rule all antimicrobial drugs should be used when possible only after sensitivity tests have been performed and for specific indications. Otherwise the development of resistance will start again or will be accelerated.

References

1. Fritsche, D., Schulz-Stübner, A.: Dtsch. med. Wschr. 97: 1963 (1972).
2. Fritsche, D., Dostal, Ch., Spieckermann, Ch.: Zbl. Bakt. Hyg. I. Abt. Orig. A 223: 513 (1973).
3. Pulverer, G., Amirlak, F.: Rhein. Ärztebl. Heft 6 (1969).
4. Schaal, K. P.: Hals-Nas.-Ohrenarzt 18: 105 (1970).
5. Spieckermann, Ch.: Med. Welt 36: 1973 (1969).

Table I.
Sensitivity patterns of penicillins and cephalosporines (region of Cologne, W. Germany, 1972/73)

Percentages of sensitive strains

Bacteria	penicillin G 5 U		penicillin G 100 U		ampicillin 5 mcg		carbenicillin 100 mcg		cephaloridin 15 mcg	
	1972	1973	1972	1973	1972	1973	1972	1973	1972	1973
E. coli	0,0	0,6	57,9	61,8	50,9	60,1	75,4	70,5	79,6	87,3
Klebsiella	0,0	0,4	13,8	36,0	3,8	5,4	21,5	35,0	47,7	68,0
Enterobacter group	0,0	0,3	32,4	26,0	19,8	29,3	79,4	77,0	20,6	27,0
Serratia marcescens	0,0	0,0	0,0	0,0	3,3	1,9	16,7	19,8	0,0	1,9
Proteus group	2,9	6,6	41,2	45,6	38,4	42,4	58,3	67,6	42,3	45,0
Pseudomonas aeruginosa	0,6	0,7	3,9	2,8	1,7	1,7	54,1	62,3	1,7	2,1

Table II.

Sensitivity patterns of bacteriostatic antibacterial drugs (region of Cologne, W. Germany 1972/73)

Percentages of sensitive strains

Bacteria	sulphonamides 100 mcg		sulfamethoxazole+trimethoprime 25 mcg		chloramphenicol 10 mcg		tetracycline 10 mcg		nitrofuranzone 100 mcg	
	1972	1973	1972	1973	1972	1973	1972	1973	1972	1973
E. coli	25,1	60,4	70,7	95,2	77,2	79,3	54,5	54,7	100,0	98,0
Klebsiella	15,4	57,2	37,7	86,5	36,9	70,0	36,2	74,5	95,4	98,8
Enterobacter group	41,2	72,3	61,8	94,6	52,9	83,3	70,6	88,0	88,2	97,0
Serratia marcescens	26,7	45,5	80,0	75,7	10,0	41,0	23,3	57,1	86,7	90,3
Proteus group	24,7	53,4	65,0	65,6	42,6	61,6	10,9	22,6	89,1	88,5
Pseudomonas aeruginosa	35,9	51,0	28,2	46,0	6,1	7,8	9,9	34,6	1,1	4,7

Table III.

Sensitivity patterns of streptomycin, aminoglycoside antibiotics and colistin (region of Cologne, W. Germany, 1972/73)

Percentages of sensitive strains

Bacteria	streptomycin 10 mcg		kanamycin 5 mcg		gentamycin 10 mcg		neomycin + bacitracin		colistin	
	1972	1973	1972	1973	1972	1973	1972	1973	1972	1973
E. coli	59,5	63,5	73,7	68,2	97,6	96,8	88,0	84,7	98,8	98,5
Klebsiella	31,5	68,3	35,4	64,5	73,3	91,3	46,2	79,7	95,4	97,6
Enterobacter group	50,0	76,7	61,8	80,0	91,2	97,0	79,4	88,7	91,2	87,7
Serratia marcescens	16,7	28,8	16,7	26,3	80,0	78,2	13,3	52,0	0,0	3,2
Proteus group	69,7	70,8	51,3	60,8	65,5	79,9	80,2	83,1	0,5	3,3
Pseudomonas aeruginosa	11,6	35,3	3,3	6,8	87,8	93,0	82,3	88,2	96,1	98,0

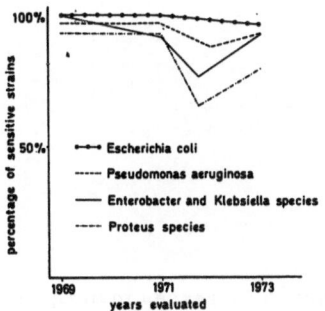

Fig. 1. Changes in sensitivity to gentamycin (10 mcg) in the region of Cologne (W. Germany) from 1969 to 1973 .

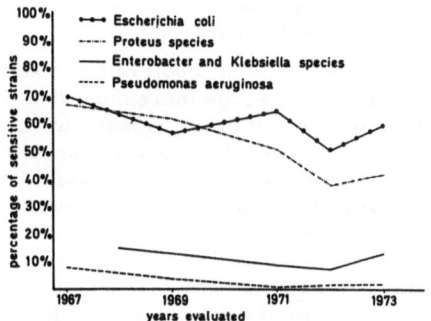

Fig. 2. Changes in sensitivity to ampicillin (5 mcg) in the region of Cologne (W. Germany) from 1967 to 1973.

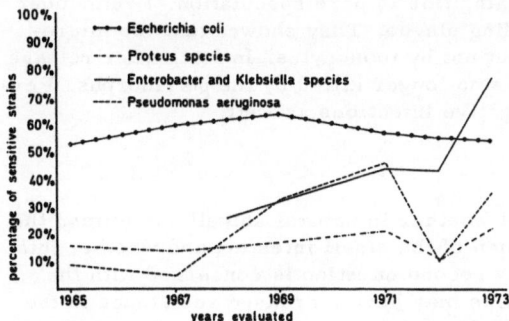

Fig. 3. Changes in sensitivity to tetracycline (10 mcg) in the region of Cologne (W. Germany) from 1965 to 1973 .

Discussion

NETER

SMITH

Springer: Dr. Smith, I would like to ask whether using mucin increases toxicity as well as decreases it.

Smith: I am aware that mucin increases the virulence of organisms, but I am not aware that it will increase toxicity.

Wundt: Dr. Smith are experimental studies known indicating in which way endotoxins are released in vivo besides the Herxheimer reaction following high doses of antibiotics at the beginning of antibiotic treatment? Is anything known about the toxicologic action of enterotoxins on organ systems other than the gut?

Smith: If I could deal with the second question first, I believe that cholera toxin and E. coli toxin only react in the gut, but I am open to correction, by increasing the normal secretion due to activating adenylcyclase. With regard to the release of endotoxin within the tissues, in my opinion, this is one of the most important aspects of the pathogenicity of gram-negative organisms. I do not know of any work elucidating the factors which determine its release. I do remember studies by Elizabeth Work, some years ago, showing that E. coli lipopolysaccharide is released in vitro into the medium or being retained on the microorganisms according to the particular growth conditions. Surely, if release can occur in vitro, it must occur also in vivo. It might be the nutritional conditions which determine its release under the latter conditions, but this is pure speculation.

Rácz, Hamburg/Germany: Regarding Salmonella, intracellular parasitic bacteria, we have to differentiate between leukocytes and macrophages, because the former can kill the bacteria at the beginning of the infection and the latter only if they have been activated.

Smith: Yes, I think it is possible, but again, this is pure speculation. I remember studies by Cavanaugh and Randall regarding plague. They showed that the microorganism may be killed by polymorphs but not by monocytes. In fact, after release from the monocytes the microorganism is no longer killed by the polymorphs. Perhaps, this may happen in other gram-negative infections as well.

FORMAL

Spaulding: I would like to ask Dr. Formal whether in natural shigellosis of man the microorganisms first multiply in the lumen of the small intestine and whether this is a prerequisite for clinical disease. My second question is concerned with the problem whether invasive types of shigellae may have increased resistance to the inhibitory action of the normal flora of the large bowel.

Formal: I do not believe that muliplication of shigellae in the small intestine is a necessary prerequisite to produce classical dysentery (multiple loose stools of small volume which contain blood, mucous, and inflammatory cells), for one can observe this syndrome following the inoculation of virulent bacilli directly into the caecum of monkeys a natural host. However, multiplication in the small bowel may be a necessary factor if watery diarrhea is a part of the clinical picture. I am not aware of any evidence which indicates that virulent shigellae are resistant to the inhibitive action of the normal flora of colon. It is my impression that in many cases, multiplication in the lumen of the colon is limited and the large numbers of shigellae which may be present in the dysenteric stool results from multiplication within the colonic epithelial cells.

Chedid: According to Rowley, the absence of virulence in smooth gram-negative strains could be due to a very minor chemical defect of the cell wall. In view of these theories and of the very interesting tables presented by Dr. Formal, I would like to ask him if the hybrid produced with a smooth mannose-determinant was sensitive to the bactericidal activity of normal serum?

Formal: We have not compared the sensitivities of our hybrid strains to the bactericidal action of serum. For the purposes of our discussion we have used the ability to penetrate epithelial cells as a measure of virulence. Since differences in this can be detected using HeLa cell monolayers, sensitivity to serum is not an important factor. It obviously may be important in the intact host following invasion of the bowel wall.

Clowes: Do non-invasive strains which still produce endotoxin or exotoxin produce fever in animals and/or volunteers?

Smith: May I add to this question? Unless I am mistaken, the mutant which produced invasiveness but not toxicity was less virulent than the strain which produced toxicity and mucosal infection. Does Dr. Formal envisage that the enterotoxin has an effect in this process or is the difference due to some other factor determining virulence?

Formal: Non-invasive mutants of dysentery bacilli do not produce significant rises in body temperature when fed to volunteers. In regard to Dr. Smith's question: the exotoxin elaborated by Shiga's bacillus must play a part in the disease process, but its precise role in pathogenesis is not known. Although Shiga toxin has been classically thought of as a neurotoxin, it probably does not function as such in man. It could act as a cytotoxin as suggested by Vicari, or perhaps as an enterotoxin as suggested by Keusch.

Neter: In the illustration I gave on the family incidence of infection, including shigellosis, it is quite obvious that we are dealing with the identical strain and not with mutants. The clinical manifestations range from severe illness to subclinical infection. Yet, within a few days an antibody response can be shown to the O antigen in both groups of subjects. Therefore, enough bacteria or their antigen must have passed the mucous membrane and reached the immune apparatus to account for this antibody response. How do you explain, then, the differences in the clinical manifestations?

Formal: I cannot explain this. Differences in clinical severity could be explained on the extent to which invasion of the mucosa occurs, and secondly, on the extent of multiplication in the mucosa. The antibody response in an individual with mild disease might be of the same order of magnitude as in one whose illness was more severe.

Berry: The role of enterotoxin in cholera is not as simple as has sometimes been presented. Dr. Neil Guentzel and I have evidence that a highly toxigenic strain of Vibrio cholerae has low virulence in the baby mouse, a good test animal for cholera. Poorer toxin-producing strains that adhere to the wall of the intestinal mucosa are more virulent. We should keep in mind that enterotoxin alone is not the total story in cholera.

Formal: I think that Dr. H. Williams Smith pointed this out with the K 88 antigen. Obviously there are two factors: toxin production and perhaps more important multiplication in the small bowel which would be adherence.

Berry: Toxin-producing avirulent organisms do multiply, so toxicity is not the total story. It may be that the proximity of the vibrios to the mucosa itself plays an important role in pathogenesis.

Smith: The reason why information on the genetics of Vibrio cholerae is not as far advanced as that of E. coli explains our present inability to analyze the pathogenesis incisively.

Springer: Dr. Neter, are the O antibodies of the IgM, IgA or IgG class?

Neter: IgM.

Irwin: Dr. Formal, are your strains starting an inflammatory response?

Formal: We have followed infections with dysentery bacilli using both histological and fluorescent antibody techniques. It is our impression that whenever invasion occurs, an inflammatory process follows.

Neter: I think it is quite obvious that many questions remain to be elucidated in the future. Perhaps, Dr. Urbaschek should organize another symposium to provide answers to these moot questions.

WIEDEMANN

Fey: I would like to ask Dr. Wiedemann regarding the present concept concerning the difference between fimbriae and sex pili. Is it still considered to be correct that fimbriae are responsible for attachment to cells and sex pili for the transfer of genetic material?

Wiedemann: There is a difference, particularly regarding the number; there are only a few sex pili and many fimbriae on a cell. There is a difference in function as well. Sex pili apparently are necessary for the contact during conjugation.

Fey: Since plasmids are relatively easily lost and colicins are controlled by plasmids, how often does it happen that colicin types are changed in vivo affecting the results of epidemiologic investigations?

Wiedemann: First, in my judgment, colicins are not easily lost and, second, what I demonstrated was the linkage between colicin production and colicin genes to other genes in the plasmids. I have no experience regarding the value of the determination of the plasmid character of colicins in epidemiologic studies.

Lowbury: We found that the virulence of Pseudomonas aeruginosa for mice persisted after the acquisition of carbenicillin resistance through transfer of the R-factor RP_1. The continued presence of these organisms in our Burns Unit was apparently due to selection pressure by several antibiotics to which Enterobacteriaceae were commonly sensitive when they were not carrying the R-factor.

Wundt: As is possible with E. coli, do you know if plasmids can be transferred to other Enterobacteriaceae, for instance to salmonallae?

Wiedemann: To the best of my knowledge, definitive information on the transfer to other species is not yet completely elucidated.

Formal: I think that Dr. Smith has transferred enterotoxin to Salmonella and observed that it did not increase the virulence of the organism.

Jachertz: How do you prove that R-factor-bearing bacteria are more virulent than R-factor-free bacteria?

Wiedemann: We know that, for example, tetracycline-resistant mutants have a longer generation time than the susceptible parent strains. Introduction of R-factors into sensitive strains provides additional genetic information without alteration of the chromosome. The generation time of such strains does not differ markedly from the sensitive parents. In continuous culture experiments we observed that resistant strains due to R-factors survive as well in mixed cultures as sensitive strains, in contrast to strains with chromosomal resistance, which survive less well.

von GRAEVENITZ , SPAULDING

Bloch: Dr. von Graevenitz, do you have any evidence that some strains isolated from patients are more pathogenic than others in experimental animals, or do you consider host factors to be of incisive importance?

von Graevenitz: I certainly agree that host factors largely account for the greater frequency of these infections in certain patients, but these microorganisms do not occur exclusively in patients with underlying disease. I have done mouse pathogenicity studies only with Herbicola-Lathyri bacteria in which very large inocula - 10^9, but not 10^6 - caused death. Human infections with bacteria that I have mentioned are seen mostly in compromised hosts or at sites with deficient defense mechanisms, i. e., in situations unrelated to those prevailing with traditional animal pathogenicity studies. Low resistance, no doubt, is a factor determining "pathogenicity"; and most of our patients had low resistance. "Normal" hosts or host sites are either not affected by these bacteria, or react only to large inocula.

Fey: Dr. von Graevenitz, have you studied the immune response of your patients?

von Graevenitz: No.

Neter: We have done rather extensive studies on patients with underlying diseases, such as subjects with cystic fibrosis. Investigating the immune response to Pseudomonas and coliform bacteria isolated from these subjects, we observed that some patients mounted an antibody response and others did not. The clinical importance of colonization may be reflected in the specific immune response.

Clowes: Serratia has been encountered with undue frequency in subjects with organ transplantation and related immunosuppression.

von Graevenitz: I know that in renal homotransplant recipients under immunosuppression, Pseudomonas aeruginosa and Serratia marcescens have been the most frequently encountered gram-negative bacteria causing septicemia. We have only seen a few of those patients and I cannot conclude that they had a particular predilection for Serratia. The infecting organism may reflect the flora prevailing in the feces or on the ward.

Irwin: Dr. von Graevenitz, you mentioned an observation that males are affected more frequently than females in a ratio of 4: 1. Do you have an explanation for this observation?

von Graevenitz: The ratios are about 4: 1 for Proteus rettgeri and 2.5: 1 for Serratia marcescens. On the other hand, urinary tract infections with other Enterobacteriaceae are more frequent in females, even after the age of 50. I have no explanation for the particular sex ratios.

McCabe: Dr. von Graevenitz, your discussion of clusters of patients with Serratia and Proteus infections were not related to age and sex specific infection rates. Is this correct? Could the observation be due to the fact that these infections occur particularly often in the elderly and males are the ones who have indwelling catheters so frequently?

von Graevenitz: While the incidence of urinary tract infection rises in males after the age of 50, it is still higher in females even then. And in most hospitals, including ours, there are no more male than female patients, and the same holds true for patients with indwelling catheters. Yet, Proteus rettgeri and Serratia marcescens infections occur - at least after 50, data for the earlier age periods are too skimpy to be evaluated for significance - more frequently in males. In Providencia stuartii, the preference for males is evident, even at ages under 40.

Fritsche: For years we studied all clinical samples at our Institute in Cologne for anaerobes. I doubt that Bacteroides fragilis is responsible for hospital infections. The natural habitat of Bacteroides fragilis and Bacteroides thetaiotaomicron, the two pathogenic species, is the large bowel. Forty-five per cent of our large bowel flora is Bacteroides, including both pathogenic species Bacteroides fragilis and Bacteroides thetaiotaomicron. In general one is concerned with facultative anaerobes which grow aerobically and one believes to have isolated the organisms responsible for the infectious process. We are certain that hematogenous spread from the large bowel occurs and that Bacteroides fragilis can seed traumatized tissues via the blood stream. We have a case, for example, where osteomyelitis after an old tibia fracture was caused by Bacteroides fragilis. Also we had an old gun shot wound of the knee, said to be healed for many years, which developed an empyema of the joint. The organism responsible in this case was Sphaerophorus necrophorus, the pathogenic strain in this species. I would like to emphasize that we do not believe that Bacteroides fragilis is responsible for hospital infection. Rather it spreads hematogenously from its natural location, the large bowel, to tissues which it infects.

von Graevenitz: I am in firm agreement with Dr. Fritsche; it is a matter of how one defines "nosocomial". According to the definition of the Center for Disease Control, which most American hospitals use, nosocomial is referred to as to any hospital acquired infection which does not exist before entering the hospital. In reference to this definition most of the cases of Bacteroides fragilis septicemia should be defined as hospital infections - possibly with the exception of those infections arising from nonmedical abortion. In contrast, however, if one uses the definition of Louis Weinstein, which maintains that only those hospital infections of exogenous origin are nosocomial (which are far more difficult to determine) then Bacteroides fragilis septicemia does not belong to hospital infections.

Fritsche: It follows then that every post-operative infection falls under the definition of nosocomial disease.

von Graevenitz: Yes, according to the definition of the Center for Disease Control; with the exception of those cases which were already infected before entering the hospital and which were operated thereafter.

Mergenhagen: I would like to ask Dr. Spaulding regarding his experience with oral infections, particularly in view of the fact that Bacteroides are frequently present in the chronic inflammatory diseases around the teeth, notably Bacteroides melaninogenicus.

Spaulding: Yes, Bacteroides melaninogenicus is one of the common isolates from the mouth and throat and outnumbers Bacteroides fragilis in frequency.

MARGET

Werner, Berlin/Germany: During urinary tract infections, have you noted any change in the secretory IgM system in the urine? I am referring to secretory piece-conjugated IgM which occurs in IgA deficient patients.

Marget: I have not studied this.

Springer: Dr. Marget, would you please comment on the relationship between urinary isolates and blood group antibodies?

Marget: We have absorbed blood group antisera with urinary isolates and observed that these bacteria frequently absorb these antibodies.

Braude: I call your attention to a natural immune disorder in the medulla of the kidney that may be responsible for the difference in age-susceptibility. This is related to concentrating ability. Hypertonicity of the medulla makes the kidney an immunologic desert, so to speak. Phagocytosis and bactericidal reactions cannot occur there because of tremendous tonicity, with osmolality higher than one molar. For this reason I wonder whether or not in children older than one year, concentration in the medulla may already be operative, whereas in younger children and infants concentration of urine is not as good and therefore the medulla not hypertonic so that immune reactions can prevent infection. What is the relative concentrating ability and tonicity of the medulla in children older and younger than one year?

Marget: Of course we are aware of this physiological behavior without an explanation in this field. But until now we have had no experience with children in the age group younger than one year and in the case of development of UTI it must be of considerable interest. In older children urine osmolality decreases when there is infection as shown by our co-worker Wichman, and indeed this may favor the relapses without affecting the other immunologic problems as shown.

FRITSCHE

SCHAAL

von Graevenitz: With regard to Dr. Fritsche's presentation, I would like to draw attention to Serratia rubidaea, first described by Stapp in 1941, and recently reviewed by Ewing et al.. We have isolated 7 strains of this bacterium from patient material. Most strains produce prodigiosin, but, in contrast to Serratia marcescens, they are lactose-positive and ornithine-negative. I am sure that some pigmented Serratia strains belong to that species. My question for Dr. Schaal is whether you observed any differences in the resistance between the various Proteus species. If we tabulate our Proteus isolates according to the number of "gram-negative" antimicrobials to which the average strain is sensitive, resistance increases in the order: P. mirabilis, P. morganii, P. vulgaris, P. rettgeri.

Schaal: Of course, there are distinct differences in the antibiotic sensitivity patterns of different Proteus species in our material, too. In spite of this we grouped them together in our statistics because we wanted to comprise not only sensitivity changes in the single species but also changes in the average sensitivity within a group of closely related species. Such changes might additionally be caused by a shifting from more to less sensitive species within that group, for instance from Proteus mirabilis to P. morganii or P. vulgaris.

Gröschel: I would like to comment on the modern possibilities of diagnosing Enterobacteriaceae using miniature tests. In our Institute we have had excellent experience with API 20 and also with r/b 2 and 4 tube-system. Using these systems many of the steps mentioned can be eliminated and in a matter of 2 days exact identification can be made. I think this is necessary, especially in situations, i. e. in a hospital for cancer patients.

Fritsche: We have had good experience with differentiation of Enterobacteriaceae also using the new entero-tube-system. We believe that this is an essential simplification of the routine diagnostic procedure.

Springer: What is the pathogenic potential of <u>Bacterium anitratum</u>?

Fritsche: We have isolated this microorganism in spinal fluid and blood from patients with meningitis and sepsis. When isolated in pure culture in these diseases, its pathogenic role is clearly established. When isolated together with other microorganisms in other diseases, the problem of its role is not easily resolved.

Neter: When Dr. Urbaschek developed the program, one particular area could not be included for discussion, namely, the question of the microbial persisters and their pathogenic role. Dr. Braude, would you kindly comment on this topic?

Braude: Persisters might be defined as non-pathogenic forms of bacteria, that can remain in the tissues after treatment or immune reactions and later cause an exacerbation. Most of the interest connected with persisters has been in the urinary tract because the hypertonic renal medulla is appropriate for survival of cell wall defective organisms that tend to undergo lysis in isotonic body fluids. There is no question that the tonicity of the renal medulla is high enough to allow the persistence of spheroplasts and protoplasts. As I indicated during my question to Dr. Marget the medulla of the kidney has a tonicity over a thousand millimoles per kg mainly due to urea. It has been difficult, however, to prove that spheroplasts or L-forms are present in the human renal medulla. Nevertheless, there is sufficient indirect circumstantial evidence, and theoretical basis for postulating that the urinary tract is a very suitable place for gram-negative organisms treated with antibiotics (that act on the cell wall) to persist, revert to bacterial forms and reactivate disease. As far as persistence in other parts of the body is concerned, I think that there is much less theoretical, circumstantial or direct information to support any real possibility that they are important. One reason for this I think is that the kidney medulla is unique in providing the hypertonic conditions that cell wall defective organisms require for such persistence. There are other forms of organisms that may be thought of as potential persisters outside of the gram-negatives. For example, G-colonies, the gonidial forms of staphylococci, are resistant to antibiotics and to immune mechanisms, they are not pathogenic in the tissues, but when antibiotics are removed, they can revert to pathogenic bacterial forms and cause disease. But of course, these organisms are not gram-negatives and are not appropriate subjects for the Symposium today. Finally there are organisms that probably persist in the tissues not in some unusual form as the gonidial colony or as a spheroplast or protoplast, but simply as non-growing non-multiplying forms of typical organisms that cause no trouble because of their low pathogenicity. Under some circumstances, when immunity is suppressed by immuno-suppressive drugs for transplantation and for treatment of cancer, these organisms multiply and produce disseminated infection. I think the best example of this is in patients with pyelonephritis who are a poor risk for transplantation because organisms that cause the pyelonephritis seem to cause no systemic infection until they are given these immuno-suppressive agents. It is very likely that these organisms then fall into the category of persisters.

Chapter II

Immunology

Opposing Effects of Immunity to Endotoxin: Hypersensitivity Versus Protection

A. I. Braude

Immune reactions to endotoxin may be either beneficial or harmful. In non-immunized subjects most, but not all, reactions attributed to immune mechanisms appear to be deleterious. Active or passive immunization, on the other hand, may lead either to protection or hypersensitivity (increased susceptibility). In fact, the two may be present together as the following review will attempt to demonstrate.

Terminology

The terms "immunized" and non-immunized are used here to connote artificial immunization, either active or passive, by the use of vaccines or antisera. Exposure to endotoxins or related antigens also occurs naturally and it is possible that such exposure contributes to endotoxin reactions. Such "natural immunity" is probably a universal condition. In the discussion that follows it is conceded that all subjects, even though not artificially immunized, probably have natural immunity to endotoxin.

1. Reactions to endotoxins in nonimmunized subjects

Beneficial or protective reactions

The first evidence that previously nonimmunized subjects can protect themselves against endotoxin was the demonstration that radioactive lipopolysaccharide was rapidly removed from the blood into the liver (1). Over 95 % of a sublethal dose disappears from the blood of mice or rabbits in a few minutes and 60 % or more appears in the liver (1), presumably in the Kupffer cells (2). If this protective clearance is overcome by increasing the intravenous endotoxin to a lethal dose, then the amount removed by the liver falls to 20 % and at least 1/3 remains in the blood for over two hours. Phagocytosis by reticuloendothelial cells in the liver is probably responsible for this clearance and large doses of endotoxin most likely overwhelm this phagocytic mechanism so that lethal amounts continue to circulate in the blood.

Harmful immune reactions

Stetson was the first to suggest that the inflammatory reaction in the skin of non-immunized rabbits was a form of hypersensitivity (3). He felt that it was analogous to the delayed hypersensitivity elicited by tuberculin. In man, however, we found that intradermal injections of endotoxin produced a more accelerated reaction, and did so in everyone examined among more than 200 volunteers. Although both tuberculin and endotoxin evoked a focal perivascular reaction in human skin*, the reaction to endotoxin was intense during the first 3 hours whereas that to tuberculin was just beginning (4). The perivascular response to endotoxin contained many polymorphonuclear leukocytes and resembled the Arthus reaction as described by Cochrane (5). At 20 hours, the tuberculin reaction had an intense perivascular response, almost entirely of mononuclear cells, whereas the perivascular response to endotoxin had greatly subsided.

A striking difference between the reaction of man to tuberculin and endotoxins was also noted by intradermal injection of each subject's own erythrocytes after the cells had been coupled with tuberculin or endotoxin. In tuberculin positive men, tuberculin-coupled red cells injected intradermally disappeared from the skin within 24 hours.

---

*As noted in small biopsies in volunteers.

In subjects reacting to endotoxin, on the other hand, the endotoxin-coupled red cells persisted at the site of injection for 48 hours, producing a small area of ecchymosis like that seen where untreated red cells were injected. This marked difference between tuberculin and endotoxin, in the presence of red cell markers, is explained by the early intense vascular injury and thrombosis from endotoxin during the developing Arthus reaction. The vascular occlusion blocks removal of the red cells that is later accomplished by the delayed reaction to tuberculin. This delayed reaction removed red cells coupled to tuberculin.

These differences between humans and rabbits in the time course, histology, and effect on sensitized erythrocytes of the tuberculin and endotoxin skin reactions can be explained by the state of natural immunity. Later studies by Lee and Stetson (6) demonstrated an accelerated dermal reactivity to E. coli endotoxin in rabbits that had been exposed previously to endotoxin. The accelerated reactions in rabbits resemble those to endotoxin in human beings. It is possible, therefore, that man, in contrast to rabbits, can develop an accelerated reactivity to endotoxin through natural exposure to the endotoxin of his intestinal flora. Rabbits, on the other hand, generally have no coliform bacteria to provide a comparable sensitization to endotoxin.

Shock and death

After finding that lethal endotoxin shock required a heat-labile factor in serum, Spink and Vick (7) suggested that complement may be essential for the production of shock by endotoxin. The production of Arthus reactions upon intradermal injections of endotoxin in man suggested that complement may also be involved in local tissue reactions to endotoxin (4). Pillemer (8) had already reported that endotoxin inactivated complement in vitro and we had found that normal serum invariably contained antibody to endotoxin (4). In view of these observations, it seemed reasonable to consider that local and systemic injury caused by endotoxin may be mediated by an immune reaction involving complement.

To test this idea, we sought evidence that complement was consumed in vivo during systemic reactions to endotoxin in rabbits. We found that only lethal doses of endotoxin consistently produced a rapid and sustained fall of serum complement levels (9). Complement levels in the serum fell 70 % below normal by one hour after intravenous injection of lethal doses. Since the level of antibody to endotoxin usually fell simultaneously with complement, it is possible that reactions in vivo between endotoxin and its antibody caused "complement-fixation" in vivo similar to that seen after injection of other antigens into immunized animals. Such a reaction would involve all components of complement, beginning with C1, C4, and C2 and later C3 - C9. In the absence of demonstrable antibody, endotoxin can preferentially consume the terminal components C^3 - C^9 with little detectable inactivation of C1, C4 or C2 (6). Hence there are two potential routes for complement inactivation by endotoxin, which may or may not require antibody, and both may be involved in the reaction that occurs in vivo.

Whatever the mechanism, mounting evidence seems to implicate complement as a mediator of toxic reactions to endotoxin in experimental animals and in patients. In animals complement appears to be required for endotoxin to initiate coagulation, chemotaxis, and shock (11). In patients, with gram-negative bacterial septicemia, complement consumption can be related to the occurrence of shock and death as in the experimental animal given endotoxin. McCabe (12) reported a highly significant depression of C3 levels only in patients with lethal bacteremias or in bacteremia complicated by shock. It is conceivable that lethal endotoxemia is equivalent to the systemic Arthus reaction ("Protracted Anaphylaxis"), since there is a remarkable similarity in their physiologic disturbances, serum complement depression, and pathologic findings (13, 14). In both endotoxemia and the systemic Arthus reaction, animals must first survive an acute period resembling anaphylaxis before going on to die from a slowly progressive form of shock. If a sensitive guinea pig is given antihistamines to protect against death from true acute anaphylaxis the animal will die later of a systemic Arthus reaction (15).

2. Immune reactions to endotoxin after immunization

Active or passive immunization can produce either hypersensitivity to endotoxin or increased resistance.

Active immunization

Hypersensitivity is best elicited in mice given small intraperitoneal doses of endotoxin on alternate days for two weeks (16). When challenged with intravenous endotoxin four days after the last dose the mice die within two hours, in contrast to non-immunized controls whose deaths occur after 12 hours. These hypersensitivity deaths resemble anaphylaxis. In addition to accelerated deaths, hypersensitive animals also die from smaller doses of intravenous endotoxin than controls so that the LD_{50} is cut in half in hypersensitive mice. In contrast to hypersensitivity, immunization can also induce increased resistance to endotoxin in mice. Such immunity (tolerance) is achieved by reducing the number of immunizing doses or by increasing the interval between immunization and challenge.

Passive immunization: Lethality

Passive immunization can also confer either protection against endotoxin or hypersensitivity to endotoxin. Rabbit antiserum against the endotoxins in smooth and rough bacteria, as well as rabbit antiserum against the endotoxins in heterologous organisms, prevent death when given intraperitoneally to mice challenged with lethal intravenous doses of endotoxin (17, 18). When similar protection is attempted in mice with rat antisera, instead of protection such sera sensitize mice so that otherwise sublethal doses of intravenous endotoxin produced deaths from anaphylaxis in 1 hour (18, 19). These small doses of endotoxin did not produce anaphylactic deaths in mice given immune rat serum prepared against heterologous endotoxins. The differential effects of protective rabbit antibody and anaphylactic rat antibody are thus attributed to reactions with different antigenic sites on the endotoxin molecule: anaphylactic antibody with oligosaccharide "0" determinants, and protective antibody with a core antigen (19).

Passive anaphylaxis to endotoxin in mice can be prevented by Isuprel, epinephrine and reserprine and by physiocochemical alteration of the serum by dilution or ultra-violet irradiation. When anaphylaxis was avoided by any of these maneuvers, the antisera prevented death from endotoxin (19). Separation of immune globulin fractions on Sephadex columns also eliminated the anaphylactic property and the individual 19S and 7S fractions of immune serum each prevented death from endotoxin.

A striking feature of passive anaphylaxis is a massive clearance of endotoxin into the lungs, that is not seen in protected mice. Up to 20 % of the injected lethal dose localized in the lungs of mice given sensitizing antisera within a few minutes and is followed by anaphylactic collapse. It appears that endotoxin-antibody complexes are trapped in the lung because the peripheral release of vasoactive amines are presumably released after complement is fixed by the aggregates. Both epinephrine and isoproterenol were successful in preventing anaphylaxis through dilatation of pulmonary vessels.

Overall, these studies demonstrate that a given antiserum to endotoxin possesses the remarkable property of mediating both anaphylaxis and protection toward the same toxic antigen, endotoxin.

Passive immunization: The dermal Shwartzman reaction

Protection by passive immunization is not limited to prevention of lethality. We also found that antiserum prevented the local Shwartzman reaction (20). Passive prevention of the Shwartzman reaction resembles passive protection against lethality in several respects. The most important similarity is the ability of antisera, prepared against one species of bacteria, to protect against endotoxins of an unrelated species. By extending the phenomenon of heterologous protection to the Shwartzman reaction we were able to provide further support for the concept that the toxic moiety of bacterial lipopolysaccharides is a molecule common to multiple species of gram-negative bacteria. Conversely, it indicates that antibody to the oligosaccharide units responsible for the "0" antigenic activity is not necessary for protection against endotoxin.

Further insight into the problem of heterologous protection came from the remarkable observations that E. coli antiserum, lacking "0" specific determinants, gave strong passive protection against Shwartzman reactions due either to E. coli or S. typhimurium endotoxins. This showed conclusively that "0" antibody is not necessary for protection and that common protective antigens are present in rough mutants of unrelated bacterial species.

Another similarity between protection against endotoxin death and the local Shwartzman reaction is that the 19S immunoglobulin fraction of antiserum can prevent either of these manifestations of endotoxin.

A third, and especially significant similarity between the protective sera lies in their effect on the clearance of chromate-labelled endotoxin from the circulating blood. Despite protection against both homologous and heterologous endotoxins, only homologous endotoxin is cleared more rapidly from the blood by antiserum. In other words, prevention of the lethal and Shwartzman reactions to endotoxin does not depend on accelerated clearance from the blood.

Passive immunization: Disseminated intravascular coagulation (DIC)

The ability of antiserum to prevent local intravascular coagulation in the skin (Local Shwartzman reaction) suggested that such serum might also prevent DIC and the generalized Shwartzman reaction. Accordingly we did experiments to prevent DIC and renal cortical necrosis with antiserum to endotoxin. Rabbit antiserum to endotoxin injected intravenously 2 hours before the second dose of endotoxin (treatment of DIC) or 96 hours before the first dose (prevention) decreased the frequency of bilateral renal cortical necrosis from 90 % to 18 % (21). DIC was also prevented by antiserum to both heterologous endotoxins and to side-chain-deficient mutants, by the 19S immunoglobulins of antisera, and by hyperimmune globulin treated with mercaptoethanol and iodoacetate.

Studies of the clotting mechanism showed that antiserum to endotoxin prevented the precipitous drop in fibrinogen and platelets that accompanies fibrin deposition in the glomeruli and elsewhere. Thus antiserum to endotoxin can prevent not only renal cortical necrosis but also consumptive coagulopathy during the evolution of the generalized Shwartzman reaction.

Passive immunization: Bacteremia

After showing that antiserum could prevent death, the local Shwartzman reaction, and DIC (the generalized Shwartzman), the next step was to test such antiserum for its ability to treat overwhelming bacteremia due to gram-negative bacilli (22).

In order to produce antiserum with broad protection against different gram-negative bacteria we immunized rabbits with a UDP galactose-deficient mutant (J5) of E. coli 0:111 that is unencumbered by "0" antigen since it cannot incorporate galactose into core lipopolysaccharide. Without "0" side chains core is accessible for producing antisera to a wide range of bacteria with similar cores. Such antisera had previously been shown to prevent unrelated endotoxins from producing dermal and generalized Shwartzman reactions and death.

Bacteremia was produced in rabbits made granulocytopenic with nitrogen mustard and fed in their drinking water either K. pneumoniae, E. coli 017 (multiple antibiotic

resistant), or E. coli 04. Since most rabbits are coliform-free the organisms colonized the gut and overwhelming bacteremia occurred as granulocytes disappeared.
When non-immune serum was given i. v. at the onset of bacteremia with any of these
3 organisms only 3.1 % to 5.7 % survived. In contrast, survival rates in animals
given antiserum to the J5 mutant were 33.3 % from E. coli 04, 40 % from Klebsiella, and 69.9 % from the multiple antibiotic resistant E. coli 017. (Combined P =
.0005). An impressive demonstration of the therapeutic power of the J5 antiserum
is its ability to increase survival beyond that of gentamicin, the only antibiotic to
which E. coli 017 was sensitive. Gentamicin doses equivalent to the largest given to
patients produced a survival rate of only 50 % (22).

In contrast to the J5 antiserum, antiserum to the parent E. coli 0:111 (with "0"
antigen intact) provided no significant protection. These experiments with living bacteria, taken together with those cited above with their endotoxins, offer considerable
evidence that "0" antigenic determinants are not only unnecessary, but even detrimental to the development of protective antisera against endotoxin and bacteremia.

Conclusions

From these considerations it would appear that natural and acquired immunity to
endotoxin may take part in either harmful or protective reactions to endotoxin. The
harmful reactions may produce either tissue damage or systemic injury. The tissue
damage is characterized mainly by an Arthus reaction, while the systemic disturbances produce intravascular coagulation, shock, and death. Complement has been
implicated in each of these and it is likely that the complement system reacts with
endotoxin both through an antigen-antibody reaction or by direct activation of C3 without antibody. Acquired immunity, conferred either by active or passive immunization may increase susceptibility to endotoxin so that death occurs faster and from
smaller doses. These reactions may also involve complement in the course of aggregate anaphylaxis in which endotoxin-antibody complexes, trapped in the lungs, release vasoactive amines.

This form of acquired hypersensitivity may mask the protective power of antiserum against endotoxin, unless the hypersensitivity factor is removed by heat-in-
activation or sephadex fractionation, or blocked by Isuprel or epinephrine. When
these steps are taken, a lethal serum may become strongly protective against endotoxin.

The hypersensitivity antibody appears to be directed against the "0" antigenic side
chains, so that it can be avoided completely by immunization with rough mutants deficient in "0" antigen (18). Antiserum prepared against rough organisms provide excellent protection against virtually all effects of endotoxin, including fever, the local
Shwartzman reaction (20), disseminated intravascular coagulation (21), and death
(18). Such antiserum also protects against overwhelming bacteremia caused by unrelated species of gram-negative bacteria (22). It is against gram-negative bacterial
infections in man that the protective property of antiserum has its most important
potential application.

Summary

Immune reactions to endotoxin may be either harmful or protective. This dual
effect is seen most clearly in certain antisera that possess the remarkable property
of mediating both anaphylaxis and protection toward endotoxin.

References

1. Carey, F. J., Braude, A. I., Zalesky, M.: J. clin. Invest. 37:441 (1958).
2. Cremer, N., Watson, D.: Proc. Soc. exp. Biol. Med. 95:510 (1957).
3. Stetson, C. A. Jr.: J. exp. Med. 101:421 (1955).

 4. Sell, S., Braude, A.: J. Immunol. 87: 119 (1961).
 5. Cochrane, C. G., Weigle, W. O., Dixon, F. J.: J. exp. Med. 110: 481 (1959).
 6. Lee, L., Stetson, C. A. Jr.: J. exp. Med. 111: 761 (1960).
 7. Spink, W. W., Vick, J.: J. exp. Med. 114: 501 (1961).
 8. Pillemer, L., Schoenberg, M. D., Blum, L., Warz, L.: Science 122: 545 (1955).
 9. Gilbert, V. E., Braude, A. I.: J. exp. Med. 116: 477 (1967).
10. Gewurz, H., Shin, H. S., Mergenhagen, S. E.: J. exp. Med. 128: 1049 (1968).
11. Mergenhagen, S. E., Snyderman, R., Phillips, J. K.: J. infect. Dis. 128: S86 (1973).
12. McCabe, W. R.: New Engl. J. Med. 288: 21 (1973).
13. Williamson, R.: J. Hyg. 36: 588 (1936).
14. Weil, M. H., Spink, W. W.: J. Lab. clin. Med. 50: 501 (1957).
15. Weiser, R., Myrvik, Q., Pearsall, V.: Fundamentals of immunology, p. 160 (Lea and Febiger, Philadelphia 1969).
16. Braude, A. I., Siemienski, J.: Bull. N. Y. Acad. Med. 37: 448 (1961).
17. Freedman, H. H.: Proc. Soc. exp. Med. 102: 504 (1959).
18. Davis, C. E., Brown, K. R., Douglas, H., Tate, W., Braude, A. I.: J. Immunol. 102: 563 (1969).
19. Brown, K. R., Douglas, H., Braude, A. I.: J. Immunol. 106: 324 (1971).
20. Braude, A. I., Douglas, H.: J. Immunol. 108: 505 (1972).
21. Braude, A. I., Douglas, H., Davis, C. E.: J. infect. Dis. 128: S157 (1973).
22. Ziegler, E. J., Douglas, H., Sherman, J. E., Davis, C. E., Braude, A. I.: J. Immunol. 111: 433 (1973).

Infections in Immunodeficiency States

M. W. Hess, H. Cottier, J. Schaedeli, and S. Barandun*

Immunodeficiency states in man are characterized by the inability of the organism to mount a normal immune response following antigenic stimulation. Since Bruton's first description in 1952 (6) of a young boy who suffered from recurrent severe infections and who was found to be agammaglobulinemic - a disease now called "infantile X-linked agammaglobulinemia"-, a broad spectrum of immunodeficiency states with defects of humoral and/or cell-mediated immunity has been discovered (reviews: 3, 9, 12, 14). It should be recognized that, in addition to faulty or absent production of humoral antibodies and/or sensitized lymphocytes ("primary specific immunodeficiency"), lowered resistance to infectious disease may also be due to 1. immunoglobulin hypercatabolism or loss (e.g. protein-losing enteropathy); 2. exogenous physicochemical causes, e.g. cytotoxic or immunosuppressive agents, ionizing radiation; 3. concomitant diseases, e.g. metabolic disorders, such as Diabetes mellitus, viral diseases, or neoplasia of the lymphoreticular or hemopoietic system; or 4. dysfunction of phagocytic cell systems (e.g. chronic granulomatous disease). The present review is largely limited to findings pertaining to the group of "primary specific immunodeficiency states". These rare but well-studied cases offer an opportunity for comparison of basic functional and morphological defects with current views and hypotheses of pathogenetic mechanisms.

It may be recalled that, according to current immunological notions, development and maintenance of immune reactivity in mammals is considered to be regulated by two independent populations of immunologically competent lymphocytes: "thymus-derived" (T-) and "bone-marrow-derived" (B-) cells (reviews: 21, 22, 24). This view holds that T-cells, in addition to their role in classic delayed-type hypersensitivity, are instrumental in immunologic memory, and in various cell-mediated immune reactions (e.g. "killer" cells in graft rejection and elimination of neoplastic cells as well as the source of lymphokines) and exert an important "helper" function in antibody formation against most antigens. In contrast, the B-cell population is exclusively linked with humoral immunity inasmuch as these elements are considered to differentiate into antibody-forming cells. The most recent classification of immunodeficiency states is based on this T- and B-cell model (9). Although some diseases appear to conform to such a classification, "the majority of patients with immunodeficiency cannot yet be unequivocally classified, and are therefore grouped under the heading of 'variable immunodeficiency' " (9).

It may be worthwhile to digress shortly and consider some aspects of this hypothesis in the light of experimental and clinical findings. Evidence for the thymus-independent development of antibody-forming elements (derived from B-cells) rests on experimental data obtained in birds and in the genetically athymic nu/nu mouse as well as on the clinical observation of so-called "thymic aplasia" (di George syndrome). The role of the bursa of Fabricius in birds as a central, thymus-independent source of B-cells has to be reevaluated in the light of recent experimental evidence indicating that this organ is a highly specialized site of contact with cloacal (antigenic) material: in the avian species, uptake, degradation and storage of particulates derived from the gut lumen appear to be limited to this organ (26). Gut-associated lymphoid tissue (GALT) in mammals, hypothetically identified as "bursa-equivalent", apparently serves a similar function. The origin of the first bursal

*Work supported by the Swiss National Foundation for Scientific Research

lymphoid cells is as yet unknown. The majority, if not the totality of, lymphocytes in Peyer's patches and mesenteric lymph nodes of newborn mice are thymus-derived as demonstrated by stable labeling techniques (17). Since kinetic data demonstrating migration of lymphoid elements from the bone marrow to developing peripheral lymphoreticular structures are still lacking, the hypothesis of an antigen- and thymus-independent development of the B-cell population remains speculative. The demonstration of plasma cells and antibody production in the thymus-less nude mouse and in cases of di George-syndrome is not necessarily strong evidence in favor of the above hypothesis. In both the nude mouse and in cases of so-called "thymic aplasia" upon careful examination remnants of thymic tissue have been detected in some instances. In addition, variable numbers of peripheral lymphocytes in nude mice carry thymus-specific surface allo-antigens.

Because of these incertainties alternatives to the current hypothesis of the ontogeny of the immune system should not be dismissed. Lymphoid cells found in the thymus at early stages of ontogeny most probably are of extrathymic origin. The possibility that, under physiological conditions, precursor cells cease to immigrate into the thymic cortex at later developmental stages is not excluded. Mechanisms governing the proliferative activity of lymphoid cells in the thymic cortex are unknown, but may include antigenic stimulation. Newly generated progenitor cells leave the thymus at the cortico-medullary junction, populate peripheral lymphoreticular organs, including those of the intestinal tract, and participate, in part, in lymphocyte circulation and recirculation; a large number of thymic emigrants may get lost in the gut lumen. Whether an immunocompetent cell differentiates into an antibody-producing element or whether it becomes a sensitized lymphocyte may be determined not by its origin from different precursor populations in different anatomical sites but by other factors. Among these one might consider, for example, the developmental stage of the organism as well as route, amount and physical form of the stimulating antigen. Different forms of immunodeficiency diseases would correspond to defects of the lymphoreticular system at different stages of development: the most severe form of immunodeficiency might be expected as a consequence of deficient or absent immigration of stem cells into the thymus anlagen, while defects in the "periphery", e.g. qualitatively and/or quantitatively faulty production of antibodies or sensitized lymphocytes, would produce deficiencies with a less severe clinical course. According to this hypothesis, one might expect different patterns of intermediate defects between those extremes with correspondingly intermediate clinical symptoms.

Following these theoretical considerations, we will concentrate on the clinical picture some of these immunodeficiency states present. Particular emphasis will be placed on the types of infectious agents found in combined or isolated deficiencies of humoral and/or cell-mediated immunity.

Combined defect of humoral and cell-mediated immunity

Swiss-type agammaglobulinemia (review: 16) (SAG) - now termed "severe combined immunodeficiency" according to the proposition by the nomenclature committee of the WHO (9) - represents the most extensively studied nosological entity of this group of deficiencies. In the first reports, the severe lymphopenia in the circulating blood and in lymphoreticular organs of these children was attributed to infection, in particular generalized candidiasis, and the thymic disorder as well as a-gammaglobulinemia were not recognized (11). Later, the finding at autopsy of an undescended, hypoplastic thymic rudiment with profound structural defects in two siblings of the same family led to the suggestion that thymic dysplasia may be responsible for the observed underdevelopment of peripheral lymphoid tissues (7). The finding of total agammaglobulinemia in the 7-year old boy, the older of the two children, corresponded well with the total lack of plasma cells and germinal centers in lymphoid organs. Since these early observations close to 100 typical cases of SAG have been described and documented with additional morphological, immunological and clinical findings. A series of cases with a less severe clinical course and atypical morphology has also been reported over the years (15), but will not be dealt with here.

The total absence of humoral and cell-mediated immune reactivity in tests in vivo and in vitro reflects itself clinically in overwhelming, generalized infections. During the first week of life children with SAG usually develop rhinitis, bronchopneumonia, and diarrhea caused by a variety of gram-positive and gram-negative bacteria, and a host of viral agents. In addition, infection with fungi, particularly Candida albicans, often develops in the skin and in mucous membranes. Vaccination with live vaccines, such as vaccinia or BCG, is most often followed by generalized, usually fatal, disease. Patients with untreated SAG invariably died during the first year of life with pneumonia, meningitis and/or sepsis. Conventional treatment with antibiotics or immunoglobulin reconstitution is without lasting effect. Attempts have been made more recently to correct the cellular defect by transfusion of bone marrow cells (review: 4, 5) and/or transfer factor therapy (review: 18).

Defects of humoral immunity

Deficiencies of the antibody-producing immune system vary from near-total agammaglobulinemia to lack of single immunoglobulin-subclasses (27) or antibody specificities (1). Antibody deficiency syndromes may present themselves as familial or sporadic, transient or progressive, primary or secondary diseases and are observed in children or in adults (2).

Infantile X-linked agammaglobulinemia was first described in 1952 (6). This rare form is characterized by its occurrence in boys only; by a general inability to produce immunoglobulins of all classes; and by a relatively intact system of cell-mediated immunity. No morphological defect of the thymus has been detected. Peripheral lymphoid organs appear to be of near-normal cellularity, but plasma cells and germinal centers are absent. Clinical consequences include recurrent pyogenic infections: conjunctivitis, otitis media, pneumonia, and gastrointestinal infections and infestations, particularly with Giardia lamblia, are often observed.

The large spectrum of selective immunoglobulin disorders ("dysimmunoglobulinemias") or "adult acquired forms" of antibody deficiency syndromes may present a similar clinical picture. Although these patients usually present no common morphological characteristics which would help to identify their individual functional defect, one might hypothetize that these disorders could be due to a variety of derangements in the differentiation of the plasma cell line. Recently we had the opportunity to study the lymph node morphology of a case of selective IgG deficiency with increased levels of serum IgM, a particular type of dysimmunoglobulinemia reviewed by Rosen et al. (25). The presence of unusually large germinal centers combined with the absence of plasma cells might suggest a differentiation arrest of lymphoid germinal center cells; increased IgM production could be interpreted as a consequence of an insufficient feedback mechanism mediated by circulating IgG.

Clinical consequences of antibody deficiency show more similarities among themselves and are less severe as compared to those encountered in cases of combined immunodeficiency. Typically, these patients suffer from recurrent bacterial infections, especially pneumonia, acute pansinusitis, otitis media, and gastrointestinal disorders with massive diarrhea. Infections of the respiratory and gastrointestinal tract are particularly frequent in cases with selective IgA deficiency (8). If the condition is not diagnosed and treated adequately, repeated septicemic episodes, and recurrent meningitis, osteomyelitis or arthritis can occur. Late manifestations may include torpid, chronic inflammatory processes, hepatosplenomegaly, bronchiectasis, pulmonary fibrosis or a malabsorption syndrome with all its consequences. It should be emphasized that patients with antibody deficiency syndromes usually show normal resistance to infection with mycobacteria, viruses or fungi.

Defects of cell-mediated immunity

Primary, selective deficiencies of cell-mediated immunity with an intact antibody-producing cell system have been postulated to exist on the basis of the hypothesis of the T- and B-cell systems of the immune apparatus (12). Two disorders are
often cited in the literature as substantiating both hypothesis and prediction: the di
George syndrome ("Thymic and parathyroid aplasia") and autosomal recessive thymic aplasia with lymphopenia (Nézelof syndrome). The case Nézelof described in
1964 (23) is presently interpreted as a milder form of di George syndrome (18). Derivatives of the third and fourth pharyngeal pouches during embryogenesis include
epithelial components of the thymus, the parathyroid glands, and part of the aortic
arch and of the esophagus. A developmental defect resulting in thymic and parathyroid aplasia or hypoplasia in combination with defects in the aortic arch, mandible
and ear lobes has been termed di George syndrome (10, review: 19). Patients with
this disorder require parathyroid hormone substitution to prevent and treat otherwise fatal neonatal tetany. In survivors increasing susceptibility to bacterial, viral
or fungal infection develops, and most patients die during the first year of life.

Only a few cases of this complex developmental defect have been examined systematically. The most important among contradictory findings is the observation
that in carefully autopsied cases thymic tissue may be found (19, 20). Peripheral
lymph nodes have a severely reduced content of lymphocytes only in those cases that
died of infection. The number of plasma cells may be normal or reduced, but the
capacity to form humoral antibodies appears to be impaired in almost all cases. The
whole picture could be interpreted as the consequence of early commitment of a reduced pool of immunocompetent cells to environmental antigens, in particular, intestinal microorganisms.

These examples may suffice to demonstrate that views on the pathogenesis of immunodeficiency states remain hypothetical because of our relative ignorance with
regard to regulatory mechanisms of the morphological and functional development
of the lymphoreticular system. In the face of all these incertainties, immunodefi-
cient patients are treated with varying success. What cells, cell products or both
are beneficial upon transfer to individuals suffering from recurrent infections? Substitution of humoral defects, associated with infection by "extracellular" bacteria,
is readily achieved by treatment with immunoglobulin preparations. Defects of cell-
mediated immunity can be corrected only by transfer of tissue and/or immunocompetent cells (review: 4). Although clinical and immunological "correction" has been
achieved in more than 20 cases of immunodeficiency involving cell-mediated immunity (13), it has to be considered that the population of "purified lymphoid stem cells"
which is obtained by density gradient centrifugation from bone marrow (review: 4)
may contain precursors of the erythrocytic, monocytic and granulocytic series beside potentially immunocompetent elements. The relative importance of monocytes
and granulocytes in defense against facultative or obligatory intracellular bacteria,
viruses and fungi, microorganisms exquisitely associated with cellular defects,
should not be underestimated.

Summary

Immunodeficiency states in man are characterized by the inability of the organism to mount a "normal" immune response following antigenic stimulation. Lowered
resistance to infectious disease may be due to 1. faulty or absent production of humoral antibodies and/or sensitized lymphocytes ("primary specific immunodeficiency"); 2. immunoglobulin hypercatabolism or loss; 3. exogenous causes, such as cytotoxic or immunosuppressive agents; 4. concomitant systemic disorders, e.g. virus diseases, neoplasia of the lymphoreticular or hemopoietic system; or 5. dysfunction of phagocytic cell systems.

A combined defect of both humoral and cell-mediated immune responsiveness
("severe combined immunodeficiency") is characterized by severe, generalized in-

fection with various bacterial, viral or mycotic agents very early after birth. Deficient humoral immunity is associated with high susceptibility to infection with "extracellular" bacteria while disease caused by viruses or fungi is rarely observed. Patients with defective cell-mediated immunity usually exhibit increased susceptibility to infection with fungi, viruses and "obligatory or facultative intracellular" bacteria.

Current hypotheses on the pathogenesis of immunodeficiency disease and therapeutic measures are briefly discussed.

References

1. Barandun, S.: Bibl. haemat. (Basel) 17: 85 (1964).
2. Barandun, S., Riva, G., Spengler, G.A.. In: Bergsma, D., Good, R.A.: Immunologic Deficiency Diseases in Man (U.S. Natl. Foundation Original Articles Series IV, 40, 1968).
3. Barandun, S., Hess, M.W. Cottier, H.: Verh. dtsch. Ges. inn. Med. 78: 741 (1972).
4. Van Bekkum, D.W.: Transpl. Rev. 9: 3 (1972).
5. Biggar, W.D., Park, B.H., Good, R.A.: Ann. Rev. Med. 24: 135 (1973).
6. Bruton, O.C.: Pediatrics 9: 722 (1952).
7. Cottier, H.: Trans. 6th Congr. Europ. Soc. Haemat., Copenhagen 1957, p.41 (Karger, Basel-New York, 1958).
8. Crabbe, P.A., Heremans, J.F.: Gut 7: 119 (1966).
9. Fudenberg, H., Good, R.A., Goodman, H.C., Hitzig, W., Kunkel, H.G., Roitt, I.M., Rosen, F.S., Rowe, D.S., Seligmann, M., Soothill, J.R.: Pediatrics 47: 927 (1971).
10. Di George, A.M.: J. Pediat. 67: 907 (1965).
11. Glanzmann, E., Riniker, P.: Wien. med. Wschr. 100: 35 (1950).
12. Good, R.A.: Amer. J. Path. 69: 484 (1972).
13. Good, R.A., Kersey, J., Spector, B., Hansen, J., Gajl, K.: Proc. Schering Symposium on Immunopathology, Cavtat 1973 (in press).
14. Hess, M.W., Schaedeli, J., Cottier, H.: Verh. dtsch. Ges. Path. 55: 175 (1971).
15. Hitzig, W.H.. In: Handbuch der Inneren Medizin, VII/1, 1973 (in press).
16. Hitzig, W.H., Barandun, S., Cottier, H.. In: Ergebnisse der inneren Medizin und Kinderheilkunde, p. 80 (Springer, Berlin-Heidelberg-New York 1968).
17. Joel, D.D., Hess, M.W., Cottier, H.: J. exptl. Med. 135: 907 (1972).
18. Levin, A.S., Spitler, L.E., Fudenberg, H.H.: Ann. Rev. Med. 24: 175 (1973).
19. Lischner, H.W.: J. Pediat. 81: 1042 (1972).
20. Lischner, H.W., di George, A.M.: Lancet 2: 1044 (1969).
21. Miller, J.F.A.P.. In: New Concepts in Allergy and Clinical Immunology, p. 3, Proc. VII th Internatl. Congr. Allergology, 1970.
22. Miller, J.F.A.P., Mitchell, G.F.: Transpl. Rev. 1: 3 (1969).
23. Nézelof, C., Jammet, M.L., Lortholary, P., Labrune, B., Lamy, M.: Arch. franç. Pédiat. 21: 897 (1964).
24. Osoba, D.: Med. clin. North Amer. 56: 319 (1972).
25. Rosen, F.S., Craig, J.M., Vawter, G.,Janeway, C.A.. In: Bergsma, D., Good, R.A.: Immunologic Deficiency Diseases in Man (U.S. Natl. Found. Original Articles Series IV, 67, 1968).
26. Schaffner, T., Cottier, H., Hess, M.W.: (in preparation 1974).
27. Schur, P.H., Borel, H., Gelfand, E.W., Alper, C.A., Rosen, F.S.: New Engl. J. Med. 283: 631 (1970).

Species Differences in the Effectiveness of Intestinal Barriers against Penetration of Inert Particulates and Bacteria

H. Cottier, Th. Schaffner, A. D. Chanana, D. D. Joel, B. Sordat, B. J. Bryant, and M. W. Hess

Host and microflora interactions occur on all surfaces of the body exposed to the environment. The microbial ecology of the gastrointestinal tract may be regarded as the most important site of interaction if numbers of microorganisms and diversity and magnitude of antigenic stimulation involved are considered. Because of suppression of, or competition with, potential invaders, possibly also by virtue of its ability to block access of invaders to the tissues, the autochthonous flora contributes to the defense of the host. It may also be related to the mosaic-like distribution of bacterial species along the gastrointestinal tract. It has been recognized in recent years that intimate symbiotic associations between microorganisms and the inner surface of the gastrointestinal tract in small rodents are instrumental in maintaining a state of health. An unexpected finding was that large numbers of hitherto unknown, highly oxygen-sensitive bacteria form dense layers covering the mucosal epithelium (1, 2, 3). Little is known with respect to the relative importance of secretory IgA, lysozyme, complement components and other humoral factors for defense against microorganisms in the neighbourhood of, or within, intestinal epithelial layers. A so-called chemical barrier resembling the intracellular digestive apparatus, as observed in mammalian tissue culture cells (4), has been proposed to exist near the brush border of epithelial cells where activities of enzymes such as acid phosphatase and nonspecific esterase have been demonstrated. However, this defense mechanism does not provide sufficient protection against such pathogenic microbes as Salmonella, Shigella, Cryptosporidia and other organisms which penetrate and/ or reside and multiply within the brush border or the entire epithelial layer (review: 5); moreover, it is sensitive to the action of certain enterotoxins such as those from E. coli and Vibrio cholerae (6, 7, 8). Our relative ignorance with regard to species differences in host defense mechansims associated with the gastrointestinal epithelium emphasizes the need for further experimentation to clarify this problem.

Although the integrity of the gastrointestinal epithelium can be regarded as a prerequisite for hindering bacterial invasion, it is not the only and perhaps not the most important defense apparatus of the host. Studies on healthy, conventional mice have revealed that small numbers of bacteria normally pass through the gut wall and can be cultured from mesenteric lymph nodes, rarely also from liver and spleen (9). It has been proposed recently that decreased trapping of intestinal bacteria and/or nonliving antigens in a diseased liver may result in more of this material reaching the spleen and thus eliciting more marked immune responses. It remains to be examined if this mechanism is primarily responsible for the well-known hyperglobulinemia observed in derangements such as liver cirrhosis (10). Strong evidence in favor of the assumption that cell systems other than the gut epithelium are the most important in preventing enterogenic septicemia came from radiobiolgical studies. Whole-body X-radiation of mice with a dose of 550-700 R is followed by a maximum incidence of fatal enterogenic bacteremia around day 11 post-exposure, i. e. at the time of maximum tissue leukopenia and not when the radiation-induced gastrointestinal epithelial

This work has been supported by the Swiss National Foundation for Scientific Research and the U. S. Atomic Energy Commission. BJB is a Leukemia Society of America Special Fellow.

damage is at its height (11, 12). Species differences with regard to M. L. D. of whole-body radiation may largely be due to disparate degrees of radiosensitivity of cellular defense mechanisms such as those provided by granulocytes, macrophages and immunocompetent cells, probably in that order of relative importance (review: 13).

Hence, in some species bacteria normally enter the intestinal wall and reach the regional lymph nodes; but how these microorganisms penetrate the epithelial layer is not fully understood. It has often been assumed that the epithelial layer of the gastrointestinal tract is impermeable for particulate matter and nonpathogenic bacteria. This view can no longer be maintained. Passage of particulates from the gut lumen into the intestinal wall, the lymph and the blood has been described as a physiological phenomenon as early as 1844 (14). Different species- and age-dependent mechanisms appear to be involved in the various types of persorption.

One such process, the endocytotic uptake of macromolecules and small particulates from the gut lumen by so-called vacuolated epithelial cells of the small intestine, has been studied in depth in calves wherein the intestinal persorption of maternal colostral immunoglobulins plays a dominant role in preventing Colisepticemia (15, 16, 17, 18, 19, 20). Transfer in mammals of passive immunity from mother to offspring is achieved by transplacental passage of plasma immunoglobulins and/or intestinal absorption of colostrum. Marked species differences exist in this respect. For example, rabbits receive maternal antibodies almost exclusively across the placenta, whereas intestinal persorption of colostral immunoglobulins is the principal way of passive humoral immunization of newborn piglets. Rats and mice use both routes (reviews: 21, 22). The reason why rabbits are not as efficient in intestinal persorption of antibody is not yet fully understood; however, a particular lysosomal response in epithelial cells has been implicated (21). The transport of horse ferritin or protein in rabbits from the gut lumen through epithelial cells was followed by electron microscopy and immunofluorescence. It was observed that the material did not reach the blood stream in detectable amounts, in contrast to the findings in rats (23, 24). Other macromolecules or colloidal materials are also readily taken up by vacuolated intestinal epithelial cells of suckling mammals and may be passed to the lymph and/or blood stream (25). It is of interest that this type of persorption is more pronounced in distal than in proximal parts of the small intestine (26), and that it seems to disappear at weaning. Rats and mice do so at about 18 days after birth (27), rabbits and pigs at days 20-22, and ferrets at 33-34 days of postnatal life, respectively (28, 29). After this period of time, vacuolated cells in conventional animals usually disappear within a few days. Apparently, they are replaced by epithelial cells incapable of such cytopempsic function (30). Recent studies have demonstrated that ileal epithelial cells in germ free piglets remain vacuolated beyond 3 weeks of age and that they exhibit a longer migration time for movement from the crypts were they are formed, to the tips of the villi, than cells of conventionalized non-germ free animals of the same age. It is tempting, therefore, to postulate that changes in the autochthonous intestinal microflora related to weaning may bring about a higher replacement rate of epithelial cells and thus prevent epithelial cells from reaching an advanced state of differentiation corresponding to that of vacuolated cells (31). Comparative studies on rates of epithelial migration in the small intestine of newborn, suckling and weaned rats (32) are consistent with this hypothesis. It should be emphasized that transport of macromolecules such as horseradish peroxidase through intestinal epithelium occurs beyond the age of weaning (33, 34). In adult rats, the magnitude of this process appears to be even more important than in neonatal intestine, particularly in the jejunum (35). After parenteral and especially following oral immunization with horseradish peroxidase or bovine serum albumin, passage of these macromolecules through the intestinal epithelium diminishes, suggesting that this phenomenon may be influenced by local immune reactions (36, 37). Whether macromolecular uptake and transport is more pronounced in so-called tuft cells (38) than in ordinary brush-border epithelia, and the species dependency of such uptake, need evaluation.

Another type of particle persorption in the intestine is found not only in suckling but also in adult mammals. Small particles such as starch globules, yeast cells and others measuring up to more than 50 μm in diameter, readily pass from the gut lumen into the lymph and blood stream (39). In humans, it was demonstrated that persorbed food particles even pass the placenta and enter the blood stream of the fetus (40). It is noteworthy that this type of persorption seems to be enhanced by drugs which increase intestinal motility and contractions (41) suggesting a pressure effect. Histological examination of intestinal biopsy specimens obtained from individuals fed with cakes indicated that starch particles entered the gut wall at the sites of maximum epithelial desquamation, i.e. at the tips of the villi (42). To our knowledge, this route of entry has not so far been examined quantitatively with regard to the passage of enteric bacteria into the lymph and blood stream.

In our laboratory, we became interested in these problems while studying the importance of antigenic stimulation for the development and functions of gut-associated lymphoid tissue (GALT). In early periods of postnatal life of mice, at least the vast majority of lymphocytes entering GALT is of thymic origin (43, 44). As these animals start suckling and developing an intestinal microflora, large numbers of thymic lymphocytes accumulate in Peyer's patches and mesenteric lymph nodes, in contrast to peripheral lymph nodes remote from the gut and to both GALT and peripheral organs of the lymphoreticular system in germ free mice. It was, therefore, of interest to examine the question 1. as to how intestinal antigens gain access to Peyer's patches; and 2. if these structures represent a preferential site for entry of particulates from the gut. Quantitative analysis of numbers of carbon particles located in Peyer's patches after introduction into the gastrointestinal tract revealed that this was in fact the case. The amount of this material passing through the epithelium at a distance from GALT was much less than in Peyer's patches (45). Similar findings were made on the rabbit appendix (46). In our studies it was also demonstrated that the major sites for particle entry into Peyer's patches were central areas of the epithelial layer covering the so-called dome. Colloidal carbon was endocytosed by subepithelial macrophages and was stored in this location for considerable periods of time.

Since inert particles freely enter into intestinal follicles such as Peyer's patches, the question arises as to how bacteria are dealt with in these tissues. Remarkable species differences seem to exist in this respect. Since the last century it has been known that Peyer's patches, the sacculus rotundus and the appendix of rabbits harbor large numbers of bacteria in the depth of the lymphoreticular tissue (for older literature, see 47), possibly of the fatty acid- and cellulose-fermenting group. In our studies on healthy, conventional mice, we have not so far detected intact bacteria within Peyer's patches, but we did find numerous microorganisms in the GALT of rabbits (48, see also 49, 50). In rats, commensal microorganisms in stages of development and dissolution have been found within epithelial cells of the ileum, but not in the lymphoreticular tissue (51). In analogy to observations made on rabbits (47) one might suspect, however, that the PAS-positive material contained in macrophages of subepithelial and deeper layers of GALT corresponds to bacterial breakdown products. More experimentation is needed to evaluate this possibility. In contrast to rabbits, mice and rats seem to possess efficient bactericidal mechanisms localized within or near the epithelial layer of the intestinal tract, particularly also in the vicinity of lymphoid tissue. In this respect, the situation in man appears to be similar to that in mice and rats.

The bursa of Fabricius in birds has been termed a central lymphoid organ responsible for the generation on non-committed B-cells (for references, see 52). Recent studies in our laboratory have revealed, however, that the avian bursa is uniquely specialized in the uptake of small particles (diameter up to at least 1 μm) from the cloacal content. This lymphoepithelial organ seems to be the only site in

the bird's intestinal tract where macromolecules and particulates from the lumen
are brought in great numbers into contact with immunocompetent cells. Furthermore, it was demonstrated that suspensions of bacteria introduced into the bursal
duct did not result in the appearance of intact microorganisms within the lymphoid
bursal tissue. Epithelial tufts covering the bursal follicles and transporting particulate matter into the lymphoepithelial tissue are distinctively equipped with enzymes
such as non-specific esterase, acid phosphatase and β-glucuronidase. These findings favor the assumption that the bursa is protected by local bactericidal mechanisms; however, they question the validity of the hypothesis that the development of
bursal lymphoid tissue is antigen-independent. Our observations indicate instead
that the bursa in birds as well as GALT in mammals could represent specialized organs which promote the contact of intestinal antigens with immunocompetent cells.
This interpretation finds support in birds in the observation that bursal lymphocytes
react with intestinal microorganisms, as shown by bacterial adherence technique
(53). We postulate that bursectomy at hatching deprives the bird not only of an important lymphoid organ, but also eliminates an essential site of contact between intestinal antigens and immunocompetent cells. The view that the growth of GALT in
mammals is largely antigen-driven, may best be documented by comparison of conventional and axenic (germ free) animals. In the latter, GALT and mesenteric lymph
nodes are poorly developed, but increase in size, form germinal centers and plasma cells when stimulated via the intestinal route. No such change was seen when
corresponding antigens were administered parenterally (54). It is of interest that in
hamsters oral administration of bovine serum albumin led to the formation of large
germinal centers in Peyer's patches while antibody-containing cells were mainly
confined to the lamina propria outside these structures (55). Various observations
favor the view that Peyer's patches are primarily involved in the generation of immunologic memory, i.e. precursors of antibody-forming cells, and are not an important site of plasma cell maturation (56, 57).

Summary

This summary review compares various mammals with respect to permissive
routes of microbial and inert particle penetration through the intestinal epithelium.
The host defense context is acknowledged by a) gastrointestinal secretions; b) symbiotic associations between microorganisms and the inner surface of the gastrointestinal tract; c) ill-defined actions of humoral factors in the vicinity of epithelial
layers; d) a postulated chemical and mechanical barrier of the epithelium; and e) systemic defense mechanisms, including leukocytes, macrophages and humoral immune systems. The routes of permissive particle penetration considered in mammals include transport 1.by vacuolated epithelial cells in the small intestine of suckling, conventional animals; 2. between desquamating cells at the tips of villi; and 3.
through epithelial discontinuities overlying dome regions of gut-associated lymphoid
tissue (GALT). The epithelia overlying lymphoid tissue of the avian bursa of Fabricius is similarly considered as specialized for uptake of particulates from the cloacal lumen. The tentative, unified explanation offered for these diverse observations is that the permissive pathways are specifically designed to promote, within
GALT, the interaction of differentiating lymphoid cells with macromolecules and
particulate antigens from the intestinal lumen. The numerous antibody forming cells
normally found in GALT may express the magnitude of this interaction in GALT under the impact of the enormous intestinal antigenic content.

References

1. Savage, D. C., Dubos, R., Schaedler, R. W.: J. exp. Med. 127: 67 (1968).
2. Lee, A., Gordon, J., Dubos, R.: Nature 220: 1137 (1968).
3. Savage, D. C., McAllister, J. S.: Microbial interactions at body surfaces and

resistance to infectious diseases. In: Dunlop, R.H., Moon, H.W.: Resistance to Infectious Disease, p.113 (Saskatoon Modern Press 1970).

4. Gordon, G.B., Miller, L.R., Bensch, K.G.: J. Cell Biol. 25: 41 (1965).
5. Takeuchi, A.: Current Topics in Pathology 54: 1 (1971).
6. Love, A.H.G.: Gut 10: 105 (1969).
7. Moon, H.W., Whipp, S.C., Baetz, A.L.: Lab. Invest. 25: 133 (1971).
8. Yardley, J.H., Brown, G.D.: Lab. Invest. 28: 482 (1973).
9. Gordon, L.E., Ruml, D., Hahne, H.J., Miller, C.Ph.: J. exp. Med. 102: 413 (1955).
10. Thomas, H.C., McSween, R.N.M., White, R.G.: Lancet I: 1288 (1973).
11. Hammond, C.W., Tompkins, M., Miller, C.Ph.: J. exp. Med. 99: 405 (1954).
12. Hammond, C.W., Colling, M., Cooper, D.B., Miller, C.Ph.: J. exp. Med. 99: 411 (1954).
13. Cottier, H.: Histopathologie der Wirkung ionisierender Strahlen auf höhere Organismen (Tier und Mensch). In: Diethelm, L., Olsson, O., Strnad, F., Vieten, H., Zuppinger, A.: Handbuch der Medizinischen Radiologie, p.3, Bd. II/2 (Springer, Berlin-Heidelberg-New York 1966).
14. Herbst, E.F.G.: Das Lymphgefäßsystem und seine Verrichtungen (Vandenhoeck und Ruprecht, Göttingen 1844).
15. Mason, J.H., Dalling, T., Gordon, W.S.: J. Path. Bact. 33: 783 (1930).
16. Aschaffenburg, R., Bartlett, S., Kon, S.K., Roy, J.H.B., Walker, D., Briggs, C., Lovell, R.: Brit. J. Nutr. 5: 171 (1951).
17. Bangham, D.R., Ingram, P.L., Roy, J.H.B., Shillam, K.W.G., Terry, R.J.: Proc. Roy. Soc. B 149: 184 (1958).
18. Fey, H., Margadant, A.: Zbl. Vet. Med. 9: 767 (1962).
19. Balfour, W.E., Comline, R.S.: J. Physiol. 160: 234 (1962).
20. Fey, H.: Die Bedeutung des Gammaglobulins für das neugeborene Kalb. Internat. Tierärztl. Arbeitsgemeinschaft Tierernährung, Symposium Salzburg, p. 17 (1967).
21. Kraehenbuhl, J.P., Campiche, M.A.: J. Cell Biol. 42: 345 (1969).
22. Brambell, F.W.R.: The transmission of passive immunity from mother to young. In: Neuberger, A., Tatum, E.L.: Frontiers of Biology, vol. 18 (North-Holland Publishing Co., Amsterdam-London 1970).
23. Kraehenbuhl, J.P., Gloor, E., Blanc, B.: Z. Zellforsch. 70: 209 (1966).
24. Kraehenbuhl, J.P., Gloor, E., Blanc, B.: Z. Zellforsch. 76: 170 (1967).
25. Clark, S.L.: J. biophys. biochem. Cytol. 5: 41 (1959).
26. Moon, H.W.: Vet. Path. 9: 3 (1972).
27. Halliday, R.: Proc. Roy. Soc. B 143: 408 (1955).
28. Clarke, R.M., Hardy, R.N.: J. Physiol. 209: 669 (1970).
29. Clarke, R.M., Hardy, R.N.: J. Anat. 108: 63 (1971).
30. Clarke, R.M., Hardy, R.N.: J. Physiol. 204: 127 (1969).
31. Moon, H.W., Kohler, E.M., Whipp, S.C.: Lab. Invest. 28: 23 (1973).
32. Koldovsky, O., Sunshine, P., Kretchmer, N.: Nature 212: 1389 (1966).
33. Cornell, R., Walker, W.A., Isselbacher, K.J.: Lab. Invest. 25: 42 (1971).
34. Warshaw, A.L., Walker, W.A., Cornell, R., Isselbacher, K.J.: Lab. Invest. 25: 675 (1971).
35. Walker, W.A., Cornell, R., Davenport, L.M., Isselbacher, K.J.: J. Cell Biol. 54: 195 (1972).
36. Walker, W.A., Isselbacher, K.J., Bloch, K.J.: Science 177: 608 (1972).
37. Walker, W.A., Isselbacher, K.J., Bloch, K.J.: J. Immunol. 111: 221 (1973).
38. Isomäki, A.M.: Acta path. microbiol. scand., Sect. A, Supplement 240 (1973).
39. Volkheimer, G., Hermann, H., Hermans, E., John, A., Al Abesie, F., Wachtel, S.: Zbl. Bakt. (Orig.) 192: 121 (1964).
40. Volkheimer, G., Schulz, F.H., John, H., Meier zu Eisen, J., Niederkorn, K.:

Gynaecologica 168: 86 (1969).

41. Volkheimer, G., Schulz, F.H., Hofmann, I., Pieser, J., Rack, O., Reichelt, G., Rothenbaecher, W., Schmelich, G., Schurig, B., Teicher, G., Weiss, B.: Pharmacology 1: 8 (1968).

42. Volkheimer, G., Wendlandt, H., Wagemann, W., Reitzig, P., Schneider, D., Böhm, C., Böhm, M., Eras, B., Gröning, H., Hauptmann, G., Hiller, R., Lorenz, H., Mandelkow, A., Strauch, S.: Path. Microbiol. 31: 51 (1968).

43. Joel, D.D., Hess, M.W., Cottier, H.: J. exp. Med. 135: 907 (1972).

44. Chanana, A.D., Schädeli, J., Hess, M.W., Cottier, H.: J. Immunol. 110: 283 (1973).

45. Joel, D.D., Sordat, B., Hess, M.W., Cottier, H.: Experientia 26: 694 (1970).

46. Hanaoka, M., Williams, M., Waksman, B.H.: Lab. Invest. 24: 31 (1971).

47. Friedenstein, A., Goncharenko, I.: Nature 206: 1113 (1965).

48. Hess, M.W., Cottier, H., Sordat, B., Joel, D.D., Chanana, A.D.: The intestinal barrier to bacterial invasion. In: Braun, W., Ungar, J.: Non-Specific Factors Influencing Host Resistance. A Reexamination, p. 447 (Karger, Basel 1973).

49. Shimizu, Y., Andrew, W.: J. Morphol. 123: 231 (1967).

50. Waksman, B.H., Ozer, H., Blythman, H.E.: Lab. Invest. 28: 614 (1973).

51. Reimann, H.A.: JAMA 192: 100 (1965).

52. Cooper, M.D., Lawton, A.R., Kincade, P.W.: Clin. exp. Immunol. 11: 143 (1972).

53. Alten van, P.J., Meuwissen, H.J.: Science 176: 45 (1972).

54. Pollard, M., Sharon, N.: Infect. Immun. 2: 96 (1970).

55. Bienenstock, J., Dolezel, J.: J. Immunol. 106: 938 (1971).

56. Cooper, G.N., Turner, K.: J. Reticuloendoth. Soc. 6: 419 (1969).

57. Craig, S.W., Cebra, J.J.: J. exp. Med. 134: 188 (1971).

Transformation of Mouse B-Lymphocytes by the Lipid Component of Endotoxin

S. E. Mergenhagen, D. L. Rosenstreich, A. Nowotny, and T. Chused

Several investigators have suggested that there is an immunological basis for certain of the biological activities associated with endotoxin reactions. This is based upon a number of recent studies which have shown that endotoxic lipopolysaccharides (LPS), like immune complexes, are effective in activating humoral effector systems such as complement and coagulation, and are mitogenic for lymphoid cells.

Peavy and coworkers (1) first reported that endotoxins were potent mitogens for mouse spleen cells and subsequent work by Gery and others (2, 3) suggested that LPS were mitogenic for bone-marrow-derived (B) lymphocytes but not T-lymphocytes. Recently, a number of laboratories (4-7), including our own, have become involved in determining which subcomponent of the endotoxin macromolecule is mitogenic for mouse lymphocytes in vitro. Our most recent experiments are summarized in this communication.

Materials and Methods

In the experiments that follow we used the LPS derived from a smooth strain of Salmonella minnesota and from Serratia marcescens. In addition, we employed the O-polysaccharide deficient endotoxic glycolipid prepared from the rough mutant of S. minnesota, R595 (8). This glycolipid has been found to be essentially free of protein, nucleic acid and sugars other than glucosamine and KDO. In addition, several biological activities of this glycolipid have been studied and found to be similar or better in potency compared to the smooth strain LPS.

Mouse (C57/B1) spleen cells were cultured in standard tissue culture media containing 2 % fetal calf serum with and without stimulants (7). After 72 h the cultures were pulsed with tritiated thymidine for the last 4 h of culture. Cultures were harvested and TCA insoluble radioactivity was measured by standard techniques. Results of spleen cell stimulation experiments are expressed as the stimulation ratio (experimental (E)/control (C)) where E/C is defined as the ratio of mean counts per min of incorporated tritiated thymidine in the stimulated cultures divided by mean counts per min in the unstimulated cultures.

Results and Discussion

Fig. 1 shows the proliferative response of mouse spleen lymphocytes induced by a broad concentration range of the smooth strain of S. minnesota LPS and the rough mutant glycolipid. LPS produced barely significant stimulation at concentrations of 10^{-4} and 10^{-3} µg/ml. From 10^{-3} to 1 µg/ml the response increased linearly and then remained constant up to concentrations of 100 µg/ml. In contrast, the glycolipid produced significant stimulation at a concentration of 10^{-5} /µg/ml. Thus it is clear that bacterial glycolipid, free of O-polysaccharide, is fully mitogenic.

Salmonella minnesota LPS was heated at $100°$ C for 30 min in 1.0 N HCl or in 0.1 N NaOH or in phosphate buffered saline (PBS), and then tested for mitogenic activity using a broad dose range of stimulant. The maximum stimulation obtained is presented in Table I. When LPS is heated in HCl, it is fractionated into an insoluble lipid ("lipid A") and a soluble polysaccharide. The lipid still retained mitogenic activity. More recent experiments, which equilibrated more closely the concentrations of lipid and LPS, have shown that they are equally active if used in the same concentrations. In contrast to the activity of lipid, the haptenic polysaccharide has no mitogenic activity.

Both acid and alkaline hydrolysis result in degradation of the LPS molecule and produce molecular fragments of marked heterogeneity. However, it is possible to detoxify LPS using much milder chemical procedures. One such procedure, de-acylation, involves treatment with potassium methylate, which only cleaves certain long chain carboxylic acids from the lipid moiety and produces a molecule that is devoid of toxicity but is antigenic. This compound is referred to as an "endotoxoid". Intact LPS from S. marcescens was subjected to deacylation and then tested for mi-togenicity. As seen in Table II, S. marcescens LPS alone at a concentration of 10 µg/ml produced a maximum stimulation of 12.8. The endotoxoids, on the other hand, prepared with two different concentrations of potassium methylate, were al-most devoid of mitogenic ability when tested at concentrations up to 300 µg/ml. These results indicate that specific ester-bound long chain fatty acids must be pre-sent in order for the lipid moiety to exert its mitogenic effect.

In order to demonstrate that the lipid was a specific B-cell mitogen, all active preparations were tested for their effect on isolated B or T lymphocyte populations (Table III). B-cells were prepared by selectively killing the T-lymphocytes in the spleen suspension with anti-theta antiserum plus complement. T-lymphocytes were prepared by selectively killing the B-lymphocytes with a heterologous anti-immuno-globulin antiserum plus complement. The use of sodium azide in this procedure has been found to enhance the specific cytotoxicity of anti-immunoglobulin antiserum presumably by preventing the modulation of membrane immunoglobulin from the surface of the B-cell.

As shown in Table IV, the purity of each cell population is demonstrated by the response to the T-cell mitogen, Conconavalin A. Thus the B-cell population did not respond to Con A while the response of the T-cell population was equivalent to that of the whole spleen cell population. Each of the lipid containing preparations, LPS, glycolipid and "lipid A" were as stimulatory for the pure B-cell population as for the unfractionated spleen cells. In contrast, none of the lipid preparations had any effect on the T-cell population.

Illustrated in Fig. 2 are some of the possible mechanisms whereby LPS stimu-lates B-lymphocytes. One possibility is that LPS binds directly to the lymphocyte surface membrane and this binding initiates transformation of the lymphocyte. Our data, that demonstrates the essential requirement of intact fatty acids in the lipid moiety suggests that this binding may take place in the lipid layer of the surface membrane, since these fatty acids are the lipophilic component of the molecule. However, we presume that in order to be mitogenic the fatty acids must be bound to the glucosamine core since isolated fatty acids which are present in both the endo-toxoids and the alkaline hydrolyzate of LPS appear to be inactive. On the other hand, recent work by Bitter-Suermann and coworkers (9) has suggested that B-lympho-cytes may be stimulated in vitro by endotoxin or cobra venom factor through the activation of the third component of complement. Since only B and not T lympho-cytes possess a C3 receptor, this theory would explain the B-cell specificity of LPS. In this regard, it is interesting that those preparations that we have shown to be mitogenic, LPS, glycolipid and "lipid A", have been shown to activate C3, while the nonmitogenic compounds do not (10, 11).

Summary

An analysis of which subcomponent of lipopolysaccharide is mitogenic for mouse lymphocytes has been performed. A purified glycolipid from S. minnesota (R595) is more mitogenic than an intact lipopolysaccharide derived from the smooth strain of S. minnesota. Acid hydrolysis separates lipopolysaccharide into two components. The lipid fraction is mitogenic, while the polysaccharide is not. Those procedures, which degrade or modify only the lipid moiety while preserving the antigenic inte-grity of the polysaccharide, destroy mitogenicity. The lipid preparations are fully active on highly purified B-lymphocyte populations while they have no effect on highly purified T-lymphocyte populations. These data demonstrate that the lipid moiety of endotoxin is the B-lymphocyte mitogen.

Acknowledgment

Certain of the data in this paper has been reproduced from reference 7 by permission from the Journal of Infection and Immunity.

References

1. Peavy, D.L., Adler, W.H., Smith, R.T.: J. Immunol. 105:1453 (1970).
2. Gery, I., Kruger, G.J., Spiesel, S.Z.: J. Immunol. 108: 1088 (1972).
3. Andersson, J.,Möller, G., Sjöberg, O.: Cell. Immunol. 4: 381 (1972).
4. Chiller, J.M., Skidmore, B.J., Morrison, D.C., Weigle, W.O.: Fed. Proc. 32: 1021 (1973).
5. Andersson, J., Melchers, F., Galanos, C., Lüderitz, O.: J. exp. Med. 137: 943 (1973).
6. Peavy, D.L., Shands, J.W., Jr., Adler, W.H., Smith,R.T.: J. Immunol. 111: 352 (1973).
7. Rosenstreich, D.L., Nowotny, A., Chused, T., Mergenhagen, S.E.:Infect. Immunity 8: 406 (1973).
8. Chen, C.H. et al: J. inf. Dis., in press (1973).
9. Bitter-Suermann, D., Dukor, P., Gisler, R.H., Schumann, G., Dierich, D., Konig, W., Hadding, U.: Abstr. Fifth International Complement Workshop, LaJolla, California (1972).
10. Gewurz, H., Mergenhagen, S.E., Nowotny, A., Phillips, J.K.: J. Bact. 95: 397 (1968).
11. Mergenhagen, S.E., Gewurz, H., Bladen, H.A., Nowotny, A., Kasai, N., Lüderitz, O.: J. Immunol. 100: 228 (1968).

Table I.

Effect of Acid and Alkaline Hydrolysis on the Mitogenic Activity of LPS

Treatment	Final product	Maximum lymphocyte proliferative response $(E/C)^*$
None	LPS (S. minnesota)	26.2 ± 3.0
Boiling	LPS	33.6 ± 7.7
Acid Hydrolysis:		
Insoluble precipitate	Lipid A	10.8 ± 1.3
Soluble phase	Haptenic polysaccharide	1.5 ± 0.3
Alkaline Hydrolysis	Polysaccharide + degraded lipid	2.4 ± 0.8

* Baseline proliferative response = 970 counts per min per 10^6 cells.

Table II.

Effect of Deacylation on the Mitogenic Acitivity of LPS

Treatment	Final product	Maximum lymphocyte proliferative response (E/C)[*]
None	LPS (S. marcescens)	12.8 $\pm$ 0.3
Deacylation:		
0.1 M Potassium methylate	Endotoxoid 1	4.2 $\pm$ 0.5
0.3 M Potassium methylate	Endotoxoid 2	1.9 $\pm$ 0.4

*Baseline proliferative response = 970 counts per min per 10^6 cells.

Table III.

Preparation of Bone-Marrow-Derived (B) and Thymus-Derived (T) Lymphocytes

Preparation of	Protocol
B-cells	Mouse splenocytes plus AKR anti-theta (C_3H) antiserum followed by addition of guinea pig serum (complement).
T-cells	Mouse splenocytes plus goat anti-mouse immunoglobulin and sodium azide followed by addition of guinea pig serum (complement).

Table IV.

In Vitro Lymphocyte Stimulation of Isolated Bone-Marrow-Derived (B) and Thymus-Derived (T) Lymphocytes Induced by Several Lipid Preparations

	Maximum lymphocyte proliferative response (E/C)[*]		
Stimulant	Spleen cells	B-cells	T-cells
Con A	79.5 $\pm$ 9.0	0.7 $\pm$ 0.1	84.6 $\pm$ 7.0
LPS	108.0 $\pm$ 1.5	106.3 $\pm$ 12	0.9 $\pm$ 0.3
Glycolipid	100.0 $\pm$ 9.1	126.6 $\pm$ 15	0.85 $\pm$ 0.1
Lipid A	28.5 $\pm$ 3.2	21.3 $\pm$ 3.5	0.66 $\pm$ 0.3

* Stimulants were tested at several different concentrations and the maximum response is reported. Baseline proliferative response: spleen cells, 522 counts per min per 10^6 cells; B-cells, 242 counts per min per 10^6 cells; T-cells, 200 counts per min per 10^6 cells.

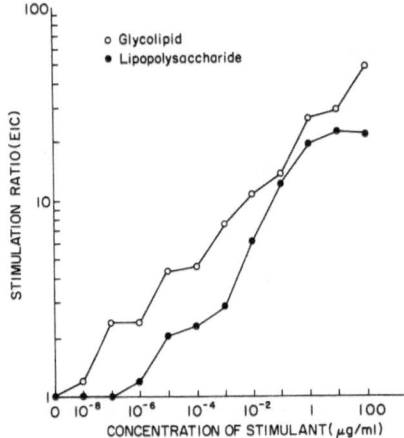

Fig. 1.   In vitro proliferation of mouse spleen lymphocytes induced by a purified glycolipid in comparison to lipopolysaccharide. Glycolipid was derived from the O-polysaccharide deficient mutant of S. minnesota, R595. Lipopolysaccharide was derived from a smooth strain of S. minnesota. Results are the mean of four experiments plotted on a log scale as the stimulation ratio (E/C).

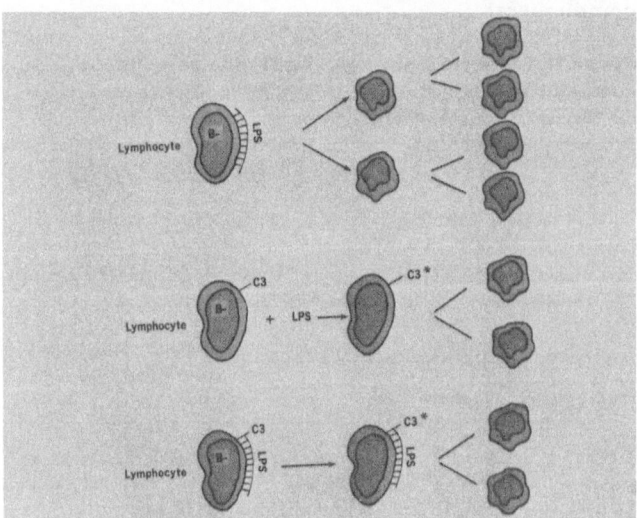

Fig. 2.   Hypothetical models of possible mechanisms whereby lipopolysaccharide (LPS) stimulates B-lymphocytes (see text for details).

Evaluation and Action of Biological Activity of Mediators Generated from Normal Serum by Interaction with Foreign Macromolecules

J. H. Wissler

1. Cell-Mediator Relation

Plasma and serum components and their metabolic products are uniquely distributed and ubiquitously present in intimate contact with the cells of the host. They inevitably participate in reactions occurring on cell membranes and in the fluid phase of the interstices. Structural alterations and the local or systemic change in their concentration, induced by internal or external factors, and the concomitant emergence of new compounds, represent the humoral signals, interplaying with the body's cellular components and triggering them in a regulatory manner to display their genetically determined functions. In response to these humoral signals, a cell may either release new substances or physically influence the fluid phase and thus further alter the normal physiological environment for itself and for other cells. These processes do not occur randomly; rather, they follow regulatory principles. Otherwise, the process of serum or plasma protein alteration and the uncontrolled accumulation of cell-generated mediators could lead to fatal shock, hypersensitivity or autoimmune reactions. Since a considerable proportion of the population is constantly suffering from these and related diseases (1), the evaluation and knowledge of regulatory principles of the process of serum protein alteration, the mechanism of biological action of the concomitantly generated products, and of the interplay of cells with their (mediator-) changed environment (recognition and response of cells) are relevant areas of research, especially from the socio-economic view point.

2. Mode of Biological Action of Foreign Macromolecules

Indirect biological action of endotoxin lipopolysaccharides, other foreign macromolecules, and generation of plasma- and serum-derived mediators

Plasma or serum components are profoundly influenced by contact with endotoxin and other naturally occurring, and synthetic, foreign macromolecules (e. g. immune complexes, endotoxin lipopolysaccharides, dextrans and their derivatives, i. e. dextran sulfates, and yeast, yeast cell walls, zymosan, inulin, agar, venoms, proteolytic enzymes, some types of polyvinylpyrrolidons, some inorganic crystals, etc.) or by complexes of these agents with cells or immunoglobulins. In addition, other external factors (burn, radiation, pressure and other mechanical forces, altered body temperature, unphysiological pH, ionic strength, and electrical field, etc.) induce their alteration. Comprehensive treatises on these subjects with special reference to endotoxin have appeared recently in two volumes edited by Kadis et al. (2, 3) and earlier by Landy and Braun (4). Among other reviews and detailed reports with reference to endotoxin interaction with cells and blood protein systems are those in the references (5-22, 65-69). A collection of data concerning the generation of numerous mediators and their biological activities by synthetic and naturally occurring, foreign macromolecules, notably endotoxins, is available in the handbooks given in references (23, 24). These data amply indicate that most of the exter-

This work has been supported by the Schweizerischer Nationalfonds zur Förderung der wissenschaftlichen Forschung, Grant No. 3.8750-72, by the Deutsche For-schungsgemeinschaft, Grant No. Wi 406/1a and Wi 406/1b and by Hoffmann-La Roche AG, Basel.

nal factors trigger cell functions through interaction with plasma or serum components by generation of specific mediators, and not by direct interaction with the cells.

Evaluation and relevance of biological activity of mediators

Concerning the evaluation of a biological activity of a mediator, the demonstration of its natural role and its participation in vivo, Miles and Wilhelm (summarized by Wilhelm (89)) and Sir Dale (summarized by Spector and Willougby (73)) have outlined the appropriate ideal criteria. But rather than giving a collection of biological phenomena to be observable by foreign macromolecule-serum interactions, I propose to summarize some findings, describing how a single mediator, generated as a result of serum activation by a particular type of foreign macromolecule, can yield an apparently divergent, but selective multiplicity of biological phenomena (activities) and how they are linked mechanistically and regulatorily. I intend to show that biological mediator substances may acquire certain activities, either regulatorily or artificially, which are not intrinsic molecular properties, residing in their primary structure (amino acid sequence) or in their native conformation *. With respect to mediator action, individual assay systems potentially reflect various, but single biological parameters, and a multiplicity of assay systems, therefore, may frequently yield apparently divergent biological results.

The mediator to be considered is the classical anaphylatoxin (CAT) (29, 30, 32, 70, 71, 84). The biological activities considered are : spasmogenic activity and chemotactic activity for neutrophils, and its relation to chemoattractive trapping activity ("migration inhibition"). All these activities in principle can be generated from serum or plasma by appropriate contact with endotoxins and an array of other foreign macromolecules (10, 14, 15, 22-24, 29-39, 53-55, 97, 99).

The activities considered here and the parameters on which they are based, have recently shown to be involved in numerous pathophysiological and pathological phenomena, among them hyperacute graft rejection (82, 83), bacterial and viral infection defense (74-78), repair of cell necrosis (43, 76), wound healing (92) and reactions versus thrombi formation (for survey see 34), as well as acute (e. g. Arthus and Shwartzman reaction) and delayed hypersensitivity phenomena (34, 35, 57, 58, 61-66, 70-72, 77, 93), Forssman and serum-mediated, anaphylaxis-like shock reactions (23, 24, 57, 79-81) some of which may well be mediated indirectly by endotoxins.

CAT activity may be generated either by the classical complement reaction sequence, by trypsinization of the complement component C5 ** (10, 15, 59, 60), or by the so-called alternate pathway of complement activation (properdin system) (for survey see 9, 12, 13, 17, 96, 97). In vitro, endotoxin and other macromolecules, such as inulin, agar, yeast cell walls, etc. are said to activate the complement system by the latter pathway (9, 17, 20, 100) which is only recently once again believed to proceed, in some instances, without the participation of immunoglobulins (9, 96, 98). However, it still remains to be established whether this activation does in fact occur without a n y participation of antibody, notably without IgM (natural antibody). It has been considered (for survey and discussion see e. g. Landy 4, 5, Nelson 13, and Lee and Stetson 66) that even relatively few molecules of IgM, an amount beyond detection by the presently available assay methods, can suffice for activation of the complement system. This possibility receives credence as the

---

* Native conformation : phase of many microscopic conformers, making up the macroscopically apparent native conformational state of a protein, as generated within the normal physiological microenvironment and having a most compact three-dimensional spatial arrangement of amino acid residues.
** complement components according to the recommendations of Bull. WHO 39 : 935 (1968), also given in Immunochemistry 7 : 137 (1970).

multivalent IgM in combination with an appropriate antigen is one of the most potent complement-activating agents known. Hence, the in vivo importance of this activation of the complement system may be questioned, since the potent IgM is ubiquitous, and in combination with an appropriate antigen, it is about two or even three orders of magnitudes more effective than other activators of complement by pathways assumingly not involving antibody (except direct trypsinization of the complement system). However, as far as the situation can be judged at present, this does not mean that different alternatives in complement activation do not exist in vivo too, and that there is no other - or in the type of pathway for complement activation. I believe and propose that different types and molar ratios, the absolute concentrations and partition probabilities (complex stability) of antigen and antibody, when interacting to form complexes, are effective in the molar ratio range of 1 :∞ to ∞: 1, and activate the protein C3 by an allosteric or other conformationally determined way (85, 86), induced through different potentially parallel running reactions forming the trigger for conformational alteration in C3. The balance between the most effective, and macroscopically obviously running pathway is then decided not by distinct structural features of the "antigen", as it is presently considered, by dividing even chemically and physiologically unrelated substances, according to their potency of activating C3 through a c e r t a i n pathway (9, 96). Rather the selection of the "major" acting pathway in C3 activation between possibly numerous existing ones is suggested to be a thermodynamic and kinetic problem, namely when C3 activation is based on a h y s t e r e t i c mechanism (for survey see 87), an idea which has not previously been considered. Some recent evidence, showing that different pathways may indeed being operative parallely (100, 101), e. g. by contact with endotoxin (100), might substantiate this idea.

Normally, if more than one biological activity is observable in a sample, it is assumed that each of the apparent activities reflects a separate "factor" entity. Hence, only a spectrum of factors would be capable of displaying even related biological phenomena ; this is presently the situation for the activities generated by activated lymphocytes (for survey see 45, 58, 61, 62). Or alternatively, it can be assumed that groups of activities are displayed by a single "factor" with multiple active chemical centers (or regions). In the latter situation, the multiple activities residing in one mediator would normally o n l y be displayed c o i n c i d e n t l y . This has been often assumed, especially for the anaphylatoxin preparations in terms of evoking ileal contraction and chemotactic activity for neutrophils (46-50). In addition, chemotactic activity for macrophages (51, 52) has been attributed to anaphylatoxin as being its intrinsic molecular property, although without mechanistic implications in the latter case. Hence, the specificity of mediator action at the cellular level remains in doubt.

Aspects on regulation and mechanisms of mediator action

Recent evidence from this laboratory (25-33, 56) has lead to the conclusion that a limited number of substances ("factors"), generated concomitantly by contact of naturally occurring, foreign macromolecules, with serum or plasma, may display various types of biological phenomena which are regulated in that each factor acts alone or by interplay with others to modulate one factor physically or chemically. That a single type of mediator can normally exert multiple biological activities and each of them separately or all together in well defined sequences or coincidently, is indeed possible, provided that multiple biological activities derive from the alteration of the physicochemical state of a mediator within its microenvironment, rather than being related to d i f f e r e n t c h e m i c a l r e g i o n s of the mediator molecule (or to multiple factors). Moreover, we have shown (25, 27, 28), at least as far as chemotactic activity for neutrophils is concerned, that this principle can be extended to other, non-mediator protein molecules. Thus, on a mechanistic basis, it explains the finding that one type of biological activity can be displayed by numerous substances under specified conditions, even when those substances are un-related chemically and physiologically to the acting native mediator molecule.

3. Classical Anaphylatoxin : A Model for Studying Mechanism of Action and Regulation of Mediatory Activities

Intrinsic molecular properties of classical anaphylatoxin

As a model to study mediator action, it is useful to survey some aspects of my work done with the anaphylatoxin-related binary leukotactic peptide system (ACL-system) (29-32) with classical anaphylatoxin (CAT) (29, 30) and cocytotaxin (CCT) (26, 29) as peptide components. The molecular properties of these two peptides, both of which have been obtained in the crystalline state after purification to molecular homogeneity from contact-activated pig, rat, and guinea pig serum (26, 28, 29, 30), are surveyed elsewhere (28, 32). Fig. 1 shows typical, double-refractive crystals of CAT in polarized light.

Molecularly homogeneous CAT displays spasmogenic activity on contractile elements of various tissues, in vitro and in vivo, as intrinsic molecular property, residing in the primary structure and being largely independent of the conformation of the peptide (30, 32, 40, 70, 71). CCT does not display this activity (26). Dependent on the function of contractile elements within the tissues, their activation by CAT morphologically leads to seemingly different biological phenomena, e. g. fatal anaphylaxis-like shock, histamine liberation and enhancement of vascular permeability in the intact guinea pig as well as ileum contraction, coronary constriction on isolated organs of a variety of species (28-32, 70,71).

CAT and CCT, either alone or together, may display various, and, in part, mechanistically dissimilar activities (28-32) from which spasmogenic activity on smooth muscle represents an intrinsic and chemotactic activity for neutrophils an acquired biological activity of CAT. Several criteria have to be considered in attributing biological activities as intrinsic or acquired molecular properties of a peptide species. These criteria are : a) The homogeneity of the CAT peptide, b) the physicochemical state (conformation)* of the peptide in solution, c) the dependence of the activity on the concentration of the peptide, and d) the apparent physical parameters underlying the assay system.

Intrinsic and acquired biological activities of classical anaphylatoxin in relation to its molecular homogeneity

Solutions of the molecularly homogeneous, crystalline CAT display spasmogenic activity on various types of tissues, in vitro and in vivo, leading to the afore-mentioned biological phenomena, but they do not attract chemotactically neutrophils and macrophages. However, when CAT is contaminated with CCT, the combination of the two peptides may attract neutrophils, but not macrophages, provided a positive concentration gradient of CAT is present in the environment of the cell (28-33, 40, 56). The acquisition of this activity is due to a conformational alteration in CAT, induced by CCT, without formation of a stable complex of the two peptides (40, 88). Fig. 3 (right side) summarizes the work published elsewhere about this subject (28, 32, 40, 88). Furthermore, Fig. 2 shows an activity phase diagram of the acquired neutrophil chemotactic activity of the peptide combination, demonstrating that this activity of the conformationally altered CAT is elicited only in a distinct range of the absolute concentration and molar ratio of the two peptides. The magnitude of the acquired activity is also strongly dependent on the two parameters. In principle, spas-

---

* Conformational state : thermodynamic, macroscopic average properties determined by many microscopic states allowed for the system at a given temperature, pressure, etc..

mogenic activity of the native CAT is not abolished by the influence of CCT (28, 32). The combination of CAT and CCT which display chemotactic activity for different granulocyte types regulatorily (28, 33), we have termed anaphylatoxin-related binary leukotactic peptide system (29, 31, 32).

Intrinsic and acquired biological activities of classical anaphylatoxin in relation to the alteration of its physicochemical state (conformation)

Emerging from the studies of the acquisition of chemotactic activity for neutrophils in CAT consequent to its contamination with CCT, this acquisition of biological activity is inherently related to phases of CAT conformers, different from the native, serum-generated ones (40, 88). Further investigations indicated an acid-base mechnism as being the underlying principle of the conformational alteration in CAT (Brønsted or Lewis base) induced by CCT (correspondent acid). Therefore, CCT has been replaced by conventional Brønsted acids and found that these acids (e. g. acetic acid) can produce a CAT chemotactically active for neutrophils (40). Acquisition of this activity by acid treatment has been demonstrated to be the consequence of unfolding of the native CAT. Thus, CAT, whether homogeneous or not, may be transformed to chemotactically active phases of conformers which also display spasmogenic activity on contractile elements of tissues. Whether contractile elements of single migratory cells are also activated by native, chemotactically inactive CAT, remains to be investigated.

Using other, non-mediator proteins and enzymes, it has been demonstrable that similar conformational alterations may lead to their acquisition of chemotactic activity for neutrophils. Fig. 4 summarizes some typical examples, evaluated in different laboratories, and having in part biological relevance, also apart from the mechanistic issue. They indicate that a migratory motor response of neutrophils is the consequence of a cellular recognition process (25, 28). This relation is especially apparent from the fact that not every conformational alteration in proteins leads to acquisition of chemotactic activity for neutrophils. We have elaborated constraints in quantitive terms under which transitions of phases of conformers are related to acquisition of chemotactic activity for neutrophils (25). These findings experimentally substantiate related considerations concerning the influence of molecular conformations on the effectiveness of anti-inflammatory drugs (102).

Aspects concerning concentration dependence of chemotactic activity. Recognition and response of neutrophils

In many cases biological activities are non-linearly dependent on the concentration of the mediator. Likewise, chemotactic activity for either leukocyte type is far from being a linear function in dependence on the concentration of the chemotactic stimulus, e. g. denatured CAT, the ACL-system (25, 27, 28, 32, 33, 40) or of casein (28). Chemotactic activity for neutrophils of various stimuli is exerted only in a relatively narrow range on their concentration axis, increasing from zero to a maximum value of activity and then decreasing asymptotically to zero activity again within a range of two to three orders of magnitude. Similar conclusions have been reached by other authors (94, 95). In addition, the absolute maximum values of the activity may alter itself from agent to agent, under comparable conditions, which may be explained on theoretical grounds (28). Fig. 5 summarizes our findings for some neutrophil attractants.

An interpretation of these phenomena has been recently advanced by our laboratory (25, 28, 56). It has been demonstrated that neutrophils respond chemotactically to a stimulus only at certain concentrations. In contrast, at higher concentrations in which neutrophils are non-responsive in terms of chemotaxis, these cells are chemo-attractively trapped (28, 56) as demonstrated by the classical migration inhibition capillary tube assay (62, 90, 91). The effects are as cell-specific as is the acticity of the chemoattractant. This shows that leukocytes may posses a dual migratory motor response consequent to sensing of a single, distinct recognition signal. The migratory motor response to be displayed depends only on the physical presentation of the recognition signal, which, for the migratory responses, is represented by particular processes of protein transconformations (25, 27, 28, 56). Furthermore, as far as the capillary tube assay system is concerned, the results demonstrate that a particular assay system may measure indiscriminately different biological parameters, namely chemoattractive trapping and/or migration-inhibitory activity (28, 56).

Conclusions

An evaluation of the mode of action of mediators and their measurement is presented. It is shown that proteins or peptides can have intrinsic and acquired biological activities, and that multiple physicochemical parameters are involved in their display. These are the homogeneity of peptides, the process of their generation, their microenvironment which influences their physical state and their concentration under non-ideal conditions. In addition, as regards the assay system by which mediators are tested, the real biological activity to be measured may be hidden in a potential array of biological parameters which are fixed expressions of the assay method.

Acknowledgement

The author gratefully acknowledges the skilful technical assistance of Mr. Bosco Koncar.

References

1. Davis, D. J. : J. Allergy clin. Immunol. 49 : 323 (1972).
2. Kadis, S., Weinbaum, G., Ajl, S. J. : Microbial toxins, a comprehensive treatise. Bacterial endotoxins, vol. 4 (Academic Press, New York 1971).
3. Kadis, S., Weinbaum, G., Ajl, S. J. : Microbial toxins, a comprehensive treatise. Bacterial endotoxins, vol. 5 (Academic Press, New York 1971).
4. Landy, M., Braun, W. : Bacterial endotoxins. (Rutgers University Press, New Brunswick 1964).
5. Landy, M. : The significant immunological features of bacterial endotoxins. In : Wolstenholme, G. E. W., Birch, J. : Pyrogens and fever, p. 49 (Churchill and Livingstone, London 1971).
6. Work, E. : Production, chemistry and properties of bacterial pyrogens and endotoxins. In : Wolstenholme, G. E. W., Birch, J. : Pyrogens and fever, p. 23 (Churchill and Livingstone, London 1971).
7. Brown, D. L., Lachmann, P. J. : Trans. Coll. Int. Allergologicum 45 : 193 (1973).
8. Agarwal, M. K. : Int. Arch. Allergy 44 : 759 (1973).
9. Gewurz, H. : Alternate pathways to activation of the complement system. In : Ingram, D. G. : Biological activities of complement, p. 56 (Karger, Basel 1972).
10. Jensen, J. A. : Anaphylatoxin (s). In : Ingram, D. G. : Biological activities of complement, p. 136 (Karger, Basel 1972).
11. Melmon, K. L., Epstein, W., Tan, M., Nies, A. L. : Kinin generation caused by two immunologic systems (human IgG-rheumatoid factor complex and endotoxin-antibody-complement complex). In : Movat, H. Z. : Cellular and humoral mechanisms in anaphylaxis and allergy, p. 224 (Karger, Basel 1969).
12. Pillemer, L., Blum, L., Lepow, I.H., Ross, O. A., Todd, E. W., Wardlaw, A. C. : Science 120 : 279 (1954).
13. Nelson Jr., R. A. : J. exp. Med. 108 : 515 (1958).
14. Snyderman, R., Shin, H. S., Phillips, J. K., Gewurz, H., Mergenhagen, S. E. : J. Immunol. 103 : 413 (1969).

15. Jensen, J. A., Snyderman, R., Mergenhagen, S. E. : Chemotactic activity, a property of guinea pig C'5-anaphylatoxin. In : Movat, H. Z. : Cellular and humoral mechanisms in anaphylaxis and allergy, p. 265 (Karger, Basel 1969).

16. Frank, M. M., Kane, M. A., May, J. E. : Biological interactions of complement with endotoxin : Studies in C4-deficient guinea pigs. In : Braun, W., Ungar, J. : Non-specific factors influencing host resistance, a reexamination, p. 323 (Karger, Basel 1973).

17. Gewurz, H., Ertel, N. : Non-immune activation of complement : Two new phenomena. In : Braun, W., Ungar, J. : Non-specific factors influencing host resistance, a reexamination, p. 340 (Karger, Basel 1973).

18. Elin, R. J., Wolff, S. M. : Endotoxin-induced non-specific resistance to infection - a possible mechanism. In : Braun, W., Ungar, J. : Non-specific factors influencing host resistance, p. 371 (Karger, Basel 1973).

19. Mustard, J. F., Evans, G., Packham, M. A., Nishizawa, E. E. : The platelet in intravascular immunological reactions. In : Movat, H. Z. : Cellular and humoral mechanisms in anaphylaxis and allergy, p. 151 (Karger, Basel 1969).

20. Gewurz, H., Shin, H. S., Mergenhagen, S. E. : J. exp. Med. 128 : 1049 (1963).

21. Davis, S. D., Ianetta, A., Wedgwood, R. J. : Bactericidal reactions of serum. In : Ingram, D. G. : Biological activities of complement, p. 43 (Karger, Basel 1972).

22. Havemann, K., Horvat, M., Sodomann, C.-P., Havemann, K., Bürger, S. : Europ. J. Immunol. 2 : 97 (1972).

23. Erdös, E. G., Wilde, A. F. : Bradykinin, kallidin and kallikrein. In : Eichler, O., Farah, A., Herken, H., Welch, A. D. : Handbook of experimental pharmacology, vol. 25 (Springer, Berlin 1970).

24. Rocha e Silva, M. : Histamine and anti-histaminics. In : Eichler, O., Farah, A. : Handbook of experimental pharmacology, vol. 18/1 (Springer, Berlin 1966).

25. Wissler, J. H., Sorkin, E. : Mechanism of cellular recognition and chemotactic activity for neutrophil leukocytes. Abstr. Commun. 9th Int. Congr. Biochem., p. 318, Stockholm 1973.

26. Wissler, J. H. : Europ. J. Immunol. 2 : 84 (1972).

27. Wissler, J. H., Sorkin, E. : Nature and mechanism of cellular recognition and regulation of leukocyte migration. Workshop report : Chemotaxis of leukocytes, Paris 1973, Nouv. rev. française d'hématol., in press.

28. Wissler, J. H. : The binary leukotactic peptide system of serum as a model for exploring cellular recognition and leukocytes-migratory behaviour. In : Sorkin, E. : Chemotaxis of leukocytes, its biology and biochemistry (Karger, Basel 1974), in press.

29. Wissler, J. H. : Experientia 27 : 1147 (1971).

30. Wissler, J. H. : Europ. J. Immunol. 2 : 73 (1972).

31. Wissler, J. H., Stecher, V. J., Sorkin, E. : Europ. J. Immunol. 2 : 90 (1972).

32. Wissler, J. H., Stecher, V. J., Sorkin, E. : Int. Arch. Allergy 42 : 722 (1972).

33. Wissler, J. H., Stecher, V. J., Sorkin, E. : Regulation of chemotaxis of leukocytes by the anaphylatoxin-related peptide system. In : Peeters, H. : Proc. XXth Colloquium Protides of the Biological Fluids, p. 411 (Pergamon Press, Oxford 1973).

34. Sorkin, E., Stecher, V. J., Borel, J. F. : Ser. Haematol., vol. 3/1 : 131 (1970).

35. Giertz, H., Hahn, F. : Makromolekulare Histaminliberatoren. In : Eichler, O., Farah, A. : Handbook of experimental pharmacology, vol. 18/1, p. 481 (Springer, Berlin 1966).

36. Bodammer, G., Vogt, W. : Arch. Pharmacol. Exp. Path. 266 : 255 (1970).

37. Snyderman, R., Mergenhagen, S. E. : Characterization of polymorphonuclear leukocyte chemotactic activity in serums activated by various inflammatory agents. In : Ingram, D. G. : Biological activities of complement, p. 117 (Karger, Basel 1972).
38. Udaka, K., Udaka, K. : Fed. Proc. 31 : 657 (1972).
39. Keller, H. U., Sorkin, E. : Immunology 9 : 441 (1965).
40. Wissler, J. H., Stecher, V. J., Sorkin, E. : J. Immunol. 111 : 314 (1973).
41. Wilkinson, P. C., McKay, I. C. : Int. Arch. Allergy 41 : 237 (1971).
42. Wilkinson, P. C., McKay, I. C. : Europ. J. Immunol. 2 : 570 (1972).
43. Bessis, M. : Necrotaxis : chemotaxis towards an injured cell. Nouv. Rev. Française d'Hématol., in press.
44. Wilkinson, P. C. : Nature 244 :512 (1973).
45. Pick, E., Turk, J. L. : Clin. exp. Immunol. 10 : 1 (1972).
46. Bokisch, V. A., Müller-Eberhard, H. J., Cochrane, C. G. : J. exp. Med. 129 : 1109 (1969).
47. Müller-Eberhard, H. J. : Annu. Rev. Biochem. 38 : 389 (1969).
48. Müller-Eberhard, H. J., Vallota, E. H., Götze, O., Zimmermann, T. S. : Mediators of the inflammatory response-complement. In : Lepow, I. H., Ward, P. A. : Inflammation, mechanisms and control, p. 83 (Academic Press, New York 1972).
49. Vallota, E. H., Müller-Eberhard, H. J. : J. exp. Med. 137 : 1109 (1973).
50. Shin, H. S., Snyderman, R., Friedman, E., Mellors, A., Mayer, M. M. : Science 162 : 361 (1968).
51. Snyderman, R., Shin, H. S., Hausman, M. H. : J. Immunol. 107 : 316 (1971).
52. Snyderman, R., Shin, H. S., Hausman, M. H. : Proc. Soc. exp. Biol. Med. 138 : 387 (1971).
53. Keller, H. U., Sorkin, E. : Int. Arch. Allergy 31 : 505 (1967).
54. Keller, H. U., Sorkin, E. : Int. Arch. Allergy 31 : 575 (1967).
55. Keller, H. U., Sorkin, E. : Int. Arch. Allergy 35 : 279 (1969).
56. Wissler, J. H., Sorkin, E., Jungi, T. W., Stecher, V. J., Arcon, A. : Mechanisms regulating leukocyte accumulation : cellular recognition and migratory behaviour of phagocytes. Proc. Int. Meeting on Future Trends in Inflammation, Verona 1973, in press. Abstr. Commun., p. 11.
57. Giertz, H. : Pharmacology of anaphylatoxin. In : Movat, H. Z. : Cellular and humoral mechanisms in anaphylaxis and allergy, p. 253 (Karger, Basel 1969).
58. Bloom, B. R., David, J. R. : Elaboration of effector molecules by activated lymphocytes. In : Lawrence, H. S., Landy, M. : Mediators of cellular immunity, p. 247 (Academic Press, New York 1969).
59. Jensen, J. : Science 155 : 1122 (1967). Ibid : Immunochemistry 3 : 498 (1966).
60. Cochrane, C. G., Müller-Eberhard, H. J. : J. exp. Med. 127 : 371 (1968).
61. Bloom, B. R., Glade, P. R. : In vitro methods in cell-mediated immunity (Academic Press, New York 1971).
62. David, J. R. : Fed. Proc. 30 : 1730 (1971).
63. Ward, P. A., Zvaifler, N. J. : J. clin. Invest. 50 : 606 (1971).
64. Zvaifler, N. J. : Arth. Rheum. 13 : 895 (1970).
65. Cochrane, C. G. : The Arthus reaction. In : Zweifach, B. W., Grant, L., McCluskey, R. T. : The inflammatory process, p. 613 (Academic Press, New York 1965).
66. Lee, L., Stetson Jr., C. A. : The local and generalized Shwartzman phenomena. In : Zweifach, B. W., Grant, L., McCluskey, R. T. : The inflammatory process, p. 791 (Academic Press, New York 1965).
67. Nowotny, A. (Ed.) : Molecular biology of gram-negative bacterial lipopolysaccharides. Ann. N. Y. Acad. Sci. 133 (1966), various review articles.
68. Cotran, R. S., Remensnyder, J. P. : Ann. N. Y. Acad. Sci. 150 : 495 (1968).
69. Markley, K. : Ann. N. Y. Acad. Sci. 150 : 922 (1968).

70. Bernauer, W., Hahn, F., Wissler, J. H., Nimptsch, P., Filipowski, P. : Arch. Pharmakol. exp. Path. 269 : 413 (1971).
71. Bernauer, W., Hahn, F., Nimptsch, P., Wissler, J. H. : Int. Arch. Allergy 42 : 136 (1972).
72. Parish, W. E. : Clin. Allergy 2 : 381 (1972).
73. Spector, W. G., Willoughby, D. A. : Chemical Mediators II. In : Zweifach, B. W., Grant, L., McCluskey, R. T. : The inflammatory process, p. 427 (Academic Press, New York 1965).
74. Miller, M. E., Oski, F. A., Harris, M. B. : Lancet I : 665 (1971).
75. Clark, R. A., Kimball, H. R. : J. clin. Invest. 50 : 2645 (1971).
76. Buckley, I. K. : Exp. Mol. Path. 2 : 402 (1963).
77. Hurley, J. V. : Ann. N. Y. Acad. Sci. 116 : 918 (1964).
78. Brier, A. M., Snyderman, R., Mergenhagen, S. E., Notkins, A. L. : Science 170 : 1104 (1970).
79. Hahn, F. : Excerpta med. Int. Congr. Ser. 162 : 145 (1968).
80. Bernauer, W., Hahn, F., Beck, E., Kury, H. : Arch. Pharmak. exp. Path. 266 : 208 (1970).
81. Seyle, H. : The mast cell, pp. 16 and 133 (Butterworth, Washington 1965).
82. Jensen, J. A., Davies, D., Linn, B. S., Snyderman, R., Franklin, L. : Circulat. Res. 30 : 332 (1972).
83. Winn, H. J., Baldamus, C. A., Jooste, S. V., Russell, P. S. : J. exp. Med. 137 : 893 (1973).
84. Sorkin, E., Stecher, V. J., Wissler, J. H. : The anaphylatoxin-related leukotactic binary peptide system. In : Braun, W., Ungar, J. : Non-specific factors influencing host resistance, a reexamination, p. 196 (Karger, Basel 1973).
85. Monod, J., Changeux, J.-P., Jacob, F. : J. molec. Biol. 6 : 306 (1963).
86. Monod, J., Wyman, J., Changeux, J.-P. : J. molec. Biol. 12 : 88 (1965).
87. Neumann, E. : Angew. Chemie, Int. Ed. 12 : 356 (1973).
88. Wissler, J. H., Stecher, V. J., Sorkin, E., Jungi, Th. : Mechanism of leukotaxis. Abstr. Commun. 4th Tagung der Gesellschaft für Immunologie, p. 61, Berne 1972.
89. Wilhelm, D. L. : Chemical mediators I. In : Zweifach, B. W., Grant, L., McCluskey, R. T. : The inflammatory process, p. 389 (Academic Press, New York 1965).
90. David, J. R., Al-Askari, S., Lawrence, H. S., Thomas, L. : J. Immunol. 93 : 264 (1964).
91. George, M., Vaughan, J. H.: Proc. Soc. exp. Biol. Med. 111 : 514 (1962).
92. Schilling, J. A. : Physiol. Rev. 48 : 374 (1968).
93. Cohen, S., Ward, P. A. : J. exp. Med. 133 : 133 (1971).
94. Keller, H. U., Borel, J. F., Wilkinson, P. C., Hess, M. W., Cottier, H. : J. Immunol. Methods 1 : 165 (1972).
95. Jensen, J. A., Williams, D. : Chemotaxis of human neutrophils induced by lymphocytoxic sera. Nouv. Rev. Française d'Hématol., in press.
96. Müller-Eberhard, H. J., Vallota, E. H. : Formation and inactivation of anaphylatoxins. In : Austen, K. F., Becker, E. L. : Biochemistry of the acute allergic reactions, p. 217 (Blackwell Scientific Publications, Oxford 1971).
97. Lichtenstein, L. M., Gewurz, H., Adkinson Jr., N. F., Shin, H. S., Mergenhagen, S. E. : Immunology 16 : 327 (1969).
98. Gewurz, H., Pickering, R. J., Snyderman, R., Lichtenstein, L. M., Good, R. A. : J. exp. Med. 131 : 817 (1970).
99. Gewurz, H., Snyderman, R., Mergenhagen, S. E., Shin, H. S. : Effects of endotoxic lipopolysaccharides on the complement system. In : Kadis, S., Weinbaum, G., Ajl, S. J. : Microbial toxins, vol. 5, p. 127 (Academic Press, New York 1971).

100. Fine, D. P. : J. Immunol. 112 : 763 (1974).
101. Ferrone, S., Cooper, N. R., Pellegrino, M. A., Reisfeld, R. A. : Proc.
     nat. Acad. Sci. 70 : 3665 (1973), J. Immunol. 111 : 301 (1973).
102. Juby, P. F., Hudyma, T. W. : Non-steroidal antiinflammatory agents. In :
     Cain, C. K., p. 182 (Academic Press, New York 1971).

Fig. 1. Double-refractive crystals in polarized light of classical anaphylatoxin from yeast-treated pig serum. Reprinted from Wissler et al. (32) with kind permission of the publisher. Methods for preparing the crystals are reported in (30).

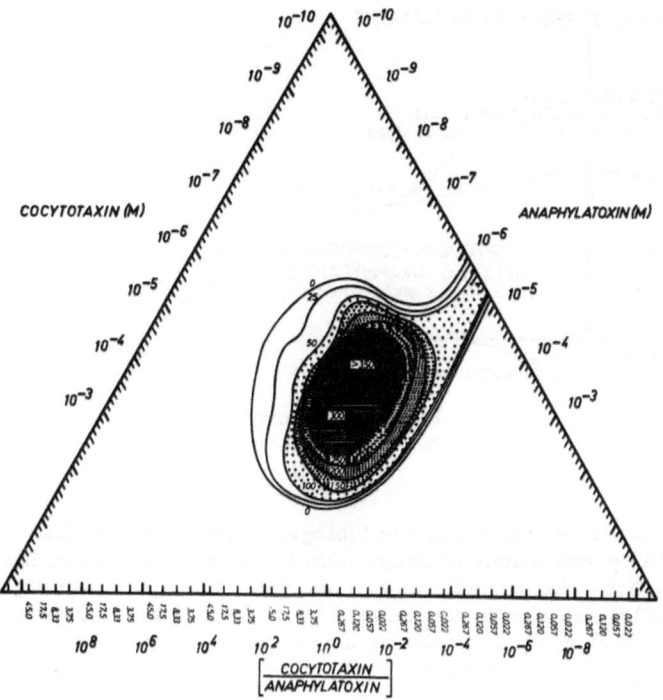

Fig. 2. Biological function of the anaphylatoxin-related leukotactic peptide system :
Relation of chemotaxis of neutrophils and homogeneity of classical anaphylatoxin.
The activity phase diagram of the binary leukotactic serum peptide system (31, 32)
represents chemotaxis for neutrophils as acquired activity of classical anaphylatoxin
(CAT) in isochemotaxis curves. Parameters from which chemotaxis activity is dependent, are the absolute concentrations (M) of CAT (30, 32) and cocytotaxin (CCT)
(26, 32) and the molar ratio of the two peptides. Isochemotaxis curves represent
constant levels of chemotactic activity in cell counts/microscopic field. Only within
the area, limited by the zero-isochemotaxis curve, the peptide system displays chemotactic activity for neutrophils. Maximum chemotactic activity is observed within
the concentration range given in the black area. Nonresponsiveness of neutrophils
to the binary peptide system in terms of chemotaxis is observed in the concentration
ranges outside of the zero-isochemotaxis curve. In contrast to the responsiveness
in terms of chemotaxis for neutrophils, the nonresponsiveness is independent of the
absolute concentrations of the two peptide components, provided that the molar ratio
CCT : CAT > 50 : 1. Note that significant chemotactic activity for neutrophils is
observed only, when the native CAT is contaminated with CCT. For explanation of
the chemotactic activity for neutrophils at very high concentrations of CAT alone,
see Fig. 5. There, the concentration dependence of chemotactic activity is shown
two-dimensionally in curves correspondent to the ones given in this Fig. Reprinted
from Wissler et al. (32) with kind permission of the publisher. Methods for assaying chemotaxis are reported in (31). Comprehensively, further details are reported
in (28).

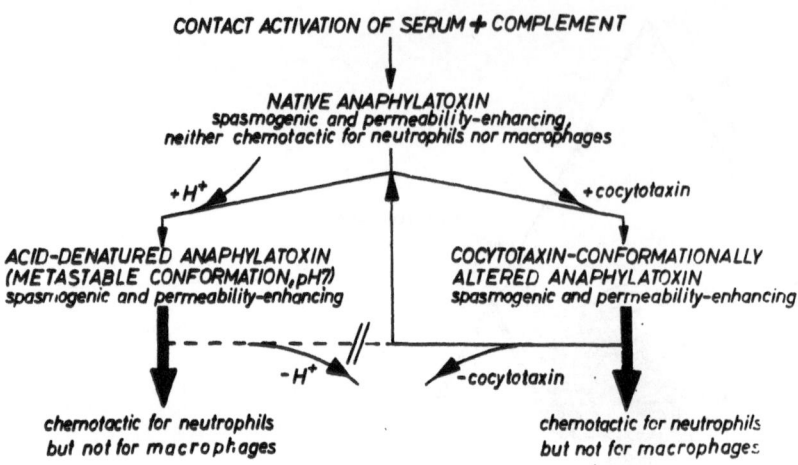

Fig. 3. Schematic representation of structural and biological activity transitions in classical anaphylatoxin (CAT) preparations of chemically identical, but physicochemically different states. It shows that a mediator can acquire a biological activity by various manipulations (40) which can be non-related to its source and derivation pathway. Reprinted from Wissler et al. (28) with kind permission of the publisher. Comprehensively, further details are reported in (28).

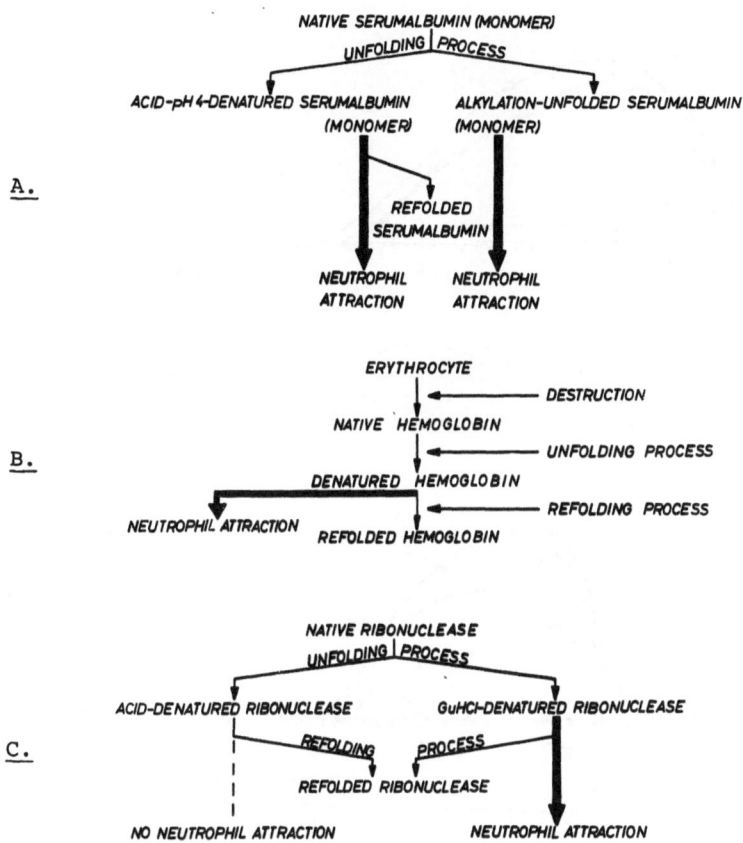

Fig. 4. Protein Transconformations and Recognition in Neutrophil Chemotaxis

A. Schematic representation for serum albumin as model.
B. Schematic representation for hemoglobin as model and of one of the possibilities by which the molecular basis of necrotaxis (43) can be explained.
C. Schematic representation for ribonuclease as model. The figures show that normal, non-mediatory cell or serum constituents can acquire a biological activity (chemotactic activity for neutrophils) by various manipulations which, by far, are non-related to their physiological activity. Note that not all transconformations are fundamental to the appearance of chemotactic activity for neutrophils. Constraints under which this activity is related to transconformations are reported elsewhere (25, 28). Qualitatively, this presentation summarizes published work of Wilkinson and McKay (41, 42) (Fig. 4 A), Wissler et al. (25, 28), Wilkinson (44) and Bessis (43) (Fig. 4 B) and Wissler et al. (25, 28) (Fig. 4 C). The behaviour of these model proteins in terms of acquiring chemotactic activity for neutrophils is comparable to the behaviour of classical anaphylatoxin (see Fig. 2). Reprinted from Wissler et al. (28) with kind permission of the publisher. There, too, further details are given comprehensively.

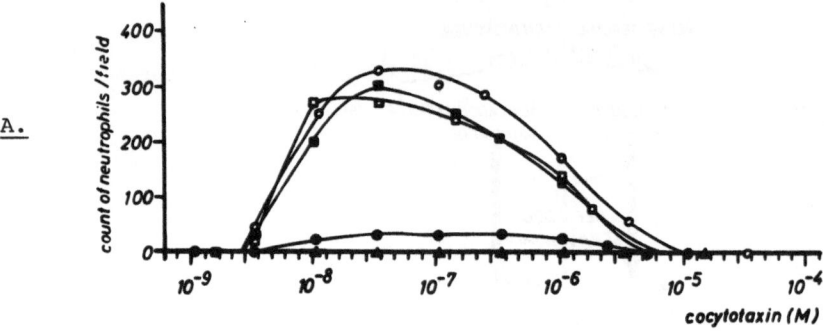

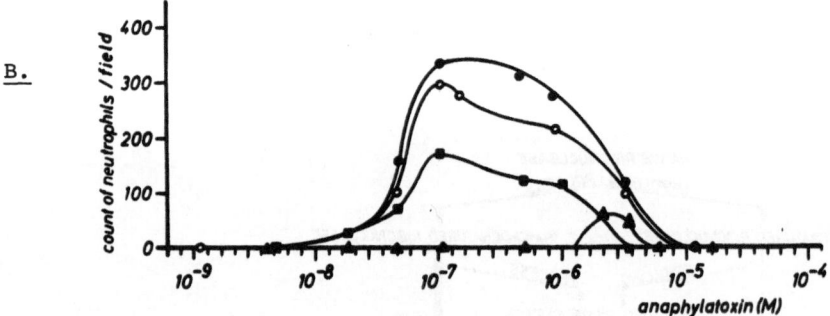

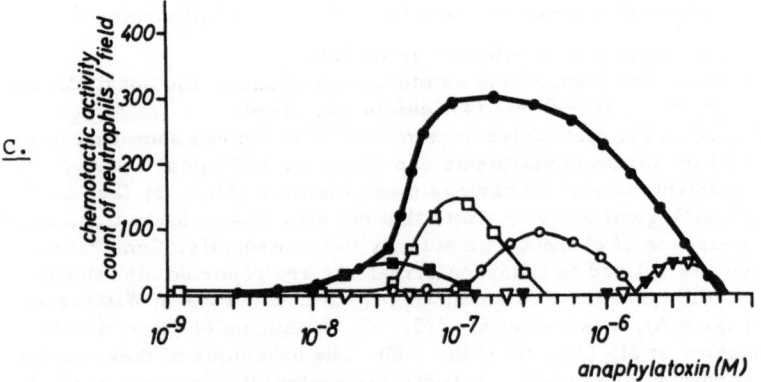

Fig. 5. Dependence of Chemotactic Activity for Neutrophil Leukocytes on the Concentration of Chemotactically Active Preparations

A. The binary anaphylatoxin-related leukotactic peptide system as chemotactically active peptide preparation (compare to Fig. 2) with varying concentrations of cocytotaxin (CCT) in combination with constant doses of classical anaphylatoxin (CAT).

Concentration of Classical Anaphylatoxin:

—▲—▲—▲—   0 M
—●—●—●— $3.50 \cdot 10^{-8}$ M
—○—○—○— $1.12 \cdot 10^{-7}$ M
—■—■—■— $2.80 \cdot 10^{-7}$ M
—□—□—□— $9.80 \cdot 10^{-7}$ M

B. The binary anaphylatoxin-related leukotactic peptide system as chemotactically active peptide preparation (compare to Fig. 2) with varying concentrations of CAT in combination with constant doses of CCT.

Concentration of Cocytotaxin:

—▲—▲—▲—   0 M
—●—●—●— $5.90 \cdot 10^{-8}$ M
—○—○—○— $1.18 \cdot 10^{-7}$ M
—■—■—■— $1.18 \cdot 10^{-6}$ M

C. Different molecularly homogenous preparations of acid-unfolded CAT

—▼—▼—▼—   native anaphylatoxin, prepared at pH 7.4
—●—●—●—   native anaphylatoxin, in $6 \cdot 10^{-8}$ M cocytotaxin (ACL)
native anaphylatoxin, treated and stored in:
—○—○—○— 0.02 N CH_3COOH, $-20^{\circ}C$, 12 h.
—□—□—□— 0.5 N CH_3COOH, $+20^{\circ}C$, 12 h.
—■—■—■—   50 % (v/v) CH_3COOH ($\sim$9N), $+20^{\circ}C$, 12 h.
—▽—▽—▽—   50 % CH_3COOH+25 % CH_3OH, $+20^{\circ}C$, 12 h.

Fig. 5 A and B demonstrate two-dimensionally the dependence of chemotactic activity for neutrophils on the concentrations of the components of the oligodispers (binary) CAT-related leukotactic peptide system (31, 32) as given in full detail in Fig. 2. The first curve in Fig. 5 C (native CAT) corresponds to the curves given in Fig. 5 B. Note that in all examples (5 A, B, and C), chemotactic activity for neutrophils is only observable in a narrow range of the concentration of the attractant. Similar behaviour show also other attractants for neutrophils, e. g., casein, etc. (28). In the chemotactic nonresponsiveness phase, neutrophil migration toward the attractant is not observable in the assay system used (31, 33). In Fig. 5 C, chemotactic activity for neutrophils of the various CAT preparations which are molecularly homogeneous in chemical terms, is displayed in different concentration ranges, indicating different kinetic stability at neutral, physiological conditions (Gey's solution (31)) of the thermodynamically unfavoured, nearly metastable conformers of CAT. The low chemotactic activity for neutrophils of the native CAT alone at very high concentrations is explainable to the unavoidable presence of about 0.5% of denatured CAT, an amount which is inherent to any preparation of purified protein. Fig. 5 A and B are reprinted from (33) and Fig. 5 C is reprinted from (28) with kind permission of the publishers. There, too, technical details are reported on which these results are based.

Purification of PO, Antigenic Subfraction of the Endotoxin Extracted from S. typhi O 901

M. Raynaud, J.-C. Chermann, M.-J. Navarro, and B. Kouznetzova

Introduction

In studying toxicity we have demonstrated (1, 2) that the purified endotoxins extracted by the method of Boivin (3) or by sodium chloride-sodium citrate hypertonic solution (4) are more toxic than those obtained by the corresponding lipopolysaccharide phenol water procedure of Westphal et al. (5) or by the technique of Galanos (6). According to some authors, lipid A solubilized with bovine serum albumin exhibits equal endotoxic activity as the original LPS in lethal toxicity or pyrogenicity (Lüderitz et al., Intern. Symp. on toxic products of bacteria, Prague, September 1973). According to other authors, the content of lipid A from LPS is not related to the level of toxicity (D. Mlynarcik, same symposium). To try to solve this problem, we have previously proved (1, 2) that a Triton X100 treatment could reveal an antigenic subfraction enriched in nitrogen from the endotoxin extracted by the trichloracetic method of Boivin (3) or by the sodium chloride-sodium citrate procedure (4).

The antigenic subfractions thus obtained from S. typhi O 901, S. typhi R_2, and S. minnesota R_{595} are immunologically identical. This special fraction is not found in the lipopolysaccharides extracted by the phenol water procedure of Westphal et al. (5) or by the method of Galanos et al. (6).

We present a description of a simple method to prepare this subfraction. It is based on the direct treatment of the germs by Triton X100 (5 %) in aqueous solution. A very heavy suspension of germs (1 g of wet germs in 5 ml) is put into contact overnight at $0^{\circ}C$. The extract contains the subfraction PO as well as various other antigens, polysaccharide subfractions we have called Ox and Oy, proteins of the cell-wall Barber type, and common heterologous antigens called r_1, r_2, r_3. The purifying process consists of a zonal centrifugation in sucrose gradient at low speed on the transparent rotor Z 15 (I. E. C.).

Material and Methods

1. Strain

S. typhi O 901 (smooth form) is cultivated in 20 1 fermentor (Biolaffitte) by a method previously described (7).

2. Extraction Method

Two methods of extraction have been used. The first one is the hypertonic solution 1 M sodium chloride-0. 1 M sodium citrate (4) ; the second one is a direct contact between the bacteria and the Triton X100 at $0^{\circ}C$ overnight as follows :

150 g of bacteria (wet weight: pellet of centrifugation 3,000 rpm during one hour on the I. E. C. PR6 centrifuge) are washed twice in saline solution (NaCl 9 p. 1000) then suspended with the Sorvall omnimixer in an aqueous solution of Triton X100 at 5 % (vol./vol.) (1 g of wet weight bacteria in 5 ml) and left one night between 0° and $4^{\circ}C$. Another centrifugation is performed (3,000 rpm 60 min) and the supernatant is retained. The sediment is extracted by the method of NaCl citrate as previously described (4).

This work has been supported by a grant from the "Institut National de la Santé et de la Recherche Médicale" : contract number 72 4 016.

3. Sera

The sera used are those described previously (2). Briefly these sera are obtained after immunization of horses with heat-killed whole germs (90 307, 483, 13 475). Only the serum of the horse 207 (bledding 13 525 and 13 455) has been obtained by immunization with a crude endotoxin treated with alkali (pH 12) at a cold temperature (-5°C) and adsorbed on calcium phosphate.

4. Centrifuges and Rotors

The centrifuges, rotors, and the digital speed integrator are from the International Equipment Company "I. E. C. ".The centrifuges are the B60 and the PR6. The rotors used in this study are the B30 titanium, the transparent Z15 (and the continuous flow rotor CF6).

5. Zonal Centrifugation

a) Rotor B30 titanium
Sample:: 30 ml of Triton X100 extract in M/1 NaCl-0.1 M sodium citrate, or the sample can be a partially purified Triton extract on Sepharose 4B (K50-100) column (Pharmacia) and introduced into a B30 rotor containing a linear gradient of 10 to 40 % of sucrose also containing 1 M sodium chloride-0.1 M sodium citrate. The sample which was introduced at the top of the gradient is followed by an overlay of 80 ml or 150 ml (as indicated in results section) of M/1 sodium chloride-0.1 M sodium citrate. Centrifugations were carried out at 40,000 rpm for 30 minutes or $\int w^2 dt = 3,800 \times 10^7$ at 4°C. The unloading of the rotor was accomplished by water injection through the center channel of the rotor of 20 to 30 ml/min. The effluent was collected in 10 ml fraction and the absorbance was determined by an ISCO-UV analyser and recorded with a Philips recorder. Each fractions was analysed immunologically against specific antisera to detect the endotoxin fraction or subfraction.
b)Rotor Z15
Sample: 20 ml of 5 % Triton X100 extract from S.typhi O 901 containing M/1 sodium-0.1 M sodium citrate and introduced into a Z15 rotor containing a linear gradient of 10 to 45 % of sucrose also containing sodium chloride-sodium citrate. The sample which was introduced at the top of the gradient is followed by an overlay of 440 ml of sodium chloride-sodium citrate. The rotor is then accelerated to 5,000 rpm in a B60 centrifuge for 55 minutes corresponding to $\int w^2 dt = 938 \times 10^6$. The unloading of the rotor was realized by a 50 % sucrose solution through the edge channel at the rate of 30 ml/min. The effluent was collected in 15 ml fractions and the sorbance was determined as described above.

Results

Treatment of endotoxin with Triton X100 has shown that the disruption realized between the PO subfraction and the remaining molecule was incomplete, leaving some whole endotoxin molecules in addition to PO subfraction. For this reason we have preferred direct treatment of the germs with Triton X100 hoping the undamaged parts of the endotoxin molecules would stay on the cell-wall, whereas the freed PO subfraction would go into solution.

The immunological examination by precipitation in a gelified medium reinforces this hypothesis and the Fig. 1b shows that, besides the PO revealed subfraction, proteins from the cell-walls (PW) lipopolysaccharide subfractions (Ox, Oy), as well as heterologous antigens r_1, r_2, r_3 are common to the Enterobactericeae.

The so-treated residual germs are resuspended in a 1 M sodium chloride-0.1 M sodium citrate solution and the residual endotoxin is found undamaged in the supernatant in addition to the PO subfractions.

Purification of the Triton Extract by Zonal Centrifugation

a) Rotor I. E. C. B30 titanium

Fig. 2 shows the profile of a typical run with the S. typhi Triton X100 extract. Immunodiffusion analysis indicates that the heterologous antigens r_1, r_2, r_3, the protein of the wall, and some modified endotoxins remain in the sample zone (1st peak in Fig. 1c), while the PO subfraction moves into the gradient.

b) Rotor I. E. C. Z15 (transparent)

Fig. 3 shows the elution pattern of a typical centrifugation with the same extract. The purification obtained with this rotor is shown in Fig. 1b (starting material) and in Fig. 1d (peak 2).

A single antigen can be seen in the PO subfraction with serum 13 455, while the proteins of the wall remain in the sample zone (Fig. 1c)

Discussion

Endotoxins extracted by the technique of Boivin or Raynaud contain two kinds of molecules as revealed by antisera prepared against heat-killed bacteria (Fig. 1a). After treatment of purified endotoxin with Triton X100, a new fraction, PO, appears. This subfraction of the endotoxin is common to the smooth (S. typhi O 901), rough (S. typhi R_2) or extreme-rough-rough forms (S. minnesota R_{595}) (1, 2). This subfraction is absent in the lipopolysaccharide extracted by the method of Westphal (5) or Galanos (6).

As shown by the elution profiles described in this paper, the PO subfraction moves into sucrose density gradients and has a high molecular weight. The centrifugation applied to the Triton X100 extract is a rate zonal centrifugation. If the run is followed for a longer time, the second peak (PO subfraction) sediments to the wall of the rotor.

The PO subfraction (second peak) of the endotoxin is toxic for ten days old chick embryos inoculated i. v. and is pyrogenic for rabbits while the material remaining in the sample zone is less toxic than the starting material.

Acknowledgement

We wish to thank Mrs. A. Carlin-Subrenat and Mr. F. Garcia for their excellent technical assistance.

References

1. Chermann, J. - C., Digeon, M., Kouznetzova, B., Raynaud, M. : Symp. Series immunobiol. Standard. 15 : 123 (1971).
2. Raynaud, M., Kouznetzova, B., Navarro, M. - J., Chermann, J. - C., Digeon, M., Petitprez, A. : J. infect. Dis., Supplement 128 : 35 (1973).
3. Boivin, A., Mesrobeanu, I., Mesrobeanu, L. : C. R. Soc. Biol. 114 : 307 (1933).
4. Raynaud, M., Digeon, M. : C. R. Acad. Sci. Paris 229 : 564 (1949).
5. Westphal, O., Lüderitz, O., Bister, F. : Z. Naturforsch. 7 b : 148 (1952).
6. Galanos, C., Lüderitz, O., Westphal, O. : Europ. J. Biochem. 9 : 245 (1969).
7. Mendiola, L. R., Kouznetzova, B., Chermann, J. - C., Sinoussi, F., Digeon, M., Raynaud, M. : Infection and Immunity 6 : 27 (1972).

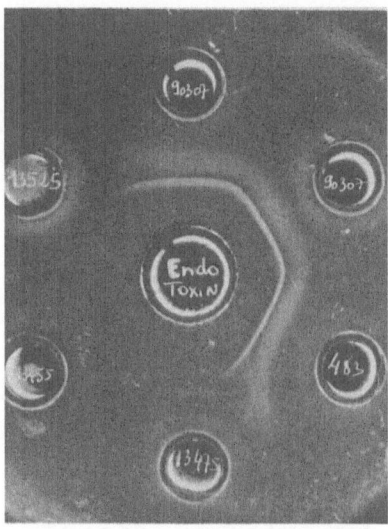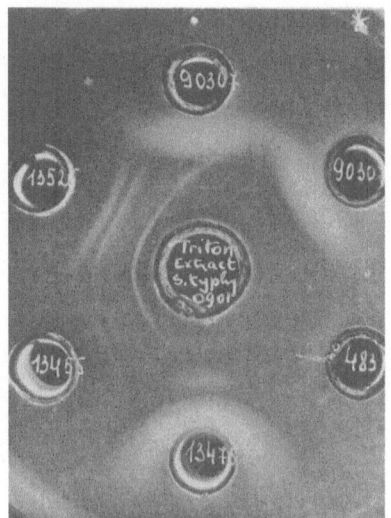

Fig. 1. Immunodiffusion plate
a) Center well: purified endotoxin extracted from S.typhi O 901 1.5 mg/ml. Note two lines of precipitation against the sera 90 307 and 483 (anti-S. typhi), and the absence of reaction with 13 525.
b) Center well: crude S. typhi O 901 Triton X100 extracted (sample material for the run described in Fig. 2. Note the heterogeneity of the preparation: the r_1, r_2, r_3 antigens are revealed by the serum 13 475, the proteins of the wall (PW) and the PO subfraction are shown by the serum 13 525 and 13 455, the modified endotoxin is detected by the serum 90 307.

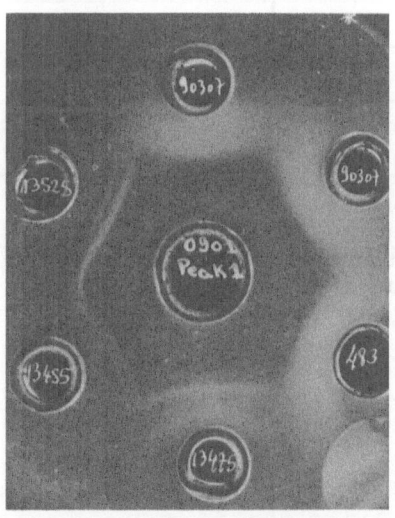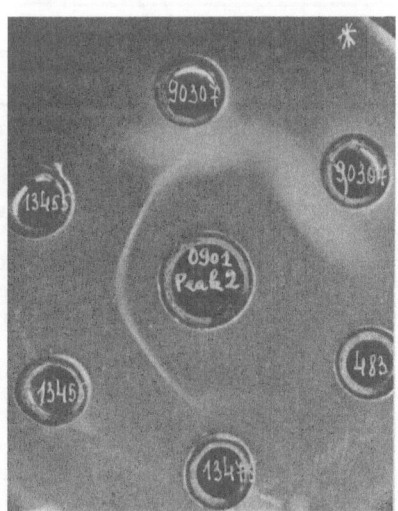

c) Center well: material remaining in the sample zone of the run described in Fig. 2. Note the presence of the residual endotoxin (serum 90 307 and serum 483) of the proteins of the wall (13 525 and 13 455) and the r_1, r_2, r_3 antigens (13 475).
d) Center well: peak 2 of the zonal centrifugation described in Fig. 2 (containing the material moving into the gradient). Note the good purification of the PO subfraction revealed by the serum 13 455.

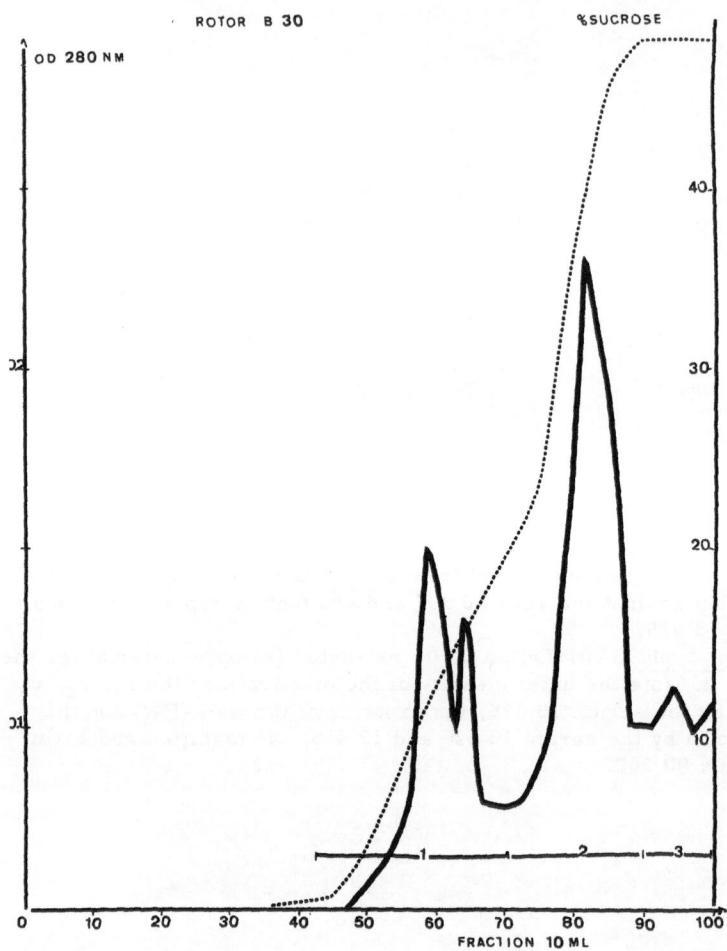

Fig. 2. Zonal centrifugation of a Triton extract from S. typhi O 901 Rotor I. E. C.
B30 operating in a B60 ultracentrifuge. Sample of 20 ml of Triton extract (5g/ml)
in 1 M sodium chloride-0. 1 M sodium citrate. Gradient 5-45 % (w/w) sucrose also
containing 1 M sodium chloride-0. 1 M sodium citrate. Overlay 1 M sodium chlo-
ride-0. 1 M sodium citrate. Run 40, 000 rpm during 30 minutes corresponding to
$\int dt = 3,800 \times 10^7$.

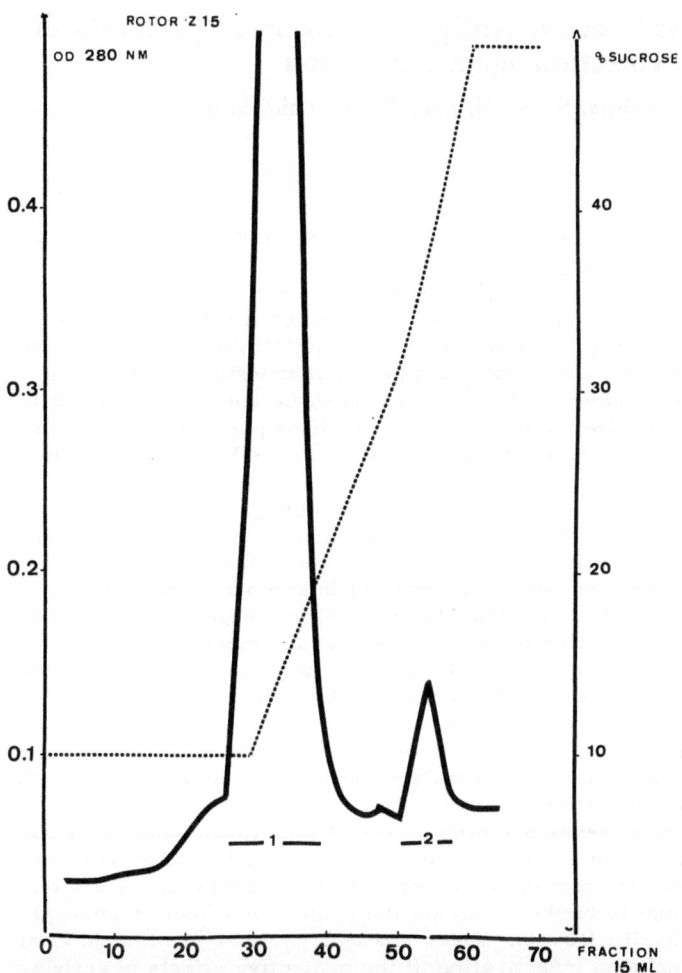

Fig. 3. Zonal centrifugation of a Triton extract from <u>S. typhi</u> O 901 Rotor I. E. C. Z15 operating in a B60 ultracentrifuge. Sample of 20 ml of Triton extract (5g/ml) in a 1 M sodium chloride-0. 1 M sodium citrate. Gradient 10-45 % (w/w) sucrose also containing 1 M sodium chloride-0. 1 M sodium citrate. Cushion 45 % (w/w) sucrose. Overlay 1 M sodium chloride-0. 1 M sodium citrate. Run 5, 000 rpm during 55 minutes corresponding to $\int \omega^2 dt = 938 \times 10^6$.

Type Specific and Cross Reactive Antigens of Gram-Negative Bacilli in Protection against Infection

W. R. McCabe, M. A. Johns, S. H. Zinner, T. DiGenio, and A. Greely

Relatively little is known about mechanisms of immunity to the enteric gram-negative bacilli in man. Antibody to capsular or K antigens or to O-specific antigens has been shown to exert protective activity in experimental animal infections (1-6), but evidence relating to human infections is lacking. Although it has been demonstrated that immunization with S. typhosa provides some protection against acquisition of typhoid fever, it is not yet clear which antigen is responsible for inducing protection (7-9). Other studies have suggested that O-specific antibody exerts little protective activity in man. Despite the fact that most humans possess "natural antibodies", these do not appear to protect against gram-negative bacterial infections. Studies in patients with pyelonephritis have demonstrated that new infections develop and established infections progress despite the presence of high titers of O antibody (10,11). Except for these observations, however, there have been no instances in which the effect of humoral antibody could be assessed in infections caused by Enterobacteriaceae. This has resulted from the nature of infections caused by gram-negative bacilli which makes such evaluations difficult. Gram-negative bacilli possess only limited capacity for invasion of the normal host and most such infections occur in patients with diminished defense mechanisms or when gram-negative bacilli are afforded direct access to sites of infection by bladder and intravenous catheters, etc. (12). This problem of evaluating immunity to "opportunistic" microorganisms in infections where host factors are of paramount importance in determining susceptibility and the outcome of infection readily explains the paucity of information concerning immunity to gram-negative bacilli in man.

The striking increase in the prevalence of nosocomial infections, especially bacteremia, caused by gram-negative bacilli, and the relative inability of control measures and antimicrobial agents to stem the increasing number of fatalities from this cause stimulated us to attempt to further evaluate the protective effect of humoral antibody to gram-negative bacilli (13-16). These studies, portions of which have already appeared in press, included investigation of the protective effects of active and passive immunization with both type-specific and shared cross-reactive antigens of gram-negative bacilli in experimental animals and evaluation of any protective effect of pre-formed antibodies to these same antigens in human gram-negative bacteremia (17-22). Since bacteremia with gram-negative bacilli so often results from operative and manipulative procedures, it was not possible to select a comparable control group to evaluate the effect of antibody in protecting against the development of bacteremia. To circumvent this problem, the effect of preformed antibody to homologous O antigen of the infecting organism and to shared, cross-reactive antigens in decreasing the severity of gram-negative bacteremia was examined. For the purposes of these clinical studies, the assumption was made that patients who developed shock or experienced a fatal outcome had more severe bacteremia and that any protective effect of pre-formed antibody would be reflected by a diminution in the frequency of these complications.

Materials and methods

Immunization and infection of animals. New Zealand white rabbits were immu-

This study was supported by United States Public Health Service research grants no. AI 09584 and AI 11116 and by training grant No. 5 T01-AI 213-10

nized with boiled Re mutants (Re) of S. minnesota and Common Enterobacterial Antigen (CA) and their serum was collected and stored as previously described (17).
Immunization with lipid A was accomplished by injection of 100, 200, 300, and
500 µg of lipid A coated acetic acid treated R 595 mutants of S. minnesota (vide in-
fra) at 3 day intervals as described by Galanos et al. (23). Serum was harvested 5
days after the last injection and stored at -20°C until used.

CF-1 female mice were immunized against Re and CA as previously described
(17, 20). Murine immunization was accomplished by weekly intra-peritoneal injection of 50, 100, 200, 200, 200, and 400 µg of lipid A coated acetic acid treated Re
bacilli.

Mice were challenged intravenously either 60 min after intravenous administration of 0.3 ml of immune rabbit serum or one week after the last injection of antigen in active immunization. Challenge was accomplished by intravenous administration of 100 LD$_{50}$'s of murine-lethal K. pneumoniae or P. morganii (erroneously i-
dentified as E. coli 107 in prior publications) as described previously (17, 20).

Clinical Material. Acute-phase serum specimens were obtained at the time bacteremia was suspected clinically or immediately after growth was noted in blood
cultures from 206 patients with gram-negative bacteremia. Blood culture isolates
were identified by standard techniques and stored until use for antigen preparation.
Since previous studies have demonstrated that the severity of the patient's underlying disease was the major determinant of the outcome of bacteremia, patients were
classified as having "non-fatal" or "ultimately-fatal underlying disease" using criteria described previously (13). Determination of shock and death being causally related to bacteremia also was based on previously described criteria. In addition,
the adequacy of antibiotic therapy in relation to the susceptibility of the infecting organism "in vitro" was assessed to insure that this did not influence results.

Antibody Determinations. Antibody to homologous bacilli, Re, and CA were determined by the passive hemagglutination test as described in earlier publications
(19-21). Antibody titers to lipid A in rabbit, mouse, and human sera were determined by the complement-dependent hemolysis technique described by Galanos et
al. with 0.025 ml of lyophilized guinea pig serum (Baltimore Biological Laboratories) demonstrated to be free of lipid A antibody being used as a source of complement (23).

Specific titers of IgG and IgM antibody to the patient's infecting organism isolated from blood cultures were determined by the indirect immunofluorescent technique with acute serum specimens from 121 episodes of bacteremia (24). All determinations were performed in duplicate and used boiled bacteria as the antigen.

The relation of height of antibody titers to the severity of bacteremia was determined by plotting the proportion of patients at each level of antibody titer who developed shock and/or death. Point biserial correlation, which allows correlation
of titers, the frequency of shock and death at each titer and the number of observations at each point, was used to detect any trend between height of antibody titers
and the severity of bacteremia (25).

Results

Antibody to Homologous Bacilli in Bacteremia. Initially, antibody titers to the
patient's infecting organism was measured by the hemagglutination method in serum
specimens obtained concomitant with or shortly after the onset of 206 episodes of
gram-negative bacteremia. Fig. 1 depicts the frequency of shock and death among
patients with ultimately fatal and non-fatal underlying disease at each level of antibody titer against each patients infecting organism. As may be seen, the frequency
of shock and death did not decrease with increasing titers of antibody against the
patients infecting organism. Shock and death did not occur less often among patients
with low antibody titers than among patients with high titers of antibody.

Since the hemagglutination technique preferentially detects IgM antibodies, titers

of IgG and IgM antibody against each patient's infecting organism were measured in
121 acute serum specimens by the indirect immunofluorescent technique. Fig. 2 illustrates the frequency of shock and death at each level of IgM antibody. The results
are quite similar to those observed with the hemagglutination technique in that increasing titers of IgM antibody were not associated with a decreasing frequency of
shock and death. In contrast, when the frequency of these complications at various
levels of IgG antibody were examined as shown in Fig. 3, a trend between increasing levels of IgG antibody and decreasing severity of bacteremia was observed. A-
mong patients with ultimately fatal underlying disease, the relation was highly statistically significant, $r_{64}=-0.519$; $p<0.001$. The relation between increasing IgG titers and decreasing frequency of shock and death approached statistical significance
for patients with non-fatal disease, $r_{55}=-0.244$; $p<0.10>0.05$, and was significant,
$r_{120}=-0.381$; $p<0.001$ for both groups of patients combined.

These findings are consistent with our earlier report of the lack of correlation
between height of hemagglutinating antibody titer and severity of bacteremia in a
smaller number of patients. A similar lack of correlation between IgM titers is consistent with its greater activity than IgG antibodies in the hemagglutination test. The
decreasing frequency of shock and death in relation to increasing IgG titers suggests
that specific IgG antibody does exert protective activity against gram-negative bacilli in man.

Antibody to Shared, Cross-Reactive Antigens of Gram-Negative Bacilli. Gram-
negative bacilli have been shown to possess several antigens which are shared by
most members of the family Enterobacteriaceae (26). One of these is the Common
Enterobacterial Antigen (CA) originally described by Kunin (27). In addition, the demonstration of the extremely similar or identical chemical composition of the core
portion of the lipopolysaccharide of all Enterobacteriaceae studied to date provides
evidence of other shared antigenic determinants. These include the antigenic determinant, probably KDO, of Re mutants and lipid A. Since these shared antigenic determinants are present in most Enterobacteriaceae, the protective activity of antibody to CA, Re, and lipid A was evaluated in experimental animal infections and
human gram-negative bacteremia.

Infections in Animals. Previously published studies have evaluated the effect of
active and passive immunization with CA against challenge with murine-lethal K.
pneumoniae and P. morganii (20). No protective activity could be detected after either active or passive immunization of mice with CA, against intravenous infection
with either of the two challenge organisms.

The results of active immunization of mice with Re or lipid A against challenge
with K. pneumoniae and P. morganii are compared with the results in control animals in Fig. 4. Immunization with Re induced antibody titers to Re of 1:320 and increased the survival rate after challenge with K. pneumoniae from 18% in controls
to 40% in Re immunized animals ($p<0.001$). Similarly, immunization with Re increased survival rates from 22% in controls to 77% in immunized animals ($p<0.001$).
Immunization with lipid A resulted in lipid A antibody titers of 1:80. Survival rates
in immunized animals, 50%, were significantly higher ($p<0.05$) than in controls,
20%, challenged with K. pneumoniae. A significant increase, $p<0.01$, in survival
was also noted in immunized mice, 75%, in comparison with controls, 30%, challenged with P. morganii.

The effects of passive immunization with rabbit antiserum to Re and lipid A are
shown in Fig. 5. Antiserum prepared by immunization with boiled Re bacteria produced Re antibody titers of 1:640 but no antibody to lipid A. Passive transfer of
0.3 ml of rabbit Re antiserum afforded significant protection against challenge with
both K. pneumoniae and P. morganii. Immunization with lipid A induced titers of
1:10, 240 to lipid A but failed to induce antibody to Re. Passive transfer of antiserum to lipid A afforded significant protection against challenge with K. pneumoniae
but failed to protect against P. morganii. The lack of protective activity of lipid A

antiserum against P. morganii was reproduced in several duplicate experiments.

Gram-Negative Bacteremia. Investigation of the effects of antibody to CA in gram-negative bacteremia failed to demonstrate any protective activity as previously reported. The frequency of shock and death was no less among patients with high titers of antibody to CA than among patients with low titers of antibody to CA.

Fig. 6 demonstrates the frequency of shock and death in patients with various titers of antibody to lipid A. Although shock and death were less frequent among patients with lipid A titers of more than 1:20, the large number of patients with undetectable lipid A antibody titers makes analysis of any relation between the frequency of shock and death and lipid A titers of questionable value.

The relation between titers of antibody to Re and the frequency of occurrence of shock and death is shown in Fig. 7. An almost linear relation, $r_{107}=-0.328; p<0.001$, is apparent between increasing titers of antibody to Re and decreasing frequency of shock and death in patients with ultimately fatal underlying disease. Although shock and death were less frequent in patients with non-fatal underlying disease who had high titers of Re antibody, the overall correlation between Re antibody titers and shock and death in this group were not statistically significant. When both groups of patients were combined, significant correlation, $r_{206}=-0.239; p<0.01$, between Re antibody titers and the frequency of shock and death was found.

Since IgG antibody against the patient's infecting organism and Re antibody both appeared to exert protective activity against bacteremia, the inter-relation between the protective effects of these two distinct types of antibody were evaluated as shown in Fig. 8. The frequency of shock and death were considerably less in patients with both non-fatal and ultimately fatal underlying diseases who had IgG titers of 1:160 or greater than in patients with lower IgG titers. However, the frequency of shock and death was always less in patients with Re titers of 1:80 or greater irrespective of the patients underlying disease or titer of IgG antibody. The decreased frequency of shock and death in patients with Re titers $\geq 1:80$ in both patients with high ($>1:160$) and low titers $\leq 1:160$ was statistically significant in both patients with non-fatal, $X^2=5.51; p<0.02$, and ultimately fatal underlying disease, $X^2=11.8; p<0.005$. Thus, the protective effect of antibody to Re is independent of any protective effect of IgG antibody and is demonstrable despite the nature and severity of the patient's underlying disease.

Discussion

The present results provide the first evidence for any protective activity of type-specific antibody to Enterobacteriaceae, other than Salmonella, in man. Although, type-specific protection is readily demonstrable in experimental animals (1-6), such protection has previously not been demonstrable in man. Our previous failure to demonstrate any protective activity of type-specific antibody in gram-negative bacteremia is undoubtedly due to the use of passive hemagglutination for detection of type-specific antibody activity. The greater activity of IgM over IgG in hemagglutination made results obtained using this technique primarily a reflection of IgM antibody. Support for this explanation is provided by the lack of correlation between protection and specific IgM titers against the patient's infecting organism. Previous demonstration that IgG antibody was more effective than IgM for opsonization of gram-negative bacilli may explain its greater protective activity (28).

Although there was suggestive evidence, it was not certain whether antibody to lipid A was associated with protection. This was partially the result of undetectable levels of antibody to lipid A in most of the patients. In addition, there appears to be some degree of species specificity for complement which may have resulted in artificially low titers of lipid A antibody.

Of greater potential importance is the demonstration of the protective activity of antibody to an antigen, Re, shared by most gram-negative bacilli against infection with heterologous bacilli in both experimental animals and man. Chedid originally demonstrated that antiserum to rough salmonellae protected mice against challenge with K. pneumoniae (29). Tate and Braude subsequently demonstrated that antiserum to rough bacilli protected against challenge with heterologous endotoxin (30). Earlier studies from our laboratory and the present investigations confirm Chedid and Tate's work and suggest that either Re or lipid A are the determinants responsible for the induction of these cross-protective antibodies (17, 22). Confirmation of the protective activity of antibody to Re in gram-negative bacteremia in man is of even greater significance since divergent effects of immunization with gram-negative bacilli in man and experimental animals have been noted. The greatest potential value of the protective effects of antibody to antigens shared by gram-negative bacilli is the possibility of increased resistance to gram-negative bacillary infections by immunologic methods.

References

1. Sanford, J. P., Hunter, B. W., Souda, L. L.: J. exp. Med. 115: 383 (1962).
2. Freter, R.: J. infect. Dis. 97: 57 (1955).
3. Markley, K., Smallman, E.: J. Bact. 96: 867 (1968).
4. Wolberg, G., DeWitt, C. W.: J. Bact. 100: 730 (1969).
5. Braude, A. I., Siemienski, J.: Bull. N. Y. Acad. Med. 37: 448 (1961).
6. Kaijser, B., Olling, S.: J. infect. Dis. 128: 41 (1973).
7. Ashcroft, M. T., Ritchie, J. M., Morrison, J., Nicholson, C. C., Stuart, C. A.: Amer. J. Hyg. 80:221 (1964).
8. Benenson, A. S.: Bull. Wld. Hlth. Org. 30: 653 (1964).
9. Ashcroft, M. T., Balwant, S., Nicholson, C. C., Ritchie, J. M., Sobryan, E., Williams, F.: Lancet II: 1056 (1967).
10. Vosti, K. L., Monto, A. S., Rantz, L. A.: J. Lab. clin. Med. 66: 613 (1965).
11. Williamson, J., Brainerd, H., Scaparone, M., Chueh, S. P.: Arch. int. Med. 114: 222 (1964).
12. McCabe, W. R.: Year Book Medical Publishers, in press.
13. McCabe, W. R., Jackson, G. G.: Arch. int. Med. 110:847 (1962).
14. DuPont, H. L., Spink, W. W.: Medicine (Baltimore) 48: 307 (1969).
15. Myerowitz, R. L., Medeiros, A. A., O'Brien, T. F.: J. infect. Dis. 124: 239 (1971).
16. Finland, M.: J. infect. Dis. 122: 419 (1970).
17. McCabe, W. R.: J. Immunol. 108: 601 (1972).
18. McCabe, W. R., Kreger, B. E., Johns, M.: New Engl. J. Med. 287: 261 (1972).
19. Johns, M. A., Whiteside, R. E., Baker, E. E., McCabe, W. R.: J. Immunol. 110: 781 (1973).
20. McCabe, W. R., Greely, A.: Infect. Immunity 7: 386 (1973).
21. McCabe, W. R., Johns, M., DiGenio, T.: Infect. Immunity 7: 393 (1973).
22. McCabe, W. R., Greely, A., DiGenio, T., Johns, M. A.: J. infect. Dis. 128: S284 (1973).
23. Galanos, C., Lüderitz, O., Westphal, O.: Europ. J. Biochem. 24: 116 (1971).
24. Reichek, N., Lewin, E. B., Rhoden, D. L., Weaver, R. R., Crutcher, J. C.: Amer. Rev. resp. Dis. 101: 238 (1970).
25. Ferguson, G.: Statistical Analysis in Psychology and Education: 2nd ed. (McGraw-Hill, New York 1966).
26. Lüderitz, O., Staub, A. M., Westphal, O.: Bact. Rev. 30: 192 (1966).
27. Kunin, C. M., Beard, M. V., Halmagyi, N. E.: Proc. Soc. exp. Biol. Med. 111: 160 (1962).
28. Smith, J. W., Barnett, J. A., May, R. P., Sanford, J. P.: J. Immunol. 98: 336 (1967).
29. Chedid, L., Parant, M., Parant, F., Boyer, F. A.: J. Immunol. 100: 292 (1968).
30. Tate, W. J., III, Douglas, H., Braude, A. I., Wells, W. W.: Ann. N. Y. Acad. Sci. 133: 746 (1966).

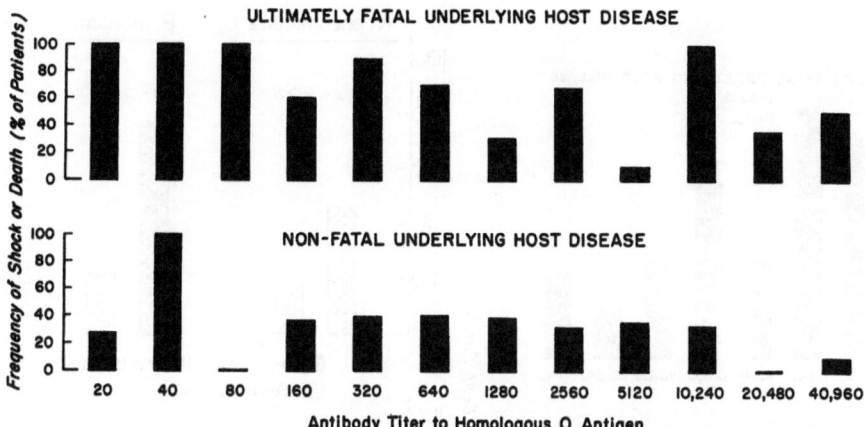

Fig. 1. The frequency of occurrence of shock and death at each titer of antibody to the patient's infecting organism in 206 episodes of gram-negative bacteremia. High titers of antibody were not associated with a decreased frequency of shock and death.

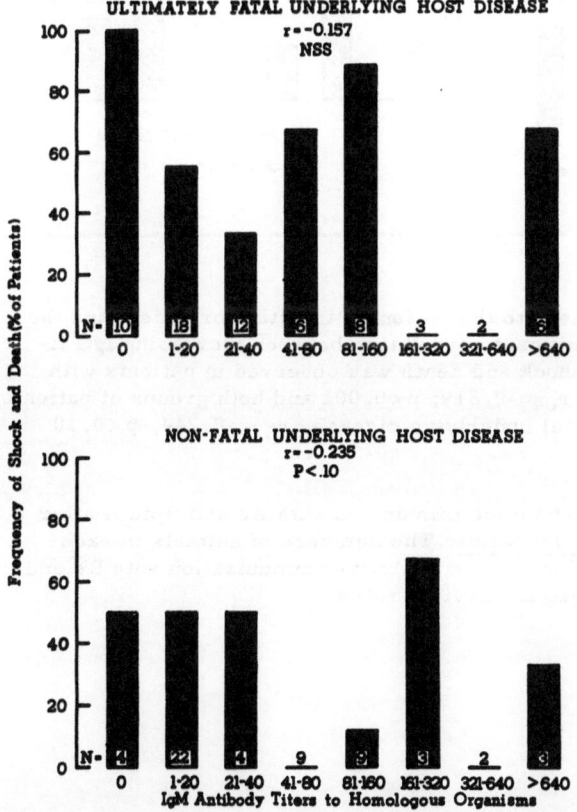

Fig. 2. Relation of specific IgM titers to the homologous infecting bacterium and the frequency of shock and death in 121 episodes of gram-negative bacteremia. No relation could be detected between IgM antibody titers and the frequency of these complications.

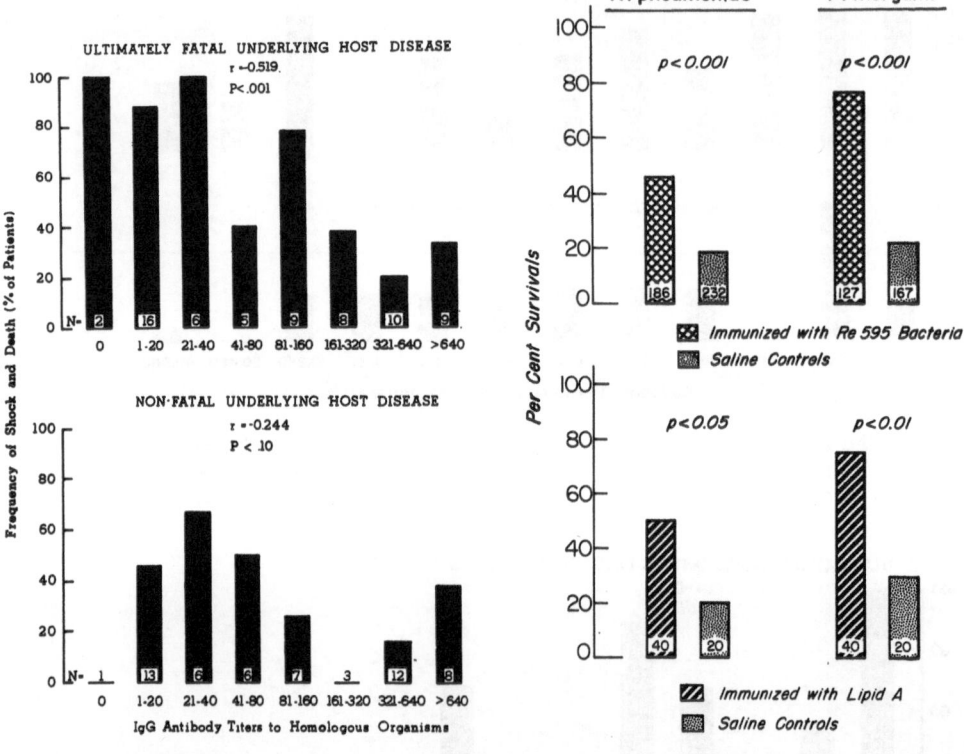

Fig. 3.   Fig. 4.

Fig. 3.   Relation of IgG antibody titers to the patient's infecting organism and the
frequency of shock and death. A significant correlation between increasing IgG titers and a decreasing frequency of shock and death was observed in patients with
ultimately fatal underlying disease, $r_{64}=-0.519$; $p < 0.001$ and both groups of patients
combined and in patients with non-fatal underlying disease, $r_{55}=-0.244$, $p < 0.10$
> 0.05.

Fig. 4.   Survival rates in controls and mice immunized with Re and lipid A after
challenge with K. pneumoniae and P. morganii. The numbers of animals in each
group are shown by the numbers within each bar. Active immunization with Re and
lipid A produced a significant increase in survival rates.

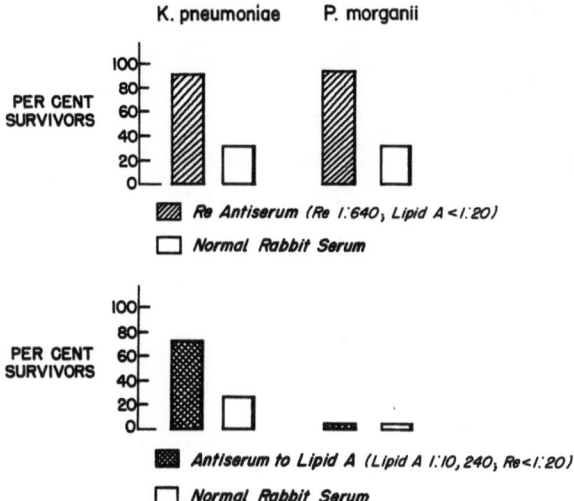

Fig. 5. Survival rates after challenge with K. pneumoniae and P. morganii in mice given 0.3 ml of normal rabbit serum or 0.3 ml of lipid A or Re antiserum. Re antiserum afforded significant protection against challenge with both K. pneumoniae and P. morganii. Lipid A antiserum protected against challenge with K. pneumoniae but failed to protect against P. morganii in several experiments.

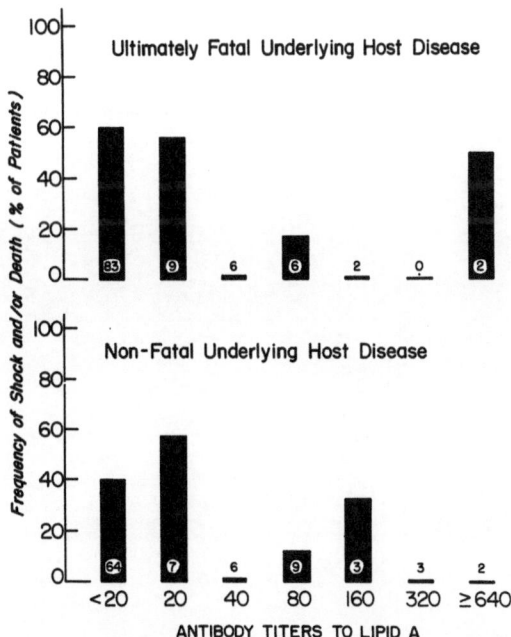

Fig. 6. Relation of frequency of shock and death to height of antibody titers to lipid A. The large number of patients with no detectable lipid A antibody limits evaluation of any protective activity of lipid A antibody.

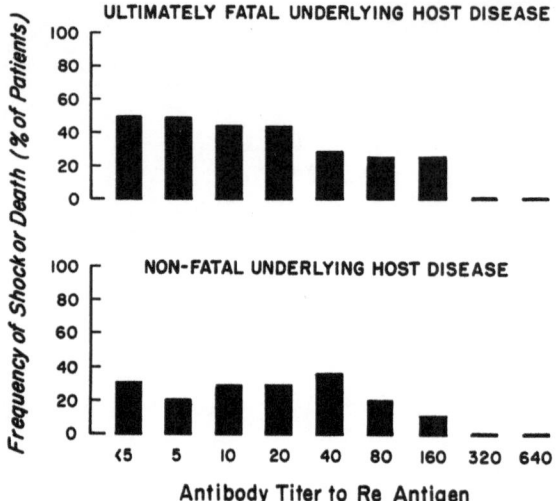

Fig. 7.   Frequency of shock and death in patients with various titers of antibody to Re. A definite correlation was detected between increasing titers of antibody to Re and decreasing frequency of shock and death in patients with ultimately fatal under-lying disease, p<0.001, and both groups of patients combined, p<0.01. No signifi-cant correlation was observed in patients with non-fatal underlying diseases.

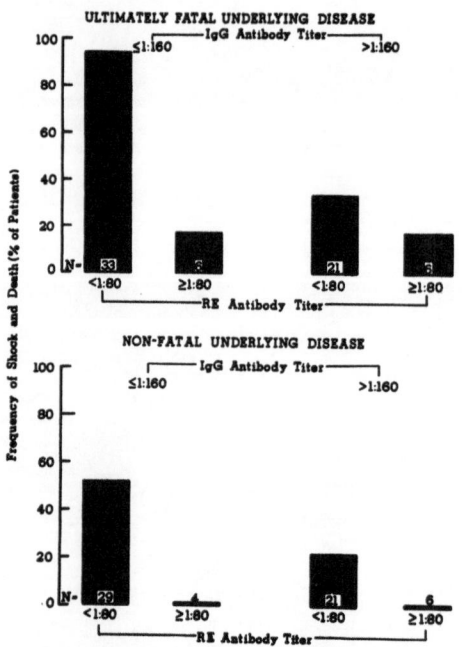

Fig. 8.   Frequency of shock and death in patients with non-fatal and ultimately fa-tal underlying diseases with high, >1:160, and low titers, ≤1:160, of IgG antibody in relation to antibody titers to Re. High titers, ≥1:80, of Re antibody exhibited protec-tive activity independent of IgG titer or the patient's underlying disease.

Discussion

BRAUDE

Chedid: Was the endotoxin you used in human skin a Boivin or LPS preparation? Also, how were the antisera prepared? We had the same experience in preventing abortion by endotoxin through antisera. We also had this dual effect: in certain cases we protected; in others we enhanced the abortive effect. Abortion is probably mediated by serotonin release from platelets and thrombocytopenia. One would imagine that antibody-antigen response would produce this thrombocytopenia, so that antiserum would favor it.

Braude: I am interested that you observed the dual effect in another system. We used both Boivin and LPS preparations for skin testing with identical results. Dr. Greisman has had this experience also. We prepare antisera by either injection of LPS or boiled cells. I have also wondered why one gets protection, sometimes, and other times hypersensitivity; and if endotoxin-antibody complexes cause thrombocytopenia why should antiserum protect? I postulate that this is a function of ratio of antibody to endotoxin so that endotoxin excess gives hypersensitivity and antibody excess protection. This is consistent with other ideas about toxicity of complexes.

Nowotny: We applied various combinations of endotoxins and corresponding antisera in different ratios and we never observed any hypersensitivity against endotoxin even giving only 3 % of optimal amount of antibody to the endotoxin, injecting the mixture prepared in vitro. We always saw a reduced pyrogenicity or reduced Shartzman reaction so we never observed a dual effect.

Braude: The difference is that the endotoxin-antibody reactions in your experiments occurred in vitro.

Gans: I would wonder if one should call this an immune reaction, because this quick disappearance of endotoxin into Kupffer cells within 1/2 hour in a non-immunized animal would be an effect of phagocytosis like India ink. Also the complement decrease would be more a bypass reaction than a specific reaction with antibody-antigen complexes. The local reaction which you called Arthus phenomenon may be morphologically similar to an Arthus phenomenon but immunologically one would demand for an Arthus phenomenon circulating antibody. However, your results in immunized animals are very interesting. Have you proof that after vaccination antibody appears?

Braude: Antibody can be demonstrated. Serum from actively immunized animals that were hypersensitive transferred this hypersensitivity. Clearance of radioactive endotoxin is an immunologic phenomenon because phagocytosis is an immunologic phenomenon involving opsonin. With regard to the Arthus reaction, the classic reaction does require both antibody and a protein antigen. Nevertheless complement activation by any device will produce a chemotactic substance and attract granulocytes with release of lysosomal reagents that can injure vessels. This is what we see with endotoxin. Whether it be by antibody-antigen reactions or by the complement bypass, we see activation of complement, chemotaxis of leukocytes, and release of lysosomal enzymes.

Gans: I know, but still I disagree that the bypass reaction could be called an immune reaction. We take away the entire basis of immunology if we call this an immune reaction.

Braude: I would be interested in Dr. Mergenhagen's response to this.

Mergenhagen: Dr. Snyderman, in our laboratory repeated some classical work of Dr. Dixon which showed that antigen-antibody complexes in antigen excess were the most damaging in serum sickness. Dr. Snyderman prepared complexes of bovine serum albumin and antibody in antigen excess, in antibody excess, and equivalence, and then activated the complement system to generate the chemotactic factor. In antibody excess and at equivalence we only got the fragments from the C5 component which degraded rapidly in the animal. In antigen excess, chemotaxis was always associated with the antigen-antibody complex. We interpreted this to mean that the fragment of complement might be carried by the immune complex and so in some way protected. If this occurs in the kidney instead of releasing these fragments of complement into the circulation it may be in a more damaging state as a complex with antigen-antibody. So I think it all goes back to the classical experiments of Dr. Dixon and I think in order to define some of these reactions we have to know what the proportion of antibody is.

Braude: Would you object to the term immune reaction to describe this involvement of endotoxin with complement?

Mergenhagen: The immune system is a big thing! If by chance we show that antibody is not involved in activation of the alternate pathway, activation of complement is then an immune reaction not involving antibody. Dr. Landy might comment on that.

Landy: We have to differentiate between the experiments done in vitro as in most of the bypass work and those that can happen inside in the host and dealing with two different antibodies. In the normal host it is IgM natural antibodies which is more efficient in complement participation than IgG by 2 or 3 orders of magnitude. In the animals you have used to prepare your antisera which are strongly provoked with antigen you have a preponderance of IgG, and must distinguish between these two. In the host it is hard for me to visualize that a bypass mechanism would work in the presence of antibodies when antigen is introduced. The complex work of Dr. Dixon is done to the best of my knowledge totally with hyperimmune sera involving IgG antibodies. These factors would perhaps help us to put this in a proper setting.

Braude: In nonimmunized animals it is possible that large amounts of endotoxin could deplete antibody, so that remaining endotoxin could activate complement by the bypass mechanism. In the immunized animals the preponderant immunoglobulin antibody against endotoxin was IgM.

Raynaud: Do you analyze your sera by immunodiffusion? When we speak of anti-endotoxin-sera, I would use an immunodiffusion plate and have 5 or 10 different systems. So depending on the condition you can have one antigen-antibody complex with x antigens giving one reaction and the other giving the other type of reaction.

Braude: We have of course analyzed it by immunodiffusion and found multiple bands but can frequently get rid of the excess bands by treatment with mercaptoethanol. In other words, most of these multiple bands are due to multiple antibody types.

Raynaud: This would not give a good explanation because if you transform IgM into IgG you do not change the antibody combining sites and you will not have one line when you get at the beginning five lines. If you have multiple bands you cannot reduce them to only one band by mercaptoethanol. So you have to explain this by some modification of mercaptoethanol on your antigen and I shall show you a picture indicating that this is the case not only for mercaptoethanol but all the substances.

HESS

Huser: The bone marrow seems to behave in many ways as if it were the sink of the body where a number of materials are dumped. It lacks lymphoid structures such as lymph vessels and lymph nodes. Since the bone marrow has no T-cells is it fair to say that the bone marrow is in a state of T-cell-deficiency?

Hess: Bone marrow does not contain T-cells and the lymphoid elements in the bone marrow are difficult to differentiate. It is difficult to tell by looking at the cell that has a lymphoid aspect whether it is going to become an erythrocyte, a granulocyte, monocyte, or a lymphocyte. With regard to your hypothesis that bone marrow functions as the sink of the body I think you are right. Bone marrow cells are constantly exposed to material in the blood and cells from peripheral lymphoid tissues probably settle in the bone marrow. As the number of plasma cells in the marrow can be impressive, and since plasma cells do not travel, they must arise in place there. So I would not be surprised if the marrow harbours many memory cells.

Bona: Do you know whether in human bone marrow there are cells which can form spontaneous rosettes with sheep red blood cells?

Hess: That is one widely used T-cell marker. As far as I know they are not present but if you treat lymphoid bone marrow cells with neuraminidase you get all the rosettes you want.

Braude: From the point of this Symposium, would you say a word about gram-negative bacterial infection in these immunodeficiency states.

Hess: I could not find anything in the literature telling whether immunodeficiency might be good or bad with regard to endotoxin. The low pathogenic or opportunistic microorganisms are more frequent, especially in cases where cell-mediated immunity is impaired. That is about all I could find.

COTTIER

Fey: When Enterobacteriaceae invade the blood stream, their portal of entry is not necessarily the intestine but the nasopharynx. In order to prove this I took three newborn calves deprived of colostrum, esophagotamised them, and infected them by the nasal route with 78:80B E. coli. These calves could not swallow the germ and inspite of this developed septicemia and pure massive culture of the organism in all organs, but not the intestine.

Cottier: The point of my presentation is that intestinal lymphoid tissue is a main site of antigenic stimulation rather than a port of entry and source of septicemia. I suspect that an intact lymphoid follicle in the gut may be an obstacle to penetration of these organisms. So, I agree that there are other portals of entry.

Chedid: What are Dr. Cottier's ideas about the penetration of endotoxin or lipopolysaccharides through the intestinal barrier? We gave radioactive endotoxin orally and could detect up to .00001 of the dose, either by toxicity or radioactivity. To our surprise, we found no endotoxin going through the gut in normal mice or rats even in germfree animals 4 or 5 days old.

Cottier: A friend of mine from Harvard University injected horse-radish peroxidase into a ligated segment of the intestine and could demonstrate that macromolecules enter the epithelial barrier and pass through. However, using horse-radish peroxidase is an amplification technique which shows minute amounts of material and it is possible by usual methods not to find such an entry. In addition, release of endotoxin within the gut wall would probably occur at the sites that you have seen

from macrophages in lymph follicles and so forth. I doubt that there is a high concentration of this material within the gut.

Braude: When you say that it passes through, do you mean that it passes through
the epithelial barrier into the blood capillaries or the lymph capillaries?

Cottier: According to this study into both.

MERGENHAGEN

Bona: Did you perform those relationships as concerned the mitogenic activity of
the haptenic polysaccharides? We get a very small effect but significant effect only
with between 10 and 50 μg for 1.5×10^6 cells. Have you some opinion concerning
the subpopulation of the B-cells stimulated by lipopolysaccharides and lipid A? We
accept that only the B-dependent fully thymic-dependent population responsible for
the sythesis of IgM in vitro is stimulated by lipid A. Have you data concerning another B-dependent population? Regarding the mechanism of transformation, it is
accepted that binding of lipopolysaccharides represents a critical event in transformation, but we find not transformation but synthesis of specific lipopolysaccharide antibody, and that only the cells which contain the antigen within the cytoplasm
synthesize specific antibody directed to lipopolysaccharides. In other words the
penetration of antigens is critical.

Mergenhagen: Your first statement is not a question. You told me you found stimulation, you told me a fact. We didn't find any stimulation with anything rich in polysaccharides, even up to 200 μg/ml. Secondly I know nothing about the mechanism.
I speculate that the lipids may bind to the lipophilic layer of the cell membrane.
The C3 receptor on B-lymphocyte may be involved although I must admit that these
stimulation experiments can be done in serum-free tissue culture media. So one
would have to postulate that the B-lymphocyte or the macrophage had to contribute
not only C3, but all components of the alternative pathway. Thirdly, I don't know
anything about the mechanism, or where the antigen is.

Springer: You mentioned that the lipid A fatty acids would most likely have to bind
lipids on the surface of the lymphocyte. We have isolated a receptor for endotoxin
which interacts with lipid A, but does not contain lipid. The interaction could be
with hydrophobic areas in the aminoacid part; but even there, our evidence is not
strong that this is the case, so one has to be careful about saying to which area on
the lymphocytes the lipidic part would bind.

Mergenhagen: I didn't say anything about the area, did I? May I ask you, you are
working with what cell as far as the receptor is concerned?

Springer: We have done most of the work on the red cell, but also some with white
cells.

Mergenhagen: This triggering of B-lymphocytes by these materials is not peculiar
to the B-lymphocyte. Lipopolysaccharides, the glycolipid, and the lipid A moiety
all stimulate macrophages equally to produce collagenase. So is this another cell
type?

Springer: I only said the receptor site on the lymphocyte has not been sufficiently
defined to make a statement as to the nature of the receptor site.

Mergenhagen: Sufficiently? I don't think it has been defined at all.

Berger, Wien/Austria: I have two questions. I observed a difference in your slide
between the chemical structure of what you call glycolipid and lipid A. The second
question is how did you administer water insoluble lipid A in your assay mixture?

Mergenhagen: It is very difficult to put these into a micro-dispersion. We exposed these materials in distilled water to ultrasound. This gave us a workable microdisperion that we could bring up with physiological saline and use. Dr. Nowotny might comment on the differences of the chemical structure of lipid A and glycolipid.

Nowotny: Glycolipid really consists of the lipid moiety of the lipopolysaccharides.

WISSLER

Springer: What is the normal conformation of anaphylatoxin and what kind of conformational changes do you provoke?

Wissler: Normal conformation is that present in the serum after generation by contact of macromolecules with serum. Unfolding is a conformation with a less compact structure.

Springer: Is it more in physical-chemical terms like alpha-helix and beta-conformation and pleated sheet or is it just a general statement?

Wissler: The active conformation is thermodynamically less favored than the native conformation in the serum. There is about 10 kilo calories difference. We are investigating if it is alpha-helix.

Nowotny: What is the specific activity of your crystallized preparation as compared to less purified materials? What is the concentration you have to use of your crystallized material to get an activity?

Wissler: Which activity? Spasmogenic activity, shock producing activity, or permeability-enhancing activity? For shock you need about 5 to 20 $\mu g/kg$ of guinea pigs, but the activity of the endotoxin may differ from batch to batch without being chemotactic and that is the important point.

Nowotny: Is your crystallized preparation more active than uncrystallized less purified material?

Wissler: When you measure the concentration by spectrophotometry it is purified 5000-fold to 6000-fold. When you measure it by Folin assay, you have about 80000-fold of purification, because the anaphylatoxin gives only a very poor color reaction with the Folin method.

Huser: Was this chemotactic activation observed in free hemoglobin, in hemoglobin-haptoglobin complex or both? I don't believe that this action is relevant to removal of aging erythrocytes, but it is relevant to purification from free hemoglobin in the plasma.

Wissler: I think that both points are important, but you don't need any. You just take normal hemoglobin and you can make it chemotactic. Dr. Bessis destroys an erythrocyte with the microlaserbeam and neutrophils migrate towards split products of the cell.

RAYNAUD

Springer: What is the molecular weight of this molecule?

Raynaud: If we accept the position on Sepharose 4 B it will be 1 million, but gel filtration is poor for determining molecular weight.

Chedid: Do you have antiserum to this common antigen and what is its protection against toxic activity or bactericidal activity?

Raynaud: It is too early to speak about it. This portion is a protein but it contains something else because we did not find 40 % nitrogen. It is absent from the LPS and it is antigenic per se. It is common to all the same species. We found it in Salmonella typhi and I know one S. typhi R. Our impression is that the polysaccharide and the lipid A are not involved, so it remains probable that it is not protein, but the complex which we have can contain other things than protein because its nitrogen content is not enough.

Chedid: What is the serum antitoxic activity towards this toxic purified antigen?

Raynaud: We have very little of this, perhaps 10 to 12 μg of the pure substance. So we have to select between chemical composition and biological activity and we know from preliminary experiments that it is very toxic, more toxic than the 0-antigen from which it is extracted.

Newsome, Surrey/England: Have you investigated KDO as an inhibitor of your antigen-antibody interaction?

Raynaud: We do not have enough KDO to see that it is inhibitory. We know that lipid A is not so and we were also quite sure that this protein contains something other than only aminoacids. In preliminary analysis we found no special aminoacids and did not find animoacid specific of the cell wall.

Braude: Have you tested it for KDO directly?

Raynaud: Not yet. We are now preparing a large batch, that will give all the answers. What we want to know before coming here was the different toxicity which is very high: 10^4 to 10^7 of the LPS.

JOHNS

Braude: It was clear from your paper that antibody to lipid A was ineffective in protecting experimentally against Proteus. Was antibody to Re ineffective?

Johns: No antibody to Re was effective in protecting against both Proteus morganii and Klebsiella pneumoniae.

Braude: How do you account for this difference between the protective effect of antibody to Re and antibody to lipid A?

Johns: We really have no idea.

Braude: Have you examined the Proteus strain for lipid A?

Johns: No, we are going to study the difference between lipid A of Proteus and Salmonella.

Neter: If I understood you correctly you have information that all 0-specific antibodies were protective if they were of the IgG immuno class rather than the IgM. Is that correct?

Johns: Actually, the IgG antibodies apparently correlate with protection, however, the tests that were done, did not differentiate specifically the 0-antigen of the organism. It's a boiled organism that is used as antigen. So that other things are present too. It could be 0, but we don't know.

Neter: Early in urinary infections IgM antibodies are largely produced, while chronic infection with the same organism results in IgG antibodies. Why was this infection not terminated, why did the infection persist, why did the antigenic stimulus persist to make the switch from IgM to IgG?

Johns: It probably indicates that the antibody against the homologous organism is not effective in protection.

Braude: The explanation for your question Dr. Neter is that IgM does not get into the urinary tract. In order to get any protective activity you have to produce IgG and this is the secret of protection against pyelonephritis.

Berry: How many cells make up the LD_{50} of these strains?

Johns: At 100 times the LD_{50} for the <u>Klebsiella</u> strain we have 10^4 organisms. For the <u>Proteus</u> strain we use 10^7.

Nowotny: How did you prepare your lipid A and how do you use it as an immunogen?

Johns: We immunize with lipid A according to the method of Galanos and Lüderitz. That is, acid treated Re 595 organisms to remove KDO and then coated with additional lipid A, which has been prepared by acid hydrolysis of the purified Re lipopolysaccharide. We immunize intraperitoneally.

Nowotny: You mean you used this lipid A preparation as an immunizing agent?

Johns: Well, it is Re 595 organisms that have the KDO removed and additional coated lipid A on their surface; not a pure lipid A.

Chapter III

Nonspecific Resistance and Endotoxin Tolerance

Recent Developments Especially Significant for Ascertaining the Mechanisms of Endotoxin-Enhanced Nonspecific Resistance

M. Landy

Of the multitude of effects on the host elicited by endotoxin none have attracted more attention and experimental work than the rapid rise in resistance to a wide variety of microbial pathogens. This enhanced nonspecific resistance to infection can to varying degree be mimicked by agents as diverse as colloidal sulfur, zymosan, heat-aggregated HSA, lipid A, starch granules and the double-stranded polynucleotide poly I C. It is suspected that while particle size and physical characteristics of these diverse agents may be as important for their effects as their composition - as yet no single characteristic of these materials has provided an acceptable explanation for their effects on host resistance. By all criteria - endotoxin displays the most striking effects quantitatively and as its composition, structure and other attributes were well-known and it was widely available it became the most extensive employed agent to study this precocious immunity. The effects of endotoxin on resistance have been shown in infections with fungi, parasites, viruses, Mycobacteria and with gram-positive and gram-negative bacterial pathogens.

The factors or variables determining this rapid nonspecific augmentation of host resistance are numerous indeed - for example characteristics of the product used to pretreat the host, dose, timing with respect to the challenge and route; with regard to the host - the species, strain, nutritional and metabolic state, age, sex, the type of enteric flora carried, or their absence etc. This has perhaps been the major burden this subject has suffered. The complexity of these factors and their interactions, suspected or unknown, has made difficult - to the point of impossibility - the definition of the mechanism by which endotoxin so quickly modulates host resistance to a broad spectrum of microbiological challenges.

In the 1950's and indeed well into the 1960's the main thrust of research in this field involved attempts to correlate humoral factors with this precocious immunity. An objective overview of these many studies leads to the conclusion that this premise could not be sustained.

That there were temporal cellular changes, notably in polymorphnuclear and mononuclear phagocytes following these stimuli was also noted early in these investigations.

A role for the granulocytosis is generally accepted - but not viewed as decisive. As regards the mononuclear phagocytes, the macrophages, by far the most attention was long given to the fixed hepatic or Kupffer cells involved in the clearance of intravenously injected carbon and other particulates from the circulation. However, the great mass of data on enhanced RES clearance in the endotoxin-stimulated host show a marked l a c k of correlation with the onset and duration of the heightened immunity.

The consensus is that the most impressive feature of the endotoxin-induced resistance is the l a c k of specificity in its expression. An analogous situation has also been demonstrated in mice infected with i n t r a c e l l u l a r pathogens - thus, mice infected with Brucella abortus were resistant to Listeria; those infected with Mycobacteria were resistant to Listeria; those infected with Mycobacteria were resistant to Klebsiella and Listeria etc. The point is that t h e s e and numerous other experiments it indicated that the acquired resistance involved a heightened nonspecific microbicidal power of macrophages. Even more striking than the nonspecificity of resistance as shown against heterologous bacteria was the finding

that t o t a l l y d i f f e r e n t types of organisms could elicit nonspecific resistance to bacteria, i. e., mice primarily infected with the protozoa Toxoplasma gondii
or Besnoitia jellisoni became strongly resistant to Listeria and Salmonella typhimurium. Parallel in vitro studies showed that macrophages from resistant animals effectively restricted the growth of the challenge organism - in the a b s e n c e of
serum. Even more dramatic was the finding that macrophages of F1 mice undergoing
a graft versus host reaction to parental spleen cells had a comparable heightened capacity to destroy Listeria and Salmonella typhimurium.

From these observations it may be concluded that a w i d e v a r i e t y of immunologic stimuli are followed by the appearance of macrophages with markedly
augmented nonspecific microbicidal activity. The stimuli referred to generally elicit cell-mediated immune responses - however my t h e s i s is that endotoxin as
the prototype of an array of agents n o t usually thought of in connection with cellular immunity a l s o brings about p r o f o u n d effects on blood monocytes and
peritoneal macrophages which could well account for the so-called nonspecific resistance studied so extensively these past 20 years. The evidence for endotoxin and
other stimulators of nonspecific resistance operating via macrophages is as follows:

1. With the possible exception of mineral or paraffin oil, the means traditionally
used by the cellular immunologist to obtain peritoneal exudates, i. e., activated macrophages is to provoke the host with Witte peptone, proteose-peptone, thioglycol-
late medium or oyster glycogen - all of which are demonstrably rich in contaminating endotoxin. I should of course add that endotoxin itself yields a comparably distinctive exudate population.

. 2. Such cells, without added serum, are intensely microbicidal, virucidal and
cytostatic for transformed or tumor cells, the latter by a cell-contact, not phagocytic process.

3. Administration of endotoxin i t s e l f yields a n a l o g o u s l y activated
peritoneal macrophages - within hours - and this state persists for several days.

4. Most convincing, however, is the fact that interaction in vitro of normal peritoneal macrophages with endotoxin (or with lipid A or with poly IC) yields within 30
to 60 minutes activated macrophages just as broadly microbicidal and just as cytostatic for tumor cells as activated macrophages generated in vivo.

"Activated" macrophages, capable of enhanced killing of a spectrum of bacteria
and viruses, differ in still other respects from normal macrophages. They a t -
t a c h immediately to glass and s p r e a d more extensively. They appear larger
than normal cells and contain more mitochondria and granules - many appearing as
lysosomes when viewed by phase contrast. The speed with which they phagocytose
particles for ingestion is increased. Ultrastructurally, activated macrophages have a
more extensive and vesicular Golgi apparatus and more lysosomes. Biochemically
they show an increase in total protein synthesis, ATP, lysosomal enzymes and glycolysis. These features are shared by cells from mice immunized with BCG or
Listeria or injected 1 to 3 days earlier with endotoxin.

The magnitude of the effects of endotoxin on macrophages is undeniable. The
question that remains is whether this is exerted directly or via an intermediate
route. I can visualize several possibilities: All mammals carry IgM natural antibodies against all principal gram-negative serotypes and macrophages are known to
have receptors for the various classes of Ig, including IgM; it therefore would not
be surprising that endotoxin would bind to these cells and thus lead to the initiation
of a variety of reactions.

Another quite separate possibility involves the fact that endotoxins are now established as the most potent mitogens for B lymphocytes, giving rise to a polyclonal
antibody response, i. e., not only specific antibody but a whole array of other antibodies as well. Rowley had shown many years ago that in mice given endotoxin there
occurred a selective major rise in spleen IgM while the amount of other classes of
Ig remained unchanged. This rise in IgM levels could not at all be accounted for by

antibody specific for the initiating endotoxin.

Recently evidence has been reported by Epstein that the elaboration in vitro by lymphocytes of the so-called soluble mediators of cellular immunity is not limited to thymic lymphocytes as had been thought previously. At least to the extent of interferon such products would also seem to be produced by B lymphocytes, the very category of immunocompetent cells so responsive to endotoxin and to lipid A.

So the circle of circumstantial evidence would seem to be complete. I have mentioned briefly some of the reasons why peritoneal and blood macrophages could be the key cells in the rapidly developing nonspecific immunity evoked by endotoxin and many other agents. In passing I have identified a number of issues that are now amenable to investigation to prove whether macrophages are called into action directly - or indirectly by generation of mediators from serum or via B lymphocytes. However that may be, there would now seem to be new opportunities for resolving the issue of how the host mounts this distinctively nonspecific and rapid resistance to infection.

Mechanisms of Endotoxin Tolerance and their Effectiveness during the Febrile Phase of Gram-Negative Bacterial Infections in Man

S. E. Greisman and R. B. Hornick

Understanding the role of endotoxin in the pathogenesis of human gram-negative bacterial infection requires the elucidation of the mechanisms by which man develops resistance to this toxin. In contrast to animal models, few critical studies are available concerning human mechanisms of tolerance acquisition to endotoxin. Even fewer studies are available concerning the effectiveness of these tolerance mechanisms during the febrile phase of gram-negative bacterial illness. This report reviews the tolerant responses of man to bacterial endotoxin as assessed by febrile and subjective toxic reactivity. Our experience is based upon studies in several hundred volunteers, both healthy and overtly ill with gram-negative bacterial infections (typhoid fever or tularemia), given repetitive intravenous injections of bacterial endotoxins from various sources. These studies were conducted for the primary purpose of developing improved prophylactic and therapeutic approaches to gram-negative bacterial infection. All volunteers were fully informed of the nature of these studies and were free to withdraw at any time. Where pertinent, results of animal studies are included to supplement the human data.

Responses of the healthy host

Man is one of the most reactive species to bacterial endotoxin. When responsiveness is compared with that of the rabbit, the threshold pyrogenic doses (μg/kg) of endotoxin are comparable, but the dose-response relationship is much steeper for man (1). By testing at various intervals after a single intravenous injection of endotoxin, the development of tolerance to the toxin can be shown to occur in two temporally distinct phases. The initial or early phase of pyrogenic tolerance, most readily seen in a man during a continuous intravenous infusion of endotoxin, possesses the following characteristics (2 - 5):

1. It appears within hours. During this early phase of tolerance, volunteers characteristically express disbelief that the endotoxin infusion is continuing.

2. It is transient, and requires closely spaced or continuing endotoxin infusion for maximum maintenance. If the initial infusion is discontinued and resumed the following morning, volunteers will have regained full responsiveness; indeed, they usually hyperreact at this time.

3. It is specific for endotoxins as a class. No tolerance occurs to other pyrogens such as staphylococcal enterotoxin, influenza virus, or tuberculin in specifically sensitized animals unless massive (shock producing) doses of endotoxin are administered. Full responsiveness also persists to preformed endogenous pyrogen.

4. No specificity of early tolerance exists between endotoxins from diverse bacterial species.

5. It is not associated with increments in circulating anti-endotoxin antibodies and can be readily induced in hosts with defective capacity to synthesize antibodies.

6. Within the sensitive dose-response range, the level of early tolerance is directly proportional to the intensity of the initial pyrogenic response.

7. It is relative, not absolute, and can be overcome by increasing the dose of endotoxin.

8. It cannot be transferred with plasma.

9. It cannot be reversed with large volumes of fresh normal plasma or whole blood.

10. It is not associated with granulocytopenia; indeed, granulocytosis regularly oc-

curs.

11. It is not associated with augmented ability of plasma or of liver homogenates to inactivate endotoxin in vitro.

Studies in vitro indicate that endotoxin is capable of releasing endogenous pyrogen from both granulocytes and macrophages (6 - 9). Early studies by Page and Good demonstrated that the febrile responses to endotoxin were not reduced in patients with agranulocytosis, and these investigators stated : "It must be concluded from this study that injury to neutrophils with liberation of endogenous pyrogen plays no important role in the development of fever following the intravenous injection of pyrogens in man". Moreover, when tolerance was tested, it was concluded that "... refractoriness to endotoxin develops in the complete absence of neutrophils in the circulating blood or in the blood forming tissues and consequently demonstrates that these cells are not essential to this defense reaction" (10). Since the hepatic Kupffer cells sequester the major portion of intravenously injected endotoxin (11), it is these macrophages that would be expected to contribute the major source of endogenous pyrogen following intravenous endotoxin administration. And, indeed, studies on both isolated hepatic Kupffer cells (7), as well as in vivo hepatic perfusion studies via indwelling hepatic vein cannulas (12), fully support this concept. These studies also indicate that pyrogenic tolerance to endotoxin is based upon inhibition of endogenous pyrogen release from the hepatic Kupffer cell. Since no detectable enhancement occurs in the endotoxin detoxifying potency of liver homogenates during this early tolerant phase (3), it appears that the hepatic Kupffer cells have become refractory to the endogenous pyrogen releasing effect of endotoxin. The mechanisms underlying this refractory state, however, are unknown. The hypothesis that depletion of the more susceptible intracellular lysosomes may be responsible for endotoxin tolerance seems pertinent in this regard (13). Alternatively (or perhaps causally related), the possibility that an immunological desensitization is involved has been considered previously, and the striking resemblance to the early tuberculin tolerant state stressed (3).

In contrast to the early phase of pyrogenic tolerance to endotoxin, the later phase, as observed by testing at various intervals after a single initial intravenous endotoxin injection, possesses very different characteristics (4, 5) :

1. It is delayed, requiring 72 or more hours to appear.
2. It is enduring, persisting for weeks.
3. It is highly, though not completely, specific for the "O" endotoxin employed for the initial injection.
4. It bears no relationship to the intensity of the initial pyrogenic response.
5. It is associated with increments in anti-endotoxin antibodies, and is reduced in hosts with defective antibody synthesizing capabilities (e. g. splenectomized man).
6. It can be transferred with serum.
7. Using hyperimmune serum, the protective serum factors reside in the IgG and IgM fractions. In addition, dissociation of the IgM fraction into non-agglutinating fragments by use of 2-mercaptoethanol does not inactivate the protective activity of this fraction. Since "O" specific endotoxin combining activity of such 2-mercaptoethanol treated IgM fractions can still be demonstrated by use of a modified Farr technique, it is apparent that the primary (combining) activity of "O" antibody is sufficient for protection (14).

While late phase tolerant serum from rabbits injected intravenously with smooth endotoxin preparations transfers tolerance primarily to its homologous "O" endotoxin test preparation, some tolerance is nevertheless conferred against heterologous endotoxins. Similar transfer of heterologous tolerance has also been demonstrated in man (15). The degree of such heterologous tolerance transferable with serum can be augmented significantly if endotoxins from the rough mutants, rather than from the smooth parent strains of the gram-negative bacteria are used to prepare the antisera. These findings, originally described by Braude and Douglas with relation to protection against the Shwartzman reaction (16), have recently been confirmed with respect to pyrogenicity (14). The findings indicate that in the absence of the "O" specific terminal antigenic side chains, common core antigens are unmasked in the endotoxin molecule and that antibodies to these common antigens can provide broad spectrum protection against endotoxins from diverse gram-negative bacteria (14, 16). It should be emphasized, however, that the degree of heterologous tolerance transferred by such anti-rough endotoxin sera is far less than that transferred by anti-"O" sera to its homologous endotoxin (14).

Anti-endotoxin antibodies appear to produce pyrogenic tolerance by directly protecting the hepatic Kupffer cells from the endogenous pyrogen releasing activity of the toxin, rather than by simply accelerating the clearance of the endotoxin from the circulation. Thus, studies by Dinarello and coworkers (7) indicate that tolerant phase serum inhibits endotoxin induced release of endogenous pyrogen from isolated hepatic Kupffer cells in vitro. In vivo studies in our laboratory indicate that presenting the toxin directly to the Kupffer cells via chronically implanted portal vein cannulae does not lessen the pyrogenic response (12). In addition, Braude and Douglas have shown that serum protection against endotoxin need not be associated with accelerated blood clearance of the toxin (16); these findings have been confirmed in our laboratory. The mechanisms whereby anti-endotoxin sera protect the hepatic Kupffer cells against endotoxin toxicity are currently under study.

The usual method of evoking endotoxin tolerance consists of daily intravenous administration of toxin. By this method, these two temporally distinct mechanisms of pyrogenic tolerance converge, so that both anti-endotoxin antibodies and direct hepatic Kupffer cellular refractoriness become superimposed during the late phase of tolerance. If endotoxins possessing "O" specific antigens from smooth bacterial strains are employed to evoke tolerance, the "O" antibodies will provide "O" specific protection while the cellular refractory state and the antibodies to common core antigens will provide broad endotoxin protection. By careful quantitative testing, it can be shown that while tolerance extends to all endotoxins, it is nevertheless most marked against the homologous "O" endotoxin preparation. Either the "O"specificity or the non-specificity of the late phase of endotoxin tolerance can then be emphasiz - ed, depending upon the prejudice of the investigator.

What is the role of the accelerated clearance of circulating endotoxin by the reticuloendothelial system in the mediation of tolerance? As indicated, pyrogenic tolerance appears to be based primarily upon failure of the hepatic Kupffer cells to produce endogenous pyrogen, and this failure can be related to a direct effect of endotoxin (cellular refractoriness) and to anti-endotoxin antibodies. Accelerated clearance of endotoxin by the hepatic Kupffer cells can be viewed as an ancillary protective mechanism that delivers the toxin more efficiently to the refractory cells. In the absence of refractoriness, more rapid uptake of endotoxin by the Kupffer cells, as occurs during portal vein infusion of toxin (12) does not result in pyrogenic tolerance.

Responses of man during the febrile phase of gram-negative bacterial illness

The responses of volunteers to bacterial endotoxin during the course of typhoid fever and tularemia have been studied intensively (17 - 19) and can be summarized as follows :
1. During active infection, human responsiveness to intravenously injected endotoxin increases markedly. This augmented responsiveness to endotoxin :
a) begins during the latter portion of the incubation period, before any overt signs or symptoms of illness can be detected;
b) increases daily in stepwise fashion during the latter portion of the incubation peri - od;
c) decreases daily in stepwise fashion during the early portion of the convalescent phase;
d) extends to endotoxins from diverse bacterial species;

e)cannot be attributed simply to non-specific effects of fever or inflammation since it begins several days before fever commences, continues into the afebrile convalescent phase, and is not seen during the febrile phase of viral infection.

2. The two temporally distinct phases of endotoxin tolerance that characterize pyrogenic tolerance, i. e. early and late phases, continue to function effectively within this hyperreactive framework. To avoid excessive toxicity, smaller doses of endotoxin than those used in healthy subjects must be employed in demonstrating the functioning of these tolerance mechanisms :

a)the effectiveness of the early phase tolerance mechanisms can be seen by administration of continuous intravenous infusions of endotoxin. After an initial hyperreactive febrile and subjective toxic response, the fever and toxicity then gradually subside just as in healthy subjects despite the continuation of the endotoxin infusion.

b)the effectiveness of the later phase tolerant mechanisms can be shown by daily intravenous administration of a constant dose of endotoxin. The resulting increments in fever and subjective toxicity decrease slowly and progressively as in healthy subjects.

c)if tolerance to endotoxin is produced b e f o r e the volunteers are infected, tolerance can still be shown to operate during overt illness; this tolerance, however, occurs within the framework of the hyperreactivity to endotoxin. Thus, subjects made tolerant by daily intravenous endotoxin injections before infection for one to four weeks, although hyperreactive to endotoxin during illness, are significantly less reactive than are similarly infected control subjects not previously rendered tolerant to endotoxin.

d)the accelerated clearance of circulating endotoxin that develops during induction of the tolerant state in healthy volunteers persists during overt illness, i. e. no inhibition occurs in clearance despite the marked hyperreactivity to the endotoxin.

3. Despite production of endotoxin tolerance before infection with Salmonella typhosa or Pasteurella tularensis, the subsequent clinical and febrile course of illness is not mitigated. Nor is the clinical or febrile course of illness mitigated by induction of endotoxin tolerance during illness. These findings indicate that in typhoid fever and tularemia c i r c u l a t i n g endotoxin is not primarily responsible for the sustained febrile and toxic course. In contrast, since tolerance to circulating endotoxin is not accompanied by tolerance to the local inflammatory activity of the toxin as seen by intradermal testing (i. e. systemic tolerance is not synonomous with diminished local tissue reactivity (15)), the inflammatory evoking activity of endotoxin at the s i t e s o f i n f e c t i o n could contribute to the sustained febrile and toxic course of illness. In addition, because of the systemic hyperreactivity to endotoxin, release of minute quantities of endotoxin into the circulation during illness could evoke acute febrile and toxic exacerbations superimposed upon the more basic mechanisms of clinical illness. Recurrences of such systemic exacerbations, however, would be limited by the development of endotoxin tolerance.

It is emphasized that the presently available evidence indicating that the endotoxin tolerance mechanisms continue to function effectively during the febrile phase of gram-negative bacterial infection in man is limited to pyrogenic and subjective toxic responses, and to two such infections, typhoid fever and tularemia. Whether such tolerance can also operate effectively against the other toxic activities of endotoxin, or during more fulminant human gram-negative bacterial infections, remains to be determined.

Summary

Certain of the mechanisms by which man develops tolerance to bacterial endotoxins have been considered. After an initial single intravenous injection of endotoxin, two temporally distinct phases of tolerance can be discerned, early and late, each with very different characteristics. Early tolerance appears to be mediated by a transiently occurring refractory state involving to a major degree hepatic macrophages (Kupffer cells). Late tolerance appears to be mediated by anti-endotoxin antibodies directed against both "O" and common core antigens. The latter antigens are masked in the presence of the former, and become effective immunogens only when the "O" specific side chains are lacking. Accelerated reticuloendothelial system clearance of circulating endotoxin is an ancillary protective mechanism that brings the endotoxin more efficiently to the cells that are refractory or protected by antibody. The tolerance mechanisms continue to function effectively in man during the febrile phase of two gram-negative bacterial infections, typhoid fever and tularemia. They do so, however, within a new and hyperreactive framework. The implication of these findings with regards to the role of endotoxin in the pathogenesis of these illnesses have been considered. Whether the human tolerance mechanisms remain functional during more fulminant gram-negative bacterial infections remains to be determined.

References

1. Greisman, S. E., Hornick, R. B. : Proc. Soc. exp. Biol. Med. 131 : 1154 (1969).
2. Greisman, S. E., Woodward, W. E. : J. exp. Med. 121 : 911 (1965).
3. Greisman, S. E., Young, E. J., Woodward, W. E. : J. exp. Med. 124 : 983 (1966).
4. Greisman, S. E., Young, E. J., Carozza, F. A. Jr. : J. Immunol. 103 : 1223 (1969).
5. Milner, K. C. : J. infect. Dis. (Supplement) 128 : 237 (1973).
6. Bodel, P., Atkins, E. : New Eng. J. Med 276 : 1002 (1967).
7. Dianrello, C. A., Bodel, P. T., Atkins, E. : Trans. Ass. amer. Physicians 81 : 334 (1968).
8. Hahn, H. H., Char, D. C., Postel, W. B., Wood, W. B. Jr. : J. exp. Med. 126 : 385 (1967).
9. Atkins, E., Bodel, P., Francis, L. : J. exp. Med. 126 : 357 (1967).
10. Page, A. R., Good, R. A. : Amer. J. Dis. Child. 94 : 623 (1957).
11. Carey, F. J., Braude, A. I., Zalesky, M. : J. clin. Invest. 37 : 441 (1958).
12. Greisman, S. E., Woodward, C. L. : J. Immunol. 105 : 1468 (1970).
13. Janoff, A., Weissmann, G., Zweifach, B. W., Thomas, L. : J. exp. Med. 116 : 451 (1962).
14. Greisman, S. E., Young, E. J., DuBuy, B. : J. Immunol., in press.
15. Greisman, S. E., Wagner, H. N. Jr., Iio, M., Hornick, R. B. : J. exp. Med. 119 : 241 (1964).
16. Braude, A. I., Douglas, H. : J. Immunol. 108 : 505 (1972).
17. Greisman, S. E., Hornick, R. B., Carozza, F. A., Jr., Woodward, T. E. : J. clin. Invest. 42 : 1064 (1963).
18. Greisman, S. E., Hornick, R. B., Woodward, T. E. : J. clin. Invest. 43 : 1747 (1964).
19. Greisman, S. E., Hornick, R. B., Wagner, H. N., Jr., Woodward, W. E., Woodward, T. E. : J. clin. Invest. 48 : 613 (1969).

The Mitogenic Activity of Lipopolysaccharides on Lymphocytes in Culture

I. Blastogenic Activity of Lipopolysaccharide in Mouse Lymphocytes Depleted of Antigen Reactive Cells

C. Damais, C. Bona, C. Galanos, and L. Chedid

Lipopolysaccharides (LPS) from gram-negative bacteria have been found to be strong mitogens for mouse spleen lymphocytes (1) but induce only a weak stimulation of human blood (1, 2), guinea pig (3) and rabbit lymphocytes (2, 4, 5). In chicken the LPS failed to activate bursa cells (B_1) but transformed a small proportion of bursa-derived cells (B_2) compared to the strong stimulation of these populations by anti-chicken Ig serum (6).

In the mouse, the LPS activates only a distinct subpopulation of B-cells which includes the cells participating in the thymus-independent IgM phase of humoral immune response in vivo (7). It was clearly demonstrated that in vitro this polyclonal stimulation (8) is paralleled by a polyclonal IgM production (9).

In a previous study the mitogenic capacity of LPS and lipid A on mice lymphocytes was compared to their effect on the cells of two primate species. The data reported here show that the cells which responded to endotoxin were B-derived and cortisone-sensitive. Experiments were also performed to evaluate the importance of a specific transformation of mouse lymphocytes by LPS as these are known to be ubiquitous and cross-reacting antigens. Accordingly the relationship between LPS-reactive cells and blast transformation was explored.

Material and Methods

In the following experiments, 2-3 month old mice were used. These mice were either C3H/He germ free mice (C.N.R.S. Orléans) or AKR highly inbred mice (Pasteur Institute).

Congenitally athymic nude mice: These mice were bred (C.N.R.S. Orléans) from an outbred wild strain homozygous for "nu" mutation originating from the Institute of Animal Genetics (Edinburg). They were used at 4-8 weeks age old.

Cortisone-resistant lymphocytes were obtained from AKR and nude mice treated 2 days previously by a single intraperitoneal injection of 5 mg of hydrocortisone acetate (Roussel).

X-irradiated and reconstituted mice: The thymus activated lymphocytes were obtained by reconstituting with $2x10^7$ AKR thymocytes either AKR or (AKRxC57Bl)F_1 hybrids which had been lethally irradiated. The spleens were recovered 5 days after reconstitution.

Separation of lymphocytes: Mouse thymus and spleen lymphocytes were prepared according to a technique previously described (10).

Culture conditions: Mouse lymphocytes were cultivated for 48 hrs at 37°C in RPMI 1640 (Eurobio) medium supplemented with 10 % foetal calf serum (Eurobio).

Mitogens: The LPS from S. enteritidis (Danysz strain) was prepared according to the technique of Boivin et al. (11). LPS from S. minnesota smooth 1111 and rough strain 595 and from S. thyphimurium (strain LT_2) were prepared according to the technique of Westphal et al. (12). Lipid A from S. minnesota rough (strain 595) was prepared according to the procedure of Galanos et al. (13).

This research was supported by D.R.M.E. grant n° 73/465 and by I.N.S.E.R.M. grant n° 73/16.

Other reagents: - Polysaccharides were prepared by acid hydrolysis of LPS extracted from S. minnesota 1111 or S. enteritidis (Danysz strain).

- N-acetylglucosamine (Koch- Light Lab.) was used in these experiments.

- Antisera: CBA anti-θ^{AKR} was prepared and purified according to a previously described technique (14), in order to eliminate the T-dependent lymphocytes. In all experiments $2x10^6$ cells were incubated for 1 hour at 37°C with 0.2 ml of a five fold dilution anti-θ^{AKR} serum and 0.2 ml of a five fold dilution guinea pig serum previously absorbed with mouse splenocytes.

Antiserum to mouse IgM was prepared in the goat (Hyland).

Estimation of LPS-reactive cells

The number of LPS-reactive cells was estimated by the rosette method. SRBC were coated with either S. enteritidis LPS or S. minnesota LPS according to the method described by Landy et al. (15). The rosette method was performed as previously described (16).

In certain cases lymphocytes were pre-incubated during 1 hour at 37°C with hapten-acid PS or with goat anti-mouse IgM serum in order to evaluate their influence on the LPS-reactive cells.

Influence of the depletion of LPS-reactive cells on blast transformation induced by LPSs and lipid A

The lymphocytes were rosetted as previously described and layered on Ficoll-triosil gradient. After centrifugation for 10 minutes at 12000 RPM the RFC were scored in:

a) the cell suspension before layering on the ficoll

b) the layer harvested after ficoll gradient

c) the pellet of ficoll-processed cells after centrifugation.

$1.5x10^6$ cells harvested from the ficoll gradient were washed 3 times and incubated with LPS and lipid A.

Influence of an immunodominant sugar on the blast transformation induced by S. minnesota LPS

Various amounts of N-acetylglucosamine were incubated with lymphocytes 1 hour before adding 10 /ug of S. minnesota acid polysaccharide.

Blast transformation

Lymphocytes were cultivated and thymidine incorporation was measured as previously described. The mean counts per minute of 3-5 tubes are presented together with the standard error of the mean, except in the case of the experiment performed on a high number of animals. In the latter case the mean of the stimulation index is given together with its standard error.

Results

1. Stimulation of mice lymphocyte subpopulations

The results which are illustrated in Table I confirm that LPS and lipid A strongly stimulate the spleen lymphocytes of normal (AKR) and nude mice. As previously shown, this stimulation was not affected by the anti-θ serum; the thymocytes were unresponsive.

Furthermore, our experiments showed that activated thymocytes in the syngeneic or allogenic system were not stimulated by LPS. They also demonstrate that LPS has a mitogenic effect on spleen lymphocytes from germ free mice and that this effect was suppressed by pretreatment with overdoses of cortisone.

2. Antigen-binding capacity of mouse spleen lymphocytes

The antigen-binding capacity of normal mouse lymphocytes was measured by scoring the rosette forming cells and measuring the binding of radioactive LPS. The specificity of this reaction was established by pretreating these cells with homologous LPS, its hapten and with goat anti-mouse IgM. As can be seen in Table II the binding of S. enteritidis LPS was inhibited by the pretreatment of the cells with anti-IgM or the hapten. These results attest to the specific binding of LPS by cells bearing an

IgM receptor.

In order to establish the contribution of LPS-binding cells to the blast transformation induced by LPS, the following experiments were performed:

3. Influence of an immunodominant sugar on blast transformation of lymphocytes induced by the hapten of S. minnesota LPS

As can be seen in Fig. 1 the hapten of S. minnesota LPS was mitogenic although its activity was considerably less than the activity of LPS or lipid A. Whereas the strong, nonspecific response is assumed to be related to lipid A, the weak response observed with the hapten could be related to a specific antigenic stimulation of LPS-reactive cells. In agreement with this possibility are the results obtained in the following experiment:

As can be seen in Table III, the stimulation induced by 10 /ug of S. minnesota hapten was strongly inhibited when the cells were incubated with 1 or 3 mg of N-acetylglucosamine which is known to be one of the main immunodominant sugars of S. minnesota LPS.

4. Influence of the depletion of LPS-binding lymphocytes on the transformation induced by LPSs and lipid A

The stimulation induced by LPS and lipid A was measured on spleen lymphocytes previously depleted of their LPS-reactive cells. Ficoll processed murine spleen lymphocytes, which had been depleted of sheep-erythrocytes rosettes forming cells, (SRBC-RFC) were used as controls (Table IV).

Following a strong depletion of S. enteritidis LPS-RFC, S. minnesota LPS-RFC and lipid A-RFC, the lymphocytes were effectively stimulated by the addition of 5 /ug of LPS or of 5 /ug of lipid A, compared to the responses obtained in the controls.

However, the presence of the LPS-coated SRBC in depleted spleen suspensions produced an elevation of the background in all cases so that the following stimulation by various endotoxic extracts is of smaller magnitude than that observed normally. Therefore the stimulation of spleen lymphocytes by endotoxin-coated SRBCs was evaluated in a separate experiment. It can be seen in Table V that incubation of spleen lymphocytes with 24×10^7 endotoxin-coated SRBCs produced a strong stimulation. The addition of 50 /ug of endotoxin to these cells gave the same index of stimulation as that obtained by incubation of lymphocytes with 50 /ug of endotoxin alone.

Discussion

Our results are in agreement with recent findings (8, 17) that the mitogenic capacity of the LPS macromolecule is vested in lipid A. Both are strong stimulators of mouse spleen B-dependent lymphocytes and not of thymocytes. Our findings have shown that even when thymocytes were activated in a syngeneic or allogenic system, they remained unresponsive. Furthermore it was observed that the mouse spleen B-dependent lymphocytes responding to LPS and lipid A are cortisone-sensitive. The precursors of antibody-forming cells are cortisone-sensitive (17), and these precursors are widely distributed throughout the T-independent lymphocytes (18).

Because of the ubiquity and the cross reactivity of endotoxins, experiments were performed in order to ascertain whether in mice the small number of LPS-reactive cells are stimulated through an immunological mechanism or by lipid A nonspecifically.

It clearly emerges from our results that the LPS-reactive cells carry an IgM receptor which can be inhibited by the hapten.

However the depletion of LPS-reactive cells did not decrease the level of ^{3}H-thymidine incorporation after stimulation by LPS or lipid A either on a non-depleted population or one depleted of SRBC-reactive cells. Nevertheless the following findings argue in favour of an immunological response by a small population represented by LPS-reactive cells. Thus:

1. Under certain conditions the hapten elicited a weak stimulation of mouse lym-

phocytes which was of the same magnitude as those observed in primate and guinea pig lymphocytes.

2. This stimulation elicited by LPS-hapten was inhibited by large amounts of an immunodominant sugar.

All these data suggest that in mice the B-lymphocytes stimulated by LPS consist of two populations: a minor population which is stimulated specifically by a LPS of a given serospecificity via an immunological mechanism and another major population stimulated nonspecifically by lipid A, the common moiety of all gram-negative bacterial LPS.

Acknowledgment

We thank Edith Bourgeois for her technical collaboration.

References

1. Peavy, D. L., Adler, W. H., Smith, R. T.: J. Immunol. 105: 1453 (1970).
2. Bona, C., Damais, C., Dimitriu, A., Chedid, L., Ciorbaru, R., Adam, A., Petit, J. F., Lederer, E., Rosselet, J. P.: Mitogenic effect of a water soluble extract of Nocardia opaca. A comparative study with some bacterial adjuvants on spleen and peripheral lymphocytes of four mammalian species. J. Immunol. (in press)
3. Bona, C.: La pinocytose de toxines bactériennes par les macrophages et les implications immunologiques (Ph. D. thesis). Arch. Biol. 82: 323 (1971).
4. Oppenheim, J. J., Perry, S.: Proc. Soc. exp. Biol. Med. 118: 1014 (1965).
5. Sell, S., Sheppard, H. W.: Science 182: 586 (1973).
6. Weber, W. T.: J. Immunol. 111: 1277 (1973).
7. Janossy, G., Humphrey, J. H., Greaves, M. F.: Bacterial lipopolysaccharide stimulation of a subpopulation of B lymphocytes due to lipid A but not to complement activation. (Manuscript sent for consultation by J. H. Humphrey).
8. Andersson, J., Melchers, F., Galanos, C., Lüderitz, O.: J. exp. Med. 137: 943 (1973).
9. Coutinho, A., Moller, G.: Nature New Biol. 245: 12 (1973).
10. Bona, C., Damais, C., Chedid, L.: Blastic transformation of mice spleen lymphocytes by a water soluble mitogen extracted from Nocardia. Proc. nat. Acad. Sci. (in press).
11. Boivin, A., Mesrobeanu, L.: Rev. Immunol. 1: 553 (1935).
12. Westphal, O., Lüderitz, O., Bister, F.: Z. Naturforsch. 7b: 148 (1952).
13. Galanos, C., Rietschel, E. T., Lüderitz, O., Westphal, O.: Europ. J. Biochem. 19: 143 (1971).
14. Bona, C., Dumitrescu, S. M., Heuclin, C., Pariset, C.: Recherches concernant le developpement ontogénique des antigènes de cytodifférenciation lymphocytaire chez le poussin et leur controle génétique chez la souris. Dans Etude phylogénique et ontogénique de la réponse immune, INSERM 1972, 229.
15. Landy, M., Sanderson, R. P., Jackson, A. L.: J. exp. Med. 122:483 (1965).
16. Bona, C., Trebiciavsky, I., Anteunis, A., Heuclin, C., Robineaux, R.: Europ. J. Imm. 2: 434 (1972).
17. Peavy, D. L., Shands, J. W., Adler, W. A., Smith, R. T.: J. Immunol. 111: 352 (1973).
18. Cohen, J. J., Claman, H. N.: J. exp. Med. 133: 1026 (1971).
19. Playfair, J. H. L.: Clin. exp. Immunol. 8: 839 (1971).

Table I.
Stimulation of Mouse Lymphocytes Subpopulations by LPS and Lipid A

Mouse Strains	Treatment of Mice	Source of Lymphocytes	S. enteritidis LPS 10 /ug	S. minnesota LPS 10 /ug	Lipid A 5 /ug
	Nil	spleen	23. 92*± 7. 8** (4)	13. 18± 3. 27 (11)	9. 9±3. 52 (11)
	Nil	thymus	1. 05 ± 0. 1 (9)	1. 1 ± 0. 3 (3)	NT
AKR	anti-θ treatment of cells	spleen	21. 27 ± 8. 3 (4)	NT	NT
	cortisone 5 mg 48 hrs before	spleen	NT	1. 13± 0. 12 (4)	1. 41+0. 16 (4)
	x-irradiation 850r and rec. syng. thym	spleen	NT	1. 48± 0. 31 (15)	NT
(AKRx C57Bl/6) F 1	x-irradiation and rec. with AKR-thymocytes	spleen	NT	1. 2+ 0. 1 (10)	NT
C3 H/He germ free	Nil	spleen	9. 16 ± 1. 6 (5)	NT	NT
Nude	Nil	spleen	9. 37 ± 2. 9 (3)	13. 77+ 2. 75 (18)	8. 56+1. 9 (18)
	cortisone 5 mg 48 hrs before	spleen	NT	3. 17+ 0. 8 (4)	1. 69+0. 35 (4)

*   Stimulation index
**   ± SD
()   Number of mice tested
NT   Not done.

Table II.
Antigen-Binding Capacity of Mouse Spleen Lymphocytes

Pretreatment of 0.5 x 10^6 lymphocytes	LPS-binding cells determined by scintillation method cpm	RFC $^o/_{oo}$
None	392 $\pm$ 4	0.52*
S. enteritidis boivin (500 μg)	28 $\pm$ 2	0.07
S. enteritidis acidic polysaccharide (500 μg)	38 $\pm$ 1.4	0.08
Goat anti-mouse IgM diluted 1 : 5	32 $\pm$ 1.6	0.06

* 4-5 x 10^4 nucleated cells counted

Scintillation method: 0.5 x 10^6 lymphocytes incubated either with 1 μg ^{14}C-LPS biosynthetically labeled.

Rosette method: 0.5 x 10^6 lymphocytes incubated with 4.10^6 SRBC sensitized previously with LPS.

Table III.
Influence of an Immunodominant Sugar on Blast Transformation

Amount of N-acetyl-glucosamine	ac-PS minnesota added	
	0	10 μg
0	704* $\pm$ 388**	1240 $\pm$ 207
0.3 mg	420 $\pm$ 201	1283 $\pm$ 88
1 mg	344 $\pm$ 158	363 $\pm$ 135
3 mg	353 $\pm$ 196	559 $\pm$ 179

* cpm
** $\pm$ SD

Table IV.
Mitogenicity of LPS and Lipid A on Lymphocyte Populations Depleted of LPS-Binding Cells

Spleen lymphocytes depleted of	%oo RFC scored after Ficoll procedure			Stimulation of lymphocytes by			
	before Ficoll procedure	in layer	in pellet	Nil	S. enteritidis LPS	S. minnesota LPS	Lipid A
Ficoll processed (control)	-	-	-	1,963*± 710**	14,648±2,817	11,832± 886	6,732±2,140
SRBC-RFC (control)	2.8	0	24.2	2,187 ± 514	16,076±3,150	10,115± 1,815	7,935± 555
S. enteritidis LPS-RFC	3.2	0	32.3	10,881 ±1,224	17,504±2,303	14,224± 1,475	12,322± 795
S. minnesota LPS-RFC	4.3	0	15.2	14,575 ±2,540	25,859±2,360	12,981± 1,490	12,496±1,505
Lipid A-RFC	1.2	1.1	6.4	1,235 ± 290	22,799±1,520	18,407± 3,980	13,974± 960

* cpm

** ± SD

Table V.

Transformation of Lymphocytes by Interaction with Endotoxin-Coated SRBCS

Preparation interacted with lymphocytes	cpm
None	13971 ± 6735*
ASED-coated-SRBC 24.10^7	37966 ± 5400
ASED in saline 50 μg	92735 ± 15650
ASED in saline 50 μg + ASED-coated-SRBC 24.10^7	88486 ± 4890

* SD

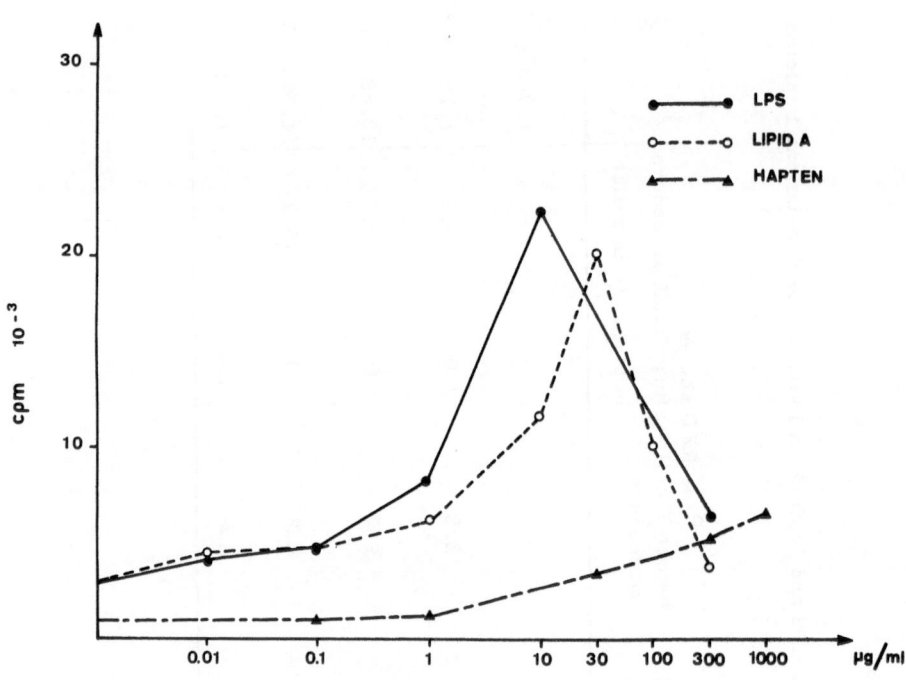

FIG.I: WEAK MITOGENIC ACTIVITY OF HAPTEN

The Role of Humoral and Cellular Factors in the Stimulation of Nonspecific Resistance Induced by Endotoxin

Zs. Pusztai-Markos

Introduction

The ability of bacterial endotoxins (ET) to alter the course of experimental infections is well-known. The transient increase in homologous and heterologous resistance that develops within hours after ET administration may be related to enhanced macrophage and reticuloendothelial activity (1, 2) and/or to enhancement of pre-existing specific antibody titres according to published observations (3, 4). Furthermore, it has been suggested that at least a part of the biological activities provoked by ET are based on a hypersensitivity state of the host (5, 6).

The investigations described in this report indicate that the importance of humoral factors in the specific host resistance is doubtful. The results of the in vivo phagocytosis experiments and the migration inhibition assays support, however, the hypothesis that ET is able to mobilize active lymphocyte factors inducing a transient nonspecific, resp. specific, cell-mediated immunity.

Material and Methods

Bacteria. Salmonella typhi 0901, E. coli 055, E. coli 026, Shigella flexneri 2a were kindly supplied by R. Rohde (Nationale Salmonella Zentrale, Hygiene-Institut, Hamburg) and S. minnesota, S. minnesota R 595 by B. Urbaschek (Hygiene-Institut, Mannheim).

The bacteria were kept at 4°C on peptone agar. For infection, 18 hours culture on DST agar (Oxoid) were washed off with 1 % casein hydrolysate solution (pH 7.2). Total bacterial counts were determined by opacity and viable counts by plate counting methods.

For immunization, the cultures were washed off with PBS and were killed by heating to 100°C in a water bath for one hour.

The LD_{50} values of the bacterial strains were : S. typhi 0901 : 1.5×10^9, E. coli 055 : 1.5×10^8 and Sh. flexneri 2a : 4.5×10^9.

Endotoxins. Purified S. typhi 0901, E. coli 055 : B6, E. coli 026 : B8 and Sh. flexneri endotoxins (Westphal-type, Difco Laboratories) were used.

For "mild" alkaline hydrolysis 1 mg/ml ET solution was kept at 37°C overnight in 0.02 n NaOH. This type of preparation was used for SRBC sensitization. For alkaline hydrolysis 2 mg/ml ET solution was treated in a water bath at 56°C in 0.25 n NaOH for one hour.

Animals. Male mice NMRI (Stolberg) and CBA (Oxfordshire Laboratory Animal Colonies, Oxon, England) weighing 18 to 22 g were used for immunization and in vivo phagocytosis experiments. For migration inhibition assay the animals weighed 28-32 g.

In vivo phagocytosis. The sensitized and control mice were infected by intraperitoneal injection of 1×10^8 bacteria in 0.5 ml. After 4 hours mice were sacrificed by cervical dislocation, 3 ml of PBS with 5 IU/ml heparin was injected intraperitoneally, the exudate collected by a syringe and filled up to 5 ml with PBS. The cell count was determined in a Buerker chamber. For viable count the exudate was immediately further diluted in PBS and 4 tenfold serial dilutions dropped out on DST agar plates in quadruplicate using calibrated micropipettes. After incubation overnight at 37°C the plates were evaluated.

For determination of the phagocytosis enhancing effect of mouse sera, two methods were employed : 1. 0.2 ml of the inactivated serum pool was injected intra-perintoneally 30 minutes before infection. 2. the infecting suspension was incubated 30 minutes at 37°C with different serum dilutions. The mice were thereafter infect-ed with these serum-suspension-mixtures.

Migration inhibition assay. The test was performed according to the method des-cribed by David (7) with slight modification : exudates were produced in experimen-tal and control mice by i. p. injection of 2 ml of sterile mineral oil (Bayol 90, Es-so) 5-8 days prior to the harvest of cells for migration studies. Each capillary tube was filled with 5×10^6 cells. Culture medium contained 15 % normal inactivated mouse serum and 100 U Penicillin/ml. Tests were run in quintuplicate at least; the data were utilized only if the deviation between the replients was less than 15 %.

Tolerance induction. For tolerance induction, LPL after "mild" alkaline hydro-lysis was used in the same dose as described by Britton (8). Tolerance state was controlled by estimating PFC count and humoral antibody titres.

Results and Discussion

In our previous experiments we were not able to demonstrate any increase in hae-magglutinating, bactericidal, or opsonizing capacity in the sera of mice 1-3 days after injection with different LPS. Investigating the plaque forming cells (PFC) and rosette forming cells (RFC) counts in the spleen of the same mice, we obtained an identical transient increase using unsensitized or with different LPS sensitized SRBC in the PFC but not in RFC number. A striking identity of PFC and RFC counts was observable after LPS stimulation, suggesting that the cells responsible for ro-sette formation in normal mice will produce C binding antibodies after LPS applica-tion.

This induction of PFC against SRBC by different LPS may reflect a nonspecific mitogenic effect of LPS molecules on the B cells or antigenic relationship between LPSs and SRBC. If the effect is based on a specific immunological mechanism, it can prove the existence of common antigenic determinants by immunization experi-ments. Injecting high doses of different LPSs (500 µg), besides an intensive specific response, induction of PFC and RFC directed also against the other LPS, while investigating the same spleen cells after 5 to 10 days. Interestingly enough, the stimulation of an immune reaction specific for SRBC is reflected only in an en-hancement of RFCs but not PFCs.

The lack of any antibacterial activity in the sera of the stimulated mice is not in accord with the idea that humoral antibodies against these common antigenic deter-minants in the LPS molecules of different gram-negative bacteria are responsible for the phenomenon of nonspecific resistance. In order to confirm the hypothesis that a cellular immunity is activated after LPS application, we investigated the mo-bilization of different cellular factors.

1. Phagocytosis Enhancing Effect of Bacterial LPS

The phagocytosis promoting activity of different LPS was studied in the mouse peritoneum. We have determined the degree of in vivo phagocytosis by injecting virulent bacteria.

In Fig. 1 is shown the correlation between LPS dose, recultivatable colony form-ing units (CFU), and peritoneal exudate cells (PEC) counts. As little as 0.1 µg LPS is highly effective in eliminating the injected bacteria (90 % of the total) without inducing a significant multiplication of PECs; the latter occurs with higher LPS doses (50-100 µg). Results similar to those depicted in Fig. 1 were obtained after infection with S. typhi 0901, or stimulating by other LPSs, as S. typhi 0901 and Sh. flexneri instead of E. coli 055 LPS. By intravenous application of the sensitizing antigens practically the same effect was evoked as by intraperitoneal route.

The degree of phagocytosis depended more on the infecting organisms than on the stimulating antigen. This could be demonstrated mostly following the enhancing ef-

fect in time as it is shown in Fig. 2. On the first three days after stimulation, independent from the applied LPS, the phagocytosis of E. coli 055 is similar but differs
from that of S. typhi 0901. The enhanced elimination of heterologous bacteria decreases or disappears in one week with the simultaneous appearance of the specific immunity.

The phagocytosis promoting effect of 100 μg LPSs remained unchanged after alkaline hydrolysis, but not the enhanced PEC count (Fig. 3). According to these data
the induction of the multiplication of the PECs, like the toxic effect and mitogenic
activity, may be a property of the lipid A component.

To check whether the enhanced elimination of bacteria following LPS administration is related to concomitant changes in serum opsonic factors, we tested the effect
of passively transferred immune sera, stimulated sera (sera of mice collected 24
hours after injection of 100 μg LPS), and normal sera on in vivo phagocytosis. Data
presented in Fig. 4 provide evidence that the in vivo phagocytosis in mice treated by
normal or stimulated sera does not differ significantly; in other words, this effect
is not transferable by serum. The assumption cannot be rejected that the enhanced
activity of RES plays an important role in the nonspecific resistance. These data
support the primary importance of macrophage activation. The increase of function
seems not absolutely connected with an increase in the number of the RES cells. The
different LPS activate these cells in the same manner, and the degree of the phagocytosis depends rather on virulence factors of the infecting organisms.

To which factors should this enhanced phagocytosis be attributed if humoral components do not take part? The interval between LPS administration and appearance
of enhanced phagocytosis, resp. nonspecific resistance, the relative minimal effective dose of LPS, as well as the course of these events, rather support the hypothesis that hypersensitivity of the delayed type - that means a cellular mechanism -
may play a primary role in stimulation of nonspecific immunity.

2. Migration Inhibition (MI) Effect of LPSs on Mouse PEC

Experimental data confirm (9) that infection-immunity can be mediated by cellular mechanisms in infections evoked by endotoxic bacteria. The inhibition of migration of specifically sensitized PEC in the presence of antigen is generally accepted
as an in vitro model for evaluating the delayed hypersensitivity state. Therefore,
the migration inhibition method was chosen to investigate the existence of cellular
immunity in the case of LPS.

Our results summarized in Fig. 5 and 6 demonstrate that S.typhi 0901 LPS exerted no MI using normal cells while E. coli 055 LPS regularly caused a pronounced
MI of normal and immune mouse PEC. Immunization of animals with LPS or killed
vaccine of S. typhi 091 and E. coli 055 bacteria resulted in a significant MI of PEC
in the presence of S. typhi 0901 LPS also.

As the inhibition of macrophages by endotoxins was described as a cytotoxic effect (10, 11), we performed control experiments on the effect of detoxified preparations. The results are shown in Fig. 7. Alkali treatment does not alter the MI activity of E. coli 055 LPS on normal cells in contrast to S. typhi 0901 LPS, which in
the detoxified form (also in the case of normal cells) displayed a dose dependent
MI capacity.

On the other hand, PEC taken from mice 24-28 hours after ET sensitization or
from tolerant mice failed to show any MI in the presence of homologous and heterologous ET.

The data obtained in this study strongly suggest that mice are sensitized against
LPS antigens in the form of delayed type hypersensitivity under natural conditions.
The alteration in the PEC reactivity subsequent to immunization or tolerance induction evidence that MI is based on an immunological mechanism. The unchanged,
resp. enhanced effect of alkali treated LPS speaks against the sole role of lipid A
as the sensitizing antigen. The mobilization of active lymphocyte factors, such as

MIF, supports the probability of RES activation in a similar way.

The interaction of sensitized lymphocytes with specific antigen results not only induction of MIF but also in a release of other factors such as mitogenic, inflammatory, macrophage activating, cytotoxic factors etc., which are correlated with the expression of cell-mediated immunity. On the other hand, it is well-known that ET exerts a "nonspecific" mitogenic effect on B lymphocytes. Other nonspecific mitogens as PHA and Con A are able to induce MIF. It seems established that lipid A is the mitogenic part of LPS molecules (12, 13). The experiments reported here stress the role of common antigenic components in different LPS, but this is probably not only in the lipid A portion in the provocation of "nonspecific" immunity. Additional experiments showed that the lipid-A-free preparation after acetic acid hydrolysis of the different LPS maintained its phagocytosis promoting activity nearly quantitatively and induced a significant MI of PEC; however, this effect seemed to depend on the serological type of LPS. Furthermore, in congenitally thymusless mice (BALB nu/nu) both cellular reactions - as in vivo phagocytosis activation and migration inhibition of PECs - are provokable by LPSs and their hydrolysed products as well. Blocking the Ig receptors by anti-mouse-Ig on the stimulated PEC population (by a concentration inhibiting completely the rosette formation) did not influence MI at all.

The findings of this study suggest that active lymphocyte factors - probably responsible for the nonspecific immunity - are mobilized primary through B lymphocytes in a nonspecific mitogenic way as well as in a specific immunologic fashion.

References

1. Biozzi, G., Benacerraf, B., Halpern, B. N. : Brit. J. exp. Path. 36 : 226 (1955).
2. Rowley, D. : Adv. Immunol. 2 : 241 (1962).
3. Michael, J. G., Whitby, J. L., Landy, M. : Nature 191 : 296 (1961).
4. Michael, J. G. : J. exp. Med. 123 : 205 (1966).
5. Braude, A. I., Siemienski, J. : Bull. N. Y. Acad. Med. 37 : 448 (1961).
6. Gingold, J. L., Freedman, H. H. : Proc. Soc. exp. Biol. Med. 128 : 599 (1967).
7. David, J. R., David, R. In : In vitro method in cell-mediated immunity, p. 249 (Academic Press, New York 1971).
8. Britton, S. : Immunology 16 : 513 (1969).
9. Collins, F. M., MacKaness, G. B. : J. of Immunol. 101 : 830 (1968).
10. Kessel, R. W., Braun, W. : Austr. J. exp. Biol. Med. Sci. 43 : 511 (1965).
11. Heilman, D. H., Bast, R. C. : J. Bact. 93 : 15 (1967).
12. Andersson, J., Melchers, F., Galanos, C., Lüderitz, O. : J. exp. Med. 137 : 943 (1973).
13. Peavny, D. I., Shands, J. W., Adler, W. H., Smith, R. T. : J. Immunol. 111 : 352 (1973).

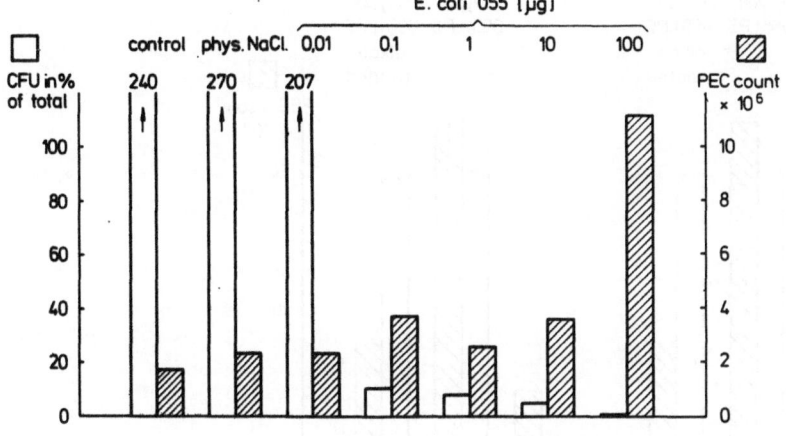

Phagocytosis enhancing effect of LPS injected i.p. 24h
before infection with 1.10^8 E. coli 055 bacteria in mice.

Fig. 1. CFU: recultivatable colony forming unit in per cent ot total injected bacteria. PEC: peritoneal exudate cell count in total exudate.

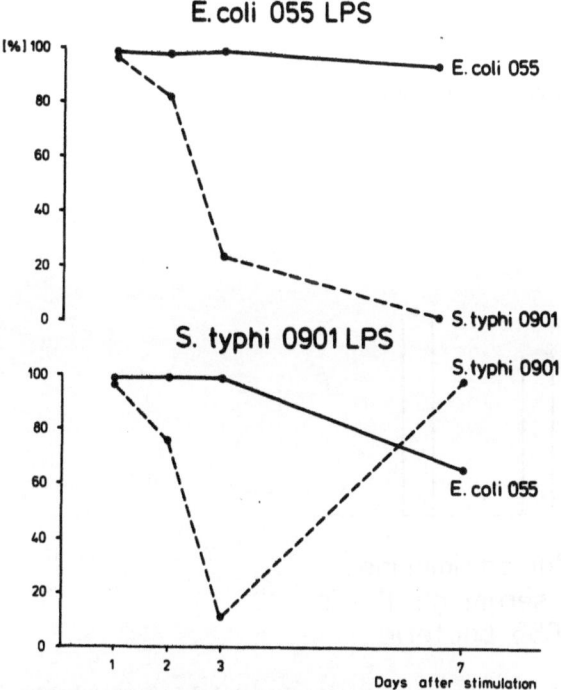

In vivo phagocytosis of i.p. injected E.coli 055 and
S. typhi 0901 bacteria (1.10^8) in mice after stimulation with
LPSs. Phagocytosis expressed in per cent to values
evaluated in non stimulated mice.

Fig. 2.

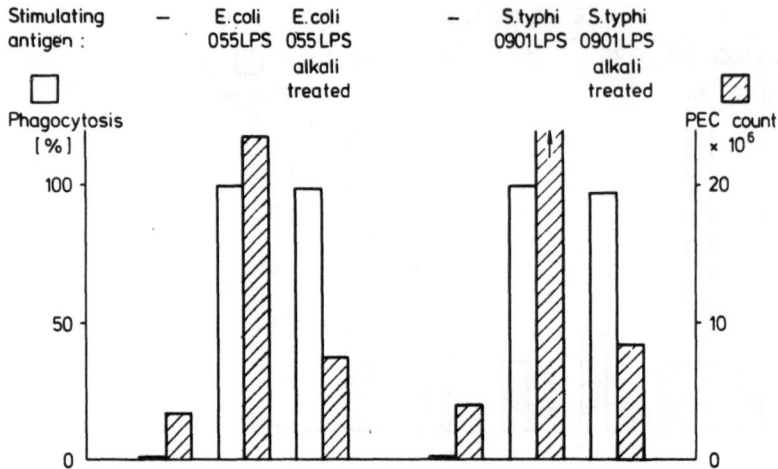

Phagocytosis enhancing effect of LPS injected i.p. 24h
before infection with 1.10^{8}E.coli 055 bacteria in mice.

Fig. 3. Phagocytosis expressed in per cent to values evaluated in control mice.
PEC: peritoneal exudate cell count in total exudate.

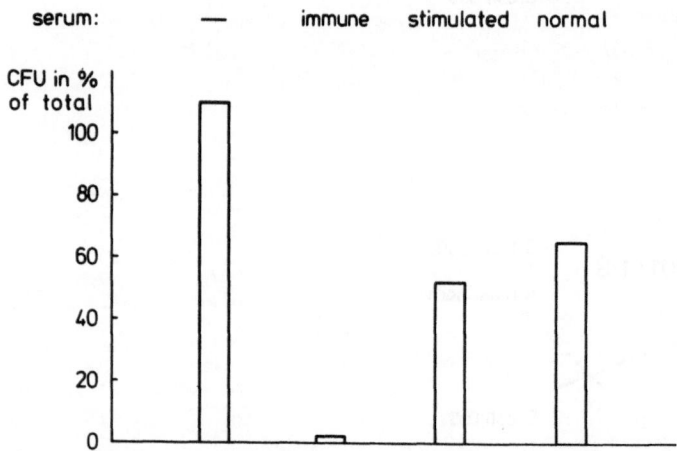

The effect of passively transferred immune,
stimulated and normal mouse serum on the in
vivo phagocytosis of E. coli 055 bacteria.

Fig. 4. CFU: recultivatable colony forming unit in per cent of total injected bacteria.

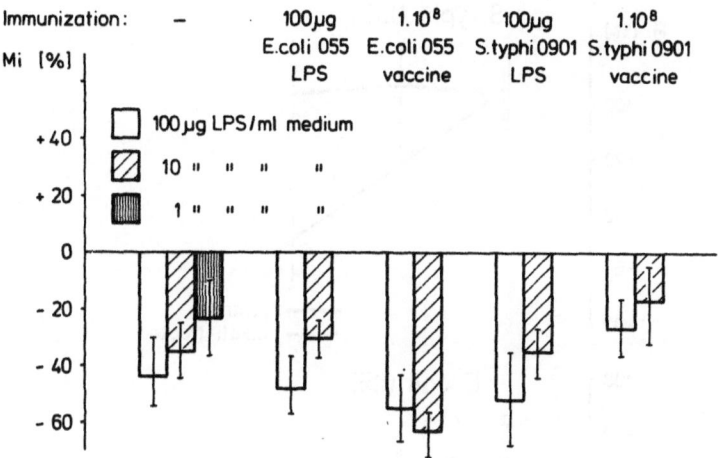

Migration inhibition effect of E.coli 055 LPS on PEC of normal and differently immunized NMRI mice.

Fig. 5.

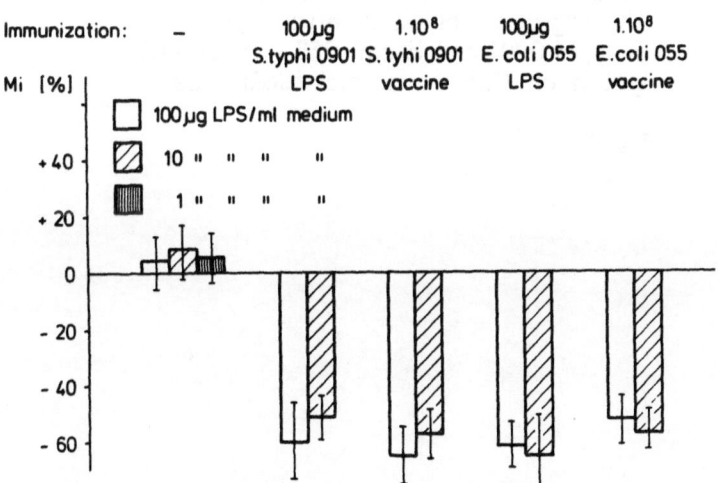

Migration inhibition effect of S.typhi 0901 LPS on PEC of normal and differently immunized NMRI mice.

Fig. 6.

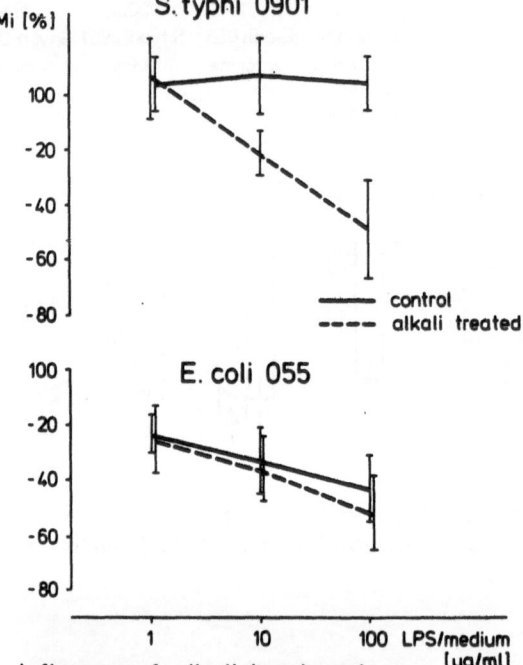

Influence of alkali treatment on
the migration inhibition effect of
S. typhi 0901 and E. coli 055 LPS investi =
gated on PEC of normal NMRI mice.

Fig. 7.

Genetic Control of Immune Responsiveness of Mice to a Soluble Brucella abortus Antigen

M. Renoux and G. Renoux

Knowledge of a genetic system in mice that controls the capacity to mount an immune response to a bacterial antigen permits further advances in investigating the function controlled by Ir genes and thus contributes to the understanding of susceptibility to bacterial diseases.

To this end, the immunogenicity of a Brucella abortus soluble antigen (ABS) was investigated as there is little likelihood that secondary responses involve prior contact with this substance; i. e. normal mice demonstrate no immunologic experience to Brucella endotoxins (1). Moreover, epidemiological studies have shown that only 18 % of serologically positive subjects develop a clinical disease, and only 65 % of exposed humans develop a positive serological test, suggesting some sort of genetic correlation with susceptibility.

B. abortus antigen (ABS) was extracted from phenol-killed antigenic suspensions for tube agglutination tests. After two spinnings each of 1 hour at maximal speed in a refrigerated centrifuge, supernatants were dialyzed against pyrogen-free distilled water until no phenol was detectable. The batch used throughout these studies, contained 8 % proteins, 35 % total sugars, 23 % lipids, and 28 % RNA. All aminoacids, including muramic acid and diaminopimelic acid, but excluding methionine were present in ABS, as well as glucosamine and galactosamine. ABS contained all the antigens of B. abortus cell walls and ABS-antisera reacted with all B. melitensis fractions that shared common antigenic structures with B. abortus (2). ABS did not react with a monospecific antiM serum (3).

To avoid the complicating effects of adjuvants, as well as to compare a pure response to ABS in responder and nonresponder mice, all suspensions or solutions were prepared in pyrogen-free water; neither complete nor incomplete Freund's adjuvant was used.

Doses of 1 μg and of 100 μg were used to immunize parental mice. Only the 1 μg dose was employed in F_1, F_2, and backcross studies. Mice were injected i. p. with antigen on days 0 and 15; they were bled from the orbital plexus on days 7 and 14 (early and late primary responses) and on days 21 and 28 (early and late secondary responses).

The anti-ABS responses were measured by titrating total and 7S antibodies in passive microtiter hemagglutination tests using sheep red blood cells (SRBC) coated by ABS via chromium chloride (4). Test sera were diluted 1:5 in veronal buffer and heated at 56°C for 30 minutes. Natural anti-sheep agglutinins were absorbed with 0.025 ml of washed SRBC per 0.5 ml of serum for 3 hours at 37°C. Serum (diluted 1:5) was added in volume of 0.025 ml to an equal volume of veronal buffer or of 0.2 M 2-mercaptoethanol (2-ME) in veronal buffer. The mixtures were incubated at 37°C for 2 hours after which time twofold serial dilutions were made in veronal buffer using the Microtiter apparatus, and 0.025 ml-amounts of the ABS-sensitized SRBC were added to each well. Each test included a serum control well in the presence of normal-uncoated SRBC. Each series of tests included an antigen control well, ABS-sensitized SRBC in presence of buffer instead of serum. The plaques covered with plastic wraps were allowed to stand overnight at room temperature. Readings ranged from +++ (complete agglutination) to - (complete absence of agglutination). The titer was taken as the reciprocal of the highest dilution of the serum giving at least a ++ reaction. Statistical analysis was performed by Student's t-test for small samples (5). Exploratory assays on 200 positive sera from various animal sources have shown that this passive hemagglutination test can detect specific antibody in as little as 1×10^{-9} grams of immunoglobulin.

To verify that nonresponder mice were not tolerant to ABS, 4 kinds of controls were utilized. a) In vitro absorption tests. Spleen, liver, and kidney of mice from the 6 strains under study were homogenized in veronal buffer. Mouse antisera to ABS were absorbed three times with an equal volume of washed, packed homogenates at $4^{\circ}C$ for 3-hr periods (6). b) Immunization of responder mice by nonresponder tissues. To test for the presence of common antigenic determinant in ABS and in cells of the nonresponder C3H/He mice, antisera were raised by injecting spleen cells from C3H/He female donors into C57BL/6 female recipients. Immunization consisted of weekly i.p. injections of 2×10^7 spleen cells in complete Freund's adjuvant for 6 weeks; animals were bled one week after the last injection. c) Cross reactivity between ABS and antisera to H-2^b and H-2^k. Anti H-2^b, anti EL4 (C57BL leukemia) pool Z1, and anti H-2^k, (BALB/c x C57BL/6) F$_1$ anti BP8 (C3H sarcome) pool E1, antisera were obtained through the courtesy of Mrs. E. Stockert (Memorial Sloan-Kettering Cancer Center, N.Y.). d) Effect of a large immunizing dose. Male and female mice of the 6 strains under study were immunized by 100 μg of ABS at days 0 and 15; sera were collected at days 7, 14, 21, and 28.

A possible relationship between H-2 antigens and mouse antibody responses was checked by agglutination tests of red cells of backcross mice with anti H-2^b and H-2^k sera.

The Centre de Sélection des Animaux de Laboratoire, Centre National de la Recherche Scientifique, provided mice of the following strains: A/Orl (H-2^a), C3H/He (H-2^k), C57BL/6 (H-2^b), C57BR/cd (H-2^k), DBA/2 (H-2^d). Random-bred CD-1 mice were purchased from Charles River, Elbeuf, France. They were maintained in plastic cages in an air-conditioned room at $24^{\circ}C$; antibiotic-free food pellets and water were available ad libitum. All mice, including F$_1$, F$_2$, and backcrosses, were 5 to 6 weeks old at the beginning of immunization (7).

Evidence for Differences in Strain and Sex Responsiveness in Parental Mice

The intraperitoneal injection of 1 μg of ABS disclosed significant differences in antibody production related to the strain and accentuated by the sex of the mice (Fig. 1). Inbred mice could be separated in 3 groups: a) two high-responding strains, A/Orl (H-2^a) and C57BL/6 (H-2^b); however, a high 7S response was sustained longer in female C57BL/6 mice than in female A/Orl mice, whereas the differences in antibody levels of males depended on a 19S production by A/Orl mice; b) two intermediate-responding strains, C57BR/cd (H-2^k) and DBA/2 (H-2^d); however, secondary 7S production was delayed in C57BR/cd mice; females were consistently higher responders than males in both strains; c) a low responding strain, C3H/He (H-2^k) in regard to female responses; male C3H/He mice were nonresponding mice. Female random-bred CD-1 mice were unable to elaborate 7S antibodies, whereas male CD-1 mice did.

Antibody Responses of Mice to Injections of 100 μg of ABS

A 100-fold higher dose of ABS was employed to determine whether low responses after injections of 1 μg of ABS were evoked by cross-reaction between ABS and tissue antigens of the low responders or whether the findings reflected a system in mice that controlled the ability for antibody responses to stimuli by one or the other antigenic determinants of ABS.

Male mice of strains C3H/He, C57BR/cd and DBA/2, and random-bred CD-1 females were immunized with 100 μg of ABS on days 0 and 15, together with males of strains A/Orl and C57B1/6 (Fig. 2). All mice responded by specific 7S and 19S antibodies at late primary bleedings. These responses were increased by a second injection of 100 μg of ABS. However, C3H/He male mice remained the lowest responding strain. Random-bred female CD-1 mice gave a high 7S secondary antibody response to 100 μg in contrast to the low 19S antibody response following a booster dose of 1 μg of ABS.

Tests for Cross-Reactivity of ABS and Mouse Tissues

Sera obtained at early and late secondary bleedings from the 6 mouse strains under study were absorbed with homologous or heterogeneous spleen, liver, and kidney homogenates and assayed by passive hemagglutination of ABS-coated sheep erythrocytes. Within the limits of dectection of our assay there was no cross-reactivity between ABS and mouse tissues.

Sera obtained after 6 weekly i.p. injections of 2×10^7 spleen cells (in Freund's adjuvant) of C3H/He female mice into C57BL/6 female recipients were examined for ABS antibodies; no evidence was obtained for a common antigenic determinant between spleen (lymphocytes) of C3H/He mice and ABS.

Finally anti H-2^b and H-2^k (also possibly used as an anti H-2^a-serum) sera did not cross react with ABS.

Antibody Responses in F_1, F_2 Hybrids and Backcrosses

The foregoing data indicate that there are two factors influencing the antibody response to ABS; one is the strain of mice, the other is sex. Reciprocal crosses between nonresponding C3H/He and highresponding C57BL/6 mice, as well as backcrosses were established to determine the number of genes controlling antibody responses to ABS and to evaluate a sex-linked effect.

Hybrid and backcross mice were immunized with 1 μg ABS at days 0 and 15. Sera were sampled for early and late primary responses at days 7 and 14, and for early and late secondary responses at days 21 and 28.

All male B6C3 or C3B6 mice gave high early and late secondary 7S and 19S responses (Tables I, II). In contrast, female B6C3 and C3B6 mice gave intermediate and low responses that differed significantly (P=0.01) from male responses. These findings suggested that a component involved in the antibody response to ABS is sex-linked, that either IgM or IgG antibodies to ABS were similarly controlled. To test this hypothesis F_2 hybrids and backcrosses were established.

No relationship between H-2 antigens and the antibody response to ABS was evident, when tested in backcross progenies (Table III) confirming findings in parent strains, where strain C3H/He (H-2^k) failed to respond, whereas C57BR/cd, with the same H-2^k allele, responded.

The antibody response produced by male or female mice in various crosses illustrate that differences in the ability of C57BL/6 and C3H/He mice to respond to ABS stimulus could be governed by an autosomal dominant trait, and that a sex-linked control is also be involved. Female F_2 mice shared their responses within three groups of nonresponders, intermediate responders, and high responders; whereas male F_2 mice were separated in two groups of, respectively, high and low responders. Male (B6xC3) F_1 x B6 backcrosses gave responses identical to that of (C3 x B6) F_1, whereas female backcross mice fell into intermediate responders and high responders. Male (B6 x C3) F_1 x C3 backcrosses shared 1/2 negative responses and 1/2 intermediate-high responses.

One may postulate a <u>B</u> responder-gene in C57BL/6 mice codominant with a <u>c</u> nonresponder-gene in C3H/He mice. One may also postulate a component involved in the antibody response is Y-linked, i.e. carried out on the Y chromosome, which influences the responses carried out on male mice by <u>B</u> or <u>c</u> hypothetical alleles. Such a hypothesis is consistent with the experimental results (Table IV).

Discussion

The present data demonstrate genetic control of antibody responses to ABS in mice, not linked to H-2 histocompatibility loci. Accordingly the ABS antibody re-

sponse system in mice is different from the branched polypeptide system (8-11) or
the copolymer system (12) described for mice, in that the responder status is not
linked to the major histocompatibility locus, and especially as regards the finding
of nonresponding animals (within the limits of detection of our assay).

The possibility that strain differences might be a manifestation of immunologic
cross reactivity between mouse antigens and some part of the ABS antigenic complex is reasonably excluded by the results of a) in vitro absorption assays of mouse
antisera, b) immunization of responder mice by non-responder spleen tissue,
c) passive hemagglutination tests for ABS antibodies in anti H-2 sera, d) immunization by a high dose of antigen.

ABS is likely to contain a number of determinants potentially antigenic for mice.
Significant differences still exist in the responses to high doses (100 μg) of ABS,
in that each inbred strain shows a distinctive pattern in the primary as well as in
the secondary responses, which resembles that obtained with 1 μg of ABS. This
may relate to a determinant moiety in ABS-complex able to induce an antibody response when given in a low amount, and thus masks most of the responses that
might be evoked by other less active determinants. However, female random-bred
mice may recognize other determinants in the ABS complex as they respond to high
doses and not to injections of 1 μg.

The clear-cut findings that emerged from our study of antibody responsiveness
to ABS in mice are conceivably related to immunological ignorance of laboratory
animals for Brucella antigens, i. e. no prior contact during their lifespan, in contrast to LPS antigens of the ubiquitous gram-negative organisms (1).

Both 19S and 7S antibody responses to immunization with ABS are under identical
genetic control. This work clearly shows also that antibody responses to ABS in
C57BL/6 and C3H/He mice are governed by at least two distinct types of genetic
control. The first type influences the ability to respond to ABS. The incidence of
responders and nonresponders among female progeny derived from various crosses
indicates that antibody responsiveness to ABS is represented by a gene (or a block
of genes), which chromosomal location is yet unknown. The second type appears to
regulate the antibody response produced by male mice. For mice identical with respect to gene dose for the postulated components, significant groupings of the results differentiate male from female responses. It may be that the involved gene
(or genes) is Y-linked and that C3H/He mice are unable to express this character.
Studies are in progress to assess the validity of this tentative hypothesis.

Summary

Strains of inbred mice varied widely in their ability to respond to a B. abortus
soluble antigen, ABS: this responsiveness was not associated with H-2 antigens.
The IgG and IgM antibody response to ABS was assessed in F_1, F_2, and backcross
progeny derived from nonresponding (C3H/He) and high responding (C57BL/6) parental strains of inbred mice. The results indicated that a major component involved in both antibody responses is an autosomal trait, whereas another factor,
probably an Y-linked component, regulates the magnitude of the antibody response
produced by male mice.

References

1. Renoux, G., Renoux, M., Tinelli, R. : Infect. Immunity 2 : 1 (1970).
2. Renoux, G., Renoux, M., Tinelli, R. : J. infect. Dis. 127 : 139 (1973).
3. Renoux, G. : Arch. Inst. Pasteur, Tunis 35 : 87 (1968).
4. Renoux, M., Renoux, G., Dubois, M. : Ann. Inst. Pasteur 115 : 978 (1968).
5. Fischer, R. A. : Statistic Methods for research works (Oliver and Boyd,
   Edinburgh 1946).
6. Gasser, D. L. : J. Immunol. 103 : 66 (1969).

7. Dickie, M. H. In: Green, E. L. : Biology of the Laboratory Mouse, p. 23 (McGraw-Hill, New York 1966).
8. McDevitt, H. O., Benacerraf, B. : Adv. Immun. 11 : 31 (1969).
9. McDevitt, H. O., Bechtol, K. B., Grumet, F. K., Mitchell, G. F., Wegmann, T. G. : In : Amos, B. : Progress in Immunology, p. 495 (Academic Press, New York 1971).
10. Benacerraf, B., McDevitt, H. O. : Science 175 : 273 (1972).
11. McDevitt, H. O., Sela, M. : J. exp. Med. 122 : 517 (1965).
12. Martin, W. J., Maurer, P. H., Benacerraf, B. : J. Immunol. 107 : 715 (1971).

Table I.

Early Secondary Responses to ABS

Total Antibodies

Strain	Males								Females										
	0	10	20	40	80	160	320	640	0	10	20	40	80	160	320	640	1280	2560	5120
B6	5																		
C3						2	6	2			2	3		6	7		3	5	2
(B6.C3) F1			1	4	13	2	6					1	5						
(C3.B6) F1.				4		10	6												
(B6.C3)(B6.C3)F2	3	4		1	3	6	1	2			3	2			2	6	1	1	
(B6.C3) F1 x B6					3	4	3	2						3	4		2	3	3
(B6.C3) F1 x C3	4	10	3		1	6					6	2		3	6	1			
C3 x (B6.C3) F1	5	1						6	3					2	2	1			

7S Antibodies

Strain	Males						Females						
	0	10	20	40	80	160	0	10	20	40	80	160	320
B6	5												
C3									3	3	5	5	
(B6.C3) F1	4		6	8	2				4	8	1		
(C3.B6) F1	8		6	8	4				4	7	2		
(B6.C3)(B6.C3)F2	9		2	3	2	2	4			3	4	2	1
(B6.C3) F1 x B6		10	4	5	2					3	7	2	
(B6.C3) F1 x C3	9	10	2	3			7		4	6	2	2	
C3 x (B6.C3) F1	10	2					1						

Table II.

Late Secondary Responses to ABS

Total Antibodies

Strain	Males								Females									
	0	10	20	40	80	160	320	640	0	10	20	40	80	160	320	640	1280	2560
B6	5																5	5
C3							3	2										
(B6.C3) F1		2	6	14	3	1				5	3	5	2	3	5		5	
(C3.B6) F1				4	8	2	3				4	4	4	1				
(B6.C3)(B6.C3)F2	9				3	1	3			4	3	1	2	3	4		5	2
(B6.C3) F1 x B6		5			3	3	2						3	8	1		1	
(B6.C3) F1 x C3	8		10	4	2					4		4	3		1			
C3 x (B6.C3) F1	6			4	2					3	1	1				3		

7S Antibodies

Strain	Males								Females									
	0	10	20	40	80	160	320	640	0	10	20	40	80	160	320	640	1280	2560
B6	5		3	5	2										5	3	2	
C3																		
(B6.C3) F1		3	14	3						5	2	2	5	4	4			
(C3.B6) F1		6	5	8	1						4	4	4	1				
(B6.C3)(B6.C3)F2	9		3	1	3					4		3	4		4			
(B6.C3) F1 x B6		6	6	1							7	3	3	1				
(B6.C3) F1 x C3	10	5	6	3						3	7	3	3	1				
C3 x (B6.C3) F1	6		2	4						4		4						

Table III.

Relationship of Secondary Late 7S Antibody Response to ABS and H-2 Antigens in F_1 x B6 F_1 x C3 and C3 x F_1 backcross Mice

Response	Sex	H-2 antigens		
		bb	bk	kk
none	male	0/10	6/25	13/14
	female	6/10	9/18	0/9
intermediate	male	1/10	9/25	2/14
	female	4/10	2/18	7/9
high	male	8/10	10/25	0/14
	female	1/10	6/18	2/9

Table IV.

Late 7S Secondary Responses to ABS. Predicted versus Actual Findings

Hypothetical Genotypes	Mice	Predicted Range	%	Found in Range
ccY3	F2	0	50	9/16
	(B6.C3)C3		50	10/24
	C3(B6.C3)		50	6/12
BBY3 + BcY3	F2	10- 80	50	7/16
BBY6 + BcY6	(B6.C3)B6	10- 80	100	13/13
BcY3	B6.C3	10- 80	100	20/20
	(B6.C3)C3		50	14/24
	C3(B6.C3)		50	6/12
BcY6	C3.B6	10- 80	100	20/20
BB	F2	320	25	4/15
	(B6.C3)B6		50	2/12
cc	F2	10	25	4/15
	(B6.C3)C3		50	3/17
	C3(B6.C3)		50	4/ 8
Bc	B6.C3	20-160	100	13/13
	C3.B6		100	13/13
	F2		50	7/15
	(B6.C3)B6		50	10/12
	(B6.C3)C3		50	14/17
	C3(B6.C3)		50	4/ 8

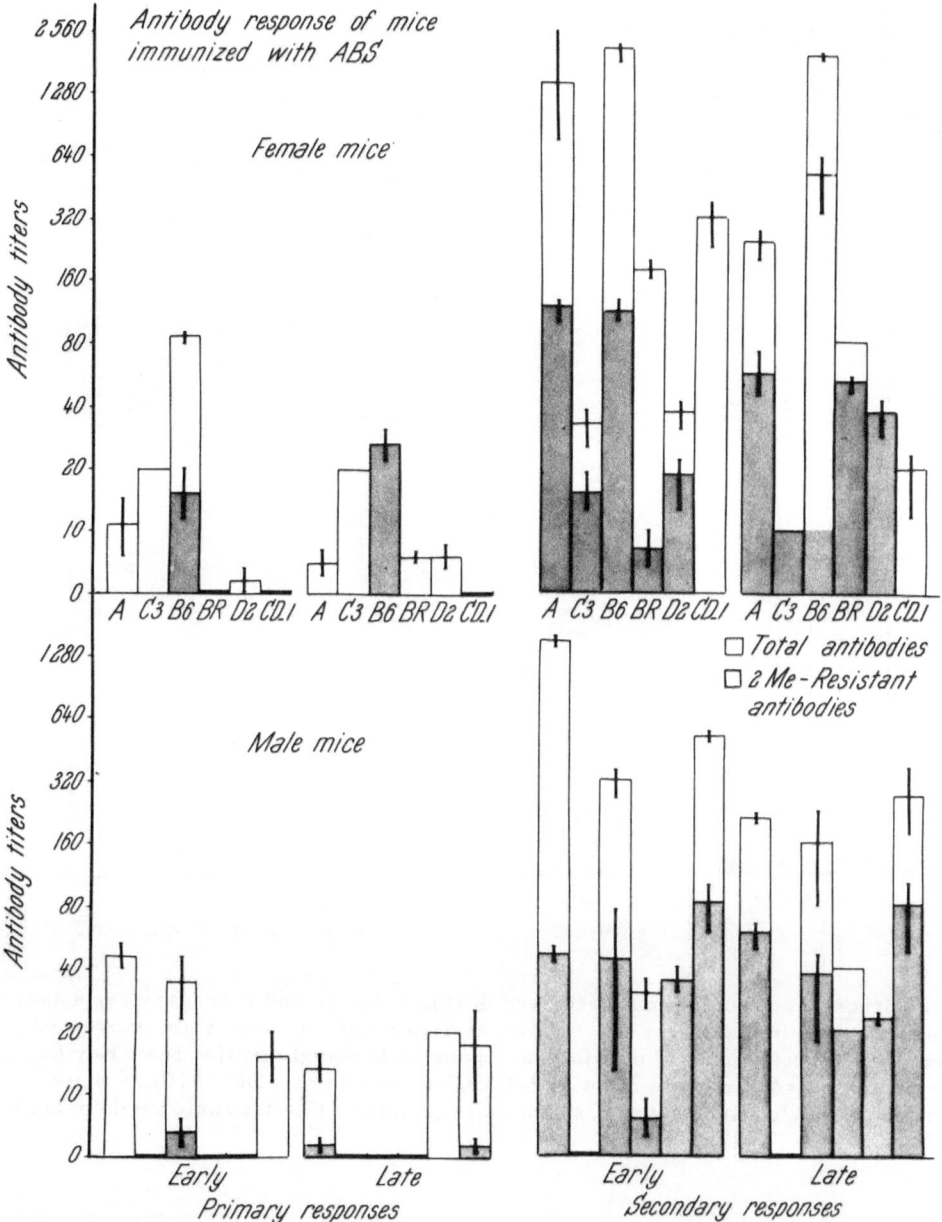

Fig. 1. <u>Brucella</u> antibodies in mouse sera during primary and secondary responses to intraperitoneal injections of 1 μg ABS at days 0 and 15. Titers are expressed as mean (± SE) reciprocals of dilutions in a passive hemagglutination test. Key for mouse strains: A=A/Orl (H-2^a) ; C3=C3H/He (H-2^k) ;B6=C57BL/6 (H-2^b) ; BR=C57BR/cd (H-2^k) ; D2=DBA/2 (H-2^d) ; CD-1=random-bred CD-1.

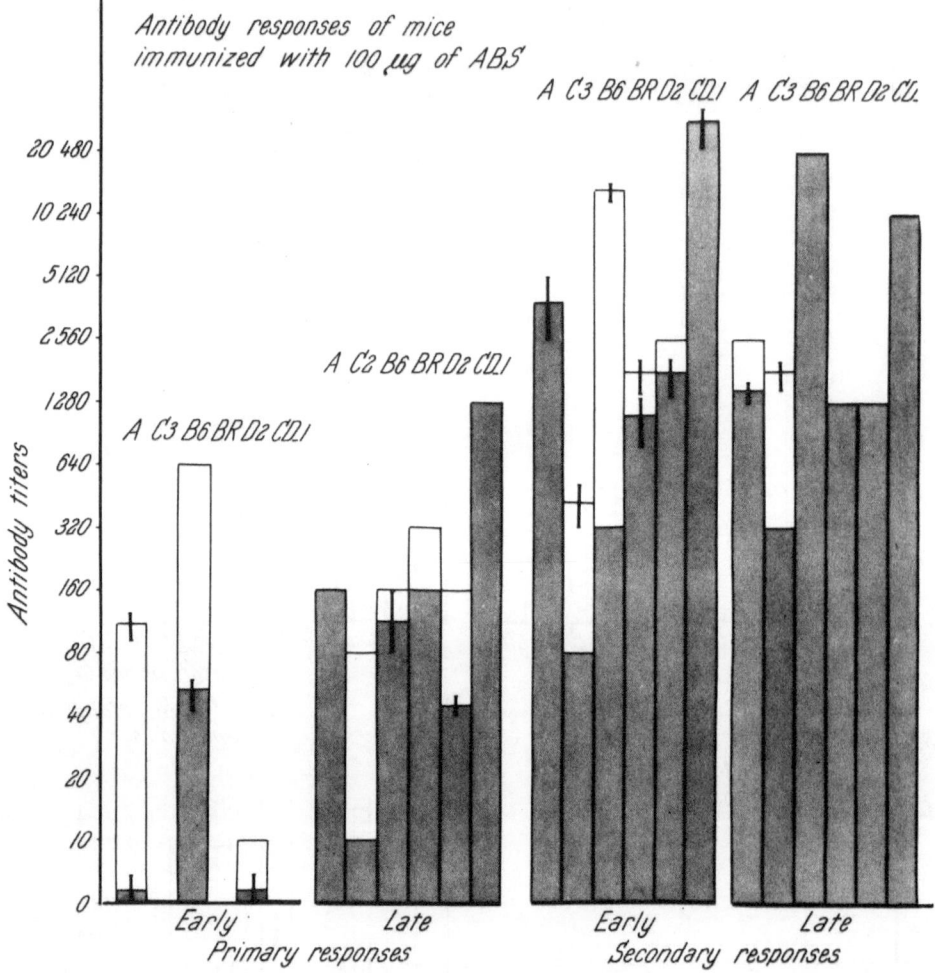

Fig. 2. <u>Brucella</u> antibodies in mouse sera during primary and secondary responses to intraperitoneal injections of 100 /ug ABS at days 0 and 15. Titers are expressed as mean (± SE) reciprocals of dilutions in a passive hemagglutination test. Key for mouse strains: A=A/Orl male mice ; C3=C3H/He male mice ; B6=C57BL/6 male mice ; BR=C57BR/cd male mice ;D2=DBA/2 male mice ; CD-1=female random-bred CD-1 mice.

Discussion

Springer: I have difficulty in finding a precise definition of what a macrophage really is.

Landy: I would think that in terms of our discussion one would regard it as a mononuclear cell with distinctive pinocytotic and phagocytic capabilities that functionally has the capacity to deal with the infectious agents we have referred to. It is a cell characterized by reaction to particulate matter by the formation of phagosomes; when appropriately activated, these fuse with lysosomes to digest and destroy bacteria, viruses, etc. This is an unconventional definition but operationally it may be a satisfactory one.

Chedid: Perhaps Dr. Springer was thinking of the hosts' populations of macrophages. There are major differences between systems using peritoneal macrophages or spleen macrophages or pulmonary macrophages; in fact there are discrepancies between experiments because of the origin of these cells. This does not in the least diminish the message; on the contrary it broadens it as we depend on several types of populations, having several kinds of activities and yielding different end-results.

Landy: Dr. Chedid's point is a most important one and may help us understand the disappointing correlates established with liver macrophages and for alveolar macrophages. The mononuclear phagocytes are a most heterogenous population of cells depending on their location and still other characteristics we know little about. The macrophages I was referring to were primarily peritoneal and to a lesser extent those in peripheral blood as reflecting the type experiments had been done.

Mergenhagen: To the list of mediators Dr. Landy identified, I would add that Rosenstreich, in our laboratory, has found now that pure B-lymphocytes stimulated with lipopolysaccharides make chemotactic factors for mouse neutrophilic leukocytes. The products of these B-cells put into chemotaxis chambers are strikingly chemotactic for mouse peripheral neutrophils.

Nowotny: For the record I would identify the experiments of Alexander which show clearly that the "arming" of normal macrophages is possible through mediators. Specific lymphocytes are exposed to tumor cells which had been used to sensitize the donors of the lymphocytes: the supernatant of this interaction is added to normal macrophages which are thereby rendered cytotoxic for these tumor cells. In quite separate experiments Alexander could render normal macrophages just as cytotoxic by exposing them to endotoxin alone.

Rietschel, Freiburg/Germany: I recall that in the differentiation of early- and late-phase tolerance in rabbits that for the induction of late-phase tolerance one needs a Boivin type preparation; a phenol-water type preparation did not work. Can Dr. Greisman tell us whether this also holds true for the human studies he has done?

Greisman: We have not tried phenol-water preparations in our human studies. In the rabbit, as you know, we have given a single dose and measured the tolerance a week later. Such phenol-water preparations as we did use were far less 0-antigenic than the Boivin products. We feel that this was the main reason for our findings. We have not done analogous experiments in man.

Chedid: Dr. Greisman did you try other pyrogens during the refractory state? Is this refractoriness broader than endotoxin refractoriness?

Greisman: No, we have not tried other agents in humans. We have tried many other pyrogens in rabbits and these animals are fully responsive to all the other pyrogens tested. We have carefully done full dose response curves, so we know we are in the sensitive range. Rabbits respond fully to all other pyrogens during the refractory stage.

Lowbury: I was wondering whether the late-phase tolerance is associated with IgM or IgG, or does it reflect a transition from one class of Ig to another type?

Greisman: We found when we hyperimmunize rabbits and look at this late phase, both IgM and IgG are associated with this tolerance and that both seem equally and highly effective.

Landy: That may be, but I would point out to Dr. Greisman that the minute doses of endotoxin he administered to his volunteers could yield only a pure IgM response.

Greisman: Yes, in man we have not studied the partition of the response into IgG and IgM. I would emphasize that in rabbits that are hyperimmunized (and only in hyperimmunized rabbits) the serum protective factors are equally divided between IgG and IgM; a single immunizing dose gives a quite different result; only IgM.

Berry: Dr. Greisman when you infect human volunteers with typhoid do you find that they have developed the early or the late tolerance to endotoxin?

Greisman: If we wait until patients convalesce from the disease, let us say a week in their convalescence, and if we have not tested them previously, they will show some degree of tolerance to endotoxin. But if we test them during the height of their illness or at any time while they are clinically sick they are always markedly, tremendously hyperreactive to endotoxin. We can, however, make them resistant by artificially giving them endotoxin.

BRAUDE: Passive Transfer of Nonspecific Resistance to Endotoxin in Gram-Negative Bacterial Infection

Landy: My question for Dr. Braude is why he has not carried out the key and seemingly obvious experiment of absorbing his protective sera with lipid A, especially as this component is now so easily prepared or readily available.

Braude: This is a very important issue and one that we have thought about a great deal. The trouble with this approach is that of contaminating test sera with lipid A or other endotoxin components. It then is no longer appropriate for our type of study because contaminating endotoxin or lipid A would itself stimulate resistance.

Landy: Would that necessarily be the case if absorption were performed with lipid A on appropriate carriers or appropriate surfaces?

Braude: That is a possibility we should consider.

Landy: I cannot see how else you will be able to resolve the specific versus non-specific character of the passively-transferred resistance.

Braude: That is true. Except that the uniformity of the protection against totally unrelated organisms possessing this common antigen, is, I think, very strong circumstancial evidence.

Rietschel: I would like to comment on this interchange. We have lipid A and anti-lipid A serum prepared by Galanos. We are able to show passive transfer with this anti-lipid A serum against the pyrogenic activity of lipid A. Pilot experiments show that it is possible to absorb the activity from the serum by lipid A or red cells coated with lipid A. In my view this is additional evidence that anti-lipid A serum has anti-endotoxin activity.

Braude: Dr. Rietschel were you able to facilitate absorption and still retain protection?

Rietschel: We have the same problems Dr. Braude mentioned since some of the lipid A may come off of the absorbing agent; there is nonetheless a definite difference between the controls and the absorbed serum despite the probability that there is some lipid A in the absorbed serum.

Müller-Berghaus: I have three questions for Dr. Braude. First, when does he inject his antiserum? Second, did he examine leukocytes, did they drop in the experiments in which he could prevent the Shwartzman reaction by antiserum? Third, did he measure complement levels since one could visualize that some complement factors are consumed in this immune reaction and that such complement factors would no longer be present for triggering intravascular coagulation.

Braude: We inject the antiserum for the Shwartzman reactions either before the preparative dose or immediately before the provocative dose; in both instances it is effective.

Müller-Berghaus: Does immediately mean ten minutes or one hour?

Braude: The interval is 24 hours, two hours before the provocative dose. We did not study the leukocytes in this system and we have not studied the complement components under these experimental conditions.

Greisman: I wonder whether Dr. Braude compared the efficacy of the antisera in these models with the mutant antiserum to ascertain the order of magnitude of differences in the protective activity with this broad spectrum protection versus the specific protection. Dr. Braude has left us with the impression that the mutant gives better protection against the homologous endotoxin than would be the case with the homologous antiserum.

Braude: I want to emphasize that the superiority of the mutant antiserum is indeed a heterologous protection. I have sought to emphasize that protection is more effective in the heterologous situation. In fact there may be no protection whatever with the smooth parent antiserum.

CHEDID

Braude: Dr. Chedid, your observation that you could not detoxify rough but could detoxify the smooth is very intriguing. One of the things I would wonder about is whether the physical state of the rough endotoxin, which is I presume more hydrophobic, may make the toxic areas less accessible to the detoxifying material in his serum-leukocyte mixture.

Chedid: I would emphasize that there are solubility problems with the rough endotoxins which may constitute a simple answer to the question, because the exposure could be incomplete. We took all reasonable precautions to use preparations which apparently were in very good suspension, and this difference was seen nonetheless.

Braude: How about hybrid endotoxin? Has it been studied from this point of view?

Chedid: Yes. And that is a rather good point because when one hooks smooth endotoxins on to rough endotoxins, they then become detoxifiable. As you know rough lipopolysaccharide is weakly immunogenic, whereas smooth lipopolysaccharide is markedly immunogenic for the rabbit. By hooking a hybrid rough endotoxin on smooth endotoxin one gets antibodies to smooth and to rough. So there must be an interesting interplay between the exposure of the molecule and perhaps even its capacity to penetrate cells.

Landy: Dr. Chedid, as you know, there is quite a literature by Rudbach and others which attests to serum detoxification being fully reversible. What is your interpretation of these divergent findings?

Chedid: My impression is this serum effect as such, is indeed reversible; but in the host this is followed by cellular changes which make the change an irreversible one.

PUSZTAI-MARKOS

Landy: Dr. Pusztai-Markos, you referred to alteration in the number of peritoneal exudate cells, but you did not indicate qualitatively what types were involved. Did these changes and numbers primarily involve macrophages or granulocytes?

Pusztai-Markos: Qualitatively there are certain differences using different endotoxins, for instance endotoxin from Sh. flexneri, E. coli or S. typhi. In general, the number of macrophages of mononuclear cells is increased and to a lesser extent the number of leukocytes. Following stimulation with endotoxin these changes disappear on the third day. Only in the case of infecting with bacteria on the third and fourth day, is there an extreme increase, primarily in the number of macrophages.

RENOUX

Landy: My question to Dr. Renoux is whether you have thought to determine if reactivity to these somatic components, in terms of their pharmacologic effects, is likewise controlled genetically? That is the urgent issue we would like to see resolved, as we still are uncertain whether immunological phenomena constitute the basis of this reactivity.

Renoux: We have not done these experiments but we are quite sure that some sex factors at least and probably other genetic factors are related to nonspecific resistance. While it is quite evident that some genetic factors do operate, most work thus far has been done on random bred animals, and there is consequently a dispersion in the results. We have indeed found that in NMRI mice, the female response is different from the male response with regard to the nonspecific immunity induced by Brucella.

Chapter IV

Pathophysiology

Nature and Specificity of Human Erythrocyte Membrane Receptor for Bacterial Endotoxins*

G. F. Springer, J. C. Adye, A. Bezkorovainy, and B. Jirgensons

Endotoxin (= lipopolysaccharide = LPS[1] or O antigen) of gram-negative bacteria in minute quantities produces numerous noxious effects (2, 11), for which attachment to host tissue is a prerequisite. LPS has the ability to fix to human red cells in vitro (cf. 14) and under extreme conditions in vivo (1, 3, 15). A receptor for LPS attachment has been isolated for the first time from the membrane of erythrocytes (17, 18, 19). It interacts with all LPS preparations tested and with the related Kunin antigen but not with other antigens of gram-negative or gram-positive bacteria and hereafter is referred to as LPS-receptor. Other compounds, such as glycolipids, lipoproteins and basic proteins also combine with LPS but are less active and so far as investigated non-specific (23).

The LPS-receptor, a major component of human red cell membranes without significant blood group ABH(0) Rh_0(D) or MN activities, was now obtained under mild conditions in physico-chemically homogenous form. It is a lipo-glycoprotein rich in N-acetyl-neuraminic acid (NANA), galactose and hexosamine, possessing 61 % peptide. Its molecular weight was 228,000 and that of its subunits 103,000 and 33,000 (20). The intact substance prevented attachment to erythrocytes of unheated and heated, smooth and rough endotoxin of all gram-negative bacteria tested. The receptor physically and reversibly blocked those lipopolysaccharide groupings which attach to red cells. It removed LPS fixed to red cells. Citraconylation of the LPS-receptor produced inactive heavy and light subunits; decitraconylation of the former restored high activity but the light subunits remained inactive. The heavy subunits had a chemical composition similar to intact LPS-receptor but the light subunits had significantly less carbohydrate. Circular dichroism spectra of the intact LPS-receptor showed β-conformation and about 15 % α-helix. The rest of the macromolecule had a somewhat flexible aperiodic conformation. Removal of lipid resulted in some disorganization. Addition of sodium dodecyl sulfate (SDS) restored the order almost completely. Receptor activity was greatly decreased by proteases, papain produced dialyzable fragments with some activity. Lipid, N-acetyl-neuraminic acid, galactose and hexosamine were not involved in the activity (20).

Isolation and characterization of LPS-receptor. Human erythrocyte stroma was prepared from donors of all blood groups and types as described previously (16). Aqueous suspensions of stroma were homogenized in a Waring blender and extracted as follows:

Packed red blood cells
washed with 0.85 % aqueous NaCl,
water hemolysis.
↓
1. Erythrocyte stroma
   1 % aqueous suspension, homogenization
   Waring blender; butanol: H_2O (1:1) extraction,

* This work was supported by a Grant-in-Aid from the American Heart Association and with funds contributed in part by the Chicago and Illinois Heart Associations 73-875, and National Institutes of Health Grant Al 11560, by the National Science Foundation Grant GB-36008 and Robert A. Welch Foundation (Houston) Grant G-051.

↓ 16 hr, 4°C, pH 8.0. Activity in aqueous phase.

2. Crude LPS-receptor (4.4 % of 1.)

> Activity increase over 1. approximately
> 1,000 %.
> Centrifugation 33,000 - 151,000 g.
> >90 % activity in syrupy interphase.
> Sepharose 4B column, 0.05 M Tris, pH 7.0;
> ↓ peak after void volume.

3. Purified LPS-receptor (1.7 % of 1.)

> DEAE-Sephadex A-25 column. Elution 0.05 M
> Tris, pH 7.0 with NaCl. Stepwise
> increase, receptor recovered at 0.2 M NaCl.
> Sucrose density gradient 5 % - 20 % in 0.01 M PO^{-3}
> 58,600 g Sephadex G-200 column, 0.05 M Tris,
> ↓ pH 7.0, receptor in void volume.

4. Highly purified LPS-receptor (0.4 % of 1.)
   Activity increase over 2. approximately 400 %.

Most of the receptor activity resided in the aqueous phase from which the receptor was isolated and purified as depicted above. The increase in activity of highly purified receptor over the stroma suspension was 40- to 50-fold.

General and Physicochemical Properties of LPS-Receptor. The highly purified receptor was a white powder readily soluble in water or buffered saline up to 6 % at 1° C, the dry receptor retained full activity for several years and in physiological solutions for at least one week at 4° C and for 72 hr at 37° C. It lost about 40 % of its activity upon incubation at 56° C for 6 hr. Heating in a boiling water bath or autoclaving at 121° C destroyed all receptor activity in <15 min. Pre-incubation with LPS did not protect it from inactivation. LPS-receptor lost 8.48 % water on drying at 80° C; it contained 0.49 % ash and <0.02 % P. The receptor possessed no human blood group A_1, B, M or Rh_0 (D) activity when tested at 10 mg/ml. It possessed traces of A_2 activity giving inhibition at 2.5 - 5.0 mg/ml under standard conditions; it had similar activities when measured with human or eel anti-human blood group H(0) sera and with rabbit but not with human anti-N sera. The receptor was non-pyrogenic at concentrations of at least 1 mg/kg body weight (21).

Physical Properties. The LPS-receptor was homogeneous and migrated with a negative charge between pH 5.0 and pH 9.0 on acetate; its isoelectric point was close to pH 3.0. The receptor also showed a single band after electrophoresis in polyacrylamide gel using the non-dissociating buffer. It appeared to be homogeneous in the ultracentrifuge (Fig. 1). Hydrodynamic parameters determined for the intact LPS-receptor are listed in Table I. The receptor is either highly asymmetric or highly hydrated as indicated by the frictional ratio f/f_0.

The molecular weight of the receptor determined by sedimentation, diffusion and partial specific volume was 256,000. Polyacrylamide gel electrophoresis in the non-dissociating buffer, a method with a known error of ± 10 % (22), was in accord with this result, since a molecular weight between 250,000 - 260,000 was found for 3 different receptor preparations.

On polyacrylamide gel electrophoresis in the mildly dissociating buffer system, the receptor separated into 2 bands (called heavy and light fragment) whose migration is indicated in Fig. 2. Molecular weights of the light and heavy fragment, measured from the leading edges of the bands, were 33,000 ± 3,000 and 103,000 ± 10,300 respectively. Densitometric tracing of the stained gels indicated that the color ratio between the 2 components was 1.1:1 in favor of the light component.

The LPS-receptor was therefore dissociated for preparative purposes by citraconylation. Ultracentrifugal analysis of the citraconylated material showed 2 fragments (called large and small). These were separated on Sephadex G-200. The 2 major fractions were ultracentrifugically homogeneous. The physical properties of both fragments are given in Table I.

Circular dichroism spectra (13) in the far ultraviolet are shown in Fig. 3, curve 1 which indicates that the LPS-receptor is a macromolecule of considerable conformational order which appears to be some pleated sheet (beta), about 15 % α-helix and some unspecified loop and bend conformation (4). The presence of α-helix and pleated sheet is indicated by the positive band at 190 - 195 nm and by the sloping plateau from 215 - 225 nm. Sialic acid may also contribute to the positive band at 190 - 195 nm (8). If 110 is taken as the mean amino acid residue weight, a residue molar ellipticity of -5,100 at 222 nm is compatible with 12 - 15 % of α-helix. In the near ultraviolet only a faint negative band could be obtained at 265 - 285 nm.

Removal of lipid resulted in significant loss of conformational order. A weak band at 190 - 195 nm and the negative band at 205 - 220 nm indicate that some order still remained. Upon addition of SDS to the delipidated receptor, the conformation was recovered almost completely (Fig. 3, curve 3).

Chemical composition of Receptor and Receptor Fragments. The composition of LPS-receptor and of its fragments are depicted in Tables II and III. Averages of 2 analyses each are given. It must be pointed out that the component fragments originated from a different receptor lot than that whose analysis is given in Tables II and III. The peptide part of the intact LPS-receptor amounted to 61 %, with mono-amino dicarboxylic acids and hydroxyamino acids predominating. There was scarcity of Trp and sulfur-containing amino acids. Cystine-S was present only in traces.

NANA predominated by a factor of about 1.5 (molar basis) over the next common sugar galactose (Gal). There were approximately 10 % more $\underline{\underline{D}}$-galactosamine (GalN) than $\underline{\underline{D}}$-glucosamine (GlcN) and in other receptor preparations there were up to 28 % more GalN.

The acetyl content was 288 moles acetyl-mole receptor which is in good agreement with the theoretical acetyl content of 255 moles/mole receptor assuming monoacetylation of all NANA and hexosamines, especially since no allowance was made for hexosamine destruction during hydrolysis. The receptor contained nearly 10 % phosphorous-free lipid (Table III) which was non-covalently bound and whose removal did not affect the activity of the residual receptor; the extracted lipid has <10 % activity of the receptor. No significant quantity of lipid was released upon hydrolysis of the receptor.

Biological Properties

Inhibition of LPS coating of red cells. The inhibitory activity of the LPS-receptor (determined by serologic means) and the range of its specificity are shown in Table IV. It can be seen in the last column of Table IV that the inhibitory activity of the LPS-receptor is quite specific, confined to O antigens and the related Kunin antigen. Closely similar receptor quantities (around 25 /ug/ml) were needed for inhibition of coating by boiled bacterial suspensions and isolated LPS (Table IV, column 3). That the ratio was lower for whole bacteria than for isolated LPS as shown in the last column is due to the contribution of inert material in coating experiments with boiled, whole bacteria.

The LPS-receptor exhibited activity not only towards all LPS preparations tested but also towards the protein-LPS antigens investigated. Although, in some instances, larger quantities of O antigen of a given bacterium were needed when protein-LPS was used instead of LPS, the amount of receptor needed to prevent coating of red cells was proportionally much smaller for the former.

The inhibition given by the receptor was physical and not enzymatic, since incubation of LPS-receptor mixtures for 38 hr and subsequent receptor destruction by autoclaving (see below) indicated no change in coating activity and serology of LPS as compared with LPS incubated and autoclaved without receptor. No decrease in the serologic activity of LPS coated onto red cells was observed after their exposure to 2 U of receptor under standard conditions for 45 min.

Table V lists the activities of highly purified LPS-receptor and its fragments as determined in the ^{32}P assay. It can be clearly seen that citraconylation led not only to fragmentation of the receptor, but also to decrease in activity of the large fragment by about 80 %, decitraconylation largely reversed this effect; the small fragment did not regain any activity.

The activities of substances that we found to possess significant coating-inhibiting activity are compared, on a strictly quantitative basis in Table VI. The inhibitors are listed in descending order of efficiency, and the high activity of the LPS-receptor as compared with all other active substances is clearly indicated. Some poorly soluble substances such as cholesterol, which had previously been reported to inhibit erythrocyte coating, are not shown in Table VI; they possessed <1 % of the activity of the LPS-receptor. Acrolein showed only a very small effect; 10,000 /ug/ml inhibited uptake of LPS by only 20 % (Table VI), even though its inhibitory effect as measured by serologic methods had appeared to be very much higher. The reason for this seeming discrepancy is that among the various inhibitors only acrolein, ganglioside, and neomycin attach to red cells and thus significantly interfere with coating by LPS thereafter. All of these results were strictly reproducible except those with neomycin, where the effects sometimes were considerably fainter or even stronger than recorded in the table.

None of more than 200 compounds, including all amino acids, proteins such as casein, γ-globulins, phospho-proteins, egg-albumin, homo- and heteropolysaccharides, blood group glycoproteins ABH(0) and MN, monosaccharides, DNA and RNA, possessed any inhibitory activity.

Mode of LPS-receptor action. It was shown in the foregoing that the receptor appeared to interact only with the LPS; its action was confined to prevention of fixation of LPS to the cell surface. Therefore, we investigated whether or not the receptor permanently blocked the combining sites on the LPS molecule. The binding of LPS to any of the three inhibitors listed is reversible, and some LPS is transferred from the inhibitors to the red cells (Fig. 4). This transfer amounted to about 40 % per 3.8 U of receptor during incubation for 5 hr. The transfer of LPS from both hemoglobin and ganglioside was much faster and exceeded 40 % after 1 hr. After 5 hr the transfer exceeded 80 % for hemoglobin but had curiously decreased for the ganglioside. This was in contrast to both LPS receptor and hemoglobin where the transfer reached a plateau. This phenomenon is explained by the fixation of ganglioside itself to the red-cell surface and the concomitant release of LPS. Additional experiments, not listed in Fig. 4, showed that under standard coating conditions, an equilibrium was reached between LPS attached to LPS-receptor and LPS transferred to red cells. Furthermore, when incubated with receptor, LPS already fixed to red cells transferred to it, again until an equilibrium was reached.

Action of Enzymes and Acrolein on the LPS-Receptor. Proteases of animal as well as plant origin such as trypsin and papain readily inactivated the receptor by >85 %, as determined serologically and by ^{32}P assays.

None of the glycosidases employed had any effect on the LPS-receptor activity even though repeated treatment with neuraminidase released >99 % of the total NANA. As was to be expected, wheat germ lipase likewise had no effect.

Lipopolysaccharides of gram-negative bacteria are among the most potent toxins known (11). As for any toxin or drug the attachment of LPS to host tissue components is a prerequisite to their action. Yet while much is known of the nature of endotoxins (10, 12) our knowledge as to the mode of action of these important substances is woefully inadequate, with the exception of the demonstration of the interaction of LPS with complement for which, however, antibody is a prerequisite (7). Furthermore, activation of the complement system is only one of the many incitements of multiple effector systems (6, 9).

The present paper reports in detail on the chemical, physicochemical and some biological properties of this lipoglycoprotein which we termed LPS-receptor and which we have now obtained in homogeneous form, as judged by a variety of physical criteria. We studied especially the effect of chemical manipulations on the receptor's endotoxin binding activity and whether or not the interaction of the receptor with LPS was irreversible.

As shown in Fig. 4 the interaction of the receptor with LPS is reversible. An equilibrium of distribution of LPS between receptor and red cells establishes itself depending on the quantity of 3 components, red cells, LPS and receptor, present.

Although the receptor is a lipoglycoprotein with a strikingly high NANA content, neither lipid nor carbohydrate appear to be involved in its activity.

It is most likely that the peptide part plays a decisive role in the inhibitory activity of the receptor. This was indicated by the receptor's heat lability, its inactivation by aldehydes as well as its susceptibility to proteases, which are known to form condensation products with the amino acids in proteins (5). The citraconylation experiments lent further support to this interpretation. In fact, these latter experiments seem to implicate lysine and, more specifically, its ε-amino groups, although it cannot be decided whether or not this effect is due to the alteration of the lysine residues or the result of an overall change of the conformation of the receptor molecule due to introduction of negative charges with resulting hydrogen bond disruption of the pleated sheet and α-helix conformations (13). This latter possibility must be seriously considered because of the considerable conformational order of the LPS-receptor.

The quaternary structure of the receptor was also revealed by the polyacrylamide gel electrophoresis in the dissociating buffer system; it indicated that two types of subunits were present: a heavy subunit with a molecular weight of near 103,000 and light subunits molecular weight near 33,000. The absence of cystine in the receptor and the lack of an effect by reduction-alkylation favor the interpretation that the subunits of the LPS-receptor are held together and the conformation is preserved by electrostatic forces.

Densitometric analysis of the receptor fragments resulting from electrophoresis in dissociating buffer revealed that, in regard to weight, both components were present in approximately equal amounts, hence it may be proposed that the native LPS-receptor consists of one heavy subunit and four light subunits with molecular weights of 103,000 and 132,000 (4 x 33,000) respectively.

The light subunit has been isolated following citraconylation, and it has been characterized after decitraconylation (Tables I - III). The large component isolated by us, however, is likely to be an incompletely desaggregated composite corresponding to 1 large and 3 small subunits as produced by electrophoresis in dissociating buffer. This interpretation is supported by the weight propositions of ca. 5:1 for the large and small components.

The high affinity of the LPS-receptor to endotoxin is remarkable because both macromolecules possess a strong negative charge; the receptor, as shown in this paper, on account of its high NANA content and LPS predominantly due to its phosphoric acid radicals (10). Removal of virtually all the NANA had a slightly, but probably insignificant, activating effect. It is equally surprising that the receptor's lipid had no effect on its LPS-binding activity and by itself possessed no acitivity.

Strong evidence has accumulated that the lipid A part of endotoxin is responsible for its attachment to tissue components (10). It is likely that endotoxin interacts with those areas of the LPS-receptor of human erythrocytes which carry hydrophobic amino acids, of which the peptide part of the receptor possesses ca. 40 %

(possibly) in clustered arrangement and favorable conformation.

The structures on the isolated receptor and those on the red cell-surface which interact with lipopolysaccharide appear to be the same in both. This is shown by the inactivating effect of acrolein on the receptor groups in both locations and by the lack of any effect on either by neuraminidase. We have proven that the attachment of LPS to red cells is reversible as is also its fixation to the receptor; the process seems to obey the laws of mass action.

We realize that LPS-receptors on other cells may differ in nature and there may be more than one kind of lipopolysaccharide-receptor. Nevertheless, we have isolated for the first time a cell-surface macromolecule which specifically prevents LPS-fixation to red cells. This macromolecule has a substantially higher affinity to endotoxin than any of the other substances investigated by us and found to be active. Our isolation of a cell-bound receptor substance which interacts with the endotoxins of gram-negative bacteria has practical clinical implications for endotoxic shock, which occurs in man after trauma, burns, septic abortion, chronic uremia, cirrhosis and radiation sickness.

From a basic scientific point of view, this receptor is likely to further the understanding of the mode of attachment of these and other toxic substances to cells and tissue components in addition to erythrocytes since any pharmacologically active substance has to attach itself first of all to cell or tissue receptors.

References

1. Boyden, S. V.: Nature (London) 171:402 (1953).
2. Braude, A.I., Carey, F.J., Zalesky, M.: J. clinical Invest. 34:858 (1955).
3. Buxton, A.: Immunology 2:203 (1959).
4. Crawford, J.L., Lipscomb, W.N., Schellman, Ch.G.: Proc. Natl. Acad. Sci. U.S. 70:538 (1973).
5. French, D., Edsall, J.T.: Advan. Protein Chem. 2:278 (1945).
6. Gewurz, H., Snyderman, R., Mergenhagen, S.E., Shin, H.S. In: Kadis, S., Weinbaum, G., Ajl, S.: Microbial Toxins, Vol. V., p.127 (Academic Press, New York, N.Y. 1971).
7. Gilbert, V.E., Braude, A.E.: J. Exptl. Med. 116:447 (1962).
8. Kabat, E.A., Lloyd, K.O., Beychok, S.: Biochemistry 8:747 (1969).
9. Kim, Y.B., Watson, D.W.: Ann. N.Y. Acad. Sci. 133:727 (1966).
10. Lüderitz, O., Galanos, C., Lehmann, V., Nurminen, M., Rietschel, E.T., Rosenfelder, E.T., Simon, M., Westphal, O.: J. Infect. Dis. 128:S17 (1973).
11. Kadis, S., Weinbaum, G., Ajl, S.J., Ed.: Microbial Toxins, Vol. V., Bacterial Endotoxins (Academic Press, New York, N.Y. 1971).
12. Milner, K.C., Rudbach, J.A., Ribi, E.: In: Microbial Toxins, Vol. IV, p 1 (1971).
13. Nakagawa, Y., Capetillo, S., Jirgensons, B.: J. Biol. Chem. 247:5703 (1972).
14. Neter, E.: Bacteriol. Rev. 20:166 (1956).
15. Springer, G.F., Horton, R.E.: J. Gen. Physiol. 47:1229 (1964).
16. Springer, G.F., Nagai, Y., Tegtmeyer, H.: Biochemistry 5:3254 (1966).
17. Springer, G.F., Wang, E.T., Nichols, J.H., Shear, J.M.: Ann. N.Y. Acad. Sci. 133(2):566 (1966).
18. Springer, G.F., Huprikar, S.V., Neter, E.: Infection and Immunity 1:98 (1970).
19. Springer, G.F., Adye, J.C., Bezkorovainy, A., Murthy, J.R.: J. Infect. Diseases 128:S202 (1973).
20. Springer, G.F., Adye, J.C., Bezkorovainy, A., Jirgensons, B.: Biochemistry, in press (1973).
21. U.S. Pharmacopoeia XVIII:886 (1970).
22. Weber, K., Osborn, M.: J. Biol. Chem. 244:4406 (1969).
23. Whang, H.Y., Neter, E., Springer, G.F.: Z. ImmunForsch. 140:298 (1970).

Table I.

Physical Properties of LPS-Receptor and Component Fragments

Parameter	LPS Receptor	Large Component [a]	Small Component [a]
$S^o_{20, w}$	6.5	6.5	1.5
$D^o_{20, w}$	2.5	2.8	5.3
V (ml/g)	0.759	0.716	0.74 [b]
dS/dC	1.6×10^{-4}		
Mol. Wt. (sed. -diff.)	256,000	202,000 [d]	28,000 [d]
Mol. Wt. (polyacrylamide)	255,000	99,000 [c] ($\pm$ 10 %)	31,500 [c] ($\pm$ 10 %)
f/f_o	2.1		
$[\alpha]^{2G}_D$	- 26.5°	- 21.5° [d]	+ 10.0° [d]
$A^{1\%}_{276 \text{ mm}}$	7.280	6.575 [d]	5.800 [d]

[a] After succinylation or citraconylation and separation on Sephadex G-200.
[b] Assumed. [c] After polyacrylamide gel electrophoresis in dissociating buffer. Intact receptor in non-dissociating buffer. [d] Decitraconylated.

Table II.

Amino Acids of the LPS-Receptor and its citraconylated Components[a]

Amino Acid	LPS-Receptor %	Moles/mole of Receptor	Large Component[b] %	Moles/mole of Component	Small Component[b] %	Moles/mole of Component
Lysine	3.59	56[c]	2.4	33	3.5	7
Histidine	2.83	41	2.3	30	1.8	3
Arginine	3.62	47	3.1	36	2.3	3
Aspartic acid	5.78	99	3.2	49	4.6	9
Threonine	4.53	87	3.4	58	2.9	7
Serine	4.51	98	3.0	58	3.0	8
Glutamic acid	7.35	114	5.2	71	7.5	14
Proline	3.68	73	2.4	42	2.2	5
Glycine	2.20	67	1.6	43	1.8	7
1/2 Cystine	trace		0.0		0.0	
Alanine	2.92	75	2.0	45	2.4	8
Valine	3.96	77	3.4	59	3.1	8
Methionine[d]	1.35	18	1.1	15	0.7	2
Isoleucine	3.03	53	2.9	45	1.8	4
Leucine	5.02	87	4.1	63	4.3	9
Phenylalanine	3.41	47	2.0	24	2.0	3
Tyrosine	2.12	28	1.4	16	1.8	3
Trytophan	1.18	13	---		---	
Total	61.07	1,080	43.5	687	45.7	100

[a]Analysis carried out on intact, exhaustively electrodialyzed receptor dried to constant weight. Citraconylated receptor not exhaustively dialyzed and only freeze-dried in order to preserve completeness of citraconylation and to avoid insolubilization. [b]After decitraconylation. [c]Nearest integer. [d]As sulfoxide.

Table III.

Composition of Citraconylated LPS-Receptor and its Components

Structural Unit	LPS-Receptor %	Moles/mole of Receptor	Large Component[c] %	Moles/mole of Component	Small Component[c] %	Moles/mole of Component
Carbohydrates						
N-Acetyl-neu-						
raminic acid	16.32[b]	135	16.22	106	9.17	8
Galactose	6.14	88	6.11	69	3.03	5
Mannose	1.46	20				
Glucose	0.11	1				
Fucose	0.92	15	0.78	10	0.92	2
Glucosamine[d]	4.78	57	3.97	45	2.27	4
Galactosamine[d]	5.38	63	6.00	68	3.49	5
Total Hexosamine	10.10	144	10.03	115	5.76	9
Acetyl[e]	2.58	153				
Total Carbo-						
hydrate	35.11		34.42			
Total peptide	61.07					
Lipids						
Non-covalent	9.61					
Covalent	< 0.33					
	108.70					

[a]Nearest integer. [b]Warren procedure, 16.67 % by resorcinol procedure. [c]Decitra-conylated. [d]Corrected for hexosamine loss during hydrolysis. [e]Exclusive of the acetyl of N-acetyl-neuraminic acid.

Table IV.

The lipopolysaccharide-receptor as inhibitor of antigen fixation to human erythrocytes.

Antigen fixed	(A) Smallest average amount of antigen affording optimal coating (/ug/ml)	(B) Smallest average amount of receptor inhibiting coating by >95 % (/ug/ml)	B:A
Gram-negative bacteria			
Suspended bacteria (9)*	22	24	1. 1
ISOLATED O ANTIGENS**(14)	3	27	9
Kunin antigens, dialyzed purified culture fluids (3)	85	125	1. 5
ISOLATED VI ANTIGENS (2)	0. 6	700	1, 170
Gram-positive bacteria			
Rantz antigens, dialyzed purified culture fluids (2)	65	4, 500	69
STREPTOCOCCUS PYOGENES ISOLATED GROUP ANTIGENS STEAROYL DERIVATIVE (2)	4	1, 750	440

*Figures in parentheses = numbers of different strains tested. **Antigens in capitals are highly purified; those in regular type are crude preparations.

Table V.

Inhibition of E. coli O_{86} ^{32}P LPS Fixation to Human Erythrocytes by Lipopolysaccharide-Receptor and its Components.

Receptor Added 5U	% Inhibition of LPS Uptake 15U
Intact	80. 46
Desialized	87. 8
Delipidized	74. 99
Large component citraconylated	17. 66
Large component decitraconylated	54. 81
Small component decitraconylated	<5

Table VI.

Inhibition of coating of red cells by ^{32}P-labeled lipopolysaccharide (LPS) from Escherichia coli O_{86}.

Inhibitor	Concentration ($\mu g/ml$)	Percentage inhibition of ^{32}P-labeled LPS uptake[x]
LPS-receptor	40	80
Serum albumin (human)	200	80
Phosphatidylethanolamine	500	80
Neomycin	1,000	75
Ganglioside	2,000	75
Asialoganglioside*	2,000	77
Human hemoglobin	2,000	82
Acrolein	10,000	20

*Prepared from bovine ganglioside (sialic acid content, 18.7 %) in this laboratory; still contained 2.0 % - 2.5 % sialic acid.
[x]25 units of LPS.

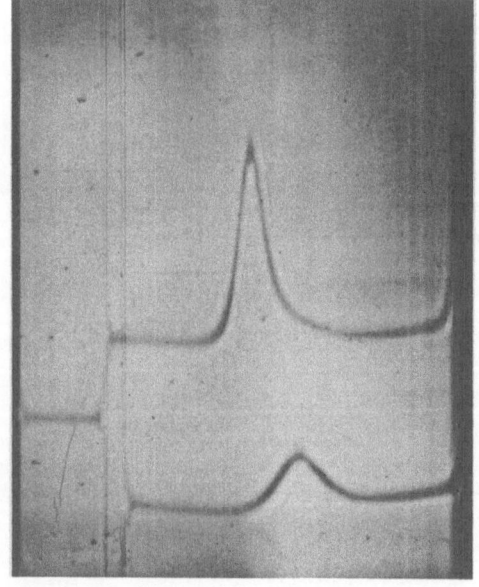

Fig. 1. 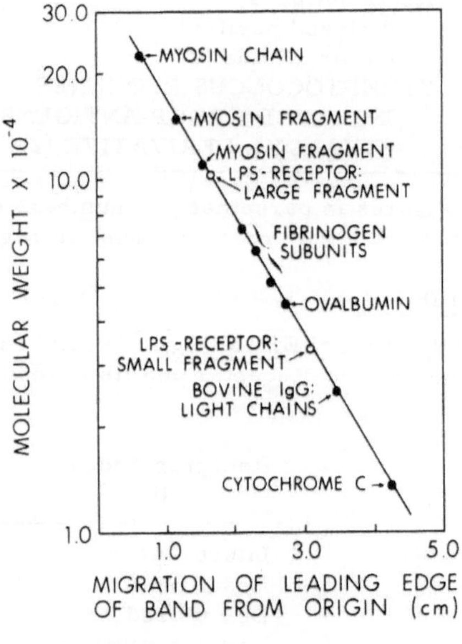   Fig. 2.

Fig. 1. Ultracentrifugation of LPS-receptor in 0.1 $\underline{N}$ NaHCO$_3$. Upper pattern - 10 mg/ml. Lower pattern - 5 mg/ml. Photo was taken 75 min after reaching speed, at a bar angle of 60°C.

Fig. 2. Migration in dissociating buffer system of various standard proteins and components of the LPS-receptor in 5 % polyacrylamide gel, 1 % SDS, 5 $\underline{M}$ Urea, 2.5 % 2-mercaptoethanol phosphate buffer, pH 7.1.

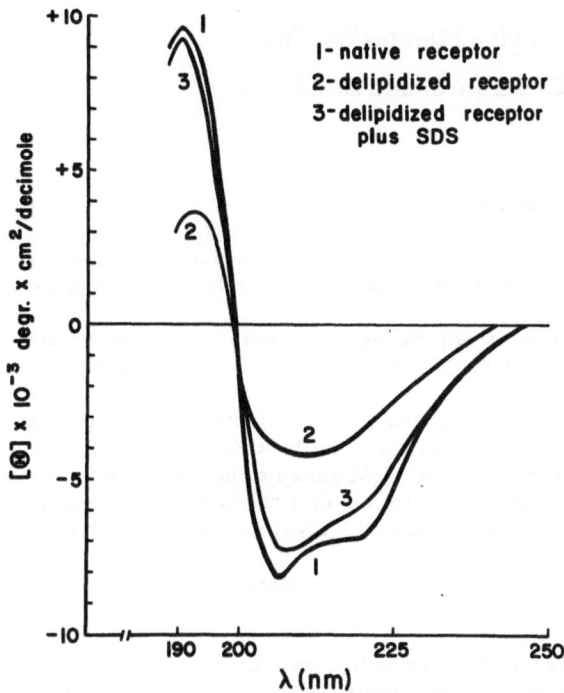

CIRCULAR DICHROISM OF LIPOPOLYSACCHARIDE-
RECEPTOR IN THE FAR UV ZONE

1- native receptor
2- delipidized receptor
3- delipidized receptor
   plus SDS

Fig. 3. Circular dichroism of LPS-receptor (lipo-glycoprotein). Curve 1, native lipo-glycoprotein 0.01 % in aqueous solution. Curve 2, delipidated glycoprotein 0.01 % in aqueous solution, pH 6.3. Curve 3, 0.0068 % delipidated material in 0.005 $\underline{M}$ sodium dodecyl sulfate, pH 6.8.

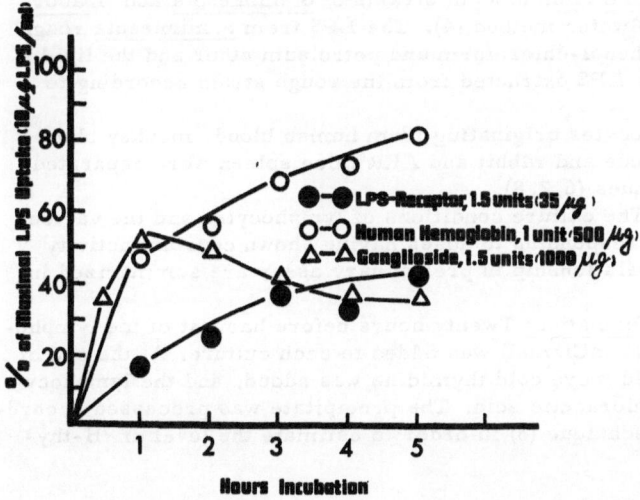

Fig. 4. Transfer of E. coli 0_{86} ^{32}P-LPS (10 /ug/ml) from inhibitors to red cells.

The Mitogenic Activity of Lipopolysaccharides on Lymphocytes in Culture

II. Comparative Study in Five Mammalian Species

C. Bona, C. Damais, C. Galanos, and L. Chedid

The administration of bacterial endotoxins to mammalian species produces various pathologic, physiologic and immunologic effects include leukocytosis, fever, the Shwartzman reaction, increase of the background of immune response against various antigens (polyclonal stimulation of B-derived lymphocytes), release of endogenous pyrogens, and death.

These endotoxins have been found to be strong mitogenic agents for mouse spleen cells in vitro (1, 2) and selective stimulants for B-derived lymphocytes. In contrast to the lectins, the endotoxins do not activate the T-lymphocytes.

According to Janossy et al. (3) lipopolysaccharide (LPS) may stimulate a subpopulation of B-dependent lymphocytes, which includes those cells which are able to take part in the relatively or fully "thymus-independent" IgM phase of humoral immune response in vitro. Actually, it was clearly demonstrated that the polyclonal, unspecific stimulation of B-cells by LPS runs in parallel with polyclonal IgM production (2).

The lipid A common moiety of all enterobacterial LPS constitutes the mitogenic part of LPS molecules (2).

In this work we made a comparative study of:

a) the mitogenic ability of LPS prepared from smooth strain of S. minnesota and B. abortus, LPS and lipid A, prepared from rough strain 595 of S. minnesota, on the lymphocytes of 5 mammalian species.

b) the ultrastructural characteristics of the human, guinea pig, and mouse lymphocytes transformed by LPS and lipid A from S. minnesota.

Material and Methods

1. Stimulating agents: The LPS from smooth strains of S. minnesota and B. abortus were prepared by the phenolwater method (4). The LPS from S. minnesota rough strain (595) was extracted by phenol-chloroform and petroleum ether and the lipid A by acetic-acid hydrolysis of the LPS extracted from the rough strain according to Galanos (5).

2. Lymphocytes: The lymphocytes originating from human blood, monkey blood and spleen, guinea pig lymph node and rabbit and AKR mice spleen were separated by previously described techniques (6, 7, 8).

3. Culture of lymphocytes: The culture conditions of lymphocytes and the various amounts of LPS and lipid A corresponding to doses having shown constant activity after establishing dose-effect relationship in preliminary assay are summarized in Table I.

4. Measuring of blast transformation: Twenty hours before harvest of the lymphocytes, 1 μCi of ^{3}H-thymidine (1 mCi/mM) was added to each culture. At the end of the ^{3}H-thymidine pulse, 100 fold more cold thymidine was added, and the lymphocytes were precipitated with trichloracetic acid. The precipitate was processed according to a previously described technique (6) in order to estimate the level of ^{3}H-thymidine incorporation.

This research was supported by D. R. M. E. grant n°73/465 and by I. N. S. E. R. M. grant n° 73/16.

The mitogenic effect of various preparations was expressed as index of stimulation (1, 5) representing the ratio of radioactivity of the lymphocytes cultivated with stimulating agents on the radioactivity of the control lymphocytes.

5. Electron microscopy: Human, guinea pig, and mice lymphocytes were harvested at the end of culture (see Table I) and prepared for electron microscopy study according to a previously described technique (9).

Results

In a preliminary study, we established the dose-effect relationship of the stimulation of lymphocytes from all 5 species investigated in this work. When the LPS or the lipid A were not mitogenic for a given species, the highest tolerated dose was chosen for all experiments. As can be seen in Fig. 1, only the mouse spleen lymphocytes were strongly stimulated by the LPS and gave a true dose-effect response.

Results seen in Table II show that the mouse spleen lymphocytes were strongly stimulated by the LPS extracted from both smooth and rough strains of S. minnesota as well as by lipid A. The lymphocytes from the other four mammalian species were stimulated very weakly by the LPS but not at all by lipid A. It is worthwhile to note that the lymphocytes from none of the 5 species studied were stimulated by B. abortus LPS.

The activation of mice, guinea pig, and human lymphocytes by LPS showed the same ultrastructural characteristics of the transformed cells, i. e. small lymphocytes, various intermediate forms between lymphocytes and immunoblasts, immunoblasts, intermediate forms between immunoblasts and plasmoblasts, plasmoblasts, and plasma cells. While the number of transformed cells was about 60 % in mouse-LPS-stimulated culture, only 2-3 % transformed cells were found in human and guinea pig lymphocyte cultures. This relative percentage of transformed cells estimated in electron microscopy corresponds to the absolute percentage of labelled cells determined by autoradiography (unpublished data). The mice-lipid A-transformed lymphocytes possess the same ultrastructural characteristics as the LPS transformed lymphocytes. In human or guinea pig lymphocyte cultures stimulated by lipid A, we did not find transformed cells. In this case numerous lymphocytes showed an unusual straying of lipic vacuoles.

Discussion

Our results are in agreement with the recent findings (2, 3) that the mitogenic property of the LPS molecule is vested in the lipid A. Both are nonspecific stimulants of mouse spleen B-dependent lymphocytes; however, the strong mitogenic effect of both the LPS and the lipid A was not observed with human or rabbit peripheral lymphocytes (6, 10).

It was of interest, therefore, to investigate whether these endotoxic preparations would stimulate the spleen lymphocytes from the rabbit and monkey (since the latter is phylogenetically close to the human). This was attempted, but the effect observed was in no way comparable to the one produced with mice spleen cells.

Several alternative hypotheses could therefore be entertained:

1. The weak mitogenic response observed in human, monkey, guinea pig, and rabbit lymphocytes could be related to the small number of B-cells in these species (11, 12, 13) as compared to the abundance of B-lymphocyte cells in mouse spleen (14).

2. The weak mitogenic effect in the four above-mentioned species may reflect specific antigenic stimulation, while in mouse the LPS and the lipid A behave as nonspecific stimulants.

3. Finally, the unique property of the mouse spleen lymphocytes to respond to the LPS could be related to the fact that the organ from this species has an haematopoietic function in the adult animal whereas this function is lacking in the other species investigated. Since the mouse lymph node lymphocytes cannot be stimulated by LPS (J. H. Humphrey's personal communication), and since the bone marrow from

the normal mouse and rabbit can be stimulated by exposure to various antigens in vitro (15), the last hypothesis is favored.

In contrast to the LPS of gram-negative bacteria, the LPS prepared from B. abortus exhibited only a very weak blastogenic ability. The absence of β-hydroxymyristic and other fatty acids from Brucella LPS indicates that the structure of its lipid A is different from that of LPS extracted from Enterobacteriaceae. However, Peavy et al. (16) have reported that β-hydroxymyristic and lauric acids made soluble by triethylamine procedure were not mitogenic when incubated in doses as high as 100 μg/ culture, indicating that the mitogenicity of lipid A is related to its structure.

Nevertheless, the cells stimulated by LPS (nonspecifically in mouse or specifically in human and guinea pig lymphocytes) exhibit a similar pattern, and this represents a true in vitro cytodifferentiation of small lymphocytes to plasma cells. Actually, we have found all the intermediate forms between small lymphocytes and plasma cells subsequent to stimulation by LPS or lipid A. It seems that the most important in vitro cytodifferentiation are as follows: small lymphocyte $\rightarrow$ immunoblast $\rightarrow$ plasmoblast $\rightarrow$ plasma cell.

It is worth noting that this cytodifferentiation takes place during a symetrical division in all the precursor cells, but only the last step – plasmoblast $\rightarrow$ plasma cell – represents a true differentiation and maturation process not involving mitosis.

These steps in vitro are similar to the sequence of cellular events in lymphoid organs from animals which received antigen in vivo (17, 18, 19, 20).

Acknowledgment

We thank E. Bourgeois for her technical collaboration.

References

1. Peavy, D. L., Adler, W. H., Smith, R. T.: J. Immunol. 105: 1453 (1970).
2. Andersson, J., Milchers, F., Galanos, C., Lüderitz, O.: J. exp. Med. 137: 943 (1973).
3. Janossy, G., Humphrey, J. H., Greaves, M. F.: Bacterial lipopolysaccharide stimulation of a subpopulation of B-lymphocytes due to lipid A, but not to complement activation (sent for personal consultation by J. H. Humphrey).
4. Westphal, O., Lüderitz, O., Bister, F.: Z. Naturforsch. 7b: 148 (1952).
5. Galanos, C., Rietschel, E. T., Lüderitz, O., Westphal, O.: Eur. J. Biochem. 19: 143 (1971).
6. Bona, C.: In vitro cytodifferentiation of human blood lymphocytes (submitted to Clin. Immunol. and Immunopath.)
7. Bona, C., Trebiciavsky, I., Anteunis, A., Heuclin, C., Robineaux, R.: Eur. J. Imm. 2: 434 (1972).
8. Bona, C., Anteunis, A., Robineaux, R., Halpern, B.: Clin. exp. Immunol. 12: 377 (1972).
9. Bona, C.: Arch. Biol. 82: 323 (1971).
10. Oppenheim, J. J., Perry, S.: Proc. Soc. exp. Biol. Med. 118: 1014 (1965).
11. Preud'Homme, J. L., Seligman, M.: Blood 40: 777 (1972).
12. Pernis, B., Forni, L., Amante, L.: J. exp. Med. 132: 1001 (1970).
13. Oppenheim, J. J., Rogentine, G. N., Terry, W. D.: Immunology 16: 123 (1969).
14. Raff, M. C., Sternberg, M., Taylor, R. B.: Nature 225: 553 (1970).
15. Mond, J. J., Thorbecke, G. J.: Cellular Immun. 5: 480 (1972).
16. Peavy, D. L., Shands, J. W., Adler W. H., Smith, R. T.: J. Immunol. 111: 352 (1973).
17. Feldman, J. D.: Adv. Imm. 4: 175 (1964).
18. Movat, H. Z., Fernando, N. V. P.: Exp. mol. Path. 4: 155 (1965).

19. Depetris, S., Karlsbad, G., Petris, B., Turk, J. L.: Int. Arch. Allergy 29: 112 (1966).
20. Thiery, Y. P.: Bull. Soc. Chim. biol. 50: 1077 (1968).

Table I. Culture Conditions of Lymphocytes

Lymphocytes	Origin of lymphocytes	Period of culture	Number of cells/ml	Medium	Supplemented with	Amount of LPS and Lipid A for 1 ml culture	
						LPS	Lipid A
Human	Bood	5 days	10^6	Eagle	10 % autologous serum	50 µg	50 µg
Monkey	Bood and spleen	5 days	10^6	"	10 % autologous serum	50 µg	50 µg
Guinea pig	Lyrmph node	3 days	3×10^6	"	20 % FCS	10 µg	1 µg
Rabbit	Spleen	3 days	2.5×10^6	"	15 % autologous serum	10 µg	1 µg
Mice	Spleen	2 days	1.5×10^6	RPMI-1640	10 % FCS	10 µg	5 µg

Table II. Comparison of Blastogenic Reactivity to LPS Lipid A in Lymphocytes of Five Mammalian Species.

Lymphocytes originating from	Organ	Number of subject tested	LPS-smooth strain S. minnesota	LPS-R-595 strain S. minnesota	Lipid-A- R-595 S. minnesota	LPS-smooth strain B. abortus
Human	blood	6	2.1 ± 0.52	1.0 ± 0.06	1.05 ± 0.08	0.95 ± 0.13
Monkey	blood	5	1.6 ± 0.3	1.7 ± 0.3	1.6 ± 0.4	1.7 ± 0.3
	spleen	5	1.3 ± 0.3	1.1 ± 0.19	1.06 ± 0.3	1.1 ± 0.2
Guinea pig	lymph node	6	2.9 ± 0.5	3.3 ± 0.62	1.16 ± 0.15	1.08 ± 0.17
Rabbit	spleen	5	2.5 ± 0.59	2.35 ± 0.65	1.5 ± 0.3	1.7 ± 0.6
Mouse	spleen	11	22.54 ± 3.9	12.0 ± 4.4	11.2 ± 3.5	2.0 ± 0.6

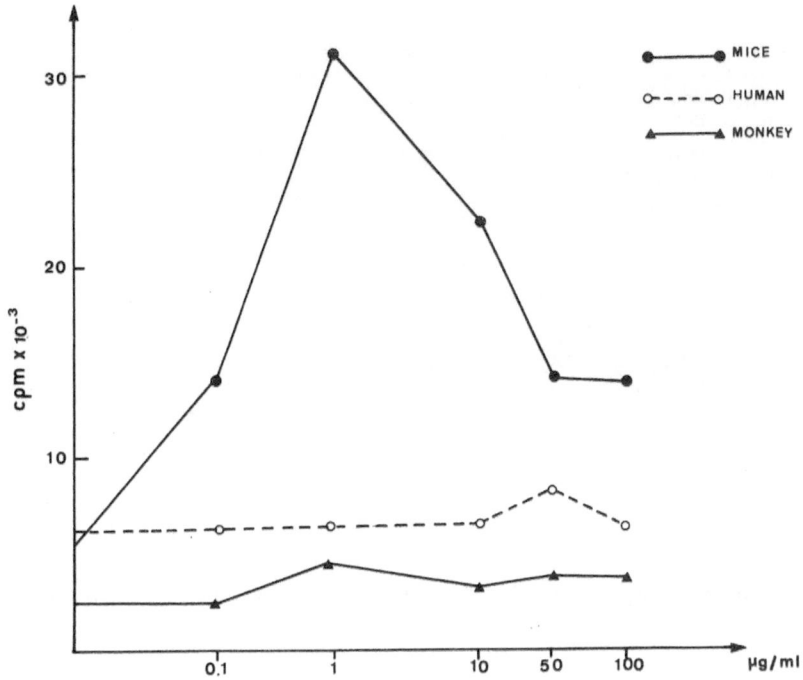

Fig. 1. Dose-response relationship of stimulation of mice, human and monkey lymphocytes by LPS.

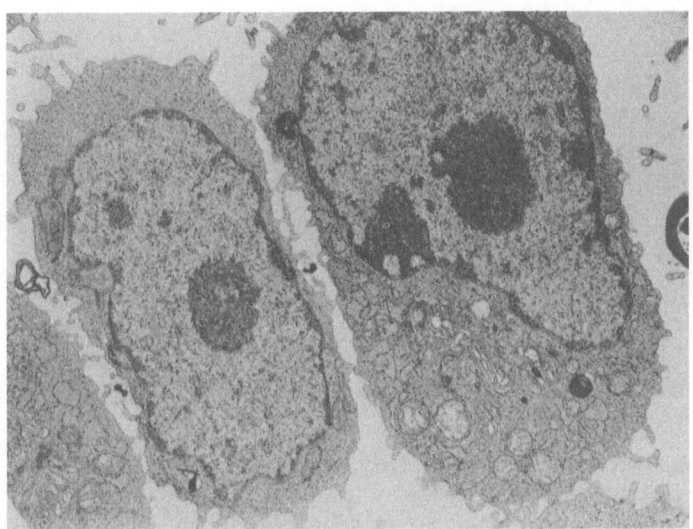

Fig. 2. Mouse spleen lymphocytes stimulated by LPS. A cluster of transformed cells constituted by immunoblast (I), intermediate forms (IF) and plasmoblasts (P). (x 9,000).

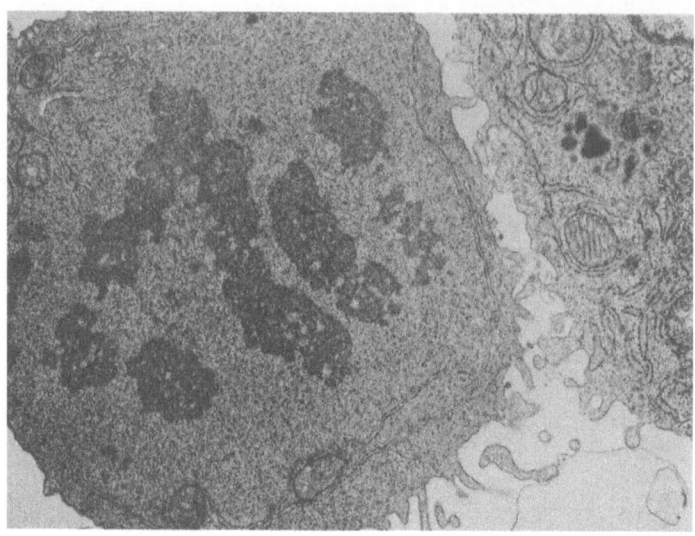

Fig. 3. Mouse spleen lymphocytes stimulated by LPS mitosis of the plasmoblast. (x 18,000).

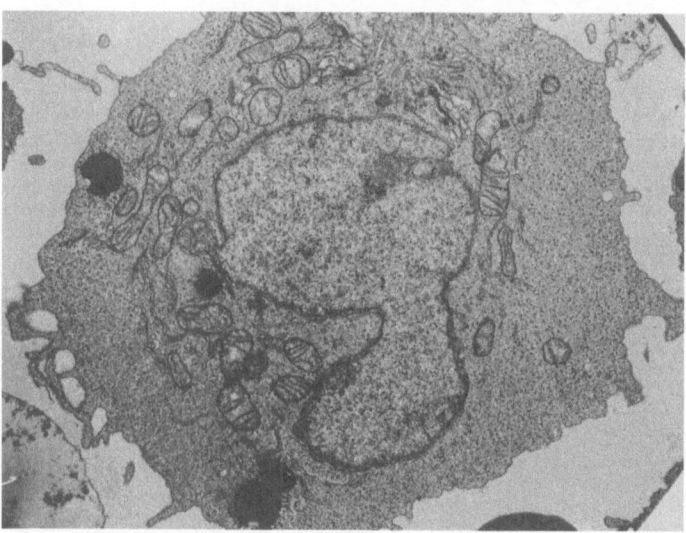

Fig. 4. Human blood lymphocyte stimulated by LPS. A transformed cells with immunoblastic characteristics. (x 18,000).

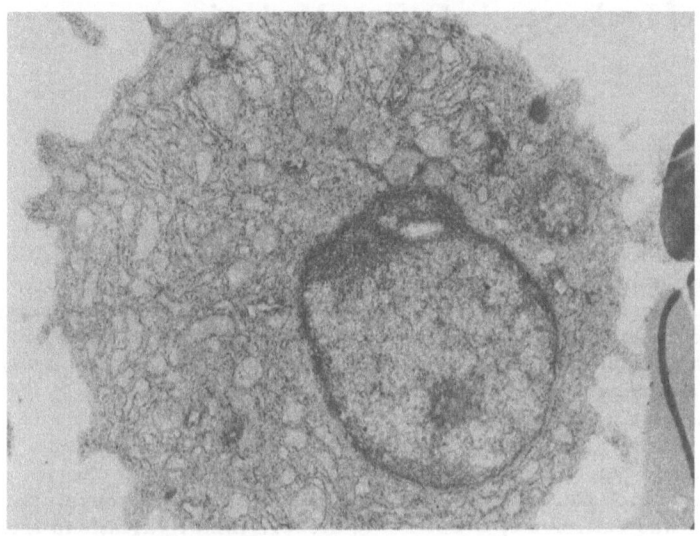

Fig. 5. Human blood lymphocytes stimulated by LPS. Plasma cell. (x 20,000).

Function of Leukocytes in Defense Mechanisms

J. W. Alexander

Neutrophils have been recognized as an essential component of host defense a-gainst infection since Metchinkoff's pioneering work at the turn of the century (1). However, critical examination of this important cell was not made until recently when renewed interest in host defense mechanisms was stimulated by a distressing awareness that sepsis continued to be one of the leading causes of morbidity and mortality. Also, increasing numbers of antibiotic resistant pathogens were emerg-ing in both nosocomial and community environments.

Studies of the antibacterial function of neutrophils started in our laboratory in 1965 after an intensive assessment of other variables of immunological defense failed to explain the marked increase in susceptibility of burn patients to the de-velopment of opportunistic infections. A more refined test for the study of this func-tion was designed which was sufficiently sensitive to discern minor variations from normal (2). Since that time, several thousand individual tests have been performed, focusing interest on normal individuals (3), patients following burn injury (3, 4), pa-tients receiving immunosuppression for renal transplantation (4), and selected pa-tients having a variety of surgical disorders (5). Studies have also been performed using several laboratory animals (6).

An attempt will be made in this review to correlate our data with data from other published investigations to show that this relatively neglected segment of host de-fense deserves more intensive investigation and to present evidence which suggests that therapeutic control of this important variable is near. The discussion will be limited to abnormalities of intracellular killing since phagocytic abnormalities ap-pear to be of lesser importance at the present time.

Inherited Abnormalities of Neutrophil Function

Berendes, Bridges, and Good (7) first described the syndrome now known as chro-nic granulomatous disease of childhood (CGD) in 1957. The initial patients had rather typical clinical features which were characterized by severe, widespread, and re-current bacterial infections during the first year of life especially associated with marked cervical lymphadenitis, pneumonias, and dermatitis (8, 9). The offending organisms were most commonly Staphylococcus, Klebsiella-Aerobacter, E. coli, and Serratia. Notably absent as etiological agents were Streptococcus, Hemophilus, and Pneumococcus. In 1966, Holmes et al. (10) discovered that patients with this disease had a profound abnormality of the bactericidal capacity of their circulating neutrophils. It was this finding, perhaps more than any other, that stimulated the intense awakening of interest in leukocyte function, especially among pediatricians. Leukocytes from patients with CGD were soon found to have an impaired ability to kill the bacteria which commonly caused their infections, but the leukocytes killed Streptococcus, Lactobacillus, and Diplococcus normally. Since the latter bacteria effectively produced their own hydrogen peroxide, a bactericidal mechanism via this pathway was suggested, and it is now believed by many investigators that it is the lack of hydrogen peroxide production within phagocytic vacuoles which is the basis for the bactericidal defect (11). Numerous studies have shown that neutro-phils from CGD patients phagocytize bacteria normally or at an increased rate and that intracellular degranulation is normal, further indicating that there is an abnor-mality of an intracellular microbicidal pathway. Hexosemonophosphate shunt acti-vity in CGD leukocytes after phagocytosis is markedly reduced. NADH oxidase activity is markedly depressed and is probably responsible for the reduced reaction with nitroblue tetrazolium. Glutathione peroxidase in some of the affected females has been normal, but in others has not (9). At first the disease was felt to be limited to males with an X-linked genetic transmission since female members of affected families had an intermediate defect with two populations of peripheral leukocytes, compatible with the Lyon effect (12). However, several female patients with similar abnormalities, were soon discovered, and in these and some of the male patients, the carrier state could not be detected in the mothers (9, 11, 13). Therefore, it would appear that several varieties of chronic granulomatous disease can occur, but the biochemical mechanism for the defect may vary. It is estimated that slightly over 100 patients with chronic granulomatous disease have been studied since its discovery.

The Chediak-Higashi syndrome in man (14, 15) begins in childhood and is characterized by photophobia, hepatosplenomegaly, partial albinism, giant lysosomal granules, and severe pyogenic infections. Infections are principally caused by the grampositive organisms. Recent studies have shown that the abnormality of granules is associated with an abnormal delay in the rate of degranulation and delayed in vitro killing of many bacteria within phagocytic cells.

Severe deficiency of glucose-6-phosphate dehydrogenase has also been described to result in fatal disease from persistent and recurring bacterial infections (16). However, the deficiency must be extreme, with only 1 % to 5 % of the normal content of enzyme before bactericidal defects occur. Leukocytes containing 20 % to 50 % of the enzyme have normal bactericidal capacity.

Myeloperoxidase deficiency has been described to be associated with increased susceptibility to infection (17), but other patients have been studied who have a complete absence of myeloperoxidase and normal antibacterial function.

Other inherited diseases have been described to have abnormal neutrophil function, but the defects have not been clearly defined.

Acquired Abnormalities of the Antibacterial Function of Neutrophils

While the inherited defects proved to be the major stimulus for closer examination of the antibacterial function of neutrophils in human disease, it is the acquired group of abnormalities which is far more important (18).

It is well recognized that malnutrition is associated with an increased morbidity and mortality from infection, but the sophistication required to study functional disorders of neutrophils was seldom available in areas plagued by malnutrition. Recently, Selvaraj and Bhat (19) showed that the malnutrition associated with kwashiorkor was accompanied by a severe abnormality of the antibacterial function of neutrophils. The abnormality showed significant improvement when the patients were treated with a high protein-high caloric diet which was supplemented with vitamins and iron. It may be of importance that they also had an iron deficiency anemia since Higashi et al. (20) demonstrated a diminished bactericidal activity in neutrophils isolated from patients with iron deficiency anemia and a concurrent neutrophilic myeloperoxidase deficiency. Other metals may be involved since plasma zinc levels in kwashiorkor have been found to be half of normal values (21). Parisi and Vallee (22) listed 29 zinc metalloenzymes and metalloproteins, some of which react with cofactors NAD and FAD. Shousha and Kamel (23) studied kwashiorkor patients and found decreased NBT dye reduction even during infection, suggesting a defect in NADH oxidase activity.

Perhaps of even greater importance in suggesting a nutritional basis for certain acquired defects in the antibacterial function of neutrophils have been our studies with thermal injury (24). Two series of patients with major burn injuries have been evaluated. The first series (A) consisted of 48 patients seen during the years 1968 to 1971, and the second series of 14 patients (B) were studied during the last year. The average age, size of burn, results of the tests, and information concerning the onset of sepsis are shown in Table I. The antibacterial function of neutrophils in this study was expressed as a neutrophil bactericidal index (NBI) which compares the function of a patient's neutrophils with the average function of neutrophils of 2 to 5 normal controls by establishing a ratio between the surviving bacteria in a patient's test and the surviving bacteria in the control tests (av.) (4). The average NBI for burn patients in series A was 5.7, indicating the presence of a rather severe deficiency. Patients in series B, by contrast, had an average NBI of 1.64, indicative of only a mild dysfunction. The average NBI was 15.9 at the onset of sepsis in series A patients, contrasted to 2.03 in series B patients. Only two of the 8 episodes of sepsis in series B patients were related to poor function of neutrophils. This dramatic difference in average neutrophil function between series A and B was associated with a significant fall of the rate of sepsis and an increase in survival of patients with these large burns. The only changes in therapy during these two periods were a more aggressive approach to excision of the burn eschar and forced nutritional supplementation by both i.v. and oral routes.

In the patients from series A, the depression of neutrophil function was significantly associated with septic episodes (Fig. 1), and patients who had septic episodes had higher values for the NBI than those who did not. In these patients, abnormalities of neutrophil function clearly preceded the onset of sepsis (Fig. 2). Abnormal function of burn leukocytes was associated with an increased rate of phagocytosis and a diminution in the ability of the neutrophils to kill ingested bacteria as measured by the addition of lysostaphin or antibiotics in the culture medium to kill extracellular bacteria, and by sequential examination of test samples by electron microscopy. Multiplication of ingested bacteria within the phagocytic vacuoles was also demonstrated by the electron microscopic studies.

Curreri et al. (25) has described a reduction in the ability of burn neutrophils to reduce NBT after stimulation with latex particles. Lennard et al. (26) have confirmed these findings and have shown that abnormal neutrophil function of burn neutrophils is associated with a decrease in NBT reduction after stimulation of the neutrophils with either latex or endotoxin. Furthermore, phagocytosis of latex appeared increased in association with higher values for the NBI. These findings are consistent with a reduction of NADH oxidase activity (11, 27, 28).

Because of the potential importance of metabolic deficiencies associated with burn injury, an attempt has been made in our laboratory to create the defect in neutrophil function by acute starvation and by protein restriction in cats. While total starvation showed no influence on the antibacterial function of neutrophils, even until the time of death, chronic protein depletion was associated with a slight abnormality in animals exhibiting significant weight loss and a decrease in serum albumin and hematocrit (29). This area of investigation obviously requires intensive study to define the important nutritional deficiencies which might be associated with abnormal neutrophil function.

Abnormal function of bactericidal capacity of leukocytes has been described in patients with diabetes mellitus, various malignancies, infections, in newborns, and following drug therapy (18, 30). A variety of isolated instances of abnormal leukocyte function have also been described which do not appear to be associated with inherited abnormalities.

Numerous drugs, including cytotoxic agents, poisons, and corticosteroids, levorphanol, sulfonamids, and others have been shown to affect the microbicidal activity of neutrophils in vitro. However, studies of their effect in vivo in clinically applicable doses have not been particularly incriminating.

Physiological Cycling of Neutrophil Function

During our early studies, the antibacterial function of neutrophils in normal individuals was noted to be somewhat variable when studied sequentially. This variability tended to occur at regular intervals with a periodicity of approximately three

weeks (3). These observations were expanded in patients with burn injury or following kidney transplantation, and the presence of cyclic variation in the antibacterial
function of neutrophils was confirmed (4, 5). The cycle was noted to occur at regular
intervals for each individual, usually between 13 and 27 days. Several animals were
also studied in a search to establish an experimental model. Cycling of neutrophil
function was found in sheep with an interval of the cycle between 11 and 30 days, in
dogs with the cycle between 14 and 28 days, in cats, between 8 and 14 days, and in
rats between 4 and 7 days (6). The regularity and length of the cycles suggested that
they might be influenced by endocrinological factors, especially since they tended to
vary in proportion with the length of the estrus cycle. Because of this, the influence
of the pituitary gland, adrenal gland, thyroid gland, gonads, and pineal body on cyclic variations on the intrinsic antibacterial function of neutrophils was studied for a
potential regulatory role. Surgical ablation of these endocrine organs and the administration of various hormones had no detectable influence on the cyclic variation
of neutrophilic function (6). It is interesting that a longer cycle has been noted in
both experimental animals and in humans with a periodicity of approximately 4 to 6
months (Figs.3 and 4).The importance of the cyclic variation is related to the change
in resistance to infection associated with the cycling (Fig. 2). In patients, nearly
all of the episodes of sepsis have had their onset during the period of the physiological cycle when a relative abnormality of antibacterial activity was present (4).

Comment

The intraleukocyte antimicrobial pathways are multiple and complex. As evidence, several investigators have shown that antibacterial activity against one organism can be normal while it is markedly abnormal for another. We have recently
shown that the physiological cycling of antibacterial activity has the same approximate periodicity for unrelated organisms (4, 29), but do not necessarily cycle in
phase. These observations provide additional evidence that several antimicrobial
pathways are operative. Whether they have a common linkage through the hydrogen
peroxide-myeloperoxidase-halide system is not determined at the present time.

Further elucidation of these pathways will be necessary before attempts for their
correction can be placed upon a sound basis. Nevertheless, the finding that similar
populations of burn patients studied during different time periods have marked differences in the overall antibacterial function of their neutrophils provides indication
that at least some of the acquired abnormalities of neutrophil function can be prevented or corrected. At the present time, it would appear that the previously described abnormality of antibacterial function in burn neutrophils was related to a nutritional deficiency which has now been prevented by more aggressive management.
Our studies in experimental animals suggest that this deficiency is not related to
caloric intake or perhaps even protein intake; instead, we believe that it is more
likely that these acquired defects were caused by abnormal vitamin or mineral metabolism. Correction of neutrophil abnormalities associated with treatment for
kwashiorkor supports this working hypothesis.

References

1. Metchnikoff, E.: Immunity in infective diseases (translated by Binnie, F.G.)
   (Cambridge University Press, Cambridge, England 1907).
2. Alexander, J.W., Windhorst, D.B., Good, R.A.: J. Lab. clin. Med. 72:136
   (1968).
3. Alexander, J.W., Wixson, D.: Surg. Gynecol. Obstet. 130:431 (1970).
4. Alexander, J.W., Meakins, J.L.: Ann. Surg. 176:273 (1972).
5. Alexander, J.W., Dionigi, R., Meakings, J.L.: Ann. Surg. 173:206 (1971).
6. Alexander, J.W., Meakins, J.L., Mitschell, M., Randall, M.J.: J. Reticuloend. Soc. In press.

7. Berendes, A.H., Bridges, R.A., Good, R.A.: Minn. Med. 40: 309 (1957).

8. Good, R.A., Quie, P.G., Windhorst, D.B., Page, A.R., Rodey, G.E., White, J.E., Holmes, B.: Semin. Hematol. 5: 215 (1968).

9. Holmes, B., Good, R.A.: J. reticuloend. Soc. 12: 216 (1972).

10. Holmes, B., Quie, P.G., Windhorst, D.B., Good, R.A.: Lancet I: 1225 (1966).

11. Karnovsky, M.L.: Fed. Proc. 32: 1527 (1973).

12. Windhorst, D.B., Holmes, B., Good, R.A.: Lancet I: 737 (1967).

13. Baehner, R.L.: Pediat. Clin. N. Amer. 19: 935 (1972).

14. Wolff, S.M., Dale, D.C., Clark, R.A., Root, R.K., Kimball, H.R.: Ann. intern. Med. 76: 293 (1972).

15. Davis, W.C., Douglas, S.D.: Semin. Hematol. 9: 431 (1972).

16. Baehner, R.L., Johnston, R.B., Jr., Nathan, D.G.: J. reticuloend. Soc. 12: 150 (1972).

17. Klebanoff, S.J., Hamon, C.B.: J. reticuloend. Soc. 12: 170 (1972).

18. Alexander, J.W., Meakings, J.L.: Int. med. Dig. In press.

19. Selvaraj, R.J., Bhat, K.S.: Amer. J. clin. Nutr. 25: 166 (1972).

20. Higashi, O., Seto, Y., Takematsu, H., Oyama, M.: Tohuku J. exp. Med. 93: 105 (1967).

21. Sandstead, H.H., Shukry, A.S., Prasad, A.S., Gabr, M.K., ElHifney, A., Mophtar, N., Darby, W.J.: Amer. J. clin. Nutr. 17: 15 (1965).

22. Parisi, A.F., Vallee, B.L.: Amer. J. clin. Nutr. 22: 1222 (1969).

23. Shousha, S., Kamel, K.: J. clin. Path. 25: 494 (1972).

24. Alexander, J.W., McClellan, M.A., Lennard, E.S., Bundeally, E.A.: Proc. of a Symp. on the Treatment of Burns, Prague. To be published.

25. Curreri, P.W., Heck, E.L., Browne, L., Baxter, C.R.: Surgery. In press.

26. Lennard, E.S., Bjornson, A.B., Petering, H.G., Alexander, J.W.: J. surg. Res. To be published.

27. Klebanoff, S.J.: Annu. Rev. Med. 22: 39 (1971).

28. Mandell, G.L., Rubin, W., Hook, E.W.: J. clin. Invest. 49: 1381 (1970).

29. Lennard, E.S., Alexander, J.W.: Unpublished data.

30. Solberg, C.O., Hellum, K.B.: Lancet II: 727 (1972).

Table I.

Relationship between Neutrophil Function and Sepsis in Burn Patients

Series	Study Period	Number Patients	Average Age	Average Size Burn % Total/30	Total and (Av) Tests Performed	Total and (Av) Days of Observation	Average NBI	Total Episode of Sepsis	Average NBI at onset of Sepsis	Incidence of Bacteremia in all Patients with Burn > 20 %. Admitted Same Approximate Period
A	1968 -71	48	15.6	53.6/39.2	573(12)	1604(33.4)	5.7	45	15.9	39.3 % of 239 Pts.
B	1972 -73	14	20.6	58.7/46.6	135(10)	368(27.3)	1.64	8	2.03	33.7 % of 83 Pts.

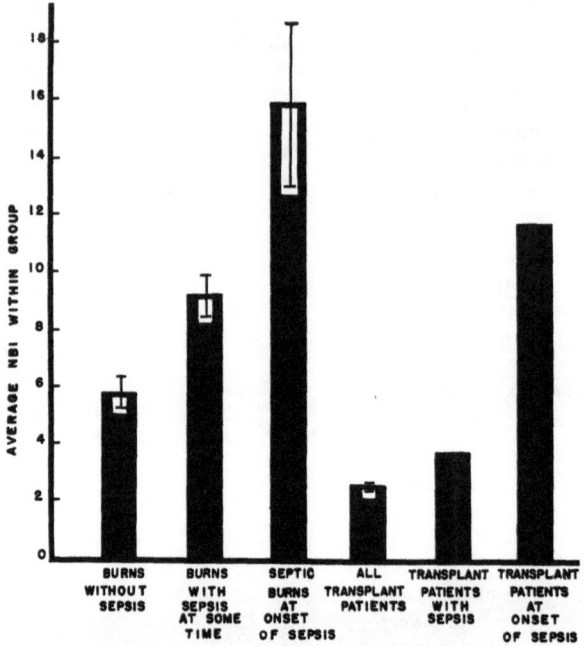

Fig. 1. Average of all determinations for the neutrophil bactericidal index (NBI) with standard errors of the mean. Twelve hundred seventy three tests were performed in 48 burn patients, and 376 tests were performed in transplant patients. Because of the small number of determinations for transplant patients with sepsis and transplant patients at the onset of sepsis, standard error of the mean was not calculated for these groups (from Alexander, J.W., Meakins, J.L.: Ann. Surg. 176:273, 1972).

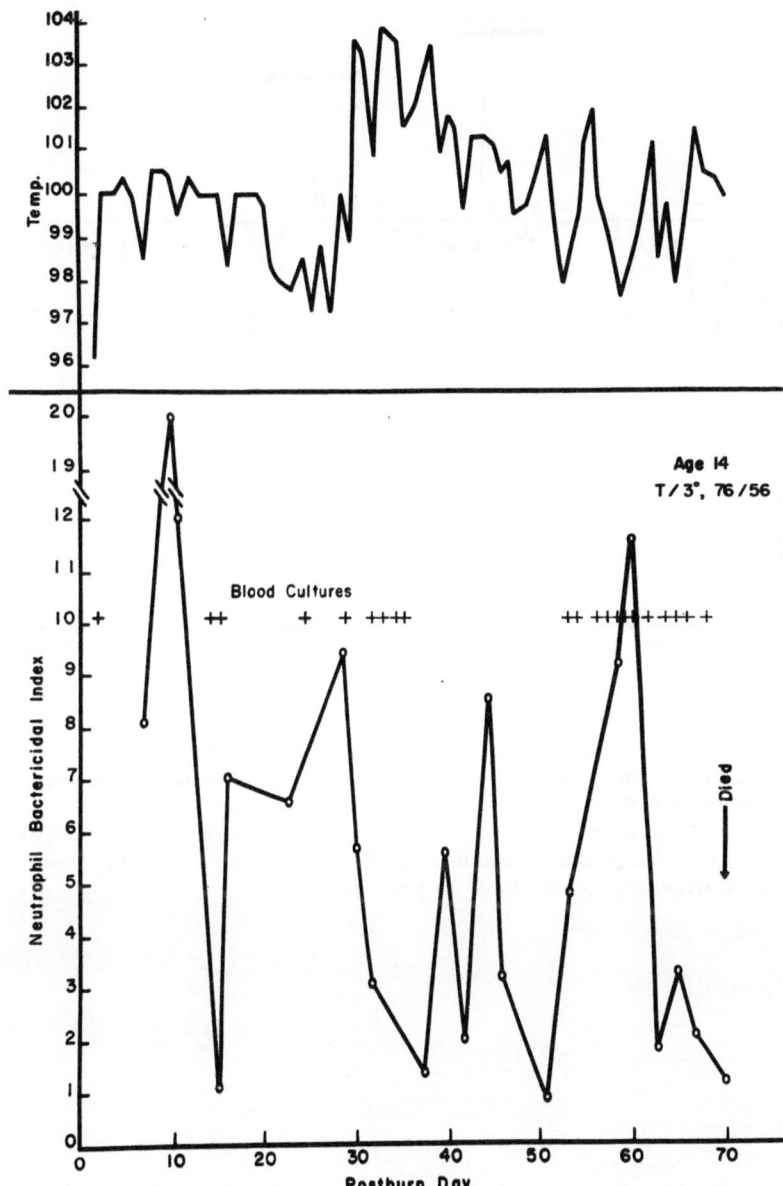

Fig. 2. Relationship between positive blood cultures and abnormalities of neutrophil function. Three distinct episodes of sepsis were observed in this 14 year old girl with a 76 % burn. Each was associated with relatively poor antibacterial function (increased NBI), but positive blood cultures were not encountered when neutrophil function was relatively better. The cyclic nature of the variation in neutrophil function is well demonstrated in this patient (from Alexander, J. W., Meakins, J. L.: Ann. Surg. 176:273, 1972).

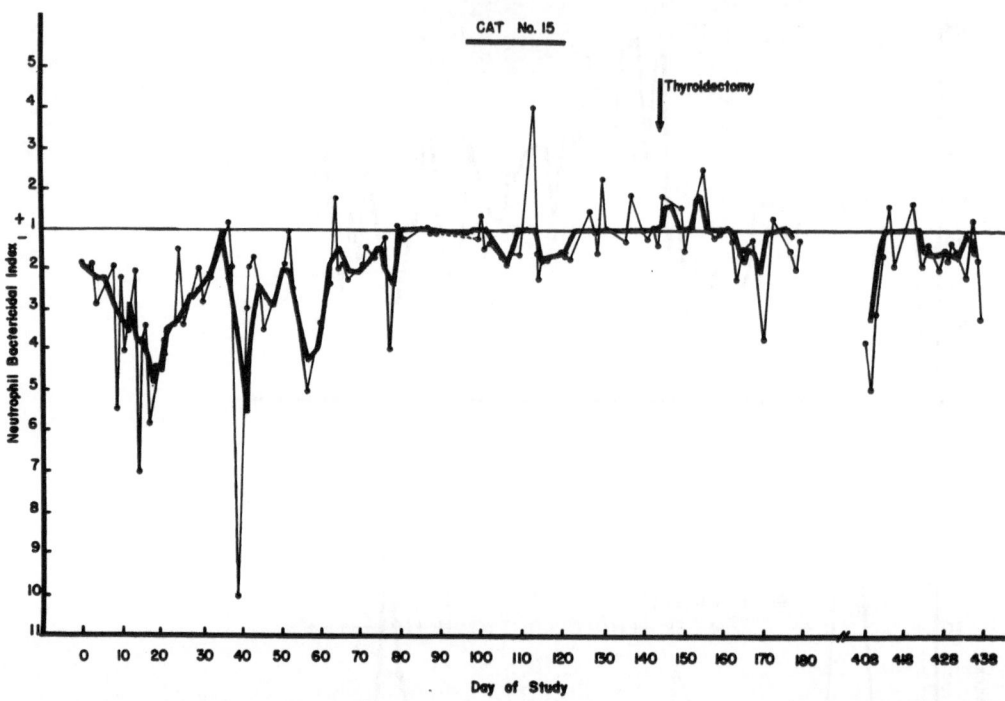

Fig. 3. Serial studies of neutrophil function from a cat. Two to nine controls were used for each data point. Individual data points are connected with a light line, and the heavy line represents a five day moving average. In addition to a cyclic fluctuation with an interval approximating the 18 days, there is a cycle with a longer trend of approximately 6 months (from Alexander, J. W., Meakins, J. L., Mitchell, M., Randall, M. J.: J. Reticuloendothel. Soc. In press).

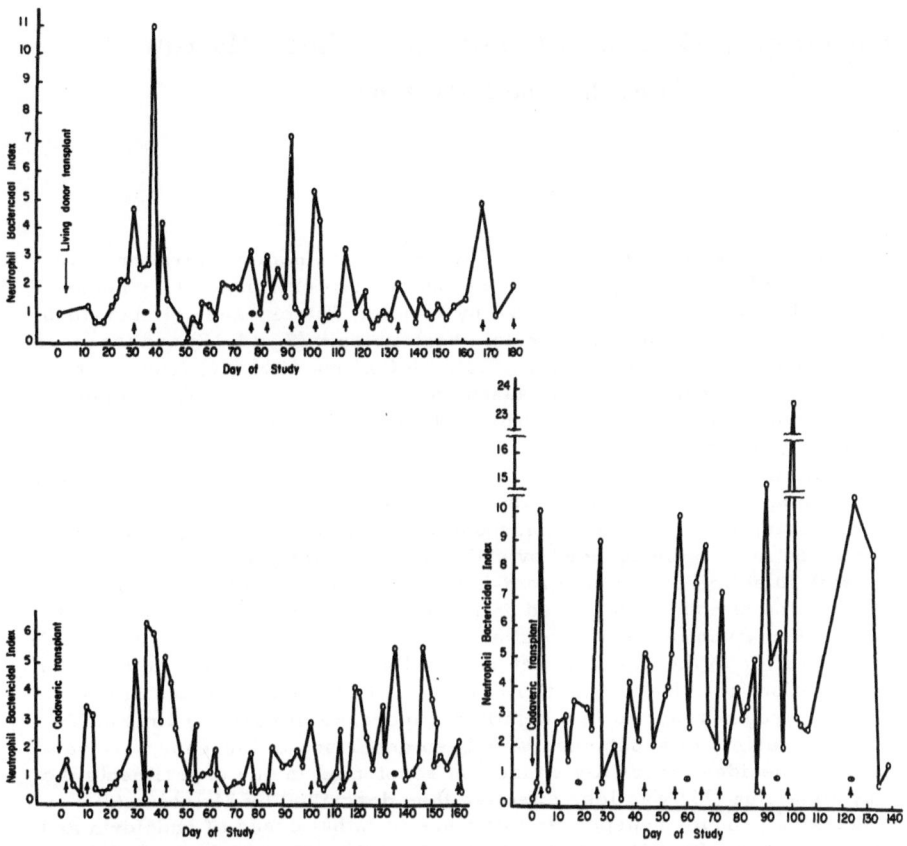

Fig. 4. Serial tests of antibacterial function of neutrophils performed on three adult patients receiving immunosuppression following renal transplantation. These patients were felt to have evidence of two cycles of different periodicity, the shorter cycle being marked by the arrows and the longer cycle by Φ. The cycle is not as clear in the patient on the right, but the other two have well demonstrated longer cycles of 60 to 100 days (from Alexander, J.W., Dionigi, R., Meakins, J.L.: Ann. Surg. 173:206, 1971).

The Effect of Detoxified Endotoxin on Bone Marrow

B. Urbaschek and R.-H. Ringert

Changes in hematological parameters following endotoxin administration have been intensively studied. As early as 1950 an induction of extramedullary hemopoiesis by bacterial pyrogens was observed by Windle, Wilcox, and collaborators (23, 24). Small doses of endotoxins stimulate hemopoiesis (7, 14-18) and increase the number of colony forming units (3, 10, 18). McCulloch et al. (6) found that multiple doses of S. typhosa endotoxin caused an increase in the number of hemopoietic stem cells present in mouse marrow and spleen that could be detected using the spleen-colony assay.

Landy and Pillemer (5) demonstrated in 1956 that injection of endotoxin produces a rapidly developing resistance to experimental infection with gram-negative bacteria in mice. Taking these findings in account Smith et al. (13) expected the mortality of irradiated mice to be reduced by endotoxin since mice are particularly prone to die with infection caused by gram-negative bacteria after irradiation.

A single injection of endotoxin caused an increase survival rate in lethally irradiated animals when given immediately after or 24 hours before irradiation (1, 13). Mice responded better to the injection before irradiation and hamsters to the injection after irradiation (13). In 1953 Mefferd et al. (7) found that repeated injections with a bacterial pyrogen resulted in a significant increase in survival time of X-irradiated mice. A variety of substances have been described to increase survival after irradiation, besides others histamine and serotonin. In our experiments the protection by histamine (11) was less effective than described by Melching (8).

Since endotoxins produce nonspecific tolerance to lethal doses of endotoxin and elicit a radioprotective effect it was obvious to study the effect of detoxified preparation of endotoxins (19) in view of a clinical application.

Pretreatment with detoxified endotoxin which was prepared with potassium methylate (9) protected animals to lethal doses of endotoxins (19, 21) and the endotoxin-caused severe disturbances in the microcirculation failed to appear (19, 20, 22). Experiments were undertaken to examine the effect of detoxified endotoxin on bone marrow and to study the radioprotective effect. Whereas in the first series of all control experiments animals died within 17 days following a whole body X-irradiation with 810 R, 75% survived after intravenous pretreatment with 100 /ug detoxified endotoxin (19). Kinosita, Nowotny and Shikata (4) demonstrated that after injection of 1 mg detoxified endotoxin mice survived a whole body X-irradiation which without pretreatment caused a lethality rate of 33% within four weeks. The results of the leukocyte count of our experiments are shown in a graph published in the paper of E. Huth in this volume. Within 20 hours the number of leukocytes dropped from an average of $8470/mm^3$ to $5110/mm^3$ after intravenous administration of detoxified endotoxin.

In two further series of experiments the radioprotective effect of endotoxin and detoxified endotoxin was studied in mice. The results of these studies concerning a comparison of the effect of different doses of endotoxin and detoxified endotoxin injected i. v. 24 hours prior to irradiation are summarized in Table I. The results using 1 mg detoxified endotoxin i. v. (prepared together with M. Frodl) at various times are listed in Table II.

Together with Ringert (11, 12) the studies on radioprotection and hematological parameters were continued using an X-ray LD 70/30 which was established prior to the experiments with 300 mice: 750 R (150 kV), 20 mA, 0.5 mm Cu, distance 30 cm, field 15 x 13.5 cm, time 9 min, 38 sec.

The following groups were compared:
1. Fourty mice, i.v. injection with 100 /ug detoxified endotoxin per mouse.
2. Fourty mice, i.v. injection with 500 /ug detoxified endotoxin per mouse.
3. Fourty mice, i.v. pretreatment with 100 /ug detoxified endotoxin per mouse, 24 hours later X-irradiation.
4. Fourty mice, i.v. pretreatment with 500 /ug detoxified endotoxin per mouse, 24 hours later X-irradiation.
5. Eighty mice, X-irradiation.
6. Fourty mice, i.v. injection with saline.

Samples of blood were taken for smears from the plexus orbitalis at day 1, 5, 14, and 30 following injection of the groups 1, 2, and 6, respectively following X-irradiation in the groups 3, 4, and 5. In group 5 no mouse survived day 30.

The statistical evaluation of the differential blood count showed a dose-dependent increase in juvenile, banded, and segmented neutrophils and monocytes after detoxified endotoxin. A decreased number of lymphocytes was observed 24 hours after detoxified endotoxin, on day 5 and 14 they were still below control values.

Results of hematologic and histologic studies of bone marrow reveal the effect of detoxified endotoxin as well as its influence on bone marrow of whole body X-irradiated mice. In the groups 1, 2, 3, 4, and 6 ten mice and in group 5 twenty mice were sacrificed at day 1, 5, 14, and 30 for bone marrow smears from the femura of the animals. The transformed mean values of the differentiated cells of the bone marrow are demonstrated in Fig. 1.

There was a decrease in number of pronormoblasts, basophilic, polychromatic, and orthochromatic normoblasts after injection with detoxified endotoxin (group 1 and 2) which was more pronounced following X-irradiation (group 5). In pretreated irradiated mice (group 3 and 4) the number of pronormoblasts and orthochromatic normoblasts was more decreased than after X-irradiation (group 5) or after injection of detoxified endotoxin (group 1 and 2). The number of promyelocytes and of neutrophilic myelocytes showed a dose-dependent increase after detoxified endotoxin which also was observed in the pretreated irradiated groups, whereas these cells decreased in number after X-irradiation. The number of metamyelocytes in irradiated mice did not at any time differ from the control values. There was a dose-dependent rise in the number of neutrophilic metamyelocytes at all times after detoxified endotoxin. Twenty-four hours after irradiation, they were lower in pretreated irradiated mice than after irradiation alone. On day 5, 14, and 30 the number of neutrophilic metamyelocytes increased markedly, compared to those of the nonpretreated irradiated mice.

The banded granulocytes were more numerous in the bone marrow after detoxified endotoxin alone as well as after irradiation alone, than in the nonpretreated irradiated animals on day fourteen. The maximum increase was observed in all groups on day five. Twenty-four hours after irradiation alone and after irradiation following pretreatment with detoxified endotoxin, a considerable increase in the number of neutrophilic segmented granulocytes was evident. On day five, the number of neutrophilic segmented granulocytes was approximately the same as the control values. On day 14 it was reduced, and on day 30 the number had increased in the pretreated irradiated mice to a greater extent than after detoxified endotoxin injection without irradiation. Irradiation induced an intense lymphopenia after 24 hours which was much less marked in the pretreated irradiated mice at this time. Whereas on day 5 and 14 a slight increase in the number of lymphocytes is noticeable after irradiation, the group of mice which received 500 /ug detoxified endotoxin before irradiation displayed values similar to those of controls. There was a large increase in monocytes after irradiation which was only moderate in irradiated mice after pretreatment with detoxified endotoxin. Detoxified endotoxin itself caused only a slight increase in monocytes at all times.

After administration of 500 µg detoxified endotoxin, the number of macrophages was increased until day 5. On day 14 and 30, these cells were reduced in number in mice after detoxified endotoxin and in pretreated irradiated mice. After irradiation alone, the number of macrophages at all times was higher than in the other groups. Irradiated mice with or without pretreatment showed an increase in megacaryocytes.

Supporting the results of bone marrow smears, in pretreated irradiated mice histological examination revealed on day 5 a more distinct cellularity than after irradiation alone. On day 14, myelopoiesis had undergone a shift to the left, while the cell count increased continuously. The observed increase in cellularity in bone marrow, increased numbers of granulocyte precursors, and increased numbers of circulating mature granulocytes implicate a stimulation of hemopoiesis and an accelerated maturation.

It should be mentioned that irradiation-induced bacteremia also occurred in mice pretreated with detoxified endotoxin. Since 70% of these animals survived, detoxified endotoxin protects these animals against the consequences of bacteremia.

The described experimental results on bone marrow are of interest in regard to the expected myelopoiesis-stimulating effect of detoxified endotoxin as pretreatment in man to polychemotherapy for instance in leukemia or morbus Hodgkin, especially since infectious complications of granulocytopenia restrict aggressive chemotherapy. The beneficial effect of endotoxin stimulation in patients with lymphoma has been described (2).

The normal maturation of myelopoietic cells in bone marrow is suppressed during proliferation of leukemic infiltrates. By stimulation with detoxified endotoxin prior to chemotherapy and during remission - as has been performed by E. Huth - it should be possible to increase the cellular defence by increasing the number of mature granulopoietic cells and monocytes.

References

1. Ainsworth, E. J., Chase, H. B.: Proc. Soc. exp. Biol. Med. 102: 483 (1959).
2. DeConti, R. C., Kaplan, S. R., Calabresi, P.: Blood 39: 602 (1972).
3. Hanks, G. E., Ainsworth, E. J.: Radiat. Res. 32: 367 (1967).
4. Kinosita, R., Nowotny, A., Shikata, T.: Blood 21: 779 (1963).
5. Landy, M., Pillemer, L.: J. exp. Med. 104: 383 (1956).
6. McCulloch, E. A., Thompson, M. W., Siminovitch, L., Till, J. E.: Cell Tissue Kin. 3: 47 (1970).
7. Mefferd, R. B., Jr., Henkel, D. T., Loefer, J. B.: Proc. Soc. exp. Biol. Med. 83: 54 (1953).
8. Melching, H.-J., Streffer, C., Ladner, H.-A., Allert, U.: Naturwissenschaften 51: 266 (1964).
9. Nowotny, A.: Nature 197: 721 (1963).
10. Reissmann, K. R., Kodetthoor, B. U., Hiroyoshi, O.: J. Lab. clin. Med. 76: 652 (1970).
11. Ringert, R.-H.: Untersuchungen zum Wirkungsmechanismus des Strahlenschutzes eines detoxifizierten Endotoxins. Dissertation, Universität Heidelberg (1971).
12. Ringert, R.-H., Urbaschek, B.: Influence of endotoxoid on myelopoiesis, lymphopoiesis and erythropoiesis in whole body X-irradiated mice. 1st meeting of the European Division of the Int. Soc. of Haematol., Milano 1971.
13. Smith, W. W., Alderman, I. M., Gillespie, R. E.: Amer. J. Physiol. 191: 124 (1957).

14. Smith, W. W. , Alderman, I. M. , Gillespie, R. E. : Amer. J. Physiol. 192: 263 (1958).
15. Smith, W. W. , Alderman, I. M. , Gillespie, R. E. : Amer. J. Physiol. 192: 549 (1958).
16. Smith, W. W. , Marston, R. Q. , Cornfield, J. : Blood 14: 737 (1959).
17. Smith, W. W. , Alderman, I. M. : Radiat. Res. 17: 594 (1962).
18. Smith, W. W. , Brecher, G. , Budd, R. A. , Fred, S. : Radiat. Res. 27: 369 (1966).
19. Urbaschek, B. : Zur Frage des Wirkungsmechanismus bakterieller Endotoxine und seiner Beeinflussung. Habilitationsschrift, Universität Heidelberg (1967).
20. Urbaschek, B. , Branemark, P. I. , Nowotny, A. : Experientia 24: 170 (1968).
21. Urbaschek, B. , Nowotny, A. : Endotoxin tolerance induced by endotoxoid. In: Chedid, M. L. : La structure et les effects biologiques des produits bactériens provenant de germes gram-négatifs, p. 357 (Editions du Centre nationale de la recherche scientifique, Paris 1969).
22. Urbaschek, B. : The effects of endotoxins in the microcirculation. In: Kadis, S. , Weinbaum, G. , Ajl, S. J. : Microbial toxins, vol. V, p. 261 (Academic Press, New York/London 1971).
23. Windle, W. F. , Wilcox, H. H. : Amer. J. Physiol. 163: 762 (1950).
24. Windle, W. F. , Chambers, W. W. , Ricker, W. A. , Ginger, L. G. , Koenig, H. : Amer. J. med. Sci. 219: 422 (1950).

Table I.

Pretreatment of mice with endotoxin or detoxified endotoxin 24 hours prior to X-irradiation.

Pretreatment	Number of dead mice until postirradiation day			dead/total
	5.	14.	30.	
Saline	-	13	1	14 / 20
Endotoxin 10 /ug	-	-	-	- / 20
Endotoxin 1 /ug	-	-	3	3 / 20
Endotoxin 0. 1 /ug	-	5	7	12 / 20
Detoxified endotoxin 1000 /ug	-	-	1	1 / 20
Detoxified endotoxin 100 /ug	1	-	2	3 / 20
Detoxified endotoxin 10 /ug	-	3	4	7 / 20
Detoxified endotoxin 1 /ug	-	3	11	14 / 20

Table II.

Injection with detoxified endotoxin (1 mg/mouse) before and/or after X-irradiation.

Groups	Time of injection			Detoxified Endotoxin	Saline
	before	after	after	Number of mice	
		X-irradiation		dead/total	dead/total
1	1 hr	-	-	10/10	10/10
2	6 hrs	-	-	8/10	10/10
3	6 hrs	6 hrs	-	10/10	10/10
4	6 hrs	24 hrs	-	4/10	9/10
5	6 hrs	6 hrs	24 hrs	10/10	10/10
6	24 hrs	6 hrs	-	10/10	10/10
7	-	1 hr	-	0/10	10/10
8	-	6 hrs	-	7/10	10/10
9	-	24 hrs	-	10/10	10/10
10	24 hrs	-	-	4/10	10/10

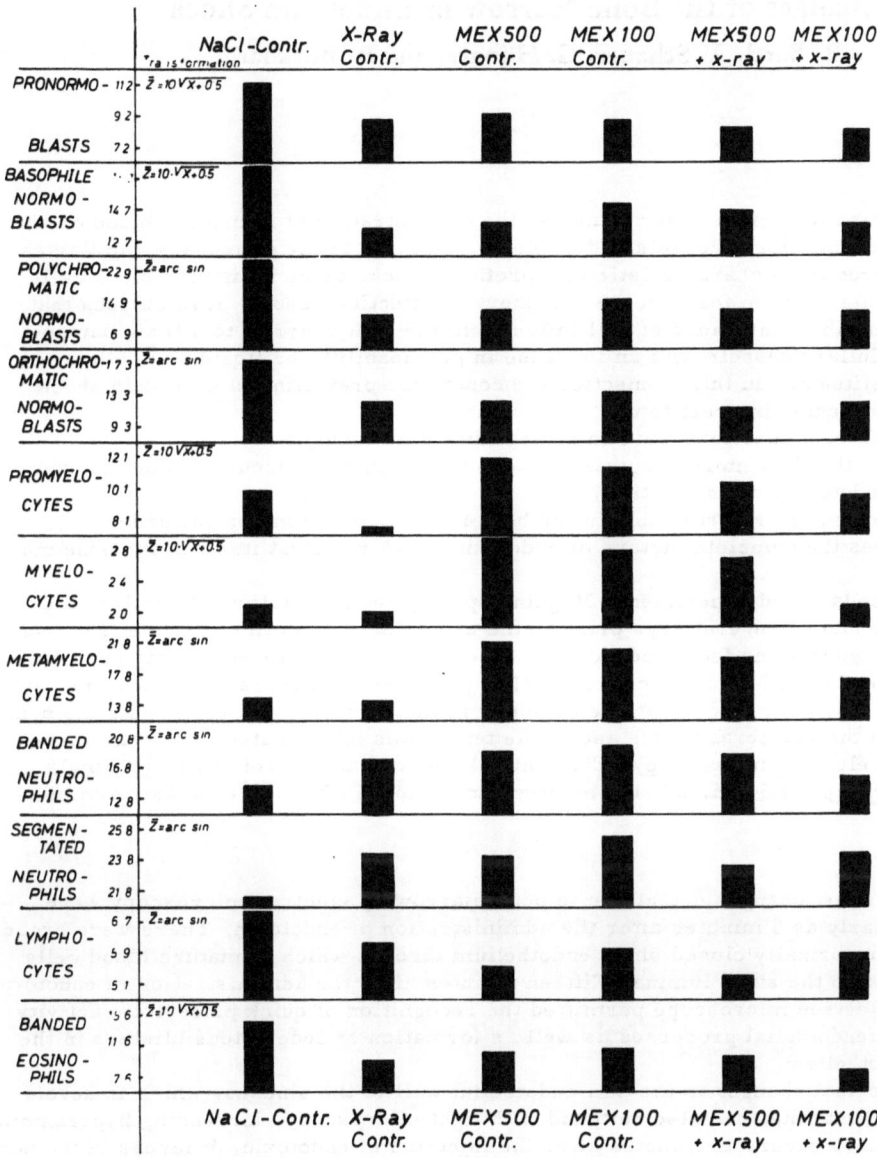

Fig. 1. Transformed mean values of differentiated bone marrow cells on day 1, 5, and 14 after injection respectively X-irradiation. MEX = detoxified endotoxin.

Changes of the Bone Marrow in Endotoxin Shock

R. Bartl, A. Schauer, G. Hübner, and R. Burkhardt

In addition to various other changes, the concentration of immature blood cells in the renal medullary vessels and of megakaryocytes in the pulmonary capillaries is considered to be characteristic of a prefinal shock. Since immature blood cells are particularly often found in the periphery in infectious shock, it is obvious that one may attribute the cause of cell influx from the bone marrow to a toxic damage of the medullary vessels with an increase in permeability for liquid and cellular blood constituents. In this connection the conditions prevailing in endotoxin shock were of particular interest to us.

The following questions were to be examined thoroughly :

1. When do the first morphologically determinable changes occur in bone marrow after endotoxin administration?
2. Which changes are detectable in the bone marrow in chronological order?
3. How does the complete picture of endotoxin shock manifest itself in the bone marrow?

In a standardized experiment, 30 guinea pigs (breeding station : Bäumler, Munich) were placed several days prior to the experiment in a climatic cage free from pathogenic germs and fed standard food. They were administered 2. 5 mg of lipopolysaccharide obtained from E. coli 0 111/B 4 (supplied by Messrs.Difco) for the initiation of an endotoxin shock. After embedding in methacrylate and epon resin the bone marrow of the vertebral bodies and of the femur was investigated with light microscopy and electron microscopy. Five animals served as control and two animals were sacrificed at 5, 15, 30, 60 minutes, and 4 hours after the endotoxin administration.

Results

The earliest changes detectable in bone marrow by electron microscopy were found as early as 5 minutes after the administration of endotoxin. There were broad gaps in the normally closed sinus endothelium through which immature blood cells penetrate into the sinus lumina. Fifteen minutes after the administration of endotoxin, the electron microscope permitted the recognition of quick pinocytotic activity in the sinuendothelial processes as well as formation of oedematous blisters in the sinus endothelium.

Pathological changes in the thin endothelial wall of the sinuses, which is devoid of basal membrane, can also be found with light microscopy. Increasing hyperaemia of the sinuses occurs 15 minutes after the injection of endotoxin. Whereas in the normal femoral bone marrow the sinuses occupy an average of 6 % by volume - not counting the central sinus - 10 % by volume can be measured after 30 minutes after injection and up to 40 % by volume 60 minutes after injection. The increasing dilation of the sinuses is paralleled by dissociation of the thin sinuendothelial layer. Immature marrow cells including megakaryocytes are found in the damaged sinus vessels after 30 minutes, whereas erythrocytes increasingly inundate the intercellular space of the bone marrow. These extravascular erythrocytes are distinguished by poor staining ability with Giemsa and Gömöri staining procedures and can often be identified merely by their contours.

The complete morphological picture of endotoxin shock in the bone marrow is present 60 minutes after the administration of endotoxin. The broad sinuses filled with blood are conspicuous. They occupy up to 40 % by volume of the bone marrow and subsequently no longer show an essential increase in lumen. The sinus destruction is so intensive that, in some areas, sinus walls can no longer be recognized. All maturation stages of the erythrocytopoiesis, granulocytopoiesis, and thrombocytopoiesis are reduced by their flooding out into the sinuses, and they can be found again in the vascular bed of the lesser and general circulation. Their place is now occupied by extended, diffuse extravasations of plasma and erythrocytes. The remaining vascular sections of the medullary circulation show only relatively small, uncharacteristic changes even in the complete picture of shock. No microthrombi could be detected in the medullary vessels of the guinea pigs.

In the peripheral blood picture a sudden diminution of leukocytes from 6.300 per cmm to 1750 per cmm is demonstrable as early as 5 minutes following the administration of endotoxin. After 60 minutes the leukocyte count amounts to 2.400 per cmm. The erythrocyte count increases from the normal rate of 5 million per cmm to 5.9 million per cmm. In summarizing the changes chronologically, the following picture is obtained : The first changes demonstrable by electron microscopy can be noted a few minutes after the beginning of the endotoxin effect. They manifest themselves as rapid and increasingly progressive sinuendothelial damage. This is paralleled by an extreme dilation of the sinuses, increasing to about 7 times the normal size after 60 minutes. A sudden diminution of leukocytes likewise occurs during the first minutes. As a late change, in bone marrow only megakaryocytosis and eosinophilia are demonstrable subsequent to the endotoxin shock.

The total findings thus characterize the bone marrow as an organ in which the circulatory system is particularly sensitive to endotoxin. The present findings do not yet reveal to what extent the first endothelial changes are haemodynamically or cytotoxically caused by endotoxin damage; however, they provide an explanation of the very rapid influx of immature blood cells into the peripheral circulation at the very first stages of endotoxin shock.

References

1. Ainsworth, E. J., Chase, H. B. : Proc. Soc. exp. Biol. Med. 102 : 483 (1959).
2. Ainsworth, E. J., Hatch, M. H. : Radiat. Res. 13 : 632 (1960).
3. Balazs, A. : Brit. J. exp. Path. 51 : 114 (1970).
4. Boggs, D. R., Athens, J. W., Cartwright, G. E., Wintrobe, M. M. : J. clin. Invest. 44 : 643 (1965).
5. Breddin, K. : Med. Welt 22 : 1165 (1971).
6. Chien, S., Dellenback, R. J., Usami, S., Gregersen, M. I. : Proc. Soc. exp. Biol. Med. 118 : 1182 (1965).
7. De Palma, R. G. : Surgery 62 : 505 (1967).
8. Fruhman, G. J. : Amer. J. Physiol. 212 : 1095 (1967).
9. Fruhman, G. J. : Blood 27 : 363 (1966).
10. Lapp, H. : Med. Welt 22 : 1180 (1971).
11. Quesenberry, P., Morley, A., Stohlman, F., Rickard, K., Howard, D., Smith, M. : New Engl. J. Med. 286 : 227 (1972).
12. Remmele, W. : Klin. Wschr. 15 : 803 (1968).
13. Remmele, W., Goebel, U. : Klin. Wschr. 51 : 25 (1973).
14. Rotter, W. : Med. Welt 22 : 1175 (1971).
15. Smith, W. W., Alderman, I. M. : Radiat. Res. 17 : 495 (1962).
16. Smith, W. W., Alderman, I. M., Cornfield, J. : Amer. J. Physiol. 201 : 396 (1961).
17. Smith, W. W., Alderman, I. M., Gillespie, R. E. : Amer. J. Physiol. 192 : 549 (1958).
18. Smith, W. W., Marston, R. Q., Cornfield, J. : Blood 14 : 737 (1959).
19. Stewart, G. J. : New Engl. J. Med. 281 : 1404 (1969).

20. Witte, S., Betke, K., Tosberg, P. : Das Knochenmark (Lehmanns, München 1968).
21. Yoshida, M., Hirata, M., Hatano, Y., Inada, K. : Jap. J. exp. Med. 38 : 335 (1968).
22. Yoshida, M., Hirata, M., Hatano, Y., Inada, K. : Jap. J. Med. Sci. Biol. 23 : 289 (1970).

Table I.

CHRONICAL CHANGES OF BONE MARROW AFTER ENDOTOXIN ADMINISTRATION

	normal	5 min	15 min	30 min	60 min
number of animals	5	2	3	3	4

ELECTRON MICROSCOPY

	normal	5 min	15 min	30 min	60 min
gaps in the sinus endothelium					
quick pinocytotic activity in the sinus endothelium					
oedematous blisters in the sinus endothelium					

LIGHT MICROSCOPY

	normal	5 min	15 min	30 min	60 min
dilation of the sinuses (vol %)	6	5	8	10	40
dissociation of the sinus endothelium					
flooding out of immature bone marrow cells					
erythrocyte extravasates					
megakaryocytes (normal 100 %)	100	100	80	63	46

PERIPHERAL BLOOD PICTURE

	normal	5 min	15 min	30 min	60 min
leucocytes (per cmm)	6350	1750	2600	2400	2650
erythrocytes (mio per cmm)	4,9	5,0	6,5	5,8	5,9

= sinus volume (vol%):6
O =megakaryocytes : 100%

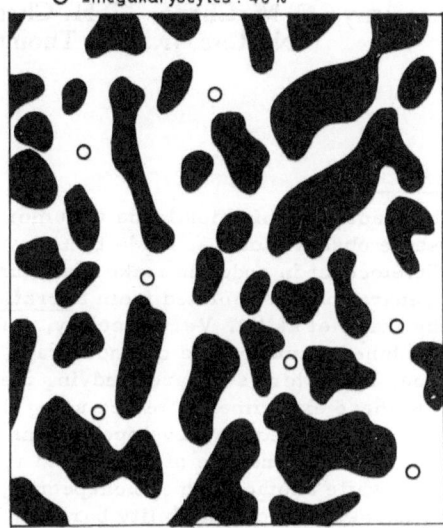

= sinus volume (vol%):40
O =megakaryocytes : 46%

Fig. 1.   Bone marrow of guinea pig
(normal).

Fig. 2.   Bone marrow of guinea pig
(60 min after endotoxin administration).

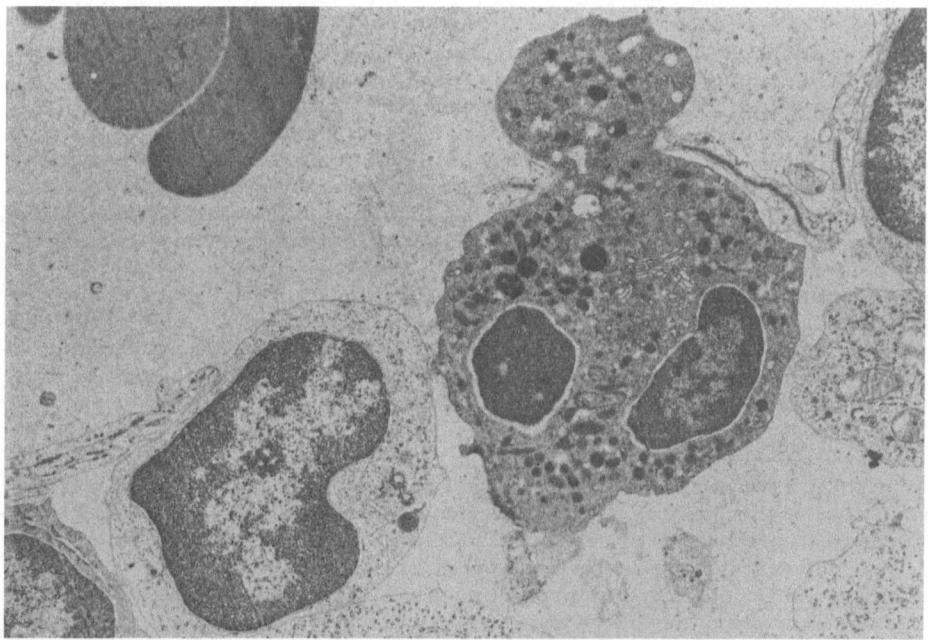

Fig. 3.   Sinus endothelium 5 min after endotoxin administration. On the right, a
neutrophil granulocyte emigrating through a small endothelial gap, on the left, the
formation of a broad endothelial dehiscence through which an immature cell is dis-
posed of.

Endotoxin-Induced Tumor Resistance

A. Nowotny, C. M. Chang, C. H. Chen, J. Grohsman, H. Liang, A. K. Ng,
N. Rote, Jr., J. J. Thompson, and C. Yang Ko*

Introduction

That bacterial infection leads to tumor regression was first reported by Busch(3).
The same observation was made by Coley (5), who artificially infected patients with
live Stretococci in order to make their tumors undergo necrosis. A tumor necrotizing preparation was isolated from Serratia marcescens by Shear (17) and from E.
coli by Ikawa et al. (8). Very recently, infections with live BCG bacteria were used
for combined immuno- and chemotherapy of leukemia by Mathe et al. (11).

In our laboratories we are studying the escape of tumors from immune recognition. In these experiments we are using the TA3 tumor line which arose in strain A
mice as a spontaneously developed mammary adenocarcinoma and was converted into ascites form. One line of this strain remained strain specific, designated as
TA3-St, while another turned nonspecific, designated as TA3-Ha. This latter line
grows across histocompatibility barriers and even in rats. These two strains, which
possess histoincompatible and possibly also tumor associated antigens, were used
for the studies of the mechanism which enables certain tumor lines to evade immune
recognition and destruction.

During these studies, the TD_{50} (= the number of viable tumor cells which will
cause death of 50% of the experimental animals) was established by us to measure
the resistance of normal or immunized animals against the TA3-Ha line. First we
observed variations in the TD_{50} values depending upon the age of the inbred allogeneic mice. Older groups were significantly more resistant to the tumor than younger
groups. Later we found that animals which underwent a transient infection and recovered from it showed a much higher resistance against TD_{50} challenge than normal animals. These variations in the TD_{50} indicated to us that bacterial infections
enhance the natural resistance of animals against the tumor.

Accordingly, 1 /ug of Serratia marcescens endotoxin (ET) was injected i. p. into
mice, followed the next day by 3 or more TD_{50} TA3-Ha tumor challenge. Significant protection could be achieved (7).

In the continuation of this work the optimal conditions for such protection were
investigated, including the effects of a) various doses of ET, b) various routes of
injection of ET, c) various time intervals between ET administration and tumor
challenge. In addition, the mechanism of the anti-tumor effects of endotoxin-induced
resistance was studied. The results obtained will be reported here.

Materials and Methods

Tumor. The maintenance of the TA3 tumor is carried out in A/J mice. The TD_{50}
determination in allogeneic systems is carried out as described earlier. The washing procedures, preparation of cell suspensions, cell viability determinations were
all given in detail in the above previous publication (7).

Endotoxin and Endotoxoid. The procedure to extract S. marcescens 08 cells with
trichloroacetic acid was described earlier (13). The detoxification with potassium
methylate was also reported (12). For better solubilization of the ET preparations,
they were brought to 0.1 N NaOH concentration at room temperature and neutralized

*Portions of this work were taken from the dissertation presented by Cynara Yang
Ko to the Graduate School of Temple University in partial fulfillment of the requirements for the M.S. Degree, 1973.

back to pH 7.2 within 5 min.

Spleen cell suspension and blast transformation were made according to the me-
thology of Rosenstreich et al. (16).

Limulus lysate clotting assay. This semi-quantitative method to detect endotoxin
was carried out by the procedure of Levin and Bang (10) using the amoebocyte lysate
generously provided by Dr. Levin.

Results

Table I shows clearly the highly significant protection achieved by the i.p. injection of 1 μg/mouse of ET, 24 h before the tumor challenge.

Regarding the optimal conditions of ET pretreatment induced tumor resistance,
Table II shows the most effective time intervals between ET pretreatment and tumor
challenge. The optimal time for low dose ET pretreatment is from -3 to 0 days before tumor challenge.

Table III shows the effect of various doses of ET. These findings indicate that
the best effects can be achieved by 1 to 10 μg ET per mouse, depending upon the
mouse strain used. If a very high dose of ET was given i.p., those animals which
survived this treatment showed a very significant resistance against tumor challenge
even after 21 days.

Regarding the route of ET injection, the i.p. application seems to be much more
effective than i.v., indicating a local phenomenon elicited by endotoxin. ET given
i.v. required a 10-fold higher dose in order to elicit a measurable protection against
the tumor.

Whether the toxicity of ET is required for the enhancement of tumor resistance
was investigated by using chemically detoxified ET (endotoxoid). Table IV summarizes the findings and indicates that after giving endotoxoid i.p., a protection comparable to the effect of toxic ET against the TA3-Ha tumors can be observed. This
is in agreement with our earlier observations, where we reported that the nonspecific resistance enhancing effect of ET against virulent gram-negative bacteria is
not dependent upon the toxicity of the preparation (9).

Regarding the mechanism of resistance enhancement, the following experiments
have been carried out so far. The question of whether endotoxin had a direct cytotoxic effect on the tumor cells was investigated by exposing tumor cells in vitro to
a concentration of 100 μg ET per 10^6 tumor cells. The cells were incubated for 90
min and the viability was determined by the Trypan blue exclusion test. The viability of the cells was not reduced by this treatment. The cells were washed free of
ET and injected i.p. into mice to check their tumorigenicity. One TD_{50} was inoculated and Table V shows that this inoculum resulted in approximately 50% mortality.

The fact that both the ET and the tumor had to be injected i.p. in order to achieve
an optimal effect indicated a local phenomenon. To investigate the role of peritoneal
exudate cells, which are known to be attracted to the peritoneal cavity in large numbers after ET injection, we injected 10 μg ET i.p. and rinsed the cavity with medium M199. 10^6 peritoneal exudate cells were transferred into normal mice i.p. and
the same animals were challenged with 3 TD_{50} tumor cells. No significant protection could be achieved.

It has also been reported earlier (7) that transfer of a spleen cell suspension
from tumor-bearing mice into normal recipients rendered them resistant against
subsequent challenge with viable tumor cells. This indicated that the tumor-bearing
mice had an immune response against the tumor they bear and that this can be transferred by immunocytes. Similarly, it has been reported by others as well as by us
that endotoxin can induce mitogenesis of spleen cell suspension in vitro. It is now
firmly established that the B cells respond to the nonspecific mitogenic effect of endotoxin (2, 6, 14, 16).

Therefore, spleen cells were exposed to endotoxin in vitro and their mitogenic
index was determined by measuring the incorporation of tritiated thymidine. It was

established that 3-day incubation reaches the maximum of mitogenic stimulation and that 10 μg ET/ml medium is the optimal concentration to stimulate 10^6 spleen cells. Accordingly, spleen cells were exposed to ET for 3 days under these conditions and 10^7 viable spleen cells were adoptively transferred into normal recipients. This transfer was followed by challenge with 4 TD_{50} of TA3-Ha tumor. No protection could be observed. On the other hand, if the incubation of the spleen cells with ET was carried out for only 5 h, adoptive transfer of these spleen cells resulted in a marginal protection.

Since ET given i.v. is known to accumulate rapidly in those organs which are rich in reticuloendothelial cells, such as spleen, liver and lung, 10 μg ET was given to mice and their spleens were taken 24 h later. Spleen cell suspension was made from these organs and washed several times with medium. Adoptive transfer of 10^7 viable spleen cells into normal recipients resulted in significant protection. Table VI summarizes these results.

The endotoxin content of the adoptively transferred spleen cells was determined by the Limulus lysate clotting assay. 10^7 spleen cells obtained from i.v. injected mice or after exposure to ET in vitro were sonicated and 0.1 ml aliquots of this were incubated with 0.1 ml hemocyte lysate in a 37° water bath. It was observed by this experiment that the amount of ET in the lysate was less than 0.02 μg/10^7 spleen cells. I.p. injection of that low amount of ET alone does not give any detectable protection in any of the mouse strains so far investigated. These results indicate that the transfer of endotoxin via spleen cell is not the sole factor responsible for the enhanced tumor resistance.

Discussion

The TA3-Ha tumor cell which can grow in allogeneic hosts in spite of the fact that it still contains transplantation antigens (H2[a]) is an ideal model to study the effect of immune stimulation or immunosuppression on the growth characteristics of the tumor and on the resistance of the host against the tumor. As we observed earlier, ascites fluid of TA3-Ha tumor-bearing mice transferred passively into normal recipients has an immunosuppressive effect which enhances the growth of the tumor. On the other hand, stimulation of the immune response by endotoxin shows its effect by inducing resistance to lethal tumor challenge dose.

Regarding the mechanism of the endotoxin-induced tumor resistance, several possibilities can be considered. Endotoxin may act on macrophages, as described by Alexander and Evans (1) and may render them cytotoxic against tumor cells. Another possibility is that since endotoxin is an exclusively B cell stimulant, and an excellent adjuvant of humoral immune responses, in our system mice treated with endotoxin will undergo B cell activation and at the same time small amounts of specific anti-tumor antibodies were also synthesized earlier than would be expected under normal conditions due to the adjuvant effect of endotoxin. These small amounts of specific antibodies could enhance the cytotoxicity of the B cells in a mechansim similar to that described by Perlmann and associates (15).

Most probably the mechanism is much more complicated. Cell cooperation as a possibility should be kept in mind, and at the present time we are involved in experiments searching for possible humoral mediators which may participate in the phenomenon described here. In another series of experiments we are separating B and T cells as well as macrophages from activated spleen cell suspensions by the methods described by Rosenstreich et al. (16). Adoptive transfer of purely B or T or macrophage cells into normal recipients, followed by tumor challenge, will hopefully provide additional clues regarding cell types involved in the enhanced tumor resistance.

Finally, we would like to emphasize that our results so far do not indicate correlation between mitogenic effect of endotoxin and its tumor resistance enhancing capacity. First, endotoxoid, which was shown here to be able to induce tumor resistance, was inactive as a mitogen. Second, an endotoxic glycolipid, isolated from rough S. minnesota R595 (4), which was shown to be a very strong mitogen, was less efficient in inducing tumor resistance than regular smooth endotoxins. Lastly, the induction of mitogenesis reaches its peak after 3 days of incubation of spleen cells with endotoxin. Transfer of such fully stimulated cells adoptively did not provide protection against subsequent tumor challenge.

Summary

TA3-Ha tumors, bearing H2^a histocompatibility antigens, can grow in allogeneic hosts. The TD$_{50}$ values of these tumor cells vary from strain to strain but remain fairly constant within the strain provided the animals are healthy and kept in carefully supervised animal facilities. Slight infections of animal colonies, as we observed, significantly enhanced their resistance against tumor after recovery from the infection.

Injection of bacterial endotoxin in 1-10 /ug quantities/mouse also enhanced their resistance against tumor challenge. Optimal conditions of the endotoxin inducible tumor resistance are described here.

Spleen cells from i.v. endotoxin injected normal mice could adoptively transfer tumor resistance to normal recipients. At the same time, in vitro exposure of spleen cells showed only marginal enhancement of tumor resistance. No correlation could be found so far between mitogenic effect of endotoxin and its capacity to induce nonspecific tumor resistance.

While the TA3-Ha tumor system growing in allogeneic hosts seems to be ideal to detect changes in the host resistance induced by immunosuppression or immunostimulation, it remains to be demonstrated whether other tumor systems could be similarly influenced by previous application of endotoxin.

References

1. Alexander, P., Evans, R.: Nature New Biol. 232: 76 (1971).
2. Anderson, J., Sjöberg, O., Möller, G.: Transpl. Rev. 11: 131 (1972).
3. Busch: Berl. Klin. Wschr. 3: 245 (1866).
4. Chen, C.H., Johnson, A.G., Kasai, N., Key, B.A., Levin, J., Nowotny, A.: J. infect. Dis. 128: S43 (1973).
5. Coley, W.B.: Amer. J. med. Sci. 105: 487 (1893).
6. Gery, J., Kruger, J., Spiesel, S.Z.: J. Immunol. 108: 1088 (1972).
7. Grohsman, J., Nowotny, A.: J. Immunol. 109: 1090 (1972).
8. Ikawa, M., Koepfli, J.B., Mudd, S.G., Niemann, C.: J. nat. Cancer Inst. 13: 157 (1952).
9. Johnson, A.G., Nowotny, A.: J. Bact. 87: 809 (1964).
10. Levin, J., Bang, F.B.: Thromb. Diath. haemorrh. 19: 186 (1968).
11. Mathé, G., Amiel, J.-L., Schwarzenberg, L., Schneider, M.: J. roy. Coll. Physicians Lond. 5: 62 (1970).
12. Nowotny, A.: Nature 197: 721 (1963).
13. Nowotny, A.M., Thomas, S., Duron, O.S., Nowotny, A.: J. Bact. 85: 418 (1963).
14. Peavy, D.L., Shands, J.W. Jr., Adler, W.H., Smith, R.T.: J. Immunol. 111: 352 (1973).
15. Perlmann, P., Perlmann, H., Wigzell, H.: Transpl. Rev. 13: 91 (1972).
16. Rosenstreich, D.L., Nowotny, A., Chused, T., Mergenhagen, S.E.: Infection Immunity 8: 406 (1973).
17. Shear, M.J.: Cancer Res. 1: 731 (1941).

Table I.

Effect of pretreatment of C57Bl/10 mice with 1 µg Serratia marcescens endotoxin (ET) administered 24 h before TA3-Ha tumor challenge

Pretreatment of mice (i.p.)	TA3-Ha Challenge (i.p.)	No. Survivors/ total	% Survival
Saline	4×10^4	3/39	7.7
ET, 1 µg	4×10^4	29/39 (p < 0.005)	74.4

Table II.

Effect of time interval between endotoxin (ET) treatment and challenge

Treatment of mice[a] (i.p.)	TA3-Ha Challenge (i.p.)	No. Survivors/ total	% Survival
None	4×10^4	2/20	10
ET			
day +1	4×10^4	4/10 (p<0.1)	40
day 0	4×10^4	17/20 (p<0.002)	85
day -1	4×10^4	14/19 (p<0.002)	74
day -3	4×10^4	8/10 (p<0.002)	80
day -7	4×10^4	5/10 (p<0.05)	50

Table III.

Dose response to endotoxin (ET) in ICR mice

Pretreatment of mice (i.p.)	Days between ET and challenge	TA3-Ha challenge (i.p.)	No. Survivors/ total	% Survival
Saline	1	3×10^3	0/10	0
ET				
0.1 µg	1	3×10^3	0/10	0
1 µg	1	3×10^3	2/10	20
10 µg	1	3×10^3	6/10 (p<0.02)	60
100 µg	1	3×10^3	1/10	10
200 µg	21	3×10^3	12/15 (p<0.01)	80
750 µg	21	3×10^3	10/12 (p<0.01)	83

Table IV.

Effect of pretreatment of ICR mice with Serratia marcescens endotoxin (ET) or a detoxified preparation (MEX B) administered i. p. 24 hours before TA3-Ha tumor challenge

Pretreatment of mice	TA3-Ha Challenge	No. Survivors/ total	% Survival
Saline	2×10^3	0/7	0
10 μg ET	2×10^3	4/7	57
10 μg MEX B	2×10^3	3/7	43
1000 μg MEX B	2×10^3	6/8	75

Table V.

Tumorigenicity of TA3-Ha cells incubated in endotoxin (ET)

Incubation of cells	Cell Challenge Dose (i. p.)	Tumor Deaths/ Total	% Mortality
Medium only 1 1/2 h	10^3	4/10	40
ET 100 μg/10^6 cells in medium 1 1/2 h	10^3	5/9	56

Table VI.

Adoptive transfer of spleen cells obtained from ICR mice injected i. v. with 10 μg endotoxin (ET)

Treatment of ICR mice (i. p.)	TA3-Ha Challenge (i. p.)	No. Survivors/ Total	% Survival
None	3×10^3	1/10	10
10^7 normal spleen cells	3×10^3	1/10	10
10^7 spleen cells from ET-injected donors [a]	3×10^3	5/9 (p< 0. 05)	56

[a] Spleen cells were taken 24 h after ET injection i. v. The ET content in 10^7 spleen cell sonicate was 0. 02 μg, as measured by the Limulus lysate clotting assay.

Discussion

SPRINGER

Landy: These are very interesting data. We did some experiments about 20 years ago on coating of red cells with endotoxins using 6 different generic specifities and coating in series, simultaneously and separately. Those experiments suggested that they might involve different sites because the eventual reactivity of the coated red cells was identical, whether we added them altogether - the antigenic products - simultaneously or in series or separately.

Springer: If you use the endotoxin, let us say endotoxin A in excess and then use endotoxin B, then endotoxin B will remove part of endotoxin A and attach to exactly the same receptor. That means the receptor sites are the same but it is extremely difficult to completely saturate them. If you succeed in doing so you can displace one LPS with the other LPS. I wish to point out that we were not the first ones to show this. The first ones to show this in 1956 without isolating any structures were Lüderitz, Westphal and Neter in elegant work with ^{32}P labelled LPS.

Landy: I would point out that accepting the concept of displacement, would it not be a remarkable coincidence that if you add a pool of 6 different products simultaneously you come up with precisely the same activity as though you put them in separately or in sequence.

Springer: I do not think so because you have an equilibrium of different endotoxins when you add them and they will attach themselves in accordance with the law of mass action. I may mention in addition to this that aldehydes specifically inactivate this receptor site. It was shown for acrolein, but others do it too, whether this site is on the intact red cell or on the isolated receptor.

Urbaschek: Just how much is known concerning the meaning of the pathophysiology of these endotoxin-specific receptors after destruction of the erythrocytes, for example an accumulation of receptors in the spleen?

Springer: I do not know.

Greisman: What does this do to the biological activity of the endotoxin when you combine the endotoxin with the receptor? What happens to its biological activity?

Springer: I can only partly answer this question. The serological specificity of the endotoxin is unchanged. Let us say you have E. coli 086 or 07 specificity, it will not change. But if you use a very large amount of receptor and preincubate it with LPS and then inject this material into rabbits you decrease the pyrogenicity. However, other substances such as gangliosides are more efficient in doing this because, as I mentioned, the union between receptor and endotoxin is a reversible one and you have to use a very large excess; it goes according to the law of mass action.

Braude: I wonder whether your experiments were all done with alkaline treated endotoxin?

Springer: The experiments were done with smooth and rough endotoxin. They were done with endotoxin which had been activated at pH 7.4 at 100°C for 3 hours or again at pH 7.4, because we consider this to approximate physiological conditions by preincubating the endotoxin for 72 hours at 37°C. It will not work very well on endotoxin which has just been extracted with phenol-water. It will work, but this endotoxin does not attach itself very well in general. You do not need strong alkali treatment, however, you should have a physiologic pH. If you want to accelerate the process you should boil. 37°C will do also, but then you have to incubate for a long time.

Braude: Have you any idea what this activation does that enables it to combine with red cell receptors?

Springer: Dr. D. A. L. Davies' feeling was at one time that acetyl groups are liberated but I think that is not certain any more.

Hadding: It seems to be a bit difficult to give the molecular weight of a part of a membrane. This will always be connected with the method you use for separation expecially since you have shown that the removal of the sialic acid will not change the functional role. I would suggest if you just give the molecular weight of the receptor you can substract the value of the sialic acid. I am not sure what you actually will give as a final size of the molecule if you start from a functional definition.

Springer: We have isolated a substance from an erythrocyte membrane like many other people have isolated substances from other membranes. This substance is fully soluble and is physico-chemically homogenous by all criteria of hydrodynamic measurements. It has a molecular weight of 256,000. It is a molecule as such and you can treat it under various conditions and it will stay like this. You can treat it with an enzyme and remove components from it and the molecular weight will decrease in proportion to the amount which you have removed. That is quite obvious. Or you can disaggregate it, but the material you extract from the red cell is reproducibly of a molecular weight of 256,000. This has nothing directly to do with the function. These are just the physico-chemical properties of this material.

Bloch: I am not sure - did you say that the smaller components were active as well?

Springer: They are practically inactive.

Bloch: So the activity of the smaller components, if you compare it on a mol per unit basis, is smaller.

Springer: It is very small indeed.

Rácz, Hamburg/Germany: Dr. Springer, you have mentioned the experiments done by P. Ehrlich. I should like to call to your attention that pathologists of the last century (for instance: Rindfleisch, E., Lehrbuch der pathologischen Gewebelehre, 2. Auflage, Wilhelm Engelmann Verlag, Leipzig 1871) had already noticed very intensive erythrophagocytic activity in macrophages of lymph nodes during the course of several infectious diseases, e. g. typhoid fever. The biological meaning of this phenomenon was not known. Regarding the facts, you have mentioned that the erythrocytes are covered with endotoxin. The macrophages take up endotoxin with erythrocytes. This may have great immunologic importance.

BONA

CHEDID

Springer: Is there anything known about the molar relationship of lipid A which would activate one lymphoblast? Another question: It is well-known from endotoxin preparations that only about 10% are adsorbed onto red cells. No matter how often you exchange your red cell population the other 90% are just not taken up. Is a similar situation prevailing in your lymphocyte experiment?

Bona: We got the following results with lipid A. We obtained stimulation of 40% of the cells with lipid A, 5 µg for 1.5 per million cells. One gets 60.4% of stimulation with 50 µg of the LPS. But this is not a problem. Also a very important problem is the avidity of the receptors for LPS. If you have high avidity a small quantity is sufficient to induce stimulation. If you have low avidity a greater quantity of LPS and/ or another mitogen is necessary to get stimulation.

Springer: How do you know about the avidity of your lymphocytes?

Bona: By the avidity of the receptor antibody for a given antigen.

Landy: Dr. Chedid, in the design of your experiments why would you have expected to get any result other than you got? The cells that bind are already presumably activated and elaborating IG. One has to assume that is the reason for the binding. So to remove them of course would make no difference in terms of the parameters you have used to follow response. One would expect that to make no difference. They are already fully functional. You could not turn them on if they are already turned on. I think the results you got are exactly what one would expect.

Chedid: We wanted to make sure that we would not activate, by additional preincubation of this special type of antigen, a very high number of cells. I quite agree with what you say, but we wanted to make sure that nothing happened. In our opinion no investigator looks at this problem as if he were dealing with a type of unique antigen which has a very broad spectrum and sometimes surprising activities. Because of these common structures it is curious that a specific mitogen is this very ubiquitious and very cross-reacting antigen. We expected to get these negative answers but we wanted to make sure we would get them. The only surprise is the experiment with polysaccharide.

Landy: Dr. Bona, you have emphasized repeatedly the selective B cell mitogenicity of LPS, but I recall experiments reported by Göran Möller in Stockholm where if the same product is coated on red cells it becomes a thymus-dependent mitogen. What is your comment on this observation?

Bona: The LPS of course is thymus-dependent and also from the mouse macrophage-independent antigen, e.g. it is very clear I think that the bypassing of the LPS throughout macrophages is not required for the stimulation of the lymphocyte and also not for the induction of antibody synthesis. This is not the case for another species, e.g. the LPS is macrophage-dependent in the case of guinea pigs. Nevertheless this thymus-independency of the LPS is very clearly demonstrated in that the nude mice respond as well as the normal mice. Unfortunately in our case, and I did not present the results concerning the nude mice, we worked with nude mice originating from an outbred white strain homozygous for new nude mutation isolated by Dr. Roberts in Genetical Animal Institute in Edinburgh. In this case we have various heterogenous response, actually in this strain we found highly responding mice with a stimulation index greater than 10, and low responding mice with a stimulation index less than 5 and non-responding mice. In the inbred nude mice strain it is very clear that the LPS is thymus-independent mitogen.

Pusztai-Markos: This thymus-independency is only then present when we evaluate the primary response. It is possible to obtain a typical secondary response if one allows a long enough interval between priming and booster. In the case of E. coli 055 LPS we could elicit a secondary response after eight to ten weeks, with IgG production, that is, mercaptoethanol-resistant antibodies, which also bring about indirect plaques. When we conducted parallel studies using nude mice, secondary response was completely absent. Furthermore, in normal mice an increase in the rosette-formation was observed, whereas they did not appear with nude mice.

Marget: I do not know the pathogenicity of Salmonella enteritidis and S. minnesota against mice. Have you done some antibody response in these mice before you made these investigations?

Chedid: You are referring to the pathogenicity of the living organisms and not to the toxicity of the endotoxin. The toxicity is the same order. The pathogenicity is quite different because the S. enteritidis strain we use at Pasteur, has something which mimics very faithfully human typhoid. It can be administered orally, gives the whole cycle, positive hemocultures, serodiagnosis, etc., whereas most other Salmonella strains except some S. typhimurium strains which are very special, must be used in mucin to give experimental septicemia. I have never worked on the virulence of S. minnesota, but I would accept it to be a non-virulent strain for the mouse. The other question is immunization by both.- I personally have no data, but I think that according to the literature, and probably some other people here would know, both immunize in a similar manner. The question nobody answered is: Why is there no stimulation of Brucella endotoxin? Is this due to the fact that mice are not exposed to Brucella? Or is it a very different chemical structure between Brucella lipid and lipid A?

Rietschel, Freiburg/Germany: It is known from the work of Asselineau, Lacave, Thiele and others that Brucella LPS, or the lipid A part of Brucella is different from the enterobacterial lipid A. All enterobacterial lipid A contain an amide-bound 3-hydroxyacid. This is not the case in Brucella lipid A. They have saturated fatty acids but no 3-hydroxy fatty acids.

Chedid: In other words this would be critical for mitogenic transformation?

Rietschel: I do not know what the core of lipid A is - whether it is in Brucella glucosamine too like in Enterobacteriaceae, but there is a definite difference in toxicity too of Brucella LPS as compared to enterobacterial LPS. According to Thiele and Asselineau there is C_{16}, C_{17} and C_{18} palmitic hexadecanoic-acid and steric acid in lipid A of Brucella. There is no 3-hydroxyacid, so it differs from Enterobacteriaceae lipid A in that there is no 3-hydroxyacid at all. The saturated acids are different too.

Mergenhagen: I would like to ask: Oppenheim and others have found that antibodies of course modulate the cellular immune response and my specific question is have either of you or do you know of anyone who has actually looked at immune antibodies in these systems to see if these O antibodies modulate the responsiveness of the B cell to the LPS?

Bona: I do not know. We have no experiment concerning this point. I have not seen any such findings in the literature.

Mergenhagen: In the presence of a gram-negative infection I suppose one would think that one would have to deal with humoral antibodies in the cellular reactions. I suppose somebody should look into it.

Smith: I think there is one other point, I would confirm the fact that the LPS from Brucella is less toxic than those from other Enterobacteriaceae. There is a difference in the lipid A, but there is also a difference in the polysaccharide structure. There are formyl groups on the polysaccharide structure of the Brucella and not as far as I know on those of the Enterobacteriaceae.

Bona: But if we consider that the lipid A is responsible for the nonspecific mitogenic activity, the differences in the structure of the lipid A is more important from this point.

Braude: Dr. Bona, I wonder if it is possible that the difference in the reactivity to LPS of rabbit and spleen lymphocytes may be related to the fact that rabbits are not likely to be colonized with these organisms, with gram-negatives, while mice are likely to be colonized.

ALEXANDER

Springer: Dr. Alexander has told us the effect, so to say, of what happens after burns, but what is the cause of the abnormality in the neutrophils?

Alexander: It would appear that the cause is probably on a nutritional basis. Unfortunately I cannot give you the exact biochemical nature of this at present.

Mergenhagen: I am wondering if you have looked at the responsiveness of neutrophils in these burn patients to any of the chemotactic factors. The reason I ask is that in a number of these patients several years ago, after having established that such materials as protein A from the Staphylococcus and endotoxin interacting with the complement system, the C5 fragment is cleaved, which is chemotactic for the neutrophils. Dr. Snyderman looked at a good number of these chronic granulomatous diseased patients as well as patients with staphylococcal infections. He immediately looked for defects in the complement system as did Dr. Miller et al. in Philadelphia. There was really no deficiency in the complement system nor in the ability of these patients to form the C5 fragment, having been activated with immune complexes or endotoxin. However, when Dr. Snyderman took the neutrophils from a number of these patients they were markedly deficient in migrating to performed chemotactic factors. So patients could form chemotactic factors, but the neutrophils would not respond to these chemotactic factors. In your burn patients have you had any experience at all with migration? The cells must get there before they can phagocytise. This would even be a more primary deficiency.

Alexander: That is certainly true and we have looked at this recently and fairly critically. There are defects in both random migration of the neutrophils and in response to chemotactic factors such as endotoxin. However, these defects seem not to be profound. They seem to have no relationship to the development of systemic sepsis in these particular individuals. The only real correlations we can make with the development of sepsis is with abnormalities in neutrophil function and with abnormalities of the alternate pathway of complement which happens early after burn injury and is usually self-correcting in about 14 days.

Mergenhagen: Do the neutrophils migrate in these patients normally to preformed chemotactic factors?

Alexander: They do not. That is, in some instances. It is very variable among the patients, but there can be demonstrated an abnormality of chemotaxis in the neutrophils.

Marget: There are some observations on PHA transformation of lymphocytes which is also decreasing in these severe burns and septicemias. Is there some correlation between your findings and the decrease in PHA?

Alexander: I cannot really answer that. Certainly there are marked deficiencies of cell-mediated immune mechanisms, i. e. lymphocyte cell-mediated immune mechanisms. These undoubtedly play a role in some of the fungal and viral infections that occur late in extensive burn injuries, but we have not really studied the sequential changes in the same patient so I cannot answer your question as to whether or not they are correlated.

Lowbury: Some years ago you spoke of the association between neutrophil dysfunction and burn disease. Today you told us about the important correction of this by control of infection. Would you say that your present results suggest that much of the adverse effect on neutrophil function during burn injury is in fact due to sepsis? Or what is the relative importance of sepsis?

Alexander: No, I think that this has been a question which has bothered us very greatly. That is, whether or not sepsis may precede a neutrophil dysfunction. In many instances we have seen severely septic patients who have had a return of their neutrophil function to normal during the time of extensive sepsis. If we look back retrospectively in all of our cases, when it has been studied well, neutrophil dysfunction always precedes the onset of systemic invasion. In think that the neutrophil function is the horse that precedes the cart, in the case of sepsis.

Greisman: I am not quite clear as to the latent period. What is the earliest time between the burn and the development of this neutrophil dysfunction, i. e. how soon is this seen?

Alexander: It is very variable from one individual to another I would say that over-all it has its greatest degree of abnormality usually during the first week, probably toward the end, and continues into the 2nd and 3rd week and the very large burns, at least in the earlier cases, extended much later. It did not seem to occur immediately after burn injury indicating that if there was immediate release of a burn toxin it did not affect the circulating neutrophils. This does not mean that it does not affect the formative neutrophils in the bone marrow.

Greisman: If you take the serum from burned patients and incubate normal leukocytes in this serum would you get any evidence of dysfunction?

Alexander: They function perfectly normally in burn serum.

Clowes: I would like to confirm Dr. Alexander's statement on the importance of nutrition and energy production on the function of these cells. We have examined, as I and others will bring out later in this session, there are significant differences and abnormalities in the pathways of energy production and in actual energy deficit in many tissues. The white cells are certainly among those that suffer worst in this situation. We have found a distinct correlation between negative nitrogen balance or proteolysis and the function of white cells both in terms of phagocytosis and also in terms of lymphocyte function of the type that was talked about yesterday. As proof of this, that it is probably the proteolysis which is responsible, the infusion of aminoacids alone has been demonstrated in our clinic by Dr. Blackburn and the rest of our colleagues to produce a situation in which fat is burned instead of aminoacids, restoring it more to the type of metabolism which one sees in starvation. Under these circumstances we can see a restoration of phagocytosis and killing in the leukocytes. Therefore I think your assumption is correct that it is probably an error in energy metabolism rather than some toxic effect that occurs in an acute fashion.

Berry: Dr. Alexander, do you feel that what you have observed in these burned patients is due to the burn and a burn toxin or do you think that this is the result of severe stress?

Alexander: May I speculate a little. I think that the majority of acquired abnormalities of neutrophil function in the burn as well as a variety of other diseases which have been described such as cancer patients, Kwashiorkor, perhaps iron deficiency anemia, are probably on a nutritional basis rather than on a toxic basis. The complex interdynamics related between hypermetabolism in the burn patient and the stress and its relationship to nutritional requirements are not very well understood at present. Certainly these requirements are markedly increased.

von Graevenitz: How great is the danger of Candida septicemia if you use hyperalimentation in those burn patients?

Alexander: The danger of Candida septicemia with hyperalimentation is related to the indwelling intravenous catheter. If you have to put it through a burn site as we do sometimes, it is very high. Mostly, however, we use oral hyperalimentation rather than intravenous with much better results.

URBASCHEK

BARTL

Springer: We have two presentations involving bone marrow, one on the influence of endotoxin and the other on the influence of the detoxified endotoxin. I wonder if Dr. Urbaschek would like to give us in a nutshell what exactly are the differences between these two substances, in their physiological effect, not in their chemical nature.

Urbaschek: Following endotoxin administration for example in miniature pigs the bone marrow smears show a shift to the left and all myelopoietic cell types are decreased in number. Endotoxoid causes an increase in the immature myelopoietic cells. The number of neutrophilic myelocytes does not change after endotoxoid whereas the number of the neutrophilic band and neutrophilic segmented granulocytes decreased. However, the decrease in the latter cells is more pronounced after endotoxin. Summarizing I can state that toxic as well as detoxified preparations cause stimulation of hematopoiesis and acceleration of myelopoietic maturation. After endotoxoid the changes begin later than after endotoxin. To further discuss the differences it would be necessary to talk about the significance of dosage and the interrelationship of the toxic and detoxified substances. Concerning your question on the differences in physiological effects I would like to refer to the coming paper in which clinical and clinico-chemical parameters are being discussed.

Hennemann: Dr. Urbaschek, you described bone marrow changes following endotoxoid injection which we know to be associated with regeneration, 5-7 days following medicamentous agranulocytosis without stimulation of the marrow itself. Therefore, my question is whether the changes described by you are indeed caused by endotoxoid.

Urbaschek: Since these effects occur in mice and pigs without X-irradiation I would ascribe them to endotoxoid.

Huser: The data summarized by Dr. Urbaschek are in surprisingly good agreement with the present knowledge on the effect of endotoxin on the bone marrow and on granulokinetics. Moreover, a very recent study by Stohlman and associates has shown the effect of endotoxin on granulopoiesis and on colony forming cells. Their study demonstrated evidence of elevated colony stimulating factor levels following the administration of endotoxin which seemed to induce differentiation of the marrow colony forming cells into the granulocytic pool and pathway. On the other hand, endotoxin also seems to effect the erythroid marrow in as much as decreased number of erythroid cells have been seen in these animals. Stem cell competition probably plays a role in the depletion of the marrow from these cells and in this context I would like to ask Dr. Urbaschek whether in his study using endotoxoid, effect on the erythroid cells in the marrow was also observed.

Urbaschek: Using endotoxoid in mice we observed the same effect on the erythroid cells in the marrow as with endotoxin.

Springer: May I ask what you would define as an absolute difference between endotoxoid and the endotoxin. You cannot make an individual tolerant against subsequent endotoxin application by endotoxoid. Is that correct or not?

Urbaschek: Endotoxoid induces tolerance against lethal doses of endotoxins.

Springer: What is exactly then the difference between endotoxoid and endotoxin? Except that it does not produce fever.

Urbaschek: Endotoxoid has no lethal effect in mice, guinea pigs, and pigs. However, in rabbits we observed together with K. Huth that when endotoxoid is given intravenously it acts to prepare for the generalized Shwartzman phenomenon. This could correlate with the behaviour of platelets in rabbits about which Dr. Lüscher earlier reported. In human volunteers pretreatment with detoxified endotoxin inhibits the subjective symptoms and vomiting. The endotoxin-caused biphasia of leukocytes is altered after pretreatment with endotoxoid in the sense that the initial drop in leukocytes fails to appear. Endotoxoid also inhibits the decrease in M_E of the thrombelastogram. The increase in temperature is less influenced and the secondary leukocytosis not at all.

Bona: Do you know if the endotoxoid keeps the adjuvant properties and the immunogenic properties?

Urbaschek: Yes, it does, but less intense than toxic preparations.

Greisman: Dr. Urbaschek, I am always worried about this endotoxoid versus endotoxin. The endotoxoid is not really completely inactive in terms of toxicity. It has some toxicity if you give a large enough dose. I think if I remember some of your data you can kill a small percentage of animals with a large enough dose of endotoxoid. In other words it is not completely inactive - there is some activity. What I wonder is if you give the untreated and free endotoxin in a dose which is very small equivalent in toxicity in other words reduce the dose of free endotoxin to a level equal to the toxicity of so-called endotoxoid can you now still induce the changes which you have described with the large doses of endotoxoid, i. e. is the endotoxoid preparation acting because there is a small amount of free endotoxin in it, rather than because it has some special properties?

Urbaschek: It is an open question. The possibility, however, could be reconciled with the pathophysiologic results thus far achieved. Unless chemically defined substances of the toxic and detoxified preparations are available this question can only be answered speculatively.

Springer: You could probably give some kind of an answer if you could make a statement as to the chemistry of endotoxoid as compared to endotoxin.

Urbaschek: According to Nowotny endotoxin is detoxified by treatment with potassium methylate which causes cleavage of the long chain fatty acids. The molecule becomes gradually more hydrophilic and thus the permeation through the cell membranes is impaired.

Graeff: May I comment on the effects of endotoxoid in animals. If pregnant rabbits are infused with endotoxin they produce a generalized Shwartzman reaction. If the same amount of endotoxoid is infused in pregnant rabbits they show no sign whatsoever of a generalized Shwartzman reaction. However, we observed a decrease in platelets corresponding to the decrease in platelets in pregnant rabbits after infusion with endotoxin. This is statistically significant, and I think it is a rather interesting feature.

Urbaschek: Your results concerning the decrease in platelets correspond with our results in guinea pigs, pigs and human volunteers, although you used a different detoxified endotoxin. Various detoxification methods have been published by several authors. I believe it should be mentioned here that the use of the expression endotoxoid should be more thought of as abbreviation since the association with the well-definded, uniform toxoid of gram-positive bacteria should be avoided.

Springer: I wonder whether one of the pioneers of endotoxoid, Dr. Nowotny, has a comment.

Nowotny: I would like to answer the comments of Dr. Greisman first. He asked whether the activity of endotoxoid is not due to some residual endotoxin in the preparation. We may assume that endotoxoid contains 1 to 10% toxic endotoxin, which is a very good possibility, since a chemical reaction is rarely complete. If you inject 10 μg endotoxoid you can render animals tolerant against lethal doses of endotoxin within 24 hours. If we assume that 1 μg of that is toxic endotoxin, and inject 1 μg toxic endotoxin we do not induce comparable tolerance. There are a number of other parameters which show quite clearly that there is more to it than the residual endotoxin in the endotoxoid which is responsible for some of the biological activity. Regarding the chemical changes in the endotoxoid molecule: We prepare this with potassium methylate cleavage of hydroxyl-bound long chain fatty acids. We are studying now the mechanism of glycolipid detoxification which is a small endotoxin molecule and we found that O-acyl-linkages are primarily involved and cleavage of palmitic acid is responsible for the loss of the toxicity. Cleaving of β-hydroxymyristic acid from the primary amino group of the glucosamines in the lipid moiety has nothing to do with the toxicity.

Springer: This would not agree with the Brucella statements made earlier. This is a very exciting subject.

Raynaud: Dr. Nowotny, which treatment do you use for detoxification?

Nowotny: We use 0.1 molar potassium methylate in anhydrous methanol for 60 minutes. If you carry out the process at 15°C, you remove a great percentage of the fatty acids especially from the β-hydroxymyristic acid polymer. This leads only to an increase in the activity of the preparation, the same treatment at 65°C leads to detoxification.

NOWOTNY

Springer: I think this is a very important and modern application of endotoxins. I did not see in your table whether you used this endotoxin in the host which was syngeneic to the tumor - i.e. A mice, or whether you used it to allogeneic mice. I think this is important.

Nowotny: We used it in allogeneic mice.

Springer: That would indicate that for the St-strain of the TA3-tumor you probably would not observe any effect.

Nowotny: We observed some effect, but it was not that remarkable.

Springer: And how was it with the Ha-strain in syngeneic mice?

Nowotny: Each experimental mouse injected with the nonspecific - or with the specific tumor-had an over 100% prolonged survival time due to endotoxin as compared to the controls. But they died from the tumor. The beauty of this system, using the Ha nonspecific tumor in allogeneic mice, is that it very sensitively indicates changes in immune capacity.

Springer: Yes, but it is of course in a very unfavourable environment already.

Mergenhagen: Could you clarify one point for me. When you stimulated your spleen cells in vitro and then passively transferred them, were these washed spleen cells, or did you transfer the soluble products from the lymphoid cells?

Nowotny: We washed the cells of course very carefully 4 times with 199 medium. Still the endotoxin may be inside of the cells and we may transfer simply endotoxin with these as you assumed. We measured the endotoxin content of the spleen cell homogenates. We sonicated control spleen cells and we measured the endotoxin content by the Limulus lysate assay (which is questionable, but in our hands it works very reproducibly). With this assay we discovered that the amount of endotoxin we transferred by transferring 10^7 spleen cells amounts to less than 0.01 /ug. This amount of endotoxin transferred does not provide any protection.

Bona: What is the time of contact of endotoxin and the spleen cells?

Nowotny: When we exposed them for 3 days which is the optimal time for lymphoblast formation we did not see any effect. When we exposed them for 5 hours we observed about 40% protection.

Schauer: Did you study the action of phythemagglutinin alone or in combination with endotoxin?

Nowotny: We administered phythemagglutinin and concanavalin A i.p. and we could not achieve any protective effect through these nonspecific mitogenic substances.

Springer: The so-called Hauschka-strain has a glycocalix, i.e. it has a very substantial glycoprotein coat, while the St-strain of TA3-tumors does not possess this. The glycoprotein coat of the Hauschka-strain has blood group N specificity. One can block the cells with anti-N-serum, especially from vicia graminea.

Neter: I wish to ask Dr. Nowotny to analyze the mode of action. Have you considered injecting your LPS or toxoid first and then to give the tumor both intraperitoneally and intrapleurally. This could then differentiate what is a local immune mechanism versus one which may lead to pleural development, if there is no protection at that site.

Nowotny: Thank you for the suggestion. The survival of this tumor in allogeneic host is very much dependent upon the route of application of the tumor. If you give it subcutaneously the tumors will always be rejected in an allogeneic system. We have to give it intraperitoneally. We also observed in other series of experiments that the development of ascites fluid protects the tumor from immune recognition. If there is no chance for the tumor to develop sufficient quantities of ascites fluid surrounding the tumor, the tumor will be recognized and rejected.

Springer: I think what needs to be stressed is, in the allogeneic systems the tumor is in a very labile position. For example if you treat it just a little with neuraminidase and remove a few sialic acid residues it will be knocked out and killed.

Bona: The system and results presented by Dr. Nowotny are very important. I wish to clarify for myself some small points. In the case of the adoptive system with immune spleen cells, this ability is lost after anti-theta serum treatment of the cells?

Nowotny: We did not check this yet. This and similar experiments are planned.

Bona: In the system as concerns the endotoxin pretreated spleen cells, did you test this in vitro? In this case you have only 5 hours of incubation. Do you have very clear cut differences between the normal spleen cells and the endotoxin pretreated cells in terms of protection?

Chedid: Do you have normal spleen cell controls?

Nowotny: Yes. We used only normal spleen cells in this experiment (not immune spleen cells) and exposed them to endotoxin. If you transfer normal spleen cells without exposing them to endotoxin you do not see any of it whatsoever.

Greisman: I think Dr. Nowotny may be familiar with this question. Endotoxin is a very potent inflammatory inducing agent and when put into the peritoneal cavity it undoubtedly invokes a good inflammatory response. Would any inflammatory inducing agent introduced into the peritoneal cavity of the mouse evoke the same tumor protection by nonspecific means of just inducing an inflammatory reaction in general?

Nowotny: As I mentioned, we tried among others peptone which is used to attract peritoneal exudate cells, none of them induced tumor resistance.

DISCUSSION: Absorption of Endotoxin from the Gastrointestinal Tract

Springer: Perhaps one should consider that the question whether or not endotoxin transgresses the intestinal barrier is a question of quantitative aspects. This is especially true if one takes into consideration that the intestines are not always completely intact but that there are disorders like diarrhea etc. Certainly all anti-enterobacterial antibodies and also the anti-blood group antibodies are most likely stimulated by the intestinal flora.

Raynaud: I want to know what happens after intraperitoneal injection.

Brobmann: This was shown very nicely by Daniele. If endotoxin is injected into the peritoneal cavity, a quite large amount remains at the site of injection - approximately 30%. A very large amount was found in the right and left thoracic ducts, approximately 15% in the right and 6% in the left - more than was found in the liver in these animals.

Chedid: Some preliminaries are required in order to somewhat clarify this. First of all according to the label which is used, the nature of the endotoxin used, and as you said the route of administration, it is normal to get different results. The endotoxin used by certain investigators is removed very rapidly, by others very slowly. This is due to the fact that if you take an antigen from something to which natural antibodies exist of course you will have opsonisation and very rapid removal. This is why an important prerequisite is to select an endotoxin which is removed very gradually such as is the case with certain very virulent strains and with these strains the tolerant response can easily be modulated. The second thing is the marker. Unfortunately certain biological incorporations appear to be bad tools of work, e.g. the ^{32}P is unlabelled very rapidly by simple incubation in serum because of phosphatases. Rowley's experiments were unfortunately not completely relevant as he himself knows, because he was following free label. It is extremely important to correlate the presence of label with toxic antigen. Certain cases in which the antigen is very well chosen, e.g. ^{14}C, can lead to a misleading situation because you are following, let us say, an oligosaccharide, which does not interest you biologically. It was very fortunate when Dr. Braude found the chromium labelling system - although this is not the most logical approach - since it is not biological labelling, it proved to be very good labelling and very well correlated to toxicity, on one condition, you must use sodium chromate and not chloride. It acts completely differently if you use chloride or chromate. Chloride does not fix to endotoxin and does not remain attached to it. As far as your answer for the placental barrier - our experiments by injection into the fetus show that endotoxin does not cross it. With gut resorption you still have the same problem. Of course endotoxin can be resorbed under pathological conditions. The question is to know if under normal conditions it is easily resorbed. One must be careful not only to use a label but also to use a relevant system. Under our conditions we injected 1 mg LPS orally in adrenalectomised mice which are killed by 1/100 μg in the circulation and we observed no deaths. This gives a ratio of 1:100,000. But this does not mean that non-toxic antigen could be perhaps recovered in organs. I therefore think that in the face of this apparent confusion it maybe much less confusing if we speak of the same antigens and the same systems with the same labels.

Urbaschek: In our opinion endotoxemia depends among others on the action of histamine and serotonin. After giving endotoxin per os in guinea pigs and injecting sublethal doses of histamine the antigen is detectable by serological methods, e. g. complement fixation or passive hemagglutination in most cases.

Greisman: If we infuse endotoxin slowly into the portal vein of rabbits we get just as good a febrile response perhaps as if you give it intravenously into the systemic circuit. In other words putting the endotoxin directly through the liver results in a good pyrogenic reaction to the toxin. Since the pyrogenic response is exceedingly sensitive - 1/1000 µg will elicite fever - it is difficult to conceive the normal animal absorbing amounts of endotoxin without becoming perpetually febrile or showing intermittent febrile spikes which a normal animal does not do. It is thus difficult to conceive that any significant amounts of toxin are absorbed through the gut without evoking a febrile response - in other words by this very sensitive technique.

Springer: I think that would depend somewhat on the animal because e. g. I think in dogs you find continuously in the portal vein gram-negative bacteria and also in the liver, so one should be careful about general statements.

Greisman: How many organisms in terms of numbers and the amount of toxin in the blood are you talking about? As you said we are talking about quantitative factors.

Gans: We did some studies in rats using the reversed Eck fistula infusing endotoxin through the liver, and found that the mortality was markedly reduced as compared to an unoperated animal in which the endotoxin reached the lung first. There is a definite clearance and detoxification by the liver.

Clowes: A word about the work of Dr. Fine who seems appropriate to have been able to show using the Limulus lysate assay that the introduction of histamine, serotonin and a great many other vaso-active peptides will elicit a positive Limulus lysate assay in the blood which is not present under normal conditions. I therefore think there may be something in the theory that under abnormal situations, shock, other types of situations which stimulate the production of vaso-active peptides that endotoxin may be absorbed from the gut. The whole problem is what does the Limulus lysate assay mean and is it a valid means for measuring endotoxin.

Gans: We found that slow infusion of sublethal doses of endotoxin into an intact dog results in a very slow turnover of ^{51}Cr-labelled endotoxin as determined by its radioactivity. T 1/2 at 3 hours was found. If, however, we determine the residual activity by the Limulus lysate test the activity is much more rapidly dissipated, the T 1/2 is 90 minutes. The question of course comes up - is this due to in vivo detoxification or are we measuring two different entities.

McCabe: I think there is a word just to emphasize the problem of the non-specificity of assays, the problem with labels etc. This relates to the Limulus assay in that we found it positive as often in patients with staphylococcal- and streptococcal bacteremia as in patients with coliform bacteremia. Finally the September issue of J. infect. Dis. contains a report by Dr. Wolff's group from the NIH demonstrating that a variety of substances - thromboplastin, thrombin, poly I, poly C give false positive Limulus assays. I think there is good evidence that one cannot consider it a very specific assay for clinical use, - only under very well controlled laboratory conditions.

Springer: What are the quantities as compared to endotoxin which gives the same effect?

McCabe: There are very small quantities under thromboplastin. They are not in terms of infinitesimal quantities of various endotoxins.

Gans: It might be conceivable that the enzymes which are supposedly released during shock may cause the positivity of the Limulus lysate test.

Chapter V

Hemodynamics and Metabolism

Part V

Hemodynamics and Metabolism

Correlation between Circulatory Status and Energy Metabolism in Various Tissues during Endotoxin Shock

P. Lundsgaard-Hansen

The living cell is a highly organized system dependent upon a continuous genera-tion of energy. For this generation, oxygen is the most essential of all substrates. It is therefore a very reasonable hypothesis that the lethal effects of endotoxins may exert themselves by inhibiting oxygen consumption. There are two basic possibili-ties to be considered : First, an interference with capillary exchange mechanisms followed by hypoxic intracellular damage, and, second, a specific toxic inhibition of cellular metabolism, due either to the large endotoxin molecule itself, or, more likely, to low-molecular humoral mediators. Since the most ominous clinical sign of endotoxinemia is shock, or an acute hemodynamic disturbance affecting cellular oxygen supply, we need to know whether the intracellular disorganization which may ultimately kill the patient originates within the circulation or inside the cell itself. The answer to this question has not only theoretical, but also eminently practical implications. As a point of departure for this session on hemodynamics and metabo-lism, I should like to present very briefly some studies we have performed to eluci-date this problem.

In rabbits subjected to an experimental endotoxin shock by the injection of E. coli 0111 endotoxin, 2.5 mg/kg body weight, we measured various blood parameters characteristic of shock and correlated the emerging pattern with the alterations of energy-rich phosphates and certain intermediate glucose metabolites in the heart, skeletal muscle, and liver. The animals were observed for 8 hours following the injection of endotoxin.

One example of a significant relationship between two blood parameters - mean blood pressure and the arterial lactate/pyruvate ratio - is shown in Fig. 1. The overall pattern is depicted in Table I.

The mean blood pressure and the veno-arterial difference of pCO_2, which corre-lated inversely with each other, were found to be the most representative values, in the sense that they correlated significantly with all other measured blood parame-ters except glucose, and showed no influence of time as such (which was the case for the arterial pH and the Base Deficit). The overall pattern conforms to expecta-tions for a deepening state of shock.

The behaviour of the representative blood parameters mean blood pressure and v-a pCO_2 was then related to the tissue pattern of energy and glucose metabolism in the heart, skeletal muscle, and liver. Immediately after withdrawing a blood sample, we secured tissue samples by the freeze-stop technique and analyzed their metabolite contents. In addition, appropriate enzyme reaction equilibrium constants were calculated.

The redox ratios lactate/pyruvate (L/P) and α-glycerophosphate/dihydroxyace-tonephosphate (α GP/DAP) reflect the state of cellular oxygenation, rising when this becomes deficient. The latter ratio is more informative, since its components re-main within the cell, in contrast to lactate and pyruvate. The relationships between the α GP/DAP ratio in heart and liver, and mean blood pressure and v-a pCO_2, is shown in Fig. 2 and 3. The α GP/DAP ratio rises when blood pressure goes down and when the v-a pCO_2 difference increases, i.e. with deepening shock. This find-

Work supported by Grant No. 3.10.68 from the Swiss National Research Foundation

ing indicates progressive tissue hypoxia. The overall pattern of relationship between mean blood pressure, v-a pCO_2, and the metabolic state of the three tissues studied, is illustrated in Table II.

First, the behaviour of the representative blood parameters correlated highly significantly and in a consistent manner with the redox ratios of the tissues. Second, they also correlated with an increased glycolytic rate, as evidenced by the drop of the phosphofructokinase reaction constant, which indicates that this rate-limiting enzyme and therefore glycolysis was progressively activated. Third, they showed a significant relationship with the behaviour of the energy-rich phosphates. In skeletal muscle, creatine phosphate went down with shock. In the heart and in the liver, there was a progressive shift within the adenyl phosphate group, i.e. from ATP to the di- and monophosphorylated compounds. Accordingly, the ATP/ADP ratio, which is one indicator of mitochondrial function, deteriorated with the progression of shock. However, as shown by the constant sum of ATP/ADP and AMP, there was no significant breakdown to non-phosphorylated compounds beyond the AMP step, the loss of which from cells into blood inhibits the restoration of a normal energy status upon resumption of an adequate cellular oxygen supply. This finding, together with the absence of a critical tissue glycogen depletion or a critical degree of hypoglycemia even at blood pressures around 40 mm of mercury, revealed that this severe degree of shock had not - or at least not yet - sufficed to produce an "irreversible" state of energy and glucose metabolism.

In the liver, there were four metabolic parameters which behaved in a manner possibly attributable to direct, intracellular endotoxin effects: The ATP/ADP ratio, glucose, and glucose-6-phosphate decreased with time in the endotoxin animals only (not in the controls), independently of the shock state, and the lactate level rose. However, from the overall results of this study, we concluded that during the first 8 hours after the injection of endotoxin, the dominating factor was the degree of shock which it produced.

In a parallel series of in vitro experiments, we have sought for any effects of the catecholamines, histamine and serotonin on hepatic glycolytic enzymes. We found that serotonin inhibits lactate dehydrogenase, the effect starting at a concentration which is only about two times the normal level in rabbit serum. Unfortunately, we have no data on the effects of this mediator on the gluconeogenetic enzymes which will be discussed later during this session, but we may conclude that serotonin is basically capable of acting as a metabolic inhibitor.

Altogether, then, it is our impression at present that the primary causes for a fatal outcome of endotoxinemia are to be sought for within the circulation, rather than inside the cell itself. However, we are of course well aware that our data cover a limited area only of this undoubtedly complex field. In conclusion, I hope that this session may bring us a little closer to understanding the lethal effects of endotoxins - the number of patients who might benefit from progress in this respect is large indeed.

References

1. Laederach, A., Urbaschek, B., Bohn, S., Büchler, A., Pappova, E., Lundsgaard-Hansen, P.: Experientia 28 : 630 (1972).
2. Lundsgaard-Hansen, P., Pappova, E., Urbaschek, B., Heitmann, L., Laederach, A., Molnes, N., Oroz, M., Wirth, U.: J. Surg. Res. 13 : 282 (1972).
3. Pappova, E., Urbaschek, B., Heitmann, L., Oroz, M., Streit, E., Lemeunier, A., Lundsgaard-Hansen, P.: J. Surg. Res. 11 : 506 (1971).

Table I.

Relationship between the "Representative" Blood Parameters Mean BP and v-a pCO_2 (cf. text) and the Remaining Blood Parameters : Trends Accompanying a Falling BP and a Rising v-a pCO_2 Difference

N = 54 BP_m ↓ (r = -0,388 **)   v-a pCO_2 ↑

Parameter	Trend	r_{BP}	r_{CO_2}
$pH_{art.}$	↓	0,309*	-0,639***
a-v pH	↑	-0,308*	0,718***
Base Def.$_{art.}$	↑	-0,458***	0,659***
a-v O_2 %	↑	-0,337*	0,674***
Lactate$_{art.}$	↑	-0,464***	0,462***
$L/P_{art.}$	↑	-0,502***	0,499***
$XL_{art.}$	↑	-0,368**	0,322*
Glucose	-	-	-
Time	-	-	-

* = 0,05 > p > 0,01   ** = 0,01 > p > 0,001   *** = p < 0,001

Table II. Relationship Between the "Representative" Blood Parameters Mean BP and v-a pCO$_2$, and Selected Metabolites in Heart, Skeletal Muscle, and Liver (Data Presented as in Table I)

N = 54

BP$_m$ ↓ (r = -0,388**) v-a pCO$_2$ ↑

Parameter	Heart			Muscle			Liver		
	Trend	rBP	rCO$_2$	Trend	rBP	rCO$_2$	Trend	rBP	rCO$_2$
α GP	↑	-0,574***	0,464***	↑	-0,524***	0,342*	↑	-0,578***	0,590***
Lactate	↑	-0,442**	0,642***	↑	-0,554***	0,635***	↑	-0,379**	0,571***
α GP/DAP	↑	-0,571***	0,538***	↑	-0,438**	0,396**	↑	-0,561***	0,600***
L/P	↑	-	0,444***	↑	-0,328*	0,320*	↑	-0,442**	0,422**
PFK	↓	0,291*	-0,333*	-	-	-	↓	0,298*	-0,325*
PCr	-	-	-	↓	0,306	-	-	-	-
ATP	-	-	-	-	-	-	↓	0,392**	-0,474***
ADP	↑	-0,435**	0,383**	-	-	-	↓	0,345*	-0,309*
AMP	↑	-0,326*	-	-	-	-	↑	-	0,302*
ATP/ADP	↓	0,324*	-0,287*	-	-	-	↓	0,354	-0,408**
ΣATP / ΣAMP / ADP	-	-	-	-	-	-	-	-	-

* = 0,05 > p > 0,01   ** = 0,01 > p > 0,001   *** = p < 0,001

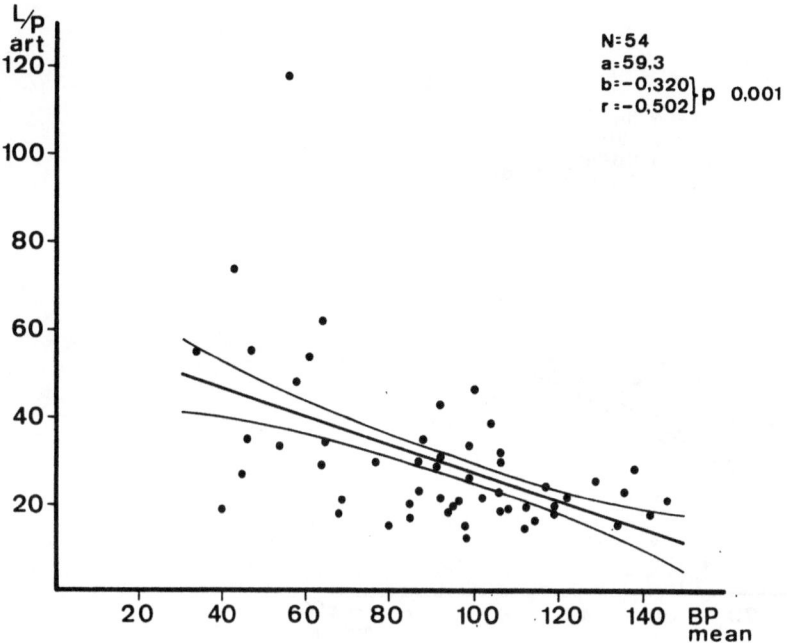

Fig. 1. Relationship between mean arterial blood pressure (BP) in mm of mercury and arterial lactate/pyruvate (L/P) ratio in 54 rabbits in endotoxin shock.

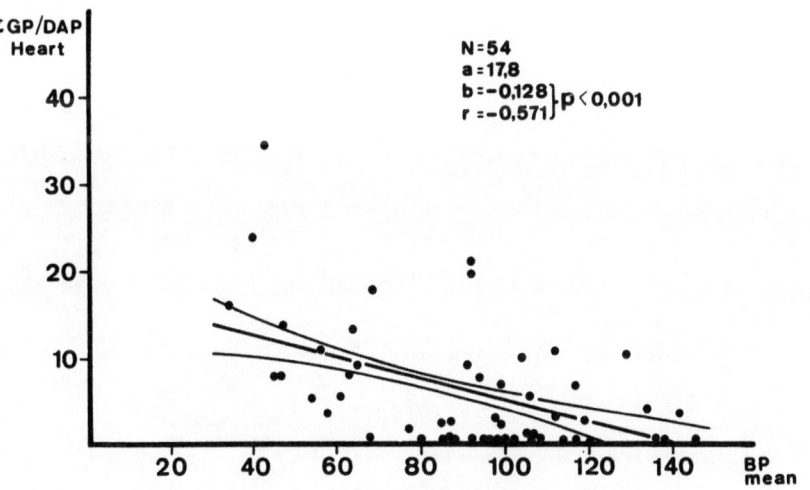

Fig. 2. Relationship between mean arterial BP and the alpha-glycerophosphate/dihydroxyacetone phosphate (α GP/DAP) ratio in heart muscle.

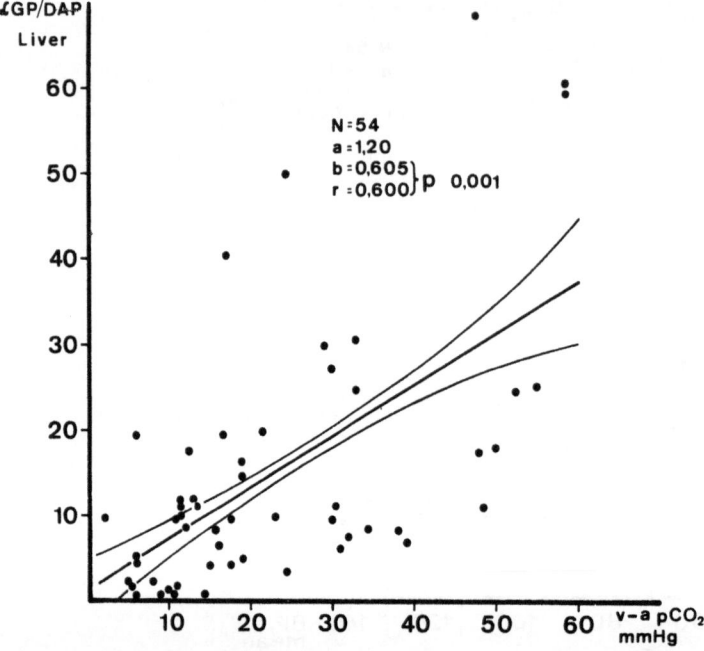

Fig. 3. Relationship between v-a pCO₂ difference in mm of mercury and αGP/DAP ratio in the liver.

Prognostic Indices for Assessment of Patients with Gram-Negative Bacteremia and Shock. An Evaluation of Hemodynamic, Respiratory and Metabolic Parameters

P. Chang*, A. A. Afifi*, V. Y. Liu, H. Nishijima, M. H. Weil, and H. Shubin

Introduction

Quantitative assessment of the severity of illness and prognosis of critically ill patients provides a basis for objective evaluation of selective aspects of patient care. In studies on patients with acute myocardial infarction and shock, linear combinations of cardiac index and lactate were most reliable for estimating severity of illness and prognosis (1). When patients with shock due to drug overdose were analyzed, two different variables emerged as the most reliable indicators, systolic pressure and arterial pH (2). The present study was undertaken to identify the most useful hemodynamic, respiratory and metabolic variables for assessing the severity of illness in patients with shock due to gram-negative bacterial infection. Prognostic indices based on combinations of these variables were derived. The effect of the time of the measurements also was evaluated to determine its influence on the reliability of the prognosis.

Method

Patient Sample

The study extended over the period from 1965 to 1971. During this time there were 96 patients treated in the Critical Care Ward, who had at least one positive blood culture. In thirty-seven of these patients only gram-positive cocci were cultured and these patients were excluded from the study. The other 59 patients had at least one blood sample from which aerobic gram-negative rods were cultured. Twenty-seven of these 59 patients also were excluded for the following reasons : 1. five were not in shock ; 2. in four patients there was some question as to whether the cultures might have been due to contamination ; 3. two patients had only minimal data ; 4. sixteen patients had other diseases which in themselves could have accounted for shock. The remaining 32 patients constituted the subjects for this study. The source of infection was in the intestinal tract in nine patients. This was most often related to peritonitis following perforation of a viscus or infarction of bowel. Twenty-three patients had nonenteric foci of infection, with the urinary tract the most common site.

The following additional criteria also were present : 1. the systolic blood pressure was less than 90 mm Hg or had decreased by more than 50 mm Hg from the pre-shock values ; 2. there was evidence of perfusion failure with an increase in lactate concentration to levels exceeding 2 mM/l. The patients ' ages ranged from 34 to 86 (mean 64) years. Twelve patients survived. Cirrhosis was found in four patients at autopsy ; nine other patients had a history of heavy alcohol intake without obvious evidence of cirrhosis. There was a history of heart disease in 10 patients

* The research programs of the Shock Research Unit are supported by grants from the John A Hartford Foundation, Inc , New York, and by the United States Public Health Service Research grants HE 05570 and GM 16462 from the National Health Institute and grant HS 00238 from the National Center for Health Services Research and Development. Computing assistance was obtained from the Health Science Computer Facility, UCLA, sponsored by NIH special research resources grant RR-3.

with either cardiomegaly, left ventricular hypertrophy and/or atrial fibrillation. Absolute hypovolemia, as reflected by a total blood volume index of less than 55 ml/kg, was present in only two patients.

Measurement

Atrial systolic pressure (SP), arterial diastolic pressure (DP) and mean arterial pressure (MAP) were measured by means of a catheter advanced into the lower abdominal aorta after percutaneous insertion into the femoral artery. Mean central venous pressure (MVP) was measured either in the superior vena cava or in the right atrium (3). Cardiac output (CO) was measured by dye dilution technique utilizing idocyanine green dye which was injected into the central venous catheter with withdrawal of blood from the arterial catheter (4). The electrocardiogram was monitored continuously. Heart rate (HR), peripheral vascular resistance (PVR), cardiac index (CI), cardiac work (CW), mean circulation time (MCT), appearance time (AT), central blood volume (CBV) and stroke volume (SV) were calculated using standard formulas (5). Rectal temperature (RT) and toe temperature (TT) were measured by means of thermistor probes. Hematocrit (Hct), hemoglobin (Hb), partial pressure of carbon dioxide (P_{co2}) and oxygen (P_{o2}) and oxygen saturation (S_{o2}) on venous and arterial blood were measured by standard techniques utilizing a Radiometer system (Model pHA 927) and an IL-Co-Oximeter (Model 182) (6). For the calculation of oxygen consumption (V_{o2}), blood was sampled simultaneously from the central venous catheter within an interval of less than 15 minutes following inscription of the indicator dilution curve. Oxygen concentration (C_{o2}) was calculated by the standard formula : Hb x Sa_{o2} x 1. 39 + (P_{o2} x 0. 0031). Oxygen transport (O_2T) was computed as the product of the Ca_{o2} and CI. Oxygen consumption was calculated from measured values of CO according to the formula : CO x C (a-v̄) DO_2, where C (a-v̄) DO_2 represents the arterial mixed venous oxygen difference. Lactate (LAC) was measured by a semiautomated adaptation of the technique of Afifi and Weil (7).

The Initial Values of a patient consisted of the first complete set of measurements which were taken within four hours after admission to the Critical Care Unit. Complete sets of "Initial Values" were available for 25 patients. The Initial Values were obtained within 12 hours after onset of shock in 12 patients and after 12 hours in eight patients. The time of the onset of shock in the other five patients was not known. The Final Values of a patient consisted of the last complete set of measurements, prior to discharge or death, exclusive of the agonal period. Complete sets of Final Values were available for 27 patients.

Statistical Analysis

The statistical technique of discriminant function analysis was used to determine the predictive values for the individual variables. A stepwise discriminant analysis procedure was used to select the best combination of variables in predicting survival. The probability of correct prediction was used to assess the reliabilities. A comprehensive description of these procedures is presented by Afifi and Azen (8).

The statistical techniques employed in this study require that the distributions of the variables can be approximated by a normal curve. Based on past experience (9), the distributions of all variables and the logarithms of CI, AT, MCT, and LAC were approximately normal. The logarithm of these measurements and the actual values of all other variables were used in the analysis.

Results

Initial Indices

The means in patients who survived and those who died were compared with the use of the t-test (Table I). The value of the Mahalanobis D^2 was determined for each variable. D^2 is the square of the distance between the means of the two groups of pa-

tients expressed in units of the pooled standard deviation. The reliability in predicting survival of each variable was computed as 100 times the estimated probability of
correctly classifying a patient into one of these two groups (Table I). This estimated
probability is the area under the standard normal probability curve to the left of D/2.
The single most reliable variables were CBV (81 %) and CW (80 %).

The stepwise discriminant analysis program BMD07M (10) was then used to determine the best Initial Values to be combined with the most reliable single variable,
CBV. The initial measurement of pHa and CW, when added to CBV, contributed most
to predicting survival. (F = 8.3, degrees of freedom 1, 22 ; p< 0.01). A linear discriminant function and a reference value were determined from the Initial Values of
these three variables (Table II). For convenience, these were referred to as the Initial Discriminant Function (IDF) and the Initial Reference Value (IRV) respectively.
A patient was predicted to survive if his value of the IDF obtained by substituting the
Initial Values into the IDF was greater than or equal to the IRV. Using this procedure to classify the twenty-five patients from whose initial measurements the IDF
was computed, 24 (96 %) were correctly classified. The reliability of this procedure
as estimated from the Mahalanobis D^2 was 92 %. The inclusion of any other single
variable did not improve the reliability significantly or increase the number of patients correctly classified.

For a given patient, the probability of survival may be computed as :

$$P(S \mid IDF) = \frac{1}{1 + e^{(IRV-IDF)}} \tag{1}$$

This probability also can be obtained from Fig. 1 by entering IDF-IRV on the abscissa and reading the probability of survival from the ordinate. The probability of
survival is then used as the prognostic index.

The same procedure was used to determine the most reliable combinations of hemodynamic and other variables separately. The resulting discriminant functions,
percentages of correct classification and reliabilities are shown in Table II. The probability of survival may be obtained from any of these IDFs using equation (1) or
Fig. 1.

Final Indices

The means and the reliabilities of the Final Values of individual variables are
given in Table I. The most reliable variables were V_{o2} (91 %), DP (90 %), MAP
(89 %), and CW (88 %).

Among all variables, the Final Values of MAP contributed most to V_{o2} (F = 12.0,
degrees of freedom 1, 24 ; p< 0.005) ; giving a combined reliability of 96 %. Using
the discriminant function based on the final measurements of these two variables, 26
of the 27 patients (96,3 %) were correctly classified (Table III). Inclusion of any additional variable did not significantly improve the reliability of the prediction or increase the number of correctly classified patients. Among hemodynamic variables,
the best combination was DP and CBV. V_{o2} and log LAC was the best combination
among the other variables. The discriminant functions and the reference values derived from the final measurements are referred to as the Final Discriminant Function (FDF) and Final Reference Values (FRV). Using Fig. 1, each FDF and its FRV
can be used to compute the probability of survival.

Time Series Index

The Initial and Final Indices were both computed from measurements taken at a
specific point in time. They reflect only the momentary status of a patient and do
not take into account the previous measurements. A more comprehensive assessment of the patient 's status was developed by examining the trend of each variable.
Twenty-six patients had three or more sets of measurements on all variables. For
each of these patients, the time trend of each variable was examined by computing

the least squares estimate of the slope (B). The individual variable whose trends showed the greatest difference between survivors and those who died was SP. Values for SP showed an upward trend in all eight patients who survived and a downward trend in all but three of the 18 patients who died.

The Time Series Discriminant Function (TDF) based on the mean (A) and the slope of SP was obtained as :

$$TDF = 0.00524 A + 0.438 B \qquad (2)$$

The reference value for this discriminant function was 4.9. Classifying the 26 patients by this TDF, 21 (81 %) were correctly classified and the reliability was 78 %. Combining the trends of any additional variables with the trend of SP did not significantly improve the reliability of prediction or increase the number of correctly classified patients. Among respiratory and metabolic variables, the trend of LAC was most reliable (60 %). Combining the trends of LAC and any other variable did not increase the reliability .

Testing of Indices on other Data

Some of the variables used in our DF were not reported in other published studies. Using the variables which were available, a DF was derived from our data for each of the studies by Wilson (11), Hopkins (12), Motsay (13), and Bell (14). The appropriate DF was then applied to the data reported in each of these studies. The results are summarized in Table IV. When the DF based on our Initial Values was applied to their data, the percent of correction classification of outcome ranged from 60 to 75 %. However, when the DF, based on our Final Values, was applied to the same data, the percent of correction classification ranged from 20 to 81 %. The lower percentage of correct classification in the patients reported by Motsay may be explained by the time of his measurements, which was described as "baseline", and reflected the patients' status early in the course of shock.

Discussion

Attempts to estimate the severity of circulatory, respiratory and metabolic abnormalities in patients with bacterial infection and shock have been based largely on the physician's clinical judgment supplemented by a limited number of quantitative physiologic measurements (15). The availability of techniques and instrumentation suitable for the bedside of the critically ill patient has made it possible to measure changes in physiological status. This quantitative information may be used to objectively assess the severity of illness and evaluate the patient's response to treatment. In patients with acute myocardial infarction and shock, linear combinations of stroke index and lactate provided the most reliable indication of severity and prognosis (1). By contrast, among patients with drug overdose, SP and pH_a or alternatively SP and Sa_{O2} were most reliable (2). Based on these studies, it became apparent that the most reliable variables differed with the underlying condition.

Measurements of V_{O2}, C_{O2} and oxygen transport, which were not available in our previous studies, have been included in the present analysis. V_{O2} emerged as the best Final Value for predicting survival. The value for V_{O2} in patients who died was less than 50 % of that in patients who survived. The higher LAC values in the patients who died also reflected their more severe perfusion failure (7). In the present study, the linear combination of the Final Values of V_{O2} and MAP emerged as the most reliable indicator of survival. When the DF was based on Initial Values only, CBV, pH_a and CW provided the most reliable linear combination.

In some instances the variables included in these best discriminant functions may not be available. Under these circumstances, a number of other combinations of hemodynamic, metabolic and/or respiratory variables may be substituted, with only a relatively small reduction in reliability (Tables II and III).

Since measurements were obtained at various times throughout the course of a patient's critical illness, the question arose as to which set of values would be most useful in assessing the patient's status at a given time. Since the Initial Values were obtained within a few hours of the patient's admission to the Critical Care Unit, it seemed appropriate to use the probability of survival based on this set of measurements to assess prognosis during the early period of the patient's acute illness. On the other hand, the probability of survival based on the Final Values, which reflected the greatest disparity between survivors and patients who died, seemed most appropriate for assessment of the severity of the patient's acute illness at a later time. These conjectures were supported by applying the IDF and FDF to our own data and to the data on patients reported by other investigators. The method of applying the various prognostic indices is detailed in the Appendix. Since patient populations may vary from one hospital to another, as well as the modes of treatment, the same DFs may not be equally applicable in different locales (Table IV). For this reason, it may be desirable to derive DFs from the data bank of individual hospitals.

The probabilities of survival which are based on Initial or Final Values are sensitive to momentary changes in the patient's status. To reduce the emphasis placed on such momentary fluctuations, the probability of survival may be obtained from all the values of selected variables throughout the patient's course. However, this has the potential of being too insensitive to sudden changes in the patient's condition, particularly as more data sets are taken into account (Fig. 2).

The current study provides indices for evaluating patients with shock complicating gram-negative bacteremia. These indices differ from those developed for patients with shock complicating myocardial infarction (1) and drug overdose (2). Once the underlying condition has been identified, an appropriate index may be selected for ongoing assessment of patient's status and response to treatment.

Summary

Thirty-two patients with shock due to gram-negative bacterial infection were evaluated to determine the most useful hemodynamic, respiratory and metabolic variables for assessing the severity of their illness. The most reliable single variable for predicting survival was oxygen consumption. The most useful prognostic index was based on the combination of oxygen consumption and mean arterial pressure. An accumulative index based on successive measurements also was derived. Validation of the indices was obtained using other published data. A procedure for use of the indices at the bedside is provided.

Appendix

Probabilities of survival obtained from the values of IDFs, FDFs and TDFs may be used as objective indices for the assessment of severity of illness and prognosis. Since the Initial Values were based on the earliest measurements obtained on these patients, probability of survival computed from the value of IDF would appear to be appropriate as a prognostic index in assessing a patient's status in the early period after the onset of shock. The probability of survival also may be obtained by substituting into the FDF the measurements taken at any later point in time. This probability may be used as an index for assessment of the momentary severity of illness at that time because the Final Values best reflect the contrasting conditions of the patients who survived and those who died. The probability of survival obtained by using the TDF which was based on the trends of SP may be used as an index of both severity and prognosis.

Computation of the probability of survival may be readily and rapidly accomplished at the bedside with the use of only a desk calculator by the following steps:

Step 1: Record the values for the selected variables as they become available.

Step 2: Choose the appropriate discriminant function based on the duration of shock. If the patient has been in shock for only a few hours, any of the three IDFs shown in Table II may be used depending on which measurements are available. If the patient has been in shock for a longer period of time, any of the three FDFs and/or the TDF shown in Table III may be used.

Step 3: Compute the selected DF using the appropriate formula (Tables II and III).

Step 4: For the selected DF (Tables II and III), compute DF - RV.

Step 5: Locate the position of DF - RV on the abscissa (Fig. 1) and read the probability of survival from the ordinate.

Step 6: When TDF is selected:

i) use the values of all available measurements of SP to compute the mean and slope by the formulas:

$$A = \frac{y_1 + y_2 + \cdots + y_n}{n}$$

$$B = \frac{(y_1 - A)(t_1 - \bar{t}) + (y_2 - A)(t_2 - \bar{t}) + \cdots + (y_n - A)(t_n - \bar{t})}{(t_1 - \bar{t})^2 + (t_2 - \bar{t})^2 + \cdots + (t_n - \bar{t})^2}$$

where y_1, y_2, ---, y_n are available values of SP, and t_1, t_2, ---, t_n are the times in hours after admission when the values of SP are determined, and $\bar{t}$ is the mean of t_1, t_2, ---, t_n.

ii) repeat Step 3 using the values A and B,

iii) repeat Steps 4 and 5 using the TDF obtained from Step 3.

The application of these indices is demonstrated for one of the patients in this study. Approximately one hour after admission to the Critical Care Unit, and four hours after onset of shock, the Initial Values were: CBV 1.84 liters, pHa 7.50 units, and CW 4145 gm m. These values give an IDF of 101.4. Entering the value IDF-IRV 1.7 into Fig. 1, the probability of survival at one hour after admission was 85%. At 35 hours after admission to the CCW, the value of V_{o2} was 187 ml/min and of MAP was 81 torr, giving a FDF 21.8. From Fig. 1, FDF-FRV 3.4 yielded a probability of survival of 97%. At this time, eight values of SP were available for this patient. The mean and slope of SP were 114 mm Hg and 1.09, respectively, giving a TDF of 6.47 and a probability of survival of 83%.

To illustrate the relationship of the three indices, the probabilities of survival computed from the measurements taken at different times were plotted (Fig. 2). The probabilities of survival based on IDF and the FDF were more sensitive to the momentary fluctuations in the status of the patient. The probabilities of survival based on the TDF were usually lower but more stable. Noteworthy is the lack of sensitivity of the IDF to the Final Values or the FDF to the Initial Values.

References

1. Afifi, A. A., Chang, P. C., Liu, V. Y., Da Luz, P. L., Weil, M. H., Shubin, H.: Prognostic indices in acute myocardial infarction complicated by shock. Amer. J. Cardiol., in press.
2. Afifi, A. A., Sacks, S. T., Liu, V. Y., Weil, M. H., Shubin, H.: New Engl. J. Med. 285: 1497 (1971).
3. Shubin, H., Weil, M. H., Rockwell, M. A. Jr.: Med. Biol. Engl. 5: 361 (1967).
4. Shubin, H., Weil, M. H., Rockwell, M. A. Jr.: Med. Biol. Engl. 5: 353 (1967).
5. Udhoji, V. N., Weil, M. H., Sambhi, M. P., Rosoff, L.: Amer. J. Med. 34: 461 (1963).
6. Zahn, R. L., Weil, M. H.: J. thorac. cardiovasc. Surg. 52: 105 (1966).
7. Weil, M. H., Afifi, A. A.: Circulation 41: 989 (1970).

8. Afifi, A. A., Azen, S. P. : Statistical Analysis : A Computer Oriented Approach, p. 252 (Academic Press, New York 1972).
9. Winkel, P., Afifi, A. A., Cady, L. D., Weil, M. H., Shubin, H. : J. chronic. Dis. 24 : 61 (1971).
10. BMD : Biomedical Computer Programs. In : Dixon, W. J., 3rd. ed., p. 233 (University of California Press, Berkeley 1973).
11. Wilson, R. F., Thal, A. P., Kindling, P. H., Grifka, T., Ackerman, E. : Arch. Surg. 91 : 121 (1965).
12. Hopkin, R. W., Sabga, G., Penn, L., Simeone, F. A. : JAMA 191 : 127 (1965).
13. Motsay, G. J., Dietzman, R. H., Ersek, R. A., Lillehei, R. C. : Surgery 67 : 577 (1970).
14. Bell, H., Thal, A. : Postgrad. med. J. 48 : 106 (1970).
15. Kwaan, H. M., Weil, M. H. : Surgery 128 : 37 (1969).

Table I. Initial and Final Mean Values and their Prognostic Reliability

VARIABLES	UNITS	INITIAL				FINAL			
		Survived	Died	P	Reliability (%)	Survived	Died	P	Reliability (%)
SP	mm Hg	90	87		53	121	73	g	83
DP	mm Hg	51	42		66	63	32	g	90
MAP	mm Hg	64	55		63	84	44	g	89
HR	beats/min	111	106		53	95	93		50
MVP	mm Hg	7.4	7.8		51	7.3	8.2		56
TR	$^\circ$C	38.2	37.0		61	37.5	35.8		64
TT	$^\circ$C	27.2	26.1		59	28.9	26.7		64
CI*	L/min/m^2	3.6	2.3	e	74	3.5	2.5	b	70
MCT*	sec	18	20		55	18	22		59
AT*	sec	9.0	9.6		53	9.2	10.3		56
PVR	dynes sec cm^{-5}	722	1049		65	1043	809		60
CBV	liters	1.97	1.30	f	81	1.91	1.31	a	72
SV	ml/beat	63	41	c	72	68	47	a	72
CW	g m	5862	3068	f	80	7311	2487	g	88
Hb$_a$	g/100 ml	10.0	10.0		50	10.4	10.2		52
Hct$_a$	%	31	30		50	31	29		54
pHa	U	7.50	7.34	c	71	7.48	7.31	b	73
Pa$_{CO2}$	mm Hg	34	36		51	32	36		59
Sa$_{O2}$	%	93	94		51	96	95		56
Pa$_{O2}$	mm Hg	94	125		57	95	92		50
LAC*	mM	3.99	6.77	a	68	2.68	8.84	e	77
Ca$_{O2}$	Vol %	13.3	13.7		51	14.2	13.7		55
C(a-$\bar{v}$)DO$_2$	Vol %	3.8	3.8		50	3.8	3.2		57
V$_{O2}$	ml/min	235	150	d	73	233	110	g	91
O$_2$T	ml/min	86	57	d	72	92	55	e	75

* Values shown are antilogs of the mean logarithmic values.

P Values:

a = < 0.05   e = < 0.005
b = < 0.025   f = < 0.001
c = < 0.02   g = < 0.0005
d = < 0.01

Table II. Initial Discriminant Functions, Reference Values and their Prognostic Reliabilities

Combination	Initial Discriminant Function (IDF)	Reference Value[*]	Actual % of Correct Classification	Reliability (%)
Hemodynamic- Metabolic- Respiratory	5.06389 CBV + 11.81079 pH_a + 0.00085 CW	99.69092	96	92
Hemodynamic	3.90008 CBV + 0.00085 CW	10.15261	92	87
Respiratory- Metabolic	0.04999 V_{O_2} + 2.16041 $C(a-\bar{v})DO_2$ + 23.11109 pH_a + 0.17628 Pa_{CO_2}	179.01953	92	92

[*] A patient is classified to survive if his value of DF is greater than the reference value.

Table III. Final and Time-Series Discriminant Functions, Reference Values and their Prognostic Reliabilities

Combination (or Variable)	Discriminant Function	Reference Value*	Actual % of Correct Classification	Reliability %
Final Discriminant Functions				
Hemodynamic-Metabolic-Respiratory	$0.05425\ V_{o2} + 0.14288$ MAP	18.38501	96	96
Hemodynamic	0.22623 DP $+ 3.37278$ CBV	16.16737	89	93
Respiratory-Metabolic	$0.0498\ V_{o2} - 4.74195$ log LAC	5.63769	93	91
Time-Series Discriminant Functions				
SP	0.05240 A $+ 0.43766$ B	4.90007	81	78

* A patient is classified to survive if his value of DF is greater than the reference value.

Table IV. Percentages of Correct Classification of Patients Reported in other Series *

Author & Reference	Initial			Final		
	IRV - IDF	Correct Classification % other studies	our study	FRV - FDF	Correct Classification % other studies	our study
Wilson (11)	$4.246 - 9.218 \log CI$	$\frac{9}{12} = 75\%$	$\frac{20}{25} = 80\%$	$9.742 - 0.199$ MAP $+ 0.0032$ PVR	$\frac{9}{12} = 75\%$	$\frac{24}{27} = 89\%$
Hopkins (12)	$1.373 - 12.291 \log CI + 5.992 \log LAC$	$\frac{7}{11} = 63\%$	$\frac{21}{25} = 84\%$	$5.080 - 0.286$ DP $+ 0.005$PVR$+6.33 \log LAC$	$\frac{9}{11} = 81\%$	$\frac{25}{27} = 93\%$
Motsay (13)	$3.548 - 12.940 \log CI -0.248$ DP $+ 0.159$ SP	$\frac{11}{15} = 73\%$	$\frac{23}{25} = 92\%$	$10.700 - 0.295$ DP $+ 0.004$ PVR	$\frac{3}{15} = 20\%$	$\frac{25}{27} = 93\%$
Bell (14)	$63.825 - 9.446 \log CI - 8.016$ pHa	$\frac{12}{20} = 60\%$	$\frac{23}{25} = 92\%$	$12.750 - 0.214$ DP $- 5.874 \log CI$	$\frac{10}{20} = 50\%$	$\frac{24}{27} = 89\%$

* The Initial and Final Discriminant functions were derived by using our Initial and Final Values to select from the available variables in each study. These discriminant functions then were used to classify the patients in each study.

PROBABILITY OF SURVIVAL %

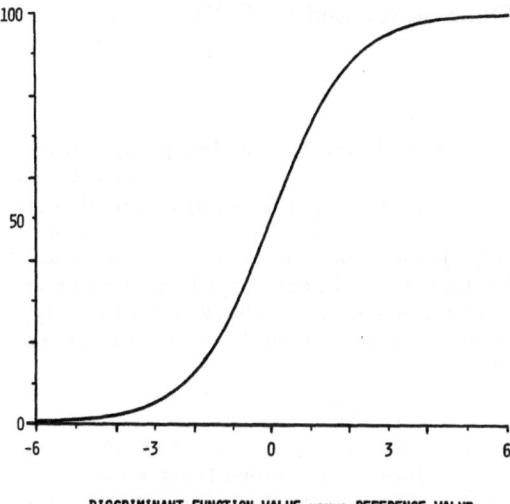

DISCRIMINANT FUNCTION VALUE minus REFERENCE VALUE

Fig. 1. Probability of survival based on discriminant function and reference values.

PROBABILITY OF SURVIVAL %

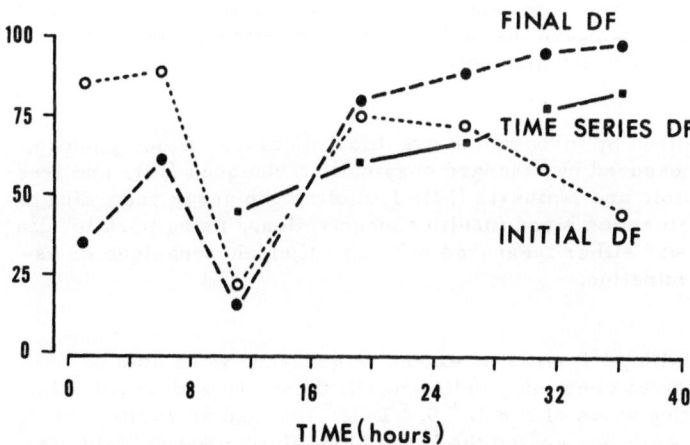

Fig. 2. Successive probabilities of survival, based on IDF, FDF, and TDF for a patient with bacterial shock. The equations for each discriminant function is given in Table II. Death appeared likely ten hours after admission when the patient's blood pressure and cardiac output were reduced. However, administration of fluids intravenously increased the cardiac output and blood pressure and the patient survived.

Abnormalities of Energy Metabolism in Sepsis and Endotoxemia

G. H. A. Clowes, Jr., T. F. O'Donnell, Jr., and N. T. Ryan

Patients who survive severe infections characteristically have low peripheral vascular resistances and maintain elevated cardiac indices in excess of 4 L/M^2/min to satisfy the high circulatory demand (1, 2, 3). On the other hand, the mortality of patients who are more than transiently hypotensive with cardiac outputs below the normal resting value is greater than 60% (4, 5). Certain well-known metabolic abnormalities accompany the septic state. Proteolysis and negative nitrogen balance far exceed the values observed in simple starvation (6, 7, 8). Blackburn et al. (9, 10) have demonstrated that in many such patients under these conditions the utilization of endogenous fat is suppressed by elevation of blood insulin secondary to glucose infusion. A pseudo diabetes and insulin resistance are typical of the seriously infected patient (11). Lactacidemia is known to accompany the low flow state, and often is found in the presence of high outputs (5). McLean et al. (12) suggest that the prognosis of the seriously infected patient is closely related to the blood lactate value.

Data obtained from the clinical observations presented in this paper suggest that a significant difference exists in the patterns of energy production in the high and low flow states. The question still exists as to the role of altered metabolism in the periphery as a determinant of the large circulatory demand versus the extent to which inadequacy of circulation and oxygen delivery may determine the utilization of metabolic substrates associated with abnormal production of metabolites by muscle and other peripheral tissues. To study this matter of peripheral metabolism in detail, experiments in pigs were designed to measure the utilization of substrates and metabolite release by a hind limb. These experimental data further indicate that not only is there a significant difference between the high and low flow state of sepsis but that endotoxemia results in a different metabolic pattern.

Methods

Cardiac output was measured by indocyanine dye dilution curves. Blood glucose, lactate and glycerol were measured by standard enzymatic techniques (13), and free fatty acid by the method of Dole and Meinertz (14). Radioimmuno assay according to Soeldner (15) was used to determine blood insulin concentrations, using pork insulin standard, urinary nitrogen was either measured by micro Kjeldahl technique or estimated by urine urea determination.

Clinical Observations

Eighteen septic patients with peritonitis or extensive gangrene were studied. All had wound or peritoneal cultures containing coliform. Of these, 10 with cardiac indices above the normal resting value of 2.8 L ± 0.4 L/M^2/min had an average of 4.6 ± 0.8 L/M^2/min. They were designated the "high blood flow" group. Eight others with an average cardiac index of 2.4 ± 0.3 L/M^2/min were designated as the "low blood flow" group. Data from these observations comparing the mean values of both groups with those of fasted normals are given in Table I.

Attention is drawn to the fact that the body temperature (100.6°F) of the low output patients who were hypotensive was significantly less than that of the high flow group (102.4°F). Other important observations were that the elevated blood glucose in these patients who were receiving 5% glucose in water intravenously was high in

The research reported in this paper was supported by Grant #GM-19951-01 of the U.S. Public Health Service.

both groups, 7.2 and 7.6 μM/ml. However, blood insulin was high (42 μU./ml) in
the former and very low in the low flow group (6 μM/ml) which compared with a resting value of 14 μM/ml in the fasting normal people. Although lactacidemia was present to some extent in both septic groups, blood lactate was significantly higher
(2.8 μM/ml) in the low flow patients. Urinary nitrogen excretion, although greater
in all septic patients than in the controls, was particularly elevated in the hypotensive septic patients, amounting on the average to 16 g/24 hrs.

Experimental Observations

Induced peritonitis: In pigs, the responses to peritonitis induced by cecal ligation
have been extensively studied by us in the past. The mortality is 47% and those who
survive, as in man, show a progressively elevated cardiac output. The average is
59% above the basal resting value after three days. In addition, the kinin and clotting systems were found to be closer to man than any of the other common laboratory animals with the exception of certain primates. The blood insulin and metabolic
responses of the pig to starvation also resemble those of man. Therefore, the pig
was selected to isolate and study the peripheral metabolism of muscle and adipose
tissue in sepsis.

In fully conscious, 30-45 kg pigs, fasted 4 days, the cardiac output, limb blood
flow and arterial-venous differences of oxygen, carbon dioxide, insulin, substrates
and metabolites were measured through previously placed inlying central venous,
femoral arterial, and femoral venous catheters. After cecal ligation performed under halothane anesthesia, measurements were repeated in three days. Animals
which died prior to this time were discarded. Each animal received a daily infusion
of 1,000 cc Ringer lactate solution, and were allowed water ad lib but no food. At
autopsy, when sacrificed, large pericecal abscesses and/or generalized serosan-
guineous peritonitis were present and cultured a mixed flora of enteric organisms.

Of the 16 pigs with peritonitis, 9 exhibited high cardiac outputs with an average
value of 5.0 $\pm$ 0.6 L/min while 7 had an average of 2.0 $\pm$ 0.3 L/min. The first group
were designated as "high flow" and the latter "low flow". The results are presented
in Table II. Attention is drawn to the fact that arterial blood insulin (41 μU/ml) is
significantly higher (p<0.001) in the group of animals which exhibited a high cardiac
output than in the basal starved or low flow group which were respectively 10 and
14 μU/ml. Despite this fact, the limb glucose uptake was slightly lower in the high
flow group than either of the others. Although the cardiac output and hind limb flow
were less in the low flow group, there was not a significant difference of limb oxygen uptake between any of the groups. Nevertheless, lactacidemia and lactate production by the hind limb of the low flow group was nearly double (p<0.001) that of
either the fasting or high flow group.

Whereas free fatty acids were taken up in starvation at a rate of .042 μM/kg/min,
FFA was released in both septic groups. The extent of lipolysis in the limb is indicated by the glycerol release. Approximately three mols of fatty acid are mobilized
from the fat deposits for each mol of glycerol. On this basis, the net fatty acid utilization by the limb in starvation was 0.22 μM/kg body wt/min, while the high flow
group had none and the low flow group 0.22 μM/kg body wt/min.

Experimental Endotoxemia

A similar preparation in conscious pigs, employing previously placed cannulae
in the jugular vein, the femoral artery, and the femoral vein, was employed to study
the effects of endotoxemia. Basal measurements were made after 4 days of starvation. Difco E. coli endotoxin, 2 mg/kg body weight, was subsequently administered
as a single dose intravenously. Observations were made at intervals thereafter during the ensuing 24 hrs. The results of these experiments are presented in Table III.

Two animals died at six hours, two between 12 and 24 hrs and two survived. Although the endotoxemia produced an immediate fall of cardiac output, the mean val-

ues at one and four hours were not significantly different from the preliminary control basal value. However, the hind limb flow declined significantly (p<0.001) to one-
half that of the initial value, with a marked rise in vascular resistance as the blood
pressure rose at four hours.

It is important to note that the arterial blood glucose progressively declined from
4.3 to 1.8 μM/ml. This remarkable hypoglycemia was accompanied by no elevation
of the blood insulin concentration. At one hour, the limb glucose uptake had fallen
significantly (p<0.001) from a normal value of 1.5 to 0.2 μM/kg/min and remained
low. However, lactate production remained unchanged despite a decline of limb oxygen uptake from 12.2 to 7.5 ml/kg body wt/min. Lipolysis, indicated by glycerol release, during endotoxemia is approximately one-quarter of that observed in simple
starvation. At the same time, free fatty acids are being released from the hind limb
in contrast to the starved state in which the limb was taking up fatty acids at a rate
of 0.25 mM/kg body wt/min. Calculation of the net fat utilization indicates that of
starvation to be 0.31 and in endotoxemia at one hour .023 and at 4 hrs .063 μM/kg
body wt/min.

Discussion

Endogenous fuel mobilization and its utilization by peripheral tissues is demonstrated by these clinical and experimental observations to differ from starvation in
high or low blood flow states of sepsis and in acute gram-negative endotoxemia. The
mechanisms may vary in each, involving alterations not only of the blood insulin concentration and its function at the tissue level, but also by the action of other hormones, particularly the corticosteroids, catecholamines, and growth hormone. Of
importance is the possibility that blocks may exist in the pathways of energy production and oxidative phosphorylation. The role of the circulation and the delivery of
oxygen probably is a major determinant, especially in the low flow states of sepsis
and endotoxemia.

Insulin has been defined by Cahill (6) as "the overall fuel control in mammals".
In the fed state it is high, suppressing mobilization of carbohydrate, fat and protein.
In the starved state, the reverse is the case as observed in the fasted humans (Table I) and the series of animals (Tables II and III). By contrast, in both septic man
and the starved septic pig, a paradoxical situation exists. Almost maximal values
of blood insulin are observed during the high output septic state. Lypolysis, as indicated by glycerol release, free fatty acids (FFA) of the blood and, secondarily,
fat uptake by the muscle of the leg are all low. Thus, it appears that fat tissue is
responding under these conditions to insulin as it should. Yet the blood insulin is
markedly elevated, while the blood glucose is in the low normal range and glucose
uptake is not significantly different from the starved state. This finding suggests
that a block exists to insulin action on muscle which promotes glucose transport
principally into muscle.

The situation regarding the blood insulin concentration in the septic patients and
animals with low cardiac output who are unable to meet the usual high circulatory
demand is very different. Here, as in the endotoxic state, the blood insulin remains
in the low range found in starvation. Possibly this is because of intense sympatho-
adrenal activity as indicated by tachycardia and a significantly elevated vascular resistance in the leg. It has been observed in shock states in which catecholamine concentrations are high that insulin secretion is low (16, 17). However, in the low flow
septic animals, the blood glucose and uptake of glucose by the hind limb remains
normal, despite the fact that more of it appears to be converted to lactate. In this
group of animals, as in the low flow septic patients, lipolysis, blood FFA, and net
fat uptake by the hind limb are all higher than in the high flow septic group, a response that might be expected in the presence of low insulin and elevated catecholamine activity.

Again, the pattern is different in the animals subjected to endotoxemia. There is obviously intense peripheral vasoconstriction in which the circulatory resistance approaches twice that of the starved control state. Not only does the blood insulin remain low, probably because of intense catecholamine activity which apparently is potentiated by endotoxin (18), but the blood glucose and glucose uptake by the limb progressively fall. The lack of glucose may well be the result of inhibition of glycogenesis in the liver by endotoxin (19). At the same time, both fat mobilization and its net utilization by the leg in the endotoxic animals progressively falls. In part, this phenomenon may be due to lack of perfusion of fat deposits and the presence of high lactate levels which are known to inhibit lipolysis (20, 21, 22, 23).

Regarding the lactacidemia in the septic patients and the production of lactate by the peripheral tissues in the low flow septic animals, it is tempting to suggest that lack of oxygen might be responsible. However, the oxygen uptake by the hind limb in the septic pigs did not differ significantly from the control state of starvation. Possibly tissue hypoxemia plays an important part in the very severe lactacidemia of the pigs given endotoxin. As the hind limb flow fell from 2.5 to 1.6 ml/kg body wt/min, the oxygen consumption of the leg fell 31%. Yet lactate production of the leg did not change.

Another possibility suggested by Drucker (24) and Schumer (25) is that an anoxic type of block occurs in the aerobic oxidation of glucose after shock. Recent studies on the enzyme activity in the energy pathways by our group (26) indicate that the insulin sensitive pyruvate dehydrogenase (PDH) is reduced in muscle during sepsis. Comparing the activity of PDH in the diaphragms of fed, starved, and septic rats with induced peritonitis, a decrease approximately to 50% was found in starvation as expected in the presence of low blood insulin values (27). However, the septic animals had PDH values of only 25% that of the starved animals despite the high insulin values observed during infection. At the same time, PDH and the glucogenic enzymes of the liver remained at a normal level of activity. Since PDH is a rate-limiting enzyme, it is suggested that there is a block to aerobic glycolysis in muscle. Thus, pyruvate is converted to acetyl Co A in tissue amounts, and lactate is released. Presumably, Cori cycle activity is then increased, accounting in part for the observation of Long et al.(8) that the glucose pool is increased in septic patients.

In considering the energy requirements, no account is taken of ketone bodies. However, ketone blood levels and utilization are proportional to the circulating free fatty acids. Data on the extent of proteolysis is also missing from these experiments, but it is apparent from Table I that the daily nitrogen excretion is nearly double that of starvation in the high flow septic patients. Those in the low flow state on the average reached the remarkably high nitrogen excretion of 16 grams per day, indicating severe proteolysis. This finding makes it tempting to speculate from crude calculations that the obvious energy deficits obtained from glucose utilization, lactate production, and fat uptake in the legs of the septic pigs must be made up by oxidation of amino acids.

Finally, it must be emphasized that the metabolic abnormalities of sepsis vary with the behaviour of the circulation. However, the high circulatory demand at rest which is satisfied in the high flow group is probably determined, in part, by the altered metabolic use of substrates and energy pathways, in great measure related to muscle insulin resistance. On the other hand, in low flow states, especially in acute endotoxemia, the pattern of metabolism probably is markedly influenced by the peripheral vasoconstriction, associated as it is with severe catecholamine activity.

In conclusion, it is apparent that the metabolic defects may lead to death by different routes. In the high flow state, sufficient energy is supplied but at the expense of muscle and other protein. In the low flow state, an acute energy deficit exists in the peripheral tissues. Obviously treatment in the two situations is different. We have found in the high flow, high insulin group that avoidance of glucose and the infusion of amino acids lowers the blood insulin which promotes endogenous fat utilization and protein sparing (9, 10). In the low flow shock group we have had success in restoring circulation and metabolism by the use of glucose, potassium and insulin (28).

References

1. Clowes, G.H.A., Jr., Farrington, G.H., Zuschneid, W., Cossette, G.R., Saravis, C.A.: Ann. Surg. 171: 663 (1970).
2. Shoemaker, W.C.: Surg. Gynec. Obstet. 134: 810 (1972).
3. Siegal, J., Greenspan, M., DelGuercio, L.R.M.: Ann. Surg. 165: 504 (1967).
4. Clowes, G.H.A., Jr., Vucinic, M., Weidner, M.G.: Ann. Surg. 163: 6 (1966).
5. Rubin, J.W., Clowes, G.H.A.: Surg. Clin. N. Amer. 49: 489 (1969).
6. Cahill, G.F., Jr.: Diabetes 20: 785 (1971).
7. Cuthbertson, D.P., Tilstone, W.J.: Adv. clin. Chem. 12: 1 (1969).
8. Long, C.L., Spencer, J.L., Kinney, J.M., Geiger, J.W.: J. appl. Physiol. 31: 110 (1971).
9. Blackburn, G.L., Flatt, J.P., Clowes, G.H.A., Jr., O'Donnell, T.F., Jr.: Amer. J. Surg. 125: 447 (1973).
10. Blackburn, G.L., Flatt, J.P., Clowes, G.H.A., Jr., O'Donnell, T.F., Jr., Hensle, T.E.: Ann. Surg. 177: 588 (1973).
11. Howard, J.M.: Ann. Surg. 141: 321 (1955).
12. MacLean, L., Mulligan, W., McLean, A., Duff, J.: Ann. Surg. 166: 543 (1967).
13. Bergmeyer, H.U., Bernt, E.: In: Bergmeyer, H.U.: Methods of Enzymatic Analysis, p. 313 (Academic Press, New York/London 1963).
14. Dole, V.P., Meinertz, H.: J. Biol. Chem. 235: 2595 (1960).
15. Soeldner, J.S., Slone, D.: Diabetes 14: 771 (1965).
16. Porte, D., Jr., Graber, A.L., Kuzuya, T., Williams, R.H.: J. clin. Invest. 45: 228 (1966).
17. Cryer, P.E., Herman, C.M, Sode, J.: Ann. Surg. 174: 91 (1971).
18. Rutenburg, S., Skarnes, R., Palmerio, C., Fine, J.: Proc. Soc. exp. Biol. Med. 125: 455 (1967).
19. Berry, L.J., Rippe, D.F.: J. infect. Dis., Suppl. 128: 118 (1973).
20. Issekutz, B., Jr., Miller, H.I., Paul, P., et al.: Amer. J. Physiol. 209: 1137 (1965).
21. Kovach, A.G.B., Rosell, S., Sandor, P.: Circulat. Res. 26: 733 (1970).
22. Fredholm, B.: Acta physiol. scand. 81: 110 (1971).
23. Zwadyk, R., Jr., Snyder, I.S.: Proc. Soc. exp. Biol. Med. 143: 864 (1973).
24. Drucker, W.R., Craig, J., Kingsbury, B., Hofmann, W., Woodward, H.: Arch. Surg. 85: 557 (1962).
25. Schumer, W.: Surgery 64: 55 (1968).
26. Ryan, N.T., Blackburn, G.L., Clowes, G.H A., Jr.: Differential tissue sensitivity to elevated endogenous insulin levels during experimental peritonitis in rats (in press).
27. Wieland, O.H., Patzelt, C., Loffler, G.: Europ. J. Biochem. 26: 426 (1972).
28. Clowes, G.H.A., Jr., O'Donnell, T.F., Jr., Ryan, N.T., Blackburn, G.L.: Energy metabolism in sepsis: treatment based on different patterns in shock and high output state (in press).

Table I.

Metabolic Substrates and Urinary Nitrogen Excretion in Fasting and Septic Man
(Values ± S. E. M.)

	Normal Fasting	Septic Semi-Starvation (Glucose 100g/24°)	
		High Flow n = 20	Low Flow n = 12
Cardiac Index (L/min/m^2)	2.8 ± 0.4	4.6 ± 0.8	2.4 ± 0.3
Mean Arterial Pressure (mmHg)	93 ± 14	110 ± 12	70.2 ± 4
Rectal Temperature (°F)	99.6 ± 0.6	102^4 ± 0.8	100^6 ± 0.6
Blood Insulin (/uU/ml)	14 ± 2	42 ± 6.4	6 ± 2
Blood Glucose (/uMol/ml)	3.94 ± .08	7.2 ± 0.5	7.68 ± 2.3
Blood FFA (/uMol/ml)	1.75 ± 15	0.41 ± .06	1.176 ± 0.9
Blood Lactate (/uMol/ml)	0.6 ± 0.1	1.15 ± .11	2.79 ± 0.3
Arterial O$_2$ Tension Breathing Air (mmHg)	90 ± 3	59 ± 6	48 ± 8
Urinary N (g/24 hrs)	4.6 ± 0.6	7.4 ± 0.9	16 ± 2

Table II.

Peritonitis (Cecal Ligation)
(Values $\pm$ S. E. M.)

	Basal n = 16	High Flow n = 9	Low Flow n = 7
Temperature ($^\circ$F)	$101^4 \pm 0.2$	$102^8 \pm 0.5$	$100^8 \pm 1.2$
Cardiac Output (L/M)	3.9 ± 0.3	5.0 ± 0.6	2.0 ± 0.3
Hind Limb Flow (ml/kg body wt/min)	2.4 ± 0.2	3.1 ± 0.1	1.9 ± 0.2
Mean Arterial Pressure (mmHg)	115 ± 3.7	103 ± 4.1	103 ± 10.8
Hind Limb Resistance (pres/flow)	49 ± 6	33 ± 2	53 ± 4
Arterial Insulin (μU/ml)	10 ± 4	41 ± 9	14 ± 5
Arterial Glucose (μM/ml)	5.6 ± 0.5	4.0 ± 0.2	4.6 ± 0.6
Glucose Uptake * (μM/kg/min)	1.6 ± 0.3	1.3 ± 0.3	1.5 ± 0.4
Arterial Lactate (μM/ml)	1.0 ± 0.1	0.9 ± 0.1	1.7 ± 0.3
Lactate Production * (μM/kg/min)	$+1.2 \pm 0.2$	$+1.2 \pm 0.3$	$+2.5 \pm 0.6$
FFA Arterial (μEq/ml)	$.715 \pm .08$	$.398 \pm .07$	$.462 \pm .03$
FFA Uptake or Release * (μMol/kg body wt/min)	$-.101 \pm .03$	$+.067 \pm .04$	$+.126 \pm .04$
Glycerol Release * (μMol/kg body wt/min)	$+.042 \pm .011$	$+.009 \pm .001$	$+.116 \pm .01$
O_2 Uptake * (ml/kg/min)	14.5 ± 3.4	13.5 ± 3.1	11.7 ± 0.7
Arterial pO2 (mmHg)	72.36 ± 2.6	69.5 ± 7.18	75.2 ± 7.2

* Net uptake (-) or release (+) by hind limb

Table III.

Endotoxin (2mg/kg)
(Values $\pm$ S.E.M.)

	Basal n = 6	1 Hour n = 6	4 Hours n = 6
Temperature (°F)	$101^8 \pm 0.6$	$102^6 \pm 0.6$	$102^6 \pm 0.6$
Cardiac Output (L/M)	3.6 ± 4	3.9 ± 1.2	3.1 ± 0.3
Hind Limb Flow (ml/kg body wt/min)	2.5 ± 0.3	1.6 ± 0.1	1.6 ± 0.2
Mean Arterial Pressure (mmHg)	112 ± 6	77 ± 10	93 ± 11
Hind Limb Resistance (pres/flow)	47 ± 7	47 ± 8	72 ± 13
Arterial Insulin (μU/ml)	3.3 ± 1.9	4.8 ± 3	2 ± 0.3
Arterial Glucose (μM/ml)	4.3 ± 0.4	3.5 ± 0.5	1.8 ± 0.4
Glucose Uptake * (μM/kg/min)	1.5 ± 0.3	0.2 ± 0.4	0.5 ± 0.1
Arterial Lactate (μM/ml)	1.1 ± 0.3	2.8 ± 0.4	2.7 ± 0.7
Lactation Production * (μM/kg/min)	$+1.4 \pm 0.7$	$+1.2 \pm 0.5$	$+1.3 \pm 0.4$
FFA Arterial (μEq/ml)	$.58 \pm .08$	$.28 \pm .05$	$.22 \pm .03$
FFA Uptake or Release * (μMol/kg body wt/min)	$-.254 \pm .109$	$+.128 \pm .086$	$+.087 \pm .057$
Glycerol Release * (μMol/kg body wt/min)	$+.02 \pm .001$	$+.05 \pm .001$	$+.05 \pm .003$
O_2 Uptake * (ml/kg/min)	11.5 ± 1.46	$7.53 \pm .65$	$7.63 \pm .94$
Arterial pO_2 (mmHg)	72.7 ± 2.5	71.3 ± 2.3	72.9 ± 3.7

* Net uptake (-) or release (+) by hind limb

Changes in Hemodynamics and Gas Metabolism after Endotoxin Injection

H. Neuhof

Hemodynamic disturbances are known to occur in mammals under the influence of endotoxins. Depending on the mode of administration and dosage, these disturbances may result in immediate death or lead to death through secondary metabolic changes. Due to anatomic and functional particularities, there are certain species-specific differences in the hemodynamics of individual partial circulations; on the whole, however, a fairly uniform course of shock is observed in primates, dogs, cats, and rabbits (1).

In our animal experiments we used rabbits whose hemodynamic reactions to endotoxin are essentially the same as those of primates. Endotoxin from E. coli was kindly prepared by Professor Urbaschek, Mannheim.

A single i. v. injection of 100 µg endotoxin/kg body weight induces severe hemodynamic disturbances in 75 % of the experimental animals. 15-20 minutes after the injection, pressure in the right ventricle and in the pulmonary artery rises due to increased acute resistance in the pulmonary vascular system (Fig. 1). Kuida and co-workers have found that this increased resistance mainly concerns the venous limb of the vessel. Due to this obstruction of blood flow, cardiac output drops acutely and - depending on the extent of the disturbance - arterial pressure also drops (2).

In dogs a different pattern is seen in this early phase. Constriction of the hepatic veins and of the small mesenterial veins causes a pooling of blood in the mesenterial circulation of this species (3, 4, 5). The decreased venous return to the heart prevents the manifestation of pulmonary hypertension, although experiments in which the venous return was constant have shown that increased resistance and increased pressure in the pulmonary vascular system occur in dogs also (2).

Oxygen uptake of the whole organism drops critically in the early phase of endotoxin shock (Fig. 1). This is primarily due to insufficient transport of oxygen to the peripheral circulation, caused by considerably reduced cardiac output. Oxygen uptake and cardiac output parallel each other (Figs. 2-5). Reduced oxygen supply to the tissues is the cause of the resulting metabolic acidosis. A hyperdynamic course of shock with relatively high cardiac output, low arterio-venous oxygen difference, low peripheral resistance, and low oxygen uptake, as occasionally seen in septic conditions in man, was not observed, either in the early or in the late phase of shock (6, 7).

50 % of our animals which showed an early phase reaction after 15-20 minutes had a very transient and markedly weaker circulatory reaction immediately (2-4 minutes) after the injection of endotoxin (Figs. 6 and 7). This indicates a two-phase liberation of vasoconstrictive substances.

In the early phase, increased right heart pressure and simultaneous deterioration of coronary perfusion were caused by falling pressure in the systemic circulation and simultaneously decreasing arterial oxygen saturation. This often results in death due to acute right heart failure. In our experiments 21 % of all animals died in this way (Fig. 8). Disturbances of excitation and conduction in the right ventricle are seen in the electrocardiogram in this phase.

The hemodynamic and metabolic changes of the early phase can regress completely, as was the case in 14 % of the rabbits used in our experiments. In 40 % of the animals the late phase of endotoxin shock develops after compensation periods of different duration : cardiac output, oxygen uptake, and arterial blood pressure decline slowly, but continuously. Progressive acidosis sets in, and these animals die.

In contrast to the early phase this renewed hemodynamic and metabolic disturbance cannot be attributed to a renewed increase in the resistance of the pulmonary vascular system.

Since severe alterations in the blood clotting system occur under the influence of endotoxin, leading to fibrin deposits and thrombocyte aggregation in the terminal vascular system, the question arises to what extent these alterations are also involved in the hemodynamic and metabolic disturbances of endotoxin shock.

In our model the blocking of coagulation with heparin had no influence on the early phase of endotoxin shock but did prevent the late phase. Likewise the lethal hemodynamic and metabolic disturbances regularly observed during and after a 10-hour endotoxin infusion occurred only rarely in the heparinized animals. Therefore intravasal coagulation processes are of pathogenetic significance in the late phase but not in the early phase of endotoxin shock.

The question as to what extent thrombocyte aggregates are involved in increasing resistance in the pulmonary vascular system was studied in rabbits which were nearly platelet-free.

For this purpose the rabbits were connected to an extracorporeal circulatory system. An average of 77 % of the platelets was removed from the circulation by filtration of the blood in a bypass over a pyrex-glass wool filter. Subsequently the animals were given i. v. injections of endotoxin (8). In control animals with a normal number of thrombocytes, the platelets decreased by $77500 \pm 62600/mm^3$ as compared to only $18600 \pm 15700/mm^3$ in the animals whose number of platelets had been reduced. Increase of systolic right ventricular pressure and the simultaneous pressure drop in the systemic circulation during the early phase reaction were not significantly different in the two groups (Fig. 9). Therefore the increased resistance in the pulmonary vascular system cannot be causally or substantially due to mechanical obstruction of the peripheral vessels by platelet aggregates.

In my opinion these results are not in contradiction to the findings of Hinshaw, Clowes and other authors who observed a rise in pulmonary arterial pressure after the administration of endotoxin only in perfusion of the isolated lung with whole blood, but not in perfusion with cell-free plasma or solutions of plasma substitutes (9). It is conceivable that biogenic amines or mediators, released from the platelets and leukocytes, under the influence of endotoxin, contribute to pulmonary vasoconstriction but that the extent of this reaction is determined by liberated vasoconstrictive substances from other organs.

References

1. Gilbert, R. P. : Physiol. Rev. 40 : 245 (1960).
2. Kuida, H., Hinshaw, L. B., Gilbert, R. P., Visscher, M. B. : Amer. J. Physiol. 192 : 335 (1958).
3. Meyer, M. W., Visscher, M. B. : Amer. J. Physiol. 202 : 913 (1962).
4. Hinshaw, L. B., Nelson, D. L. : Amer. J. Physiol. 203 : 870 (1962).
5. MacLean, L. D., Weil, M. H., Spink, W. W., Visscher, M. B. : Proc. Soc. exp. Biol. Med. 92 : 602 (1956).
6. Siegel, J. H., Greenspan, M., Del Guercio, L. R. M. : Ann. Surg. 165 : 504 (1967).
7. MacLean, L. D., Mulligan, W. G., McLean, A. P. H., Duff, J. H. : Ann. Surg. 166 : 543 (1967).
8. Neuhof, H., Kaufmann, G. : Verh. dtsch. Ges. KreislForsch. 34 : 218 (1968).
9. Hinshaw, L. B., Kuida, H., Gilbert, R. P., Visscher, M. B. : Amer. J. Physiol. 191 : 293 (1957).

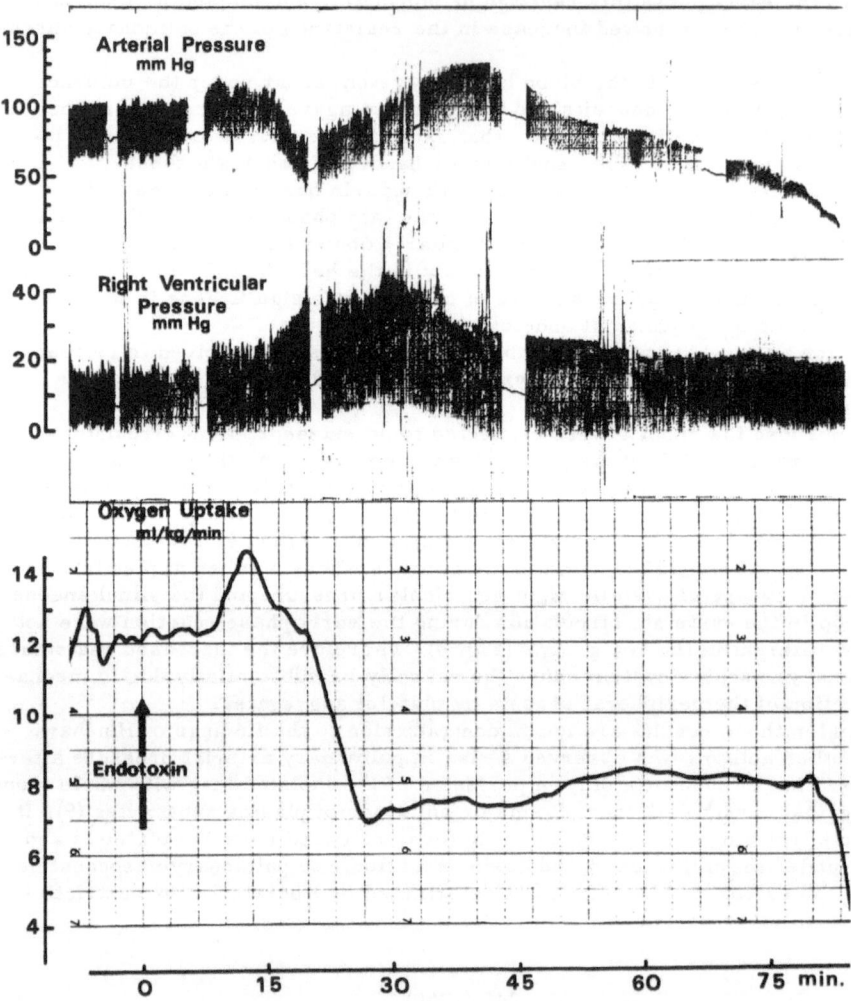

Fig. 1. Changes in arterial blood pressure, right ventricular pressure and oxygen uptake in an unanesthetized rabbit after a single i. v. injection of endotoxin (E. coli 100 µg/kg).

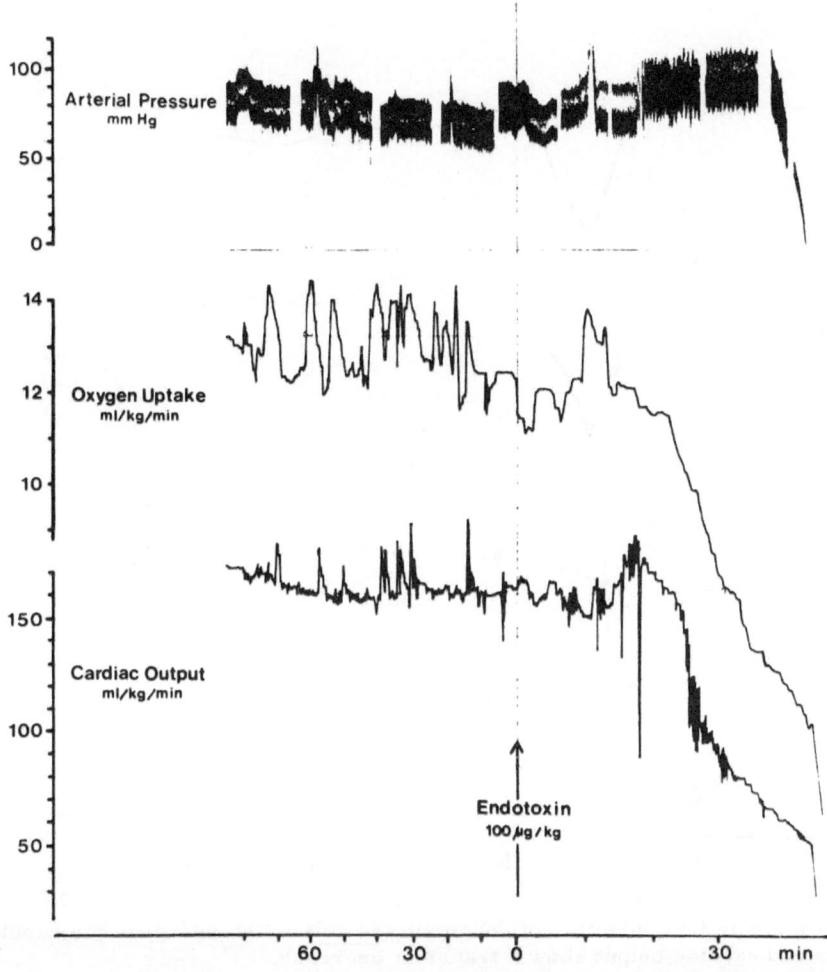

Fig. 2. In the early phase of endotoxin shock in the rabbit oxygen uptake drops critically in consequence of the decreasing cardiac output. The arterial blood pressure may be elevated above normal until short time before death.

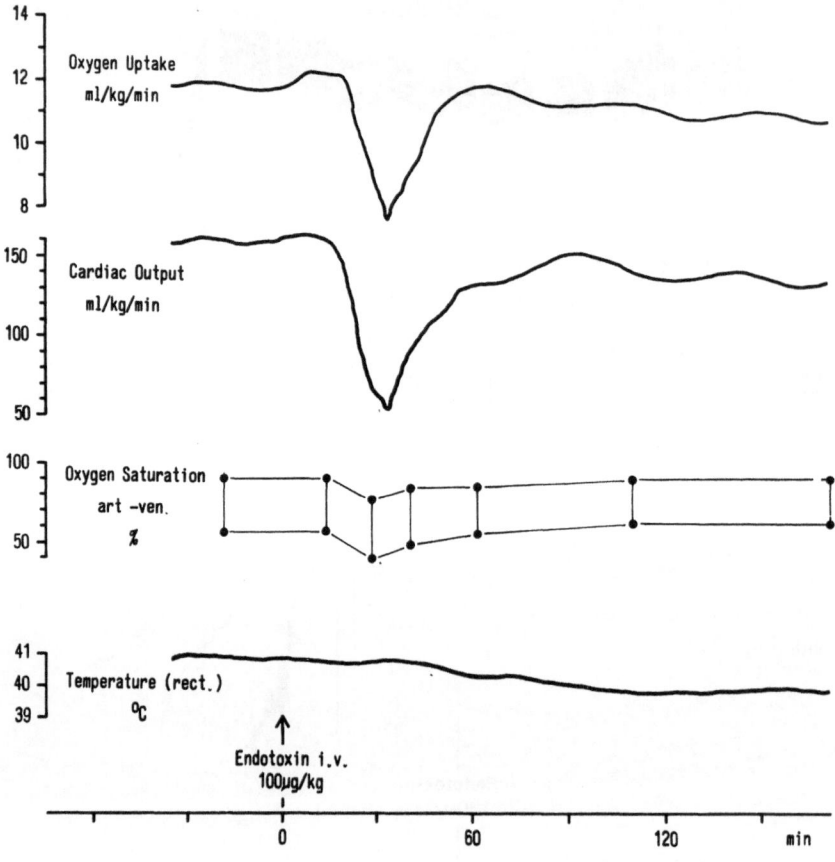

Fig. 3. After a single i.v. injection of endotoxin (<u>E. coli</u> - 100 μg/kg) in the rabbit oxygen uptake and cardiac output show a transient <u>decrease</u>.

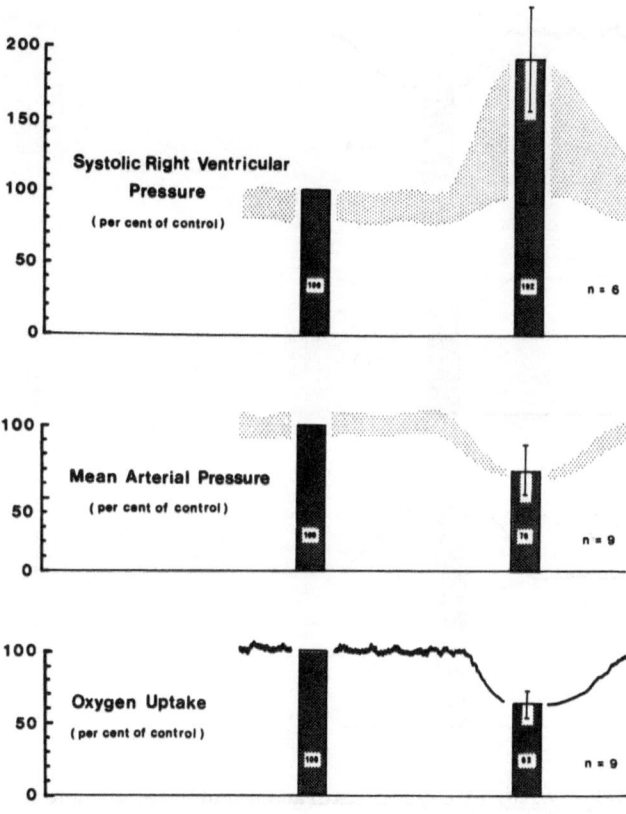

Fig. 4. Typical changes in systolic right ventricular pressure, mean arterial pressure and oxygen uptake during the early phase of endotoxin shock induced by a single i. v. injection of endotoxin (E. coli - 100 µg/kg) in unanesthetized rabbits (mean values).

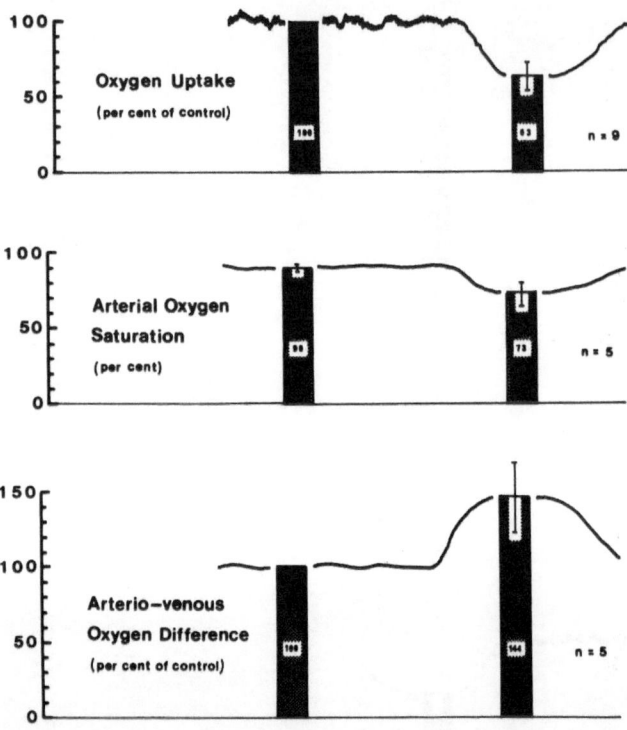

Fig. 5. Changes in oxygen uptake, arterial oxygen saturation and arterio-venous oxygen difference during the early phase of endotoxin shock induced by a single i. v. injection of endotoxin (E. coli - 100 μg/kg) in unanesthetized rabbits (mean values).

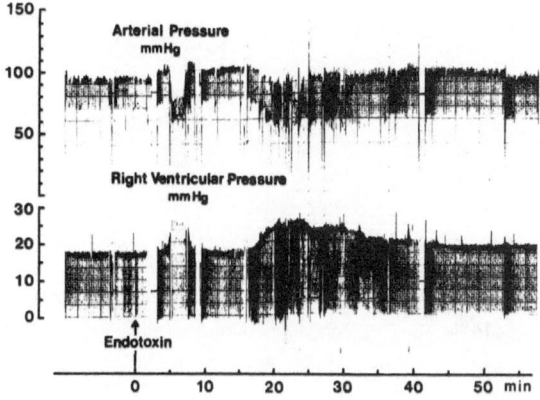

Fig. 6. Early phase reactions of arterial pressure and right ventricular pressure 6 minutes and 20 minutes after a single i. v. injection of endotoxin (E. coli - 100 μg/kg) in an unaesthetized rabbit.

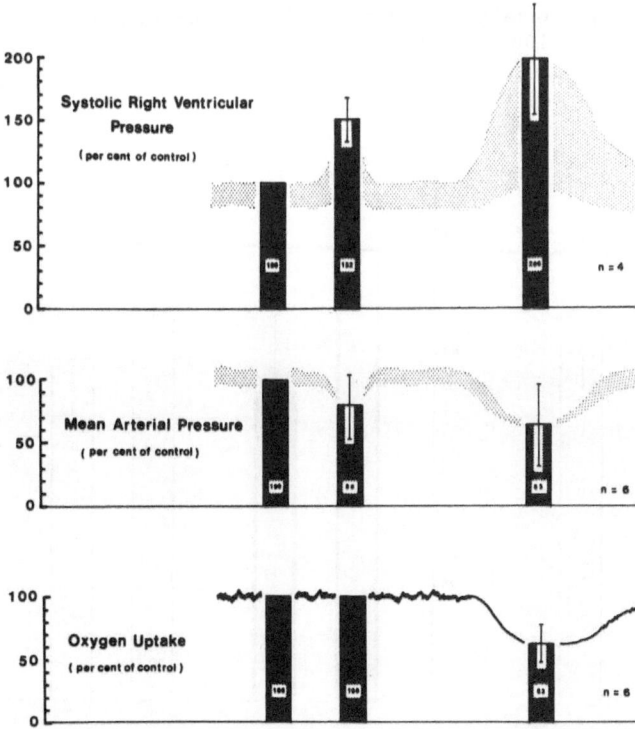

Fig. 7. Early phase reactions of systolic right ventricular pressure, mean arterial pressure and oxygen uptake 2-4 minutes and 15-20 minutes after a single i. v. injection of endotoxin (E. coli - 100 μg/kg) in unanesthetized rabbits (mean value).

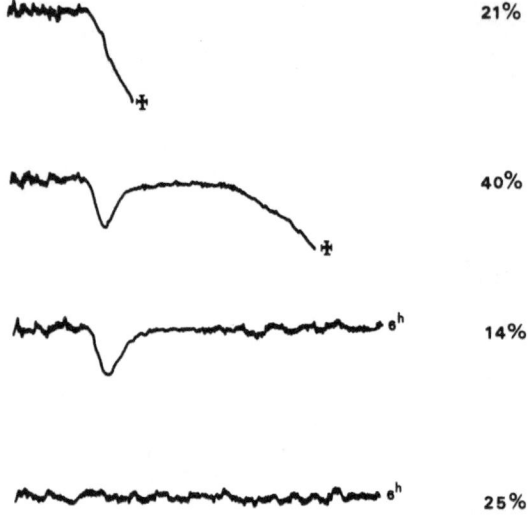

21%

40%

14%

25%

Fig. 8. Reaction patterns of oxygen uptake in rabbits (n = 28) following a single
i. v. injection of endotoxin (E. coli - 100 µg/kg).

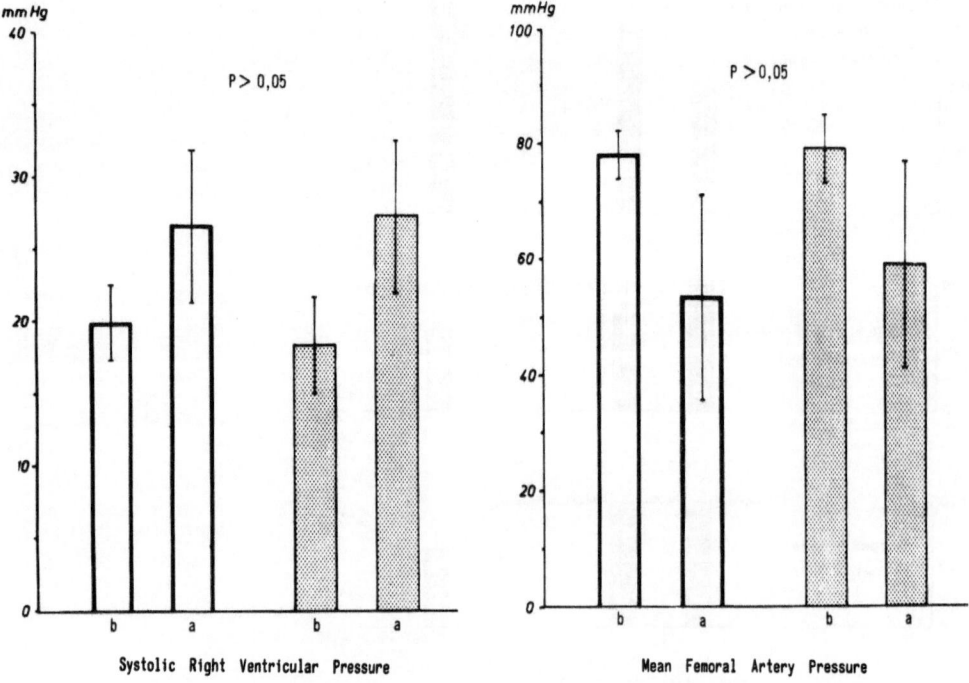

Fig. 9. Hemodynamic changes in the early phase of endotoxin shock in the rabbit.
b = before, a = after i. v. injection of endotoxin (3000 µg/kg).

normal rabbits ☐   "platelet-free" rabbits ▦

Mesenteric Hemodynamics in Endotoxin Shock

G. F. Brobmann

Introduction

In the last twenty years a widely accepted hypothesis on the pathophysiology of human septic shock has been postulated. Surveying the considerable literature one may conclude that we understand the events leading to lethal human shock. Related to the hypothesis of septic shock are the terms: adrenergic theory of shock and target organ theory. The essence of these theories is depicted in Fig. 1. Any kind of shock, whether it be hemorrhagic, septic, or due to myocardial infarction or massive trauma, leads to compensatory mechanisms to restore arterial pressure and cardiac output. One of the most important mechanisms is the stimulation of the sympathetic nervous system. Arteriolar constriction, particularly in the splanchnic region, results. Mesenteric vasoconstriction is leading to ischemic damage of critical abdominal organs, such as loss of intestinal barrier function against intraluminal contents, changes in the metabolic function of the liver, and the hepatic and splenic RES. Toxic materials enter the circulation and cannot be detoxified by the damaged RES. The subsequent endotoxemia, resulting in a vicious cycle, is responsible for what we call "irreversible" shock.

These theories were based upon the common finding of splanchnic vasoconstriction, necrosis of the bowel wall, and high concentration of endogenous catecholamines in canine endotoxin shock.

Many authors (1, 2, 3, 4, 5, 6, 7) reported on differences between the primate and the dog. Hinshaw (8) observed that the blood vessels of the perfused, denervated gut of the monkey failed to constrict upon endotoxin administration. Before extrapolating to man, the results in the shocked dog should be repeated in a species phylogenetically closer to man, such as rhesus monkeys and baboons.

Since the splanchnic hemodynamic response to endotoxin seems to be the crucial point, this study was designed to evaluate the effects of endotoxemia upon mesenteric hemodynamics in the monkey.

Methods

Nine adult rhesus monkeys, ranging in weight between 3.3 and 7.3 kg and six mogrel dogs, averaging 15 kg body weight, were selected as subjects of this study. The animals were anesthetized with intravenous sodium pentobarbital (30 mg/kg) and light anesthesia was maintained by additonal amounts of Nembutal[R].

Escherichia coli endotoxin (Difco) from a single batch was pretested for its lethality. In both monkeys and dogs a lethal dose was administered intravenously: dogs 1 mg/kg, monkeys 4 mg/kg. Fig. 2 shows the experimental preparation. The surgical procedure was identical in the two species. The anesthetized animals were placed in a supine position. An endotracheal tube was inserted into the trachea. Every 15 minutes the tube was connected to a positive pressure respirator to inflate the lungs and prevent atelectasis (especially in the monkey). Otherwise the animals breathed spontaneously. Through a midline abdominal incision the superior mesenteric artery was exposed, a precalibrated blood flow transducer (Micron, Inst.) was implanted near the artery's origin and connected to a gated sine-wave type electromagnetic blood flow amplifier (Biotronix, Inc.). A hydraulic occluder (9) was placed on the artery distal to the flow probe to obtain zero flow. Portal pressure was measured through a polyethylene catheter placed in the portal vein via a branch of the superior mesenteric vein. The abdomen was then closed. Systemic arterial pressure was measured by means of a polyethylene cannula inserted into the abdominal aorta through the left femoral artery. Both cannulas were connected to a pressure transducer (Sanborn). The left femoral vein was cannulated to inject endotoxin and additional anesthetic. The following parameter : mean systemic arterial pressure, portal venous pressure, mesenteric blood flow, and ECG were recorded simultaneously on direct-writing polygraph recorder (Sanborn). Mesenteric vascular resistance was calculated from the arterial-portal pressure gradient divided by mesenteric flow and was expressed in millimeters of mercury per milliliter per minute.

The rectal temperature was continously measured by a thermometer, and spontaneous decrease in temperature during anesthesia and endotoxin shock was prevented by means of a warming blanket.

After preparatory surgery was completed and the animal was allowed to stabilize, control measurements were obtained every 15 minutes for 1 hour. Then a lethal dose of endotoxin was injected intravenously and the animal was observed for 4 hours with measurements recorded every 15 minutes. After 4 hours the animals were sacrificed with a lethal dose of pentobarbital.

Statistical analysis was performed within each series of animals using the Sign test (10). We also compared the results of monkey versus dogs for each parameter using the Mann-Whitney U test (11).

Results

Monkeys

Three monkeys served as control. No endotoxin was injected and the animals were observed for 4 hours. No significant changes in one of the measured parameters were found. Arterial pressure ranged from 110 mm Hg at the start of control to 108 mm Hg at the end of the observation period. Comparable values for portal venous pressure are 4 and 5 mm Hg and for mesenteric blood flow 75 and 80 ml/min. The experimental group of monkeys consisted of six animals.

Systemic arterial pressure declined gradually, but progressively and reached its lowest value at 90 minutes after intravenous injection of endotoxin. The hypotensive response occurred in all animals and was significantly lower than control values from 45 minutes after injection to the remainder of the experiment. Portal venous pressure increased slightly, reaching its maximum at 45 minutes, but postinjection values were not statistically different from control until 165 minutes after endotoxin administration. There was no significant change in mesenteric blood flow at any time of the postinjection period (Fig. 3). The calculated mesenteric vascular resistance decreased after administration of endotoxin reflecting the fall in mean systemic arterial pressure and the unchanged flow. From 45 minutes after injection for the remainder of the experiment the resistance values were significantly lower than those of the control.

Dogs

The results of the canine study showed the classical effects of a lethal dose of endotoxin. This series consisted of six dogs. Within 30 seconds after endotoxin injection arterial pressure dropped markedly and reached its nadir at 3 minutes. This decrease in pressure was accompanied by a similar fall in mesenteric blood flow and a marked rise in portal venous pressure with their lowest and highest values at 3 minutes. All arterial pressure values recorded after the endotoxin injection were significantly lower than control values. Portal pressure remained elevated significantly for 30 minutes and returned to control values at 45 minutes and was not statistically significant different from control for the remainder of the experiment. Arterial pressure and mesenteric blood flow showed a partial recovery with a maximum at 30 minutes post injection and than a second more gradual, but progressive, fall for the rest of the experiment.

Mesenteric vascular resistance increased in the postinjection period. All values from 2 hours after administration of endotoxin to the end of the experiment were significantly different from control.

Canine Versus Primate Hemodynamic Responses

The initial differences in the reaction to endotoxin in the dog and the monkey are shown in Fig. 4. Arterial pressure decreased more (significantly) at 15 minutes in dogs and more in monkeys from 45 to 75 minutes after injection. Portal venous pressure elevation was greater in dogs at 3, 15, and 30 minutes and greater in monkeys from 150 to 210 minutes. There was a large decrease in flow in dogs at all times after the injection of endotoxin, whereas flow in the monkey did not change at all. All values from zero to 240 minutes were significantly different. Mesenteric vascular resistance increased in dogs and decreased in monkeys with statistically significant values from 30 minutes to the remainder of the experiment.

Discussion

The hemodynamic responses to endotoxin in the dog shown in this study have been reported often (3, 13, 14, 15, 16, 17). The dog reacts to endotoxin with hepatic venoconstriction, portal hypertension, sequestration of blood in the splanchnic viscera, and high concentrations of plasma catecholamines (14, 18). Venous return falls and a decline in cardiac output, arterial pressure and mesenteric blood flow follows. There is some recovery by 15 minutes and portal pressure returns toward control by 30 minutes after administration of endotoxin. The recovery of arterial pressure exceeds that of mesenteric blood flow, therefore calculated mesenteric vascular resistance progressively increases after endotoxin injection.

Are the plasma catecholamines released during shock (14, 18) state responsible for the mesenteric vasoconstriction in the dog model? As shown by many authors (19, 20, 21, 22, 23) prolonged sympathetic stimulation (19) or infusion of catecholamines in concentrations five times higher than measured in experimental canine shock (23) is leading only to a brief vasoconstriction in the splanchnic area followed by a rapid recovery to control levels. This ability of the mesenteric circulation to prevent underperfusion has been termed "autoregulatory escape" (19). Therefore it is very unlikely that mesenteric vasoconstriction found in the dog after endotoxin administration results from endogeneously released catecholamines. The rhesus monkey does not exhibit such hemodynamic responses to endotoxin. There is no sequestration of blood in the splanchnic area and no striking portal hypertension. Arterial hypotension is more gradual. The most significant difference between canine and monkey hemodynamic changes in endotoxin shock is the stable mesenteric blood flow. This indicates that in the rhesus monkey the mesenteric vasculature has dilated, whereas in the dog mesenteric vasoconstriction is the characteristic response to shock. Our results are in accordance with those published by Swan and associates (17) in 1971. Swan found in the baboon - phylogenetically closer to man than the rhesus monkey - similar results. Fig. 5 shows the rhesus monkey data of our study.

Swan measured the catecholamines in his baboon experiments and found a significant increase of epinephrine and norepinephrine after the administration of a lethal dose of endotoxin, but no change in mesenteric blood flow and a decrease in mesenteric vascular resistance. The above-mentioned hypothesis by Fine (24, 25, 26), Lillehei (13, 14, 15), and Nickerson (27, 28) concerning human septic shock is predicated on the existence of splanchnic vasoconstriction in the animal model. One may speculate now what form the hypothesis would have taken if primates instead of dogs as experimental animal had been used.

Which are the conclusions we should draw from these results? Since mesenteric circulation may either constrict or dilate in lethal shock states in various animal models, the intestine, the liver, or the entire splanchnic region are no more target organ areas in shock than the lung, the kidney or the heart. Death from endotoxin shock does not depend upon changes in the mesenteric circulation. In primates the intestinal and the liver (29) perfusion is maintained, no pathological changes such as necrosis of the bowel wall or hemorrhages in the gastrointestinal tract are observed, but the lethal dose of endotoxin is still lethal and the animals die. If our view were to remain confined to the gastrointestinal tract and especially to the splanchnic circulation, we probably would be unable to define the nature of refractory endotoxin shock.

From the above-mentioned experiments we understand why adrenergic blocking therapy failed in clinical septic shock. We have to raise the question again: why does a living being die after administration of a lethal dose of endotoxin?

References

1. Cavanagh, R., Rao, P. S. : Arch. Surg. 99 : 107 (1969).
2. Gilbert, R. P. : Proc. Soc. exp. Biol. Med. 111 : 328 (1962).
3. Hinshaw, L. B., Brake, C. M., Emerson, Jr. T. E., Jordan, M. M., Masucci, F. D. : Amer. J. Physiol. 207 : 925 (1964).
4. Hinshaw, L. B., Emerson, Jr. T. E., Reins, D. A. : Amer. J. Physiol. 210 : 335 (1966).
5. Vaughn, D. L., Guenter, C. A., Stookley, J. L. : Surg. Gynec. Obstet. 126 : 1309 (1968).
6. Vaughn, D. L., Peterson, E. : Obstet. Gynecol. 34 : 271 (1969).
7. Wyler, F. R., Forsyth, R. P., Nies, A. S., Neutze, J. M., Melmon, K. L. : Circ. Res. 24 : 777 (1969).
8. Hinshaw, L. B. : J. Surg. Res. 8 : 535 (1968).
9. Jacobson, E. D., Swan, K. G. : J. appl. Physiol. 21 : 1400 (1966).
10. Snedecor, G. W., Cochran, W. G. : Statistical Methods 6th ed., p. 125 (Iowa State Univ. Press, Ames 1967).
11. Siegel, S. : Nonparametric Statistics, p. 116 (McGraw, New York 1956).
13. Lillehei, R. C., Longerbeam, J. K., Bloch, J. H., Manax, W. G. : Hemodynamic changes in endotoxin shock. In : Millis, L. C., Moye, J. H. : Shock and Hypotension, p. 442 (Grune and Stretton, New York/London 1965).
14. Lillehei, R. C., Longerbeam, J. K., Rosenberg, J. C. : The nature of irreversible shock: its relationship to intestinal changes. In : Bock, K. D. : Shock, Pathogenesis and Therapy, p. 106 (Springer, Berlin 1962).
15. Lillehei, R. C., MacLean, L. D. : Ann. Surg. 148 : 513 (1958).
16. Kalas, J. P., Jacobson, E. D. : Amer. Heart J. 67 : 764 (1964).
17. Swan, K. G., Barton, R. W., Reynolds, D. G. : Gastroenterology 61 : 872 (1971).
18. Richardson, J. A. In : Catecholamines in shock. Hahnemann Symposium on Shock, p. 340 (Grune and Stretton, New York 1965).
19. Folkow, B., Lewis, D. H., Lundgren, O., Melander, S., Wallentin, I. : Acta physiol scand. 61 : 445 (1964).
20. Ross, G. : Amer. J. Physiol. 212 : 1037 (1967).
21. Greenway, C. V., Lawson, A. E. : J. Physiol 186 : 579 (1966).
22. Shehadeh, Z., Price, W. E., Jacobson, E. D. : Amer. J. Physiol. 216 : 386 (1969).
23. Swan, K. G., Reynolds, D. G. : Gastroenterology 61 : 863 (1971).
24. Fine, J. : J. Oklahoma State Med. Assoc. 59 : 419 (1966).
25. Fine, J. : Gastroenterology 52 : 454 (1967).
26. Fine, J., Minton, R. : Nature 210 : 97 (1966).

27. Nickerson, M., Carter, S. A. : Canad. J. Biochem. 37 : 1161 (1959).
28. Nickerson, M., Gourzis, J. T. : J. Trauma 2 : 399 (1962).
29. Barton, R. W., Reynolds, D. G., Swan, K. G. : Surgery 175 : 204 (1972).

THE "ADRENERGIC HYPOTHESIS"

(Fine, Nickerson, Lillehei)

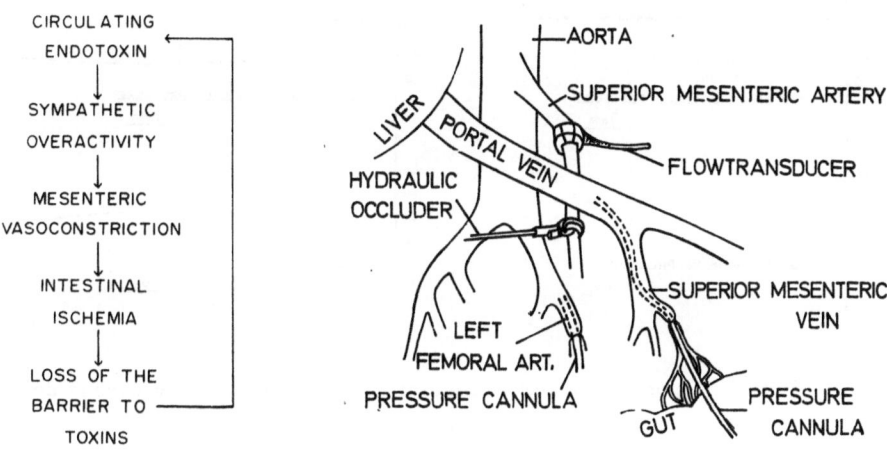

Fig. 1.   Fig. 2.

Fig. 1. Essence of the "adrenergic theory" of shock evolved as the hypothesis on the pathophysiology of lethal human shock.

Fig. 2. Schematic diagram of the experimental preparation.

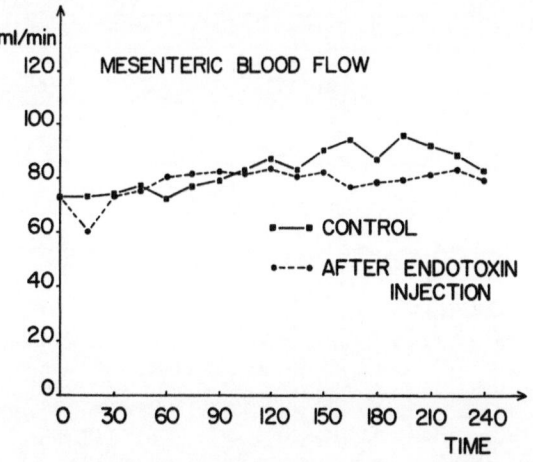

Fig. 3. Comparison of mesenteric blood flow in the three control monkeys versus the six experimental monkeys.

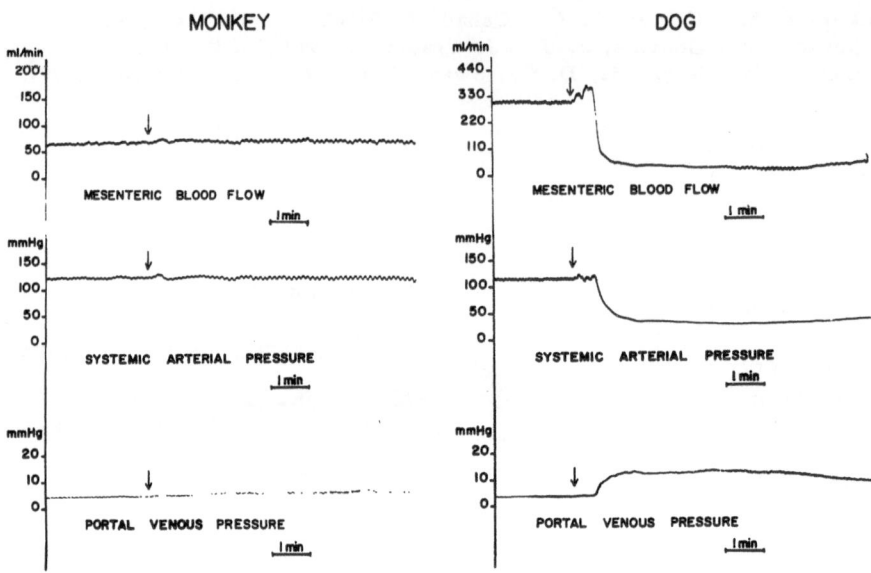

Fig. 4. Initial response to a lethal dose of endotoxin in monkey versus dog.

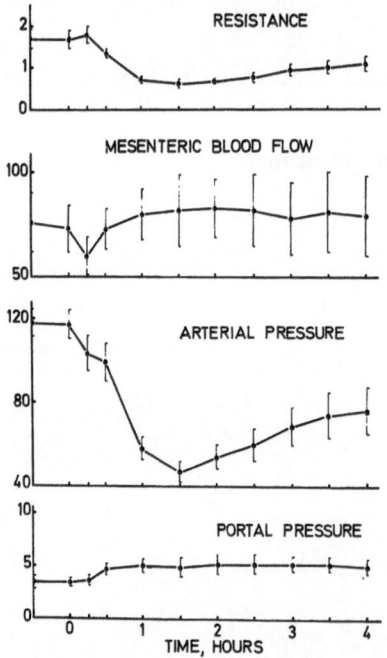

Fig. 5. Comparison of hemodynamic changes after endotoxin administration in rhesus monkeys.

Effect of Endotoxin on Metabolism and the Metabolic Changes in Bacterial Shock

L. J. Berry, M. W. Laney, and R. N. Moore

At the Airlie House conference in May 1972, I reported (2) that endotoxin inhibits the synthesis of phosphoenolpyruvate carboxykinase (PEPCK). This enzyme more than doubles in activity and in amount following an injection of hydrocortisone or the stress of fasting. The increase in amount of enzyme is blocked by either endotoxin or actinomycin D as demonstrated by radial immunodiffusion (12). More recently, Rippe and Berry (13) have found a similar effect of endotoxin on a second inducible hepatic enzyme, trytophan oxygenase (TO). Because isotopically labeled endotoxin has not been found to localize in liver parenchymal cells which contain the inducible enzymes (3, 10), it has been assumed that the observed inhibition of enzyme synthesis is mediated rather than the result of direct action.

In looking for possible target cells, those that make up the reticuloendothelial system (RES) are likely candidates. The experiments to be described permit the conclusion that neither B nor T lymphocytes serve as mediators of inhibition of enzyme induction.

As an extension of these studies, data are presented which show the sequence of changes in inducible enzymes in mice given a standard infectious dose of Salmonella typhimurium, strain SR-11. The challenge dose caused first deaths on the fifth day postinfection and 100 % mortality after about 10 days.

Methods

Mice. Specific pathogen free CD-1 and Ha/ICR inbred mice were used in the experiments. The former were employed for work with endotoxin and the latter were experimentally infected. They were produced in our animal facilities under the supervision of Mr. Mace Earls.

Enzyme Assays. PEPCK was assayed for activity by the method of Phillips and Berry (11) and for total antigenic protein by the radial immunodiffusion method of Rippe and Berry (12). The activity of TO was assayed by the method of Knox and Auerbach (8).

Glycogen Assay. Liver glycogen was determined daily in infected mice using the method of Kemp and Kits van Heijningen (7). It is expressed as per cent wet weight of liver.

Endotoxin. Endotoxin was extracted in quantity from S. typhimurium, strain SR-11, by the phenol-water method as described by Nowotny (9). The LD_{50} of this material for the mice employed was about 250 /ug.

Experimental Infections. Mice were infected intragastrically with about 10^8 cells grown overnight in brain-heart infusion broth (Difco).

Inhibitors of B and T Lymphocytes. Cyclophosphamide (Mead Johnson) was injected intraperitoneally in female mice at a dosage of 400 /ug per g body weight as recommended by Stockman et al. (14). The compound was dissolved in 0.5 ml isotonic sodium chloride solution (Travenol, Morton Grove, Ill.). Mice were used experimentally five or six days later.

Oxisuran (supplied by Dr. Henry H. Freedman, Warner-Lambert Research Institute) was dissolved in isotonic sodium chloride solution (Travenol Laboratories) and injected intraperitoneally. Each mouse received 2 mg in 0.5 ml daily for two days prior to testing.

Production of Tolerance. Mice were made tolerant to endotoxin intraperitoneally two days prior to challenge with 2LD$_{50}$.

Hemagglutination Titer. One ml of a 10 % suspension of washed sheep erythrocytes was injected intraperitoneally into mice six days after cyclophosphamide
treatment. The mice were bled five days later. Normal mice similarly immunized
with sheep rbc were bled after five days. Hemagglutination titers were determined by the tube agglutination method.

Results

Effects of Cyclophosphamide. The ability of cyclophosphamide to inhibit B cell
function as evidenced by suppressed antibody synthesis, can be seen by the data
presented in Fig. 1. All control mice had a higher hemagglutination titer than cyclophosphamide (cytoxan) treated mice. Histological sections of lymph nodes (not
shown) in treated mice were clearly deficient in cells in the area around the periphery, the B cell region. In mice known to have a deficiency in number of B lymphocytes, the effect of endotoxin on PEPCK and TO was similar to that seen in normal mice. Table I summarizes data obtained for PEPCK in treated and control
mice following 20 hours of fasting. In both groups, fasting alone resulted in more
than a doubling of the enzyme while an injection of endotoxin at the beginning of the
fast greatly reduced the level reached by the enzyme.

Results with TO are seen in Table II. An injection of hydrocortisone alone increased the amount of enzyme while endotoxin given at the time of the injection inhibited induction.

Effects of Oxisuran. The lack of effect of oxisuran on induction of PEPCK due to
fasting is seen in the data of Table III. Similarly, the absence of any altered response in TO attributable to oxisuran is evident in the results shown in Table IV.
Even though we did not confirm the ability of oxisuran to extend graft retention, as
described by Freedman and his associates (4, 6), their schedule of treatment was
followed. We did show, however, that oxisuran at the dosage employed failed to alter the hemagglutinin titer formed in response to sheep rbc administration.

Effect of Cyclophosphamide and Oxisuran on the Development of Tolerance to
Endotoxin. Since the nature of tolerance to endotoxin has not been fully elucidated,
it seemed worthwhile to determine the effect of suppressed B or T cell function on
the development of tolerance. The results are shown in Table V. Mice treated with
oxisuran become fully tolerant while those given cyclophosphamide do not. The resistance of the latter is intermediate and suggests that animals become tolerant
only when B cells are normally active. This conclusion is consistent with the recognized ability of serum from tolerant animals to passively transfer tolerance to
normal recipients (5). It does not make clear, however, why an injection of an
RES-blocking agent destroys tolerance (1).

Enzyme Levels in Infected Mice. Fig. 2 shows the number of Salmonellae in liver and spleen of mice at daily intervals postinfection. The change in bacterial
population is typically sigmoidal and indicates that pathogen proliferation occurs
during the first day, yields no net increase for about two days and then becomes
essentially logarithmic beginning at day three.

Fig. 3 shows the activity of PEPCK measured at daily intervals postinfection.
It is important to recognize that even though the animals were not eating, the enzyme remained constant for several days and then declined. This indicates that induction is blocked, presumably by endotoxin.

Fig. 4 is a graph of the liver glycogen level in mice plotted against days postinfection. Glycogen reserves decreased, possibly because PEPCK failed to increase in mice that were not feeding.

Discussion

If the ability of endotoxin to inhibit the induction of certain hormonally inducible hepatic enzymes (12, 13) is a mediated effect, then the cells responsible appear to be neither B nor T lymphocytes. It could be argued that since these cells were not completely eliminated enough remained to have been the source of adequate mediator. Were this the case, then the smallest amount of endotoxin capable of inhibiting the enzymes in normal mice should be less effective in animals given the drugs. This was not found to be true.

The amount of cyclophosphamide administered was sufficient to reduce the titer of hemagglutinin formed in response to sheep rbc. It also reduced the degree of tolerance that developed within 48 hours after a single 5 /ug injection of endotoxin. To this extent it was demonstrably effective in suppressing B cell function. Oxisuran, however, had no effect on any of the responses measured.

If mediators are responsible for impaired metabolic regulation in endotoxin-poisoned mice, then other target cells must be identified. Kupffer cells of the liver, wandering macrophages and microphages are possibly involved. Work now in progress to determine their role was not completed in time for this report but should be available soon.

Experiments with infected mice yielded results consistent with the idea that endotoxin is released in amounts sufficient to inhibit the inducible enzymes and hence impair the animal's response to stress. The data, however, are more suggestive than conclusive. Other methods of assaying for endotoxin will be used to confirm or refute these results.

References

1. Beeson, P. B.: J. exp. Med. 86: 39 (1947).
2. Berry, L. J., Rippe, D. F.: J. infect. Dis. 128: S118 (1973).
3. Braude, A. I., Carey, F. J., Zalesky, M.: J. clin. Invest. 34:858 (1955).
4. Fox, A. E., Gawlik, D. L., Ballantyne, D. L., Jr., Freedman, H. H.: Influence of oxisuran, a differential inhibitor of cell-mediated hypersensitivity, an allograft survival and humoral immunity. Unpublished.
5. Freedman, H. H.: J. exp. Med. 111: 453 (1960).
6. Freedman, H. H., Fox, A. E., Shavel, J., Jr., Morrison, G. C.: Proc. Soc. exp. Biol. Med. 139: 909 (1972).
7. Kemp, A., Kits van Heijningen, A. J. M.: Biochem. J. 56: 646 (1954).
8. Knox, W. E., Auerbach, V. H.: J. biol. Chem. 214: 307 (1955).
9. Nowotny, A.: Basis Exercises in Immunochemistry, p. 26 (Springer, New York 1969).
10. Noyes, H. E., McInturf, C. R., Blahuta, G. J.: Proc. Soc. exp. Biol. Med. 100: 65 (1959).
11. Phillips, L. J., Berry, L. J.: Amer. J. Physiol. 218: 1140 (1970).
12. Rippe, D. F., Berry, L. J.: Infect. Immunity 6: 766 (1972).
13. Rippe, D. F., Berry, L. J.: Immunological quantitation of hepatic tryptophan oxygenase in endotoxin-poisoned mice. Infect. Immunity. In press 1973.
14. Stockman, G. D., Heim, L. R., South, M. A., Trentin, J. J.: J. Immunol. 110: 277 (1973).

Table I.

Cyclophosphamide Pretreatment and TO Activity

	TO Activity in Livers of	
Treatment	Normal Mice	Mice Treated 5 Days Earlier with Cyclophosphamide
Controls	4.2 ± .3 (7)*	3.7 ± .5 (7)
4 hours after 1 mg Hydrocortisone	17.2 ± 1.1 (7)	8.7 ± .9 (7)
4 hours after 1 mg Hydrocortisone + 50 /ug Etox	11.1 ± 1.5**(7)	5.1 ± .7**(7)

* Number of mice
** Significantly less than hydrocortisone alone at P between 0.02 and 0.01

Table II.

Cyclophosphamide Pretreatment and PEPCK Activity

	PEPCK Activity in Livers of	
Treatment	Normal Mice	Mice Treated 5 Days Earlier with Cyclophosphamide
Fed Controls	146 ± 11 (10)*	127 ± 19 (6)
Controls Fasted 20 hours	332 ± 20 (10)	318 ± 35 (5)
Fasted + 150 /ug Etox	226 ± 23 (10)	169 ± 29 (5)

* Number of mice

Table III.

Oxisuran Pretreatment and PEPCK Activity

	PEPCK Activity in Livers of	
Treatment	Normal Mice	Oxisuran Treated Mice
Fed Controls	146 ± 11 (10)*	164 ± 25 (7)
Controls Fasted 20 hrs.	332 ± 20 (10)	347 ± 29 (6)
Fasted + 50 /ug Etox	226 ± 23**(10)	237 ± 14** (5)

* Number of mice
** Significantly less than fasted controls, P value 0.01

Table IV.

Oxisuran Pretreatment and TO Activity

Treatment	TO Activity in Livers of	
	Normal Mice	Oxisuran Treated Mice
Fed Controls	$4.3 \pm .3$ (7)*	8.4 ± 1.7 (7)
Controls Fasted 20 Hours	14.9 ± 1.5 (16)	13.0 ± 1.6 (7)
Fasted + 50 µg Etox	6.0 ± 1.0** (17)	7.6 ± 1.4** (17)

 * Number of Mice
** Significantly less than fasted controls, P value <0.01

Table V.

Effect of Cyclophosphamide and Oxisuran Treatment on Development of Tolerance

Treatment	Dead/Total*
Controls	16/20
5 µg Etox**	2/20
2 X 2 mg Oxisuran***	19/20
2 X 2 mg Oxisuran plus 5 µg Etox****	3/20
400 µg/g Cyclophosphamide 6 days before challenge	26/29
400 µg/g cyclophosphamide 6 days plus 5 µg Etox 2 days before challenge	11/25

 * 48 hrs after 2 LD_{50}
 ** given ip 2 days before 2 LD_{50}
 *** given ip 3 and 2 days before 2 LD_{50}
**** Etox given with 2nd oxisuran

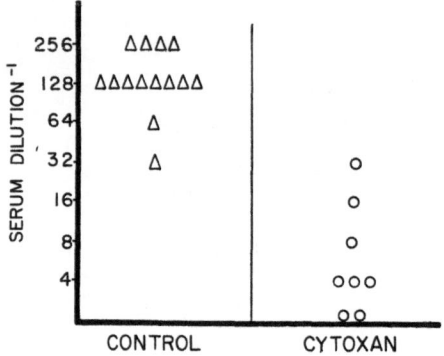

Fig. 1.   Hemagglutination titer in serum of mice vaccinated five days previously with an intraperitoneal injection of 1 ml of a 10 % suspension of washed sheep ery-throcytes. Each point is the titer of an individual animal. Mice in the group label-ed "cytoxan" were injected intraperitoneally with 400 μg per g body weight of cy-clophosphamide five days prior to immunization.

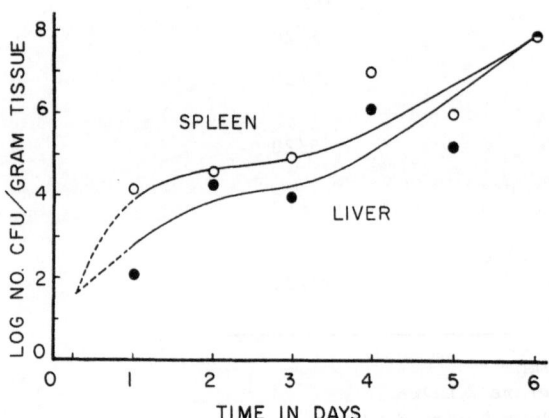

Fig. 2.   Graph of bacterial counts in spleen and liver of mice infected intra-gastrically at time zero with 10^8 colony forming units (CFU) of <u>Salmonella typhi-murium</u>, strain SR-11. The dotted lines are extrapolated back to six hours post-infection to represent the presence of fewer than 100 bacteria in both organs.

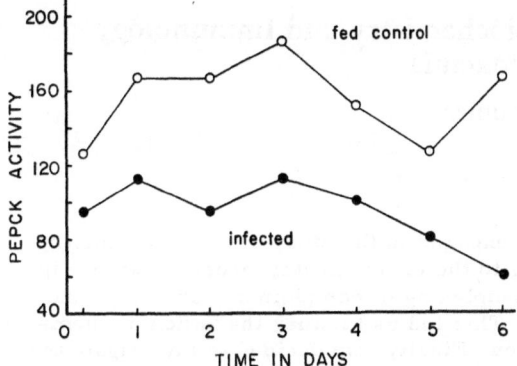

Fig. 3.   Phosphoenolpyruvate carboxykinase (PEPCK) activity in livers of control mice and in livers of mice infected intragastrically at time zero with 10^8 colony forming units (CFU) of <u>Salmonelly typhimurium</u>, strain SR-11.

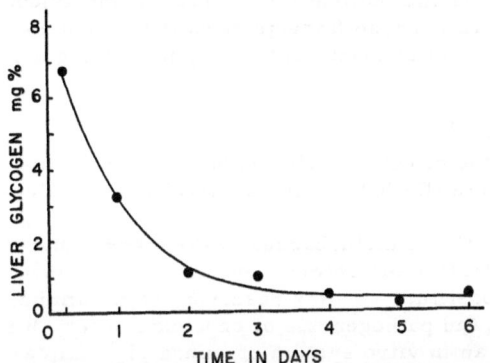

Fig. 4.   Liver glycogen in mice infected intragastrically at time zero with 10^8 colony forming units (CFU) of <u>Salmonella typhimurium</u>, strain SR-11.

Assessment of Studies in the Biochemistry and Immunology of Endotoxemia

W. Schumer

For the past four years, we have been engaged in the study of the biochemistry and immunology of endotoxemia according to the experimental model shown in Fig. 1. We studied the effect of steroids on the complexing of complement, antibody, and lipopolysaccharide (LPS) in endotoxemia. This led us to study the action of chemical mediators during this immune reaction. Finally, we decided to investigate the metabolic changes during the hypovolemic phase of endotoxic shock.

Steroid Effect on Complement Fixation

Assays utilizing specific antibodies directed against Escherichia coli LPS indicate that glucocorticoids are capable of preventing complement fixation and have a marked protective effect on the endotoxified rat. Ouabain and dehydrocholate, compounds that have steroid-like structures, were ineffective in preventing complement fixation and did not protect the endotoxified rat. These findings support the contention that steroids exert their antiendotoxic effect at least partially by preventing complement fixation.

Antiendotoxic Effect of Disodium Cromoglycate

The second phase of our studies was on the effect of antiserotonin and antihistamine agents, including disodium cromoglycate (DSCG), on the survival of adrenalectomized rats.

Histamine has been implicated in the functional disturbances of the macro- and microcirculation, which follow the administration of bacterial endotoxins to a susceptible host (1). Since tissue mast cells constitute a major reservoir of histamine, they have been suspected of participating in the pathogenesis of endotoxic shock. We have shown that histamine was liberated in an in vitro system of guinea pig complement, endotoxin, and peritoneal cells of the rat (2). This effect was attributed to anaphylatoxin, which was generated as a consequence of endotoxic activation of the complement system (3). Anaphylatoxin has been shown to affect the release of histamine from mast cells (4). Recent study has shown that disodium cromoglycate (DSCG) stabilizes mast cells and leukocytes against the immunologically induced release of slow reacting substance-anaphylaxis, histamine, and serotonin (5). In the light of these findings, we considered it worthwhile to evaluate the efficacy of DSCG in protecting against a toxic dose of endotoxin. The results of this study showed that DSCG at a dose of 10 mg/100 g body weight affords marked protection against endotoxin in the adrenalectomized rat. This suggests that in our test system endogenous shock toxins originating in mast cells and leukocytes may be a major contributing factor to endotoxin-induced lethality.

Effect of Serotonin on Cardiac Muscle Mitochondria

In a previous study we found that isolated heart mitochondria from endotoxified rats have significantly lower respiratory rates (RR) and respiratory control ratios (RCR) than those from control rats. Since in endotoxemia the tissues release a variety of endogenous chemical mediators, including histamine, norepinephrine, epinephrine, and serotonin, we investigated the problem of whether these mediators can alter respiratory function in the heart mitochondria of healthy Sprague-Dawley rats. Mitochondrial oxygen consumption was measured polarographically. The RR, RCR, and ratio of adenosine diphosphate to oxygen for both alpha-ketoglutarate (αKG) and glutamate oxidation were evaluated according to the method of Chance and Williams.

At a concentration of 1.7×10^{-4} M, only serotonin affected the performance of rat heart mitochondria. Norepinephrine, epinephrine, and histamine exerted no effect at this concentration. Also, 5-hydroxy-L-tryptophan and 5-hydroxyindole-3-acetic acid, compounds that are metabolically related to serotonin, were without effect. With serotonin the RCR for both glutamate and αKG oxidation was lowered, primarily because of decreased RR in the presence of ADP (state 3 respiration). These findings suggest that serotonin may contribute to cardiac dysfunction in endotoxemia by exerting a deleterious effect on energy metabolism in the heart.

The Effects of Endotoxin and Steroids on Gluconeogenesis

Finally, we studied the blocking effect of endotoxin on gluconeogenesis and the obviation of this block by steroids. Investigations by La Noue, Berry and Williamson have indicated that the gluconeogenic pathway in the isolated endointoxified rat liver is inhibited (6, 7, 8). We have studied the effect of endotoxin on gluconeogenesis in the intact rat liver and the effect of steroid administration on deranged gluconeogenesis.

Endotoxic shock was induced in Group A rats by infusion with 2 mg/kg body weight of E. coli endotoxin. The control Group B rats received either dexamethasone phosphate (DMP) without endotoxin or saline and endotoxin. Group C rats received endotoxin and DMP. To define the flow of fat and protein substrates to sugar production, glycolytic cycle intermediates in the liver cells were measured fluorometrically at 2 and 5 hours post-challenge. In each group of rats, these intermediates included lactate, pyruvate, phosphoenol-pyruvate (PEP), glyceraldehyde-3-phosphate (Gly-3-P), fructose diphosphate, glucose-6-phosphate (G6P), and dihydroxyacetone phosphate. Serum glucose and lactate were also measured in each group. In the Group A rats, there was a metabolic inhibition of the flow of substrate at the conversion of Gly-3-P to fructose 1,6-diphosphate. There was a 50 % decrease in fructose diphosphate, and 50 % decrease in F6P, and 80 % decrease in G6P, and a 50 % decrease in dihydroxyacetone phosphate. There was a significant increase in serum lactate and a decrease in serum glucose.

In a test of the effect of steroids on deranged gluconeogenesis, animals were given 1 mg/100 g body weight of DMP immediately after 2 mg/kg body weight of endotoxin. The DMP-treated animals showed a significant increase in serum glucose and a decrease in serum lactate, as compared to the endotoxified group. G6P and F6P were increased to control levels in the liver cell. Intracellular lactate did not change significantly.

We therefore suggest that gluconeogenesis in the intact endointoxified rat is inhibited and that this inhibition can be obviated by pharmacologic doses of steroids. Furthermore, there is an apparent need for the administration of supplemental glucose in the endointoxified rat.

References

1. Vick, J.A., Mehman, B., Heiffer, M.H.: Proc. Soc. exp. Biol. Med. 137: 902 (1971).
2. Schumer, W.: Histamine release in endotoxin shock: Effect of dexamethasone administration. In: Hinshaw, L.B., Cox, B.G.: The Fundamental Mechanisms of Shock. (Plenum Publishing Corporation, New York 1972).
3. Lichtenstein, L.M., Gewurz, H., Adkinson, N.F.Jr., Shin, H.S., Mergenhagen, S.E.: Immunology 16: 327 (1969).
4. Dias da Silva, W., Lepow, I.H.: J. exp. Med. 125: 921 (1967).
5. Orange, R.P., Austen, K.F.: Hosp. Pract. 6: 79 (1971).
6. LaNoue, K.F., Mason, A.D., Daniels, J.P.: Metabolism 17: 606 (1968).

7. Berry, L.J., Smythe, D.S.: J. exp. Med. 120: 721 (1964).
8. Williamson, J.R., Refino, C., LaNoue, K.: Effects of E. coli lipopolysaccharide B treatment of rats on gluconeogenesis. In: Porter, R., Knight, J.: Energy Metabolism in Trauma. A Ciba Symposium. (Churchill, London 1970).

PROPOSED IMMUNOLOGIC SCHEME FOR ENDOTOXIC SHOCK

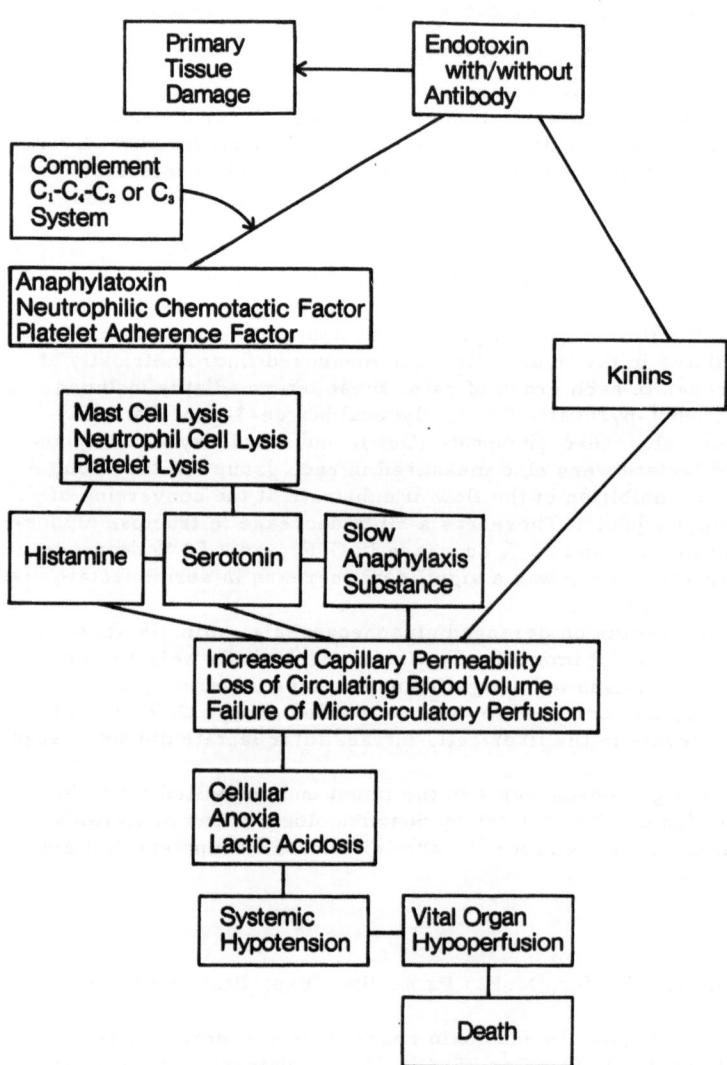

Fig. 1.

Carbohydrate Metabolism during Experimental Endotoxemia

R. E. McCallum

For many years experimental mouse typhoid has served as a model for studying host metabolism during progressive infectious bacterial disease (2). Although the intracellular parasitism of Salmonella typhimurium has been studied with this system, by far, the most work has been concerned with the exact role of endotoxin in pathogenesis of gram-negative sepsis. One of the first metabolic alterations noted to occur in experimentally infected animals as well as those given an injection of endotoxin was the rapid depletion of total body carbohydrate (2). This particular response to endotoxin has been observed by many investigators, yet its exact cause(s) remains to be elucidated.

A recent study by McCallum and Berry (8) more fully characterized gluconeogenesis, glycogen synthesis, and liver glycogen synthase (uridine 5′-diphosphate UDP-glucose : glycogen α-4-glucosyltransferase, EC 2.4.1.11) activity in endotoxin-poisoned mice. It was clearly demonstrated during liver glycogen depletion following endotoxin administration that incorporation of ^{14}C-labeled precursors into glucose and glycogen as well as liver glycogen synthase activity was significantly impaired. Evidence was also presented showing endotoxin inhibition of the increase in liver glycogen synthase due to fasting or glucocorticoid treatment, a process apparently dependent on new protein synthesis, since actinomycin D treatment inhibited it. Administration of glucose or pyruvate to endotoxin-poisoned animals failed to restore gluconeogenesis or glycogen synthesis to normal. The in vivo activation of liver glycogen synthase by a glucose load was impaired by endotoxin. These observations suggest that reduced carbohydrate synthesis was the most probable cause for rapid sugar loss during experimental endotoxemia.

This investigation was undertaken to characterize further the effects of endotoxin on liver glycogen synthase; and, more specifically, its effects on the in vivo and in vitro activation of the enzyme. It was thought that the results of such a study might indicate the molecular mode of action of endotoxin in the disruption of key-rate-limiting enzymes in the liver.

Material and Methods

Animals. Female random-bred, pathogen-free mice (ICR strain, Texas Inbred Mice, Co., Houston, Tex.) weighing 18 to 20 g were employed for these studies.

Endotoxin. The endotoxin employed in these studies was extracted and purified from S. typhimurium strain SR-11 as previously described (8). The mouse LD_{50} (mean lethal dose) of the purified preparation was determined by the method of Reed and Muench (9) using a 48 hours endpoint, and was found to be approximately 235 μg per mouse.

Fasting, injections, and sacrifice. Since the purpose of this study was to evaluate carbohydrate metabolism in mice, fasted animals were employed. It is well established that mice poisoned with endotoxin stop eating (2); therefore, fasting partially balances the experimental and control animals. In all experiments mice were

This investigation was supported by a Brown-Hazen grant from Research Corporation and General Research Support Grant No 5 SO1 RR05411 to the University of Oklahoma Health Science Center from the Division of Research Resources of the National Institutes of Health.

allowed to fast overnight (5:00 p.m. to 8:00 a.m.) prior to the initial injections and enzyme assays at 8:00 a.m. Fasting was continued throughout the course of each experiment. Injections of purified lipopolysaccharide, actinomycin D, and cortisone were carried out as described previously (8). At appropriate intervals after treatment, groups of ten mice each were sacrificed by decapitation. The livers were quickly removed, rinsed, blotted dry, weighed, and used immediately for enzyme assays.

Glycogen synthase assay. Mouse liver glycogen synthase activity was measured by modification of the filter paper method of Thomas et al. (12) as described by Hickenbottom and Hornbrook (6). A 1-g amount of liver was homogenized in three volumes of 100 mM glycylglycine buffer (pH 7.5) containing 150 mM sucrose at $0^\circ C$ for 30 seconds, and the homogenate was centrifuged at 9000 x g for 10 minutes. The resulting supernatant solution was held at $0^\circ C$, and the first assay for glycogen synthase activity was made 30 minutes after removal of the liver from the animal. Enzyme activity is expressed as millimoles of glucose transferred from uridine diphosphate glucose (UDPG) to glycogen per hour per kilogram of liver (dry weight).

In vitro activation of glycogen synthase. Activity of glycogen synthase phosphatase was determined as described by De Wulf and Hers (4) in homogenates prepared as described above incubated at $30^\circ C$ for 0, 20, 40, and 60 minutes. Phosphatase activity is reflected by the increase in glycogen synthase I during incubation at $30^\circ C$ (expressed as glycogen synthase, % I). When the effects of endotoxin added directly to the phosphatase reaction were examined, 10-fold dilutions of purified lipopolysaccharide were incubated with enzyme preparations obtained from normal fasted mice for the same time intervals indicated above.

Statistics. All results were analyzed by the rank order test (13) and Student's t-test (7) for significant differences (p $\leq$ 0.05).

Results

The results in Table I represent data obtained from several experiments designed to observe the effects of glucose, cortisone, and actinomycin D on both in vivo and in vitro activation of liver glycogen synthase. The activation reaction, catalyzed by glycogen synthase phosphatase, is reflected by an increase in glycogen synthase activity in the I-form (enzyme activity independent of added glucose-6-P). In these experiments activation of the synthase in vivo was accomplished by the administration of glucose or cortisone, whereas, in vitro activation was performed by incubating liver homogenates at $30^\circ C$.

Twelve hours after endotoxin poisoning (1 LD_{50}) liver glycogen synthase activity appeared altered. This confirms an earlier observation (8) and serves as a basis for comparison of the other results obtained in this series of experiments. Although not statistically significant (p $\leq$ 0.05), total glycogen synthase in poisoned mice was decreased from the control values. The changes seen in the I-form of synthase before incubation were more pronounced, decreased from the control values by 38 %. The in vitro activation of glycogen synthase seemed unimpaired by endotoxin treatment, although, the absolute activity of the I-form of the enzyme was significantly decreased.

As seen from the results presented in Table I endotoxin poisoning significantly impaired the in vivo activation of the enzyme due to the i.p. injection of a glucose load (36 mg) 30 minutes before sacrifice. Total glycogen synthase activity in poisoned mice given glucose was 57 % of the control values. The most significant difference, however, was seen in the I-form of the enzyme which was decreased in the poisoned mice to 26 % of the control values. The observation that 80 % of liver glycogen synthase in the controls was in the I-form (before incubation) further demonstrated the in vivo activating properties of an exogenous glucose load. Since only 29 % of the synthase in livers from poisoned mice was in the I-form, it seemed apparent that endotoxin significantly inhibited the in vivo activation reaction. This information confirms the results of a previous study (8). The phosphatase reaction in endotoxin-treated mice seemed unchanged. Since most of the enzyme in the control group was in the I-form prior to incubation (80 % I), virtually all of the synthase was independent of glucose-6-P after 60 minutes at 30°C. This was not true for the endotoxin-treated mice, although, the rate of increase in % I appeared no different from that observed for the controls.

Cortisone treatment seemed to have little effect on liver glycogen synthase in control mice, although the in vitro phosphatase reaction may have been enhanced. The synthase activity in endotoxin-poisoned mice given cortisone, however, was significantly reduced from the control values. Total activity was about 77 % of the control and the I-form was about 48 % of the control values. The in vitro phosphatase reaction did not appear reduced in endotoxin-treated mice given cortisone.

Actinomycin D treatment reduced liver glycogen synthase (both total and I-forms) to approximately 71 % and 56 % of the values observed in fasted control mice. Although the sampling time was considerably removed from the time of injection of the drug (12 hours), these data suggest that the maintenance of normal levels of glycogen synthase in mice requires DNA-dependent RNA synthesis. Previous work (8) suggested that inhibition of the above may occur in endotoxin shock, hence, contributing to the alteration in glycogen metabolism.

The results in Table I suggested that endotoxin poisoning (1 LD_{50}) had no effect on the in vitro activation of liver glycogen synthase. Preliminary experiments, however, had demonstrated some possible differences in the phosphatase-catalyzed reaction between control and endotoxin-poisoned mice. For this reason the in vitro activation of glycogen synthase by synthase phosphatase was examined more closely and the results are shown in Fig. 1. Glycogen synthase preparations were obtained from the livers of control and endotoxin-treated mice at 2, 6, and 12 hours after treatment and were incubated at 30°C. Aliquots were removed at 0, 20, 40, and 60 minutes and assayed for glycogen synthase activity. The increase in the I-form of the enzyme was plotted against the time of incubation after correction for the 15 to 20 % loss of total enzyme activity after 60 minutes at 30°C.

The rate of the glycogen synthase phosphatase reaction appeared linear with respect to time at all three sampling intervals (2, 6, and 12 hours) in fasted control mice. The rate of the reaction decreased as fasting continued, an expected phenomenon, since it had previously been shown that maximal glycogen synthase activity was observed 2 hours after treatment (8).

The total increase in glycogen synthase I activity after 60 minutes of incubation appeared no different in control and endotoxin-treated mice, although, the phosphatase reaction was not linear with respect to time in the poisoned group. This was observed for enzyme preparations obtained at 2, 6, and 12 hours after treatment. The results in Fig. 1 represent this observed difference and suggest an alteration in the kinetics of the in vitro synthase activation reaction in livers obtained from endotoxic mice.

To further test the effects of endotoxin on glycogen synthase phosphatase, purified lipopolysaccharide was added in vitro to synthase preparations obtained from normal fasted mice. The results are seen in Fig. 2. The direct addition of endotoxin (125 ng to 1.25 mg final concentration) was without effect on either glycogen synthase or the in vitro phosphatase reaction. No significant difference (p $\leq$ 0.05) in activity could be demonstrated, irrespective of the concentration of added lipopolysaccharide. These data lend further support to the hypothesis that endotoxin elicits its effect on hepatic enzymes via in vivo mediators (1).

Discussion

The results of the series of experiments reported here confirm the findings from an earlier study by McCallum and Berry (8). This includes the observation that endotoxin treatment (1 LD_{50}) significantly reduces mouse liver glycogen synthase. At

12 hours after treatment the total (I + D) activity was reduced to 70 % of the control values, whereas, the I-form activity was reduced to about 50 % of the control values. The in vivo activation of glycogen synthase by glucose was also shown to be altered in the poisoned animal. In addition, it appeared that the cortisone-induced increase in this liver enzyme was also diminished in endotoxic mice.

To extend these preliminary observations, glycogen synthase phosphatase activity was examined in endotoxin-treated mice. Although the overall phosphatase reaction in poisoned mice appeared unchanged, the observation of a non-linear increase in glycogen synthase I certainly suggests altered reaction kinetics. Possibly the substrate, synthase enzyme, for the phosphatase reaction becomes rate-limiting in the poisoned mouse. Alternate possibilities include altered substrate binding sites or catalytic sites due to conformational changes in the enzyme in addition to significant in vivo changes in ligand concentrations. Since the control and regulation of liver glycogen synthase is extremely complex, it seems likely that endotoxin may induce a number of changes which may affect the cascade-control of this enzyme. It is interesting to note that the in vitro addition of purified lipopolysaccharide from S. typhimurium had no effect on glycogen synthase or synthase phosphatase. This further suggests endotoxin elicits its effects on glycogen metabolism via mediators. The many questions posed by these results are currently under further study.

One obvious question which these data do not answer concerns the activity of liver glycogen phosphorylase during endotoxic shock. Only one bit of evidence exists which would suggest possible participation of phosphorylase enzyme in the changes noted in this study. The work of Hamosh and Shapiro (5) indicated that early after endotoxin administration (1 to 6 hours) glycogenolysis in rat liver and muscle was enhanced. This was also demonstrated with slices and homogenates of tissue from endotoxin-treated rats. They concluded that the enhanced breakdown of glycogen was due to increased phosphorylase a which, in turn, was due to enhanced activity of phosphorylase kinase. More recently, Zwadyk and Snyder (14) presented an in vitro model employing homogenates of mouse liver to study the effects of endotoxin on glycogen metabolism. Unfortunately, both of these groups sacrificed their experimental animals in a manner which causes large increases in phosphorylase a, presumably by the rapid changes which occur in nucleotide levels in tissue following iscemia or anoxia. A significant increase in glycogenolysis as a probable mechanism for rapid sugar loss during endotoxemia previously seemed unlikely. With newer concepts developing regarding the control of glycogen metabolism, however, the possibility of participation by phosphorylase cannot be ignored.

Currently, the control of glycogen metabolism in liver centers around an intimate association between phosphorylase and glycogen synthase. According to Segal (10) and Stalmans et al. (11) cyclic-adenosine 3′5′monophosphate, which is a positive effector for both phosphorylase kinase and synthase kinase, and glucose and glucocorticoids, which are positive effectors for phosphorylase phosphatase, control the operation of the glycogen cycle. It appears that phosphorylase is the most highly regulated enzyme, which, in the active form, directly binds (via protein-protein interaction) to synthase phosphatase, inhibiting its activity. In this respect, phosphorylase acts as a second messenger in the control of glycogen synthesis by glucocorticoids and by glucose.

Bitensky et al. (3) recently reported a significant selective stimulation of the epinephrine-responsive adenyl cyclase activity in livers of normal and adrenalectomized mice. They suggested that endotoxin may alter the cell membrane containing the cyclase, thereby, leading to enhanced glycogenolysis and increased sensitivity to catecholamines.

All of the possibilities discussed above remain to be clarified. There is no demonstrative proof that endotoxin affects rapid sugar loss due to glycogenolysis. There is substantial evidence, however, that reduced synthesis of carbohydrate plays a significant role in the overall biochemical lesion. From the results of this study and a previous one (8) there seems to be no question that carbohydrate homeostasis in the endotoxin-poisoned animal is disrupted.

The general theme suggested by the results of this study is that bacterial endotoxin may elicit its effects on host metabolism via a chain reaction or mediated series of events. If this be the case, then endotoxin must have a relatively selective mode of action. Complex regulatory enzymes seem likely candidates for such an effect since their activity is controlled by such a wide array of effectors.

Summary

Bacterial endotoxins have long been known to deplete an animal of its carbohydrate reserves, a phenomenon also observed to occur during mouse typhoid. Until recently, however, the mechanism(s) responsible for these changes have remained unknown. Preliminary studies have established impaired hepatic gluconeogenesis and glycogen synthesis to be directly involved in rapid and irreversible carbohydrate loss. Mouse liver glycogen synthase has also been shown to be altered early after endotoxin treatment. This study was undertaken to further characterize the nature of carbohydrate loss due to endotoxin poisoning in mice and to elucidate mechanisms responsible for the changes. Female ICR mice, fasted overnight, were injected intraperitoneally with a mean lethal dose of endotoxin extracted from Salmonella typhimurium strain SR-11. At intervals after treatment liver glycogen synthase and synthase phosphatase activities were measured. Certain regulatory properties of glycogen synthase in poisoned mice were examined in detail. The in vivo activation of the enzyme by a glucose load, cortisone, or both was impaired by endotoxin. The in vitro activation of synthase obtained from poisoned mice was also shown to be altered. The in vitro addition of endotoxin to enzyme preparations obtained from fasted normal mice had no effect on glycogen synthase or synthase phosphatase. The changes noted in liver enzyme activities in endotoxin-poisoned mice suggested possible participation of mediators. These results further suggest that impaired gluconeogenesis as well as altered synthesis and activation of liver glycogen synthase are major causes for rapid sugar loss during endotoxemia in mice.

Acknowledgement

The excellent technical assistance of M. O. Giger and L. Chavez is gratefully acknowledged.

References

1. Berry, L. J., Rippe, D. F. : J. infect. Dis., Suppl. 128 : S 118 (1973).
2. Berry, L. J., Smythe, D. S. : Ann. N. Y. Acad. Sci. 88 : 1278 (1960).
3. Bitensky, M. W., Gorman, R. E., Thomas, L. : Proc. Soc. exp. Biol. Med. 138 : 773 (1971).
4. De Wulf, H., Hers, H. G. : Eur. J. Biochem. 6 : 552 (1968).
5. Hamosh, M., Shapiro, B. : Brit. J. exp. Path. 41 : 372 (1960).
6. Hickenbottom, R. S., Hornbrook, K. R. : J. Pharmacol. exp. Ther. 178 : 383 (1971).
7. Lewis, A. E. : Biostatistics. (Reinhold Publishing Corp., New York 1966).
8. McCallum, R. E., Berry, L. J. : Infect. Immunity 7 : 642 (1973).
9. Reed, L. J., Muench, H. A. : Amer. J. Hyg. 27 : 493 (1938).
10. Segal, H. L. : Science 180 : 25 (1973).
11. Stalmans, W., De Wulf, H., Hers, H. G. : Eur. J. Biochem. 18 : 582 (1971).
12. Thomas, J. A., Schlender, K. K., Larner, J. : Anal. Biochem. 25 : 486 (1968).
13. Wilcoxon, F., Wilcox, R. A. : Some rapid approximate statistical procedures. (American Cyanimide Co., New York 1949).
14. Zwadyk, P. Jr., Snyder, I. S. : Proc. Soc. exp. Biol. Med. 142 : 299 (1973).

Table I.

In vivo and in vitro Activation of Liver Glycogen Synthase 12 Hours after Endotoxin Poisoning (1 LD_{50}) in Mice Treated with a Glucose Load, Cortisone, or Actinomycin D.

Treatment	Incubation at 30° C (min)	Glycogen Synthase Activity (mmoles/hrs/kg of liver)		%I[b]
		Total (+G6P)[a]	I-form (-G6P)	
Control	0	1449 ± 162[c]	422 ± 144	29 ± 8
	60	1090 ± 154	750 ± 170	69 ± 10
Endotoxin	0	1257 ± 130	265 ± 93[f]	20 ± 6[f]
	60	1012 ± 94	575 ± 121[f]	55 ± 8[f]
Control 30 min after Glucose Load[d]	0	1622 ± 134	1304 ± 178	80 ± 9
	60	1382 ± 156	1382 ± 150	100 ± 1
Endotoxin 30 min after Glucose Load[d]	0	1206 ± 129[f]	343 ± 53[f]	29 ± 5[f]
	60	974 ± 131[f]	612 ± 104[f]	63 ± 8[f]
Cortisone (50 mg/kg)[e]	0	1638 ± 196	374 ± 102	24 ± 6
	60	1544 ± 214	940 ± 272	72 ± 10
Cortisone + Endotoxin	0	1269 ± 145[f]	178 ± 52[f]	14 ± 3[f]
	60	1013 ± 148[f]	495 ± 98[f]	48 ± 3[f]
Actinomycin D (1 mg/kg)[e]	0	1028 ± 67	236 ± 36	23 ± 3
	60	831 ± 62	575 ± 98	68 ± 8

[a] Glucose-6-phosphate added to the incubation mixture (15 mM)

[b] I activity / Total activity X 100

[c] Mean ± standard error of values obtained from 10 mice

[d] Glucose (200 μmoles) injected i. p. 30 minutes before sacrifice

[e] Injected immediately prior to endotoxin

[f] Significantly different from control values ($p \leq 0.05$)

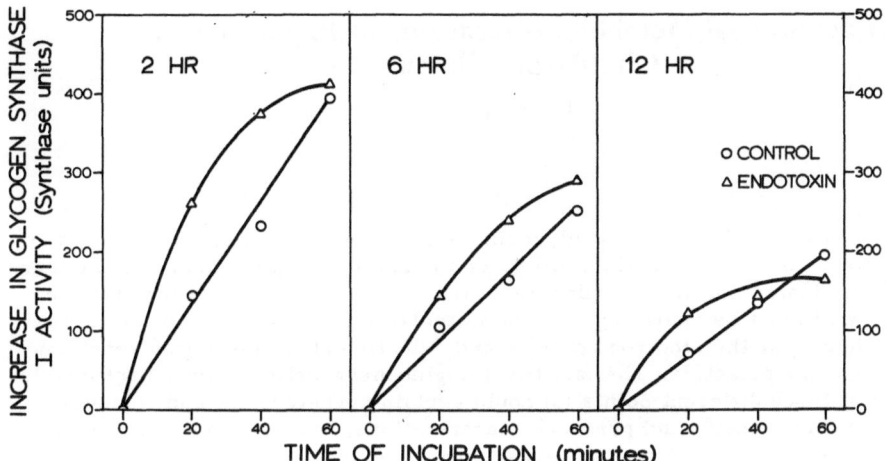

Fig. 1. Mouse liver glycogen synthase phosphatase activity in control (o) and endotoxin-treated mice (Δ) at 0, 2, 6, and 12 hours after treatment. The points represent the mean values obtained from 10 mice. Glycogen synthase I activity represents that form measured in the absence of glucose-6-P.

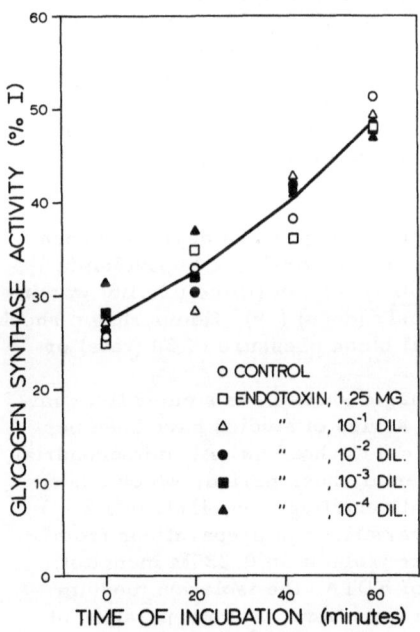

Fig. 2. Effect of the _in vitro_ addition of endotoxin on normal mouse liver glycogen synthase phosphatase activity. The points represent mean values obtained from triplicate determinations made on liver enzyme preparations pooled from 10 mice. Phosphatase activity is expressed as liver glycogen synthase activity, % I (synthase I/total synthase X 100). Total enzyme activity was measured in the presence of 15 mM glucose-6-P.

Mitochondrial Metabolic Alterations in Experimental Circulatory Shock

L. Mela

Lethal endotoxemia induces hypoglycemia, high blood lactate/pyruvate ratios and decreased tissue oxygen utilization. All these metabolic symptoms occur simultaneously. This indicates that the cytoplasmic portion of cellular glucose metabolism, glycolysis, still must be operating, but the mitochondrial reactions, the tricarboxylic acid cycle and/or the electron transfer and oxidative phosphorylation reactions could have become defective. The fact that the gluconeogenetic pathway presumably is inhibited at three different points (5) could explain hypoglycemia and somewhat higher than normal lactate and pyruvate concentrations, but could not count for decreased oxygen utilization. Thus, it was important to us, as well as to some other investigators, to attempt a clarification of possible effects of endotoxemia on mitochondrial metabolic reactions. The importance of these studies is emphasized by the central role of adequate energy production in the survival of cells and vital tissue functions.

During the last four years a number of reports have appeared in the literature providing evidence for defective mitochondrial structure and function in lethal circulatory shock (2, 3, 4, 7, 8, 10, 17, 19, 21, 22, 24, 25, 28, 29, 30, 33). Some of the biochemical and biophysical results are controversial, and in many of the reports data, unfortunately, are very scarce. Thus, today in my attempt to summarize our present understanding of the mitochondrial metabolic alterations induced by endotoxemia, I will mainly have to concentrate on reporting data from our own laboratory. And, of course, what could be easier and more pleasant than doing just that.

Experimental Procedures

In studies of mitochondria in experimental shock, both rats and dogs have been utilized. The models of experimental endotoxemia and hemorrhagic hypovolemia have been well standardized. In endotoxemia E. coli endotoxin (Difco 026:B6) was injected either intraperitoneally (rats) or intravenously (dogs) (19). Hemorrhagic shock was induced by a single bleeding to a mean arterial blood pressure of 30 (rats) or 35 mmHg (dogs) (1).

Most investigators of endotoxemic and hemorrhagic shock have studied liver mitochondrial functions. In our laboratory elaborate series of studies have been performed on liver, kidney, heart and skeletal muscle mitochondria. All mitochondrial preparations were made at the end of the experimental shock period, which varied in length, by tissue homogenization and differential centrifugation, including 2 - 3 washes of the mitochondrial preparations. Slight variations in preparations from different tissue occurred. Liver mitochondria were isolated in 0.225M mannitol, 0.075M sucrose and 100 /uM EDTA (19). Instead of EDTA, the isolation medium contained 1 mM EGTA for kidney mitochondrial preparations. For preparation of heart mitochondria, the minced tissue was incubated with a bacterial proteinase, Nagarse (Nagase, Osaka, Japan) (21). The same treatment in the presence of bovine serum albumin was utilized for the preparation of skeletal muscle mitochondria, which were isolated in a KCl-Mg^{++}- and ATP-containing medium. The isolation procedure for skeletal muscle mitochondria was adapted from Mäkinen and Lee (18).

For the studies of mitochondrial respiratory activity a Clark oxygen electrode

Supported by USPHS Grant R01-GM 19867 and the USPHS Career Development Award 5 K04 GM 50318

was used (23). Mitochondrial ATPase activity was measured in a recording pH meter by monitoring the production of H^+ occurring during ATP hydrolysis (23). The measurements were verified by direct spectrophotometric assays of ATP, ADP and AMP after their column chromatographic separation (35). The reactions of the electron transfer carriers, cytochromes, were monitored in a dual wavelength spectrophotometer (23). Ca^{++} transport activity was measured in a dual wavelength spectrophotometer with a calcium sensitive indicator murexide (20). K^+ transport was monitored with a glass - K^+ electrode. Protein concentrations of all samples were determined by the Biuret reaction.

Results

Effect of Endotoxemia on Liver and Kidney Mitochondria. Our first studies of endotoxic animals were directed to examining the liver mitochondria. The data obtained in these studies, summarized in Table I, have been reported in detail earlier (19, 21, 24). These date were obtained in liver mitochondria, isolated at the terminal stages of endotoxemia, namely 5-6 hours after an injection of an LD_{90} dose of E. coli endotoxin in the rat. Table I presents our data in three groups: The first group represents the reactions involved in electron transfer only. Oxygen utilization in State 4 (excess substrate, no ADP) is not inhibited in endotoxemia. The actual enhancement of State 4 respiration is due to loose coupling and not due to an enhancement of the electron transfer capacity per se. The reducibility of the electron transfer chain by substrates is only slightly diminished. Thus, it appears that the electron transfer reaction is relatively unaffected in lethal endotoxemia.

The second group of data in Table I presents the parameters indicative of oxidative phosphorylation reactions. All these parameters are heavily inhibited in lethal endotoxemia. State 3 (substrate and ADP in excess) respiratory activity is only slightly higher than in State 4. This also results in a large decrease in the respiratory control ratios (RCR). The rate of the ATPase activity and the amount of ATP hydrolyzed are 61 and 52% below normal. The redox response of the electron transfer carriers to ADP is completely inhibited. Similar interference of lethal circulatory shock with oxidative phosphorylation reactions in the liver has been reported by others as well (2, 7, 28, 29).

The third group of results in Table I illustrates those mitochondrial reactions involved in energy-dependent ion transport. Ca^{++} accumulation is one of the most sensitive indicators of the intactness of mitochondrial energy-linked processes. Thus, it was of great importance to us to discover that Ca^{++} uptake was one of the first mitochondrial reactions to become inhibited in endotoxemia. At the terminal stages the inhibition of Ca^{++} accumulation was often 100%. The two other important cations, which in normal mitochondria are transported very slowly and only to a small extent in the absence of specific transport inducers, are K^+ and Mg^{++}. The fact that spontaneous K^+ accumulation in liver mitochondria is doubled, in both speed and amount, after endotoxemia is of great importance. If K^+ is accumulated to this large an extent by liver mitochondria and also in vivo, the characteristic ultrastructural endotoxin effects in situ become more understandable. The tremendous swelling of liver mitochondria in situ could be due to large amounts of intramitochondrial K^+. The decrease in bound Mg^{++} concentration in endotoxic liver mitochondria is also indicative of diminished energy supply and increased membrane leakiness.

The information gathered in Table I leads to the following conclusions: 1. lethal endotoxemia damages liver mitochondria to the point where all the reactions involving energy production and utilization in mitochondria are inhibited. 2. The damage, however, is not extensive enough to destroy the cytochromes and their electron transfer function.

Fig. 1 illustrates the time course of the inhibition of mitochondrial respiratory activity in State 3. The top graph shows that after an injection of an LD_{90} dose of endotoxin the respiration has become 50% inhibited in 3.5 h. The terminal stage is reached in 5-6 h. If an LD50 dose of endotoxin is used there is no measurable inhibition of liver mitochondrial function at 3.5 h. By 18 h the respiratory activity in the animals still surviving is more than 50% inhibited: Half of the animals died before 18 h.

Qualitatively similar results were obtained in kidney mitochondria as is shown in the bottom graph of Fig. 1. Inhibition of respiration in State 3 becomes measurable slightly earlier than in the liver. At 22 h, following an injection of an LD 50 dose of endotoxin, kidney mitochondria exhibit only 25% of control respiratory activity. It should be pointed out that these data were obtained in animals surviving an LD50 dose of endotoxin for a minimum of 18-22 h.

We have, thus, shown that lethal endotoxemia induces heavy inhibition of mitochondrial energy-producing reactions in the liver and the kidney. The rate of appearance of the inhibition is dependent on the dose of endotoxin in both tissues. However, it seems that the kidney mitochondria become defective somewhat earlier than the liver.

Effect of Lethal Endotoxemia on Heart Mitochondria. Heart mitochondrial function has been studied by Schumer et al. (30) and by us (21). Our data are summarized in Fig. 2. The top graph of the Fig. shows an interesting change in the heart tissue pH during endotoxemia. It first seems, maybe, that this set of data is out of context here. The reason why I have included it is twofold: 1. the pH data indicate that the heart tissue experiences a change in endotoxemia, and 2. the change is an increase rather than a decrease in pH. All other tissues measured in the same series of experiments showed a drop in the homogenate pH. In the liver and kidney the drop was from 7.1 to 6.9 in 5-6 h; in the brain the drop occurred earlier. Thus the heart tissue shows an alkalinization rather than an acidification, so typical to other tissues in lethal endotoxemia.

Simultaneously with increasing pH, mitochondrial activity increases above control levels. In the rat heart there is an increase of about 60% by 5 h, in the dog heart the increase is very closely the same, 58%. These heart mitochondrial activities were measured at the terminal stages of endotoxemia, immediately before death. All other data obtained in the heart mitochondrial studies showed intact function. ATPase is not inhibited, ion transport activity is intact, the mitochondria are very well coupled, exhibiting respiratory control ratios above normal (15-20 in the dog, 6.5-9.0 in the rat). From these data we have to make one obvious conclusion: At the terminal stages of endotoxemia there is no inhibition of any of the mitochondrial functions in the heart. On the contrary, the mitochondrial capacity for energy production has increased significantly.

Comparison of Hypovolemic and Endotoxic Shock. We have reported earlier that hypovolemic and endotoxic shock induce similar mitochondrial alterations in the liver (19). Fig. 3 re-emphasizes these findings. It indicates that lethal hemorrhage induces an inhibition of more than 50% of liver mitochondrial State 3 respiratory activity (bottom graph) and simultaneously induces a progressive increase of State 3 activity in the heart (top). In this particular experimental series also skeletal muscle mitochondria were studied. As can be seen from the middle diagram of Fig. 3, no inhibition of State 3 respiratory activity occurred in these mitochondria. The skeletal muscle mitochondria also exhibited very tight coupling of respiration to phosphorylation with respiratory control ratios between 12-20 as compared to the control values 9-14.

Discussion

Studies of liver mitochondria isolated during lethal endotoxemia from rats and dogs have uncovered disturbed energy production (19, 21) paralleling pathologic ultrastructural findings: swelling of the mitochondria and morphologic defects in their membranes (17, 24, 33). The functional alterations indicate defective oxidative phosphorylation reactions rather than inhibition of the electron transfer chain. Other energy-linked reactions, such as ion transport across the membrane, are inhibited as well (22). The ion transport studies are specially interesting: In early endotoxemia there is a significant inhibition of energy-linked Ca^{++} accumulation by liver mitochondria. This inhibition occurs parallel with increased permeability of the mitochondrial inner membrane to K^+ which was verified by an accumulation of twice normal amounts of K^+ in the absence of specific ion transport inducers, such as valinomycin (22). These effects of the energy-dependent ion transport are evident early in circulatory shock, when the energy production is still adequate.

Changes identical to those found in liver, were also monitored in kidney mitochondria in lethal endotoxemia. The time course of the alterations suggests that kidney mitochondria become defective at a slightly earlier stage of shock than the liver mitochondria do.

As was shown above, heart and skeletal muscle do not experience mitochondrial damage in circulatory shock. In case of skeletal muscle this is easy to understand. The cells of this tissue are physiologically much more resistant to high lactates and low Po_2 and pH than other tissues. Also, during circulatory shock energy demand in skeletal muscle is quite low. As for the heart, we will return to discuss this organ later.

It should be pointed out that in vitro studies of mitochondria isolated after in vivo endotoxemia are not directly indicative of the mitochondrial activity in situ. They only reveal those mitochondrial alterations that are irreversible. In this case, irreversibility means defects acquired in vivo that are not corrected during the isolation and assay procedures, when the damaging factor or factors have been removed and the mitochondria suspended in proper osmotic environment with balanced H^+ concentrations, with plenty of substrates and phosphate acceptor available. The in vivo conditions, in actuality, might be considerably worse. Even if the mitochondria were not totally damaged yet, their acitivity might be very low, because of surrounding inhibitory factors.

Thus, knowledge of other parallel intracellular alterations becomes important, not only for the purpose of searching for etiologic causes of the mitochondrial damage but also for the understanding of the environmental conditions under which the mitochondria during circulatory shock are forced to operate. Table II will draw your attention to some of these important environmental factors.

In an organ, such as the liver, where mitochondria become quite heavily inhibited in circulatory shock, a number of other intracellular alterations occurs. Endotoxin has been shown to accumulate primarily in the liver and spleen (16, 34) and could thus, have a direct effect on the mitochondrial membranes. This is somewhat questionable though. In our earlier in vitro studies we found a specific inhibition of succinate oxidation in mitochondria incubated with endotoxin (23). The fact that this specific inhibition does not occur after in vivo endotoxemia, strongly speaks against a direct intracellular effect of endotoxin on mitochondria. However, there are other environmental changes that are known to inhibit mitochondrial function in a fashion similar to circulatory shock. These are released lysosomal enzymes (21, 24, 25), low cell pH (21) and tissue ischemia in general (6, 11, 15, 26, 31). Thus, it seems to us that the mitochondrial inhibition induced by circulatory shock could be due to tissue ischemia. The mitochondrial alterations induced by experimental tissue ischemia have been well characterized by a number of investigators (13, 14, 27, 32) and are within experimental error and laboratory differences, identical to our results on endotoxemia and hemorrhage.

There are two other facts that suggest the existence of an ischemic injury to mitochondria in shock. As can be seen from Table II, endotoxemia and hemorrhage induce similar intracellular alterations, and so do they also induce similar mitochondrial defects. The second piece of evidence is the response of the heart to circulatory shock. Heart mitochondria do not become defective. Also, as Table II shows, any indications of tissue ischemia in the heart in circulatory shock are lacking.

Our studies do not indicate any damage to the heart tissue in endotoxemia or hemorrhage. The capacity of heart mitochondrial energy production has remarkably increased after endotoxemia. This could be a compensatory response to some change that this tissue has experienced. Certainly it is an indication of a change. On the basis of our present data we cannot predict what the change might be, other than pH, which, of course, is secondary in nature. It is important to note that on the basis of our studies, possible direct effects of some toxic factors on cardiac muscle contraction and work performance cannot be excluded. Our studies only suggest that the cardiac energy-producing mechanisms, located in mitochondria, are adequate to meet even increased energy demands, assuming that substrate and ADP supplies have not been diminished.

References

1. Bacalzo, L. V., Cary, A. L., Miller, L. D., Parkins, W. M.: Surgery 70: 555 (1971).
2. Baue, A. E., Sayeed, M. M.: Surgery 68: 40 (1970).
3. Baue, A. E., Sayeed, M. M., Wurth, M. A.: Potential Relationships of Changes in Cell Transport and Metabolism in Shock. In: Kovach, A. G. B., Stoner, H. B., Spitzer, J. J.: Neurohumoral and metabolic aspects of injury, p. 253 (Plenum Press 1973).
4. Baue, A. E., Wurth, M. A., Sayeed, M. M.: Surgery 72: 94 (1972).
5. Berry, L. J.: Metabolic Effects of Bacterial Endotoxins. In: Kadis, S., Weinbaum, G., Ajl, S. J.: Microbial toxins, p. 165 (Academic Press, 1971).
6. Bounous, G., Hampson, L. G., Gurd, F.: Arch. Surgery 87: 340 (1963).
7. DePalma, R. G., Levey, S., Holden, W. D.: J. Trauma 10: 122 (1970).
8. DePalma, R. G., Holden, W. D., Robinson, A. V.: Ann. Surgery 175: 539 (1972).
9. Forsyth, R. P.: Fed. Proc. 31: 1240 (1972).
10. Hift, H., Strawitz, J. G.: Am. J. Physiol. 200: 264 (1961).
11. Hinshaw, L. B.: Release of Vasoactive Agents and the Vascular Effects of Endotoxin. In: Kadis, S., Weinbaum, B., Ajl, S. D.: Microbial toxins, p. 209 (Academic Press, 1971).
12. Hinshaw, L. B., Greenfield, L. J., Owen, S. E., Archer, L. T., Guenter, C. A.: Am. J. Physiol. 222: 1047 (1972).
13. Jennings, R. B., Kaltenbach, J. P., Sommers, H. M.: Arch. Path. 84: 15 (1967).
14. Jennings, R. B., Sommers, H. M., Herdson, P. B., Kaltenbach, J. P.: Ann. N. Y. Acad. Sci. 156: 61 (1969).
15. Kessler, M., Görnandt, L., Therman, M., Lang, H., Brand, K., Wessel, W.: Oxygen Supply and Microcirculation of Liver in Hemorrhagic Shock. In: Kessler, M., Bralley, D. F., Clark, L. C., Lübbers, D. W., Silver, I. A., Strauss, J.: Oxygen Supply, p. 252 (University Park Press 1973).
16. Levy, E., Path, F. C., Ruebner, B. H.: Am. J. Pathol. 51: 269 (1967).
17. Levy, E., Path, F. C., Slusser, R. J., Ruebner, B. H.: Am. J. Pathol. 52: 477 (1968).
18. Makinen, M. W., Lee, C. P.: Arch. biochem. biophys. 126: 75 (1966).
19. Mela, L., Bacalzo, L. V., Miller, L. D.: Am. J. Physiol. 220: 571 (1971).
20. Mela, L., Chance, B.: Biochemistry 7: 4059 (1968).
21. Mela, L., Miller, L. D., Nicholas, G. G.: Surgery 72: 102 (1972).
22. Mela, L., Nicholas, G. G., Miller, L. D.: Inhibition of Mitochondrial Energy Metabolism in Hypovolemic and Endotoxic Shock. In: Glenn, T. M.: Steroids and Shock (in press).
23. Mela, L., Miller, L. D., Diaco, J. F., Sugerman, H. J.: Surgery 68: 541 (1970).
24. Mela, L., Miller, L. D., Bacalzo, L. V., Olofsson, K., White, R. R.: Alterations of Mitochondrial Structure and Energy-linked Functions in Hemorrhagic Shock and Endotoxemia. In: Kovach, A. G. B., Stoner, H. B., Spitzer, J. J.: Neurohumoral and metabolic aspects of injury, p. 231 (Plenum Press 1973).

25. Mela, L., Miller, L. D., Bacalzo, L. V., Olofsson, K., White, R. R.: Ann. Surgery (in press).
26. Muller, W., Smith, L. L.: Surg. Gynec. Obstet. 117: 753 (1963).
27. Ozawa, K., Seta, K., Araki, H., Handa, H.: J. Biochem. (Japan) 61: 512 (1967).
28. Sayeed, M. M., Baue, A. E.: Am. J. Physiol. 220: 1275 (1971).
29. Schumer, W., DasGupta, T. K., Moss, G. S., Nyhus, L. M.: Ann. Surgery 171: 875 (1970).
30. Schumer, W., Erve, P. R., Obernolte, R. P.: Surg. Gynec. Obstet. 133: 433 (1971).
31. Slater, G., Vladeck, B. C., Bassin, R., Schoemaker, W. C.: Am. J. Physiol. 223: 1428 (1972).
32. Vogt, M. T., Farber, E.: Am. J. Pathol. 53: 1 (1968).
33. White, R. R., Mela, L., Bacalzo, L. B., Olofsson, K., Miller, L. D.: Surgery 73: 525 (1973).
34. Willerson, J. T., Trelstad, R. L., Pincus, T., Levy, S. B., Wolff, S. M.: Infection Immunity 1: 440 (1970).
35. Zalkin, H., Pullman, M. E., Racker, E.: J. Biol. Chem. 240: 4011 (1965).

Table I.

EFFECT OF TERMINAL ENDOTOXEMIA ON RAT LIVER MITOCHONDRIA

MITOCHONDRIAL REACTION	MEASURED PARAMETER	UNIT	CONTROL	ENDO-TOXIC	% CHANGE
ELECTRON TRANSFER	State 4, Succ	nmolesO_2/min/mg	7.5	10.0	+33
	State 4, Glut+Mal	nmolesO_2/min/mg	5.0	10.0	+50
	Cyto b Red., Succ	% of total	53	40	-24
	RCR Glut+Mal		7.0	1.5	-79
	State 3, Glut+Mal	nmolesO_2/min/mg	35	15	-57
OXIDATIVE PHOSPHORYLATION	ATPase, rate	nmolesH^+/min/mg	175	60	-61
	ATP hydrolyzed	nmoles/mg in 2 min.	125	60	-52
	ADP utilization, rate	nmoles/min/mg	160	0	-100
	Cyto b Oxid., ADP	% of succ red.	40	0	-100
ENERGY-LINKED TRANSPORT	Ca^{++} uptake (-Pi)	nmoles/min/mg	400	150→0	-63→100
	K^+ uptake, spont (-Pi)	nmoles/mg	85	165	+94
	Mg^{++} concentration	nmoles/mg	30	18	-40

Table II.

Tissue Effects of Circulatory Shock

	Liver			Heart	
	Endotoxic	Hemorrhagic		Endotoxic	Hemorrhagic
Mitochondrial ~ Production	↓ 19	↓ 19		↑ 21	↑ 21
Endotoxin Accumulation	+ 16,34	?		—	?
Lysosomal Enzyme Release	+ 24,25	+ 24,25		— 24,25	— 24,25
Cell pH	↓ 21	↓ *		↑ 21	?
Cell K^+	↓ 3	↓ *		?	?
Tissue PO_2	↓ 24,25	↓ 24,25		?	?
Organ Blood Flow	Stagnant 11	↓ 6,15,26,31		no change 12	↑ 9

* Silver, I.A., personal communication

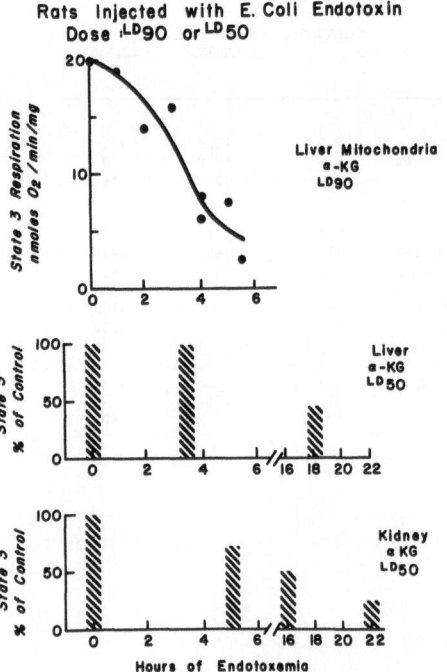

Rats Injected with E. Coli Endotoxin
Dose LD90 or LD50

Liver Mitochondria
α-KG
LD90

Liver
α-KG
LD50

Kidney
α KG
LD50

Hours of Endotoxemia

Fig. 1. Effects of E. coli endotoxemia on rat liver and kidney mitochondrial respiratory activity in State 3, using 10 mM α-ketoglutarate as substrate. The mitochondria were isolated at various times after the induction of endotoxemia, as indicated on the abscissa. State 3 respiratory activity was measured in mitochondrial suspension of 1-2 mg protein per ml in 0.225 M mannitol, 0.075 M sucrose, 10 mM Tris-Cl and 10 mM Tris-PO4, pH 7.4. ADP concentration was 520 μM.

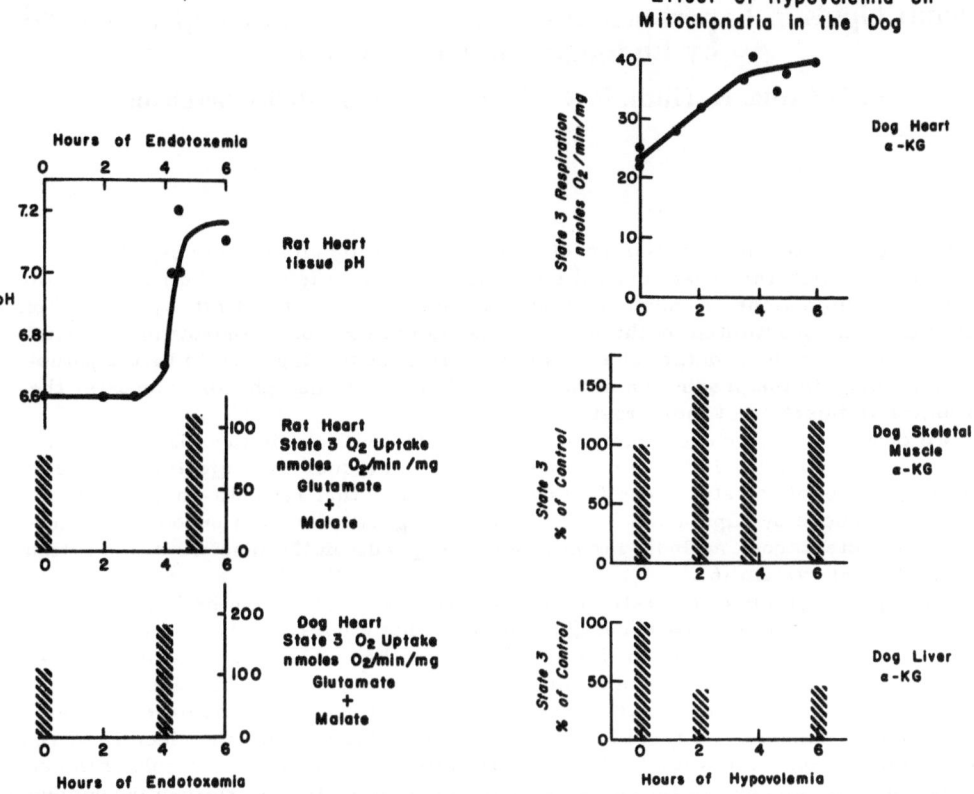

Fig. 2.   Fig. 3.

Fig. 2.   Effect of endotoxemia on heart pH and mitochondria. Rat or dog hearts
were removed prior to the death of the animal, at times indicated on the abscissa.
Tissue pH was measured immediately in a homogenate at 0°C in mannitol-sucrose
(no buffer). Assays of mitochondrial respiratory activity were done, using 10 mM
glutamate and 10 mM malate as substrates, as indicated in Fig. 1.

Fig. 3.   Effect of hemorrhagic hypovolemia on dog heart, skeletal muscle, and liver mitochondria. Experimental conditions were similar to Fig. 1.

Serum Lipids in Experimental Consumption Coagulopathy Induced by Endotoxin and Thorotrast

K. Nekarda, K. Huth, F. W. Schmahl, and G. Müller-Berghaus

The generalized Shwartzman reaction is identified by glomerular capillary thrombi and bilateral renal cortical necrosis (7). The reaction is the result of two intravenous injections of endotoxin spaced 24 hours apart. The first injection of endotoxin can be substituted by the intravenous administration of thorotrast. Impairment of the reticulo-endothelial system by thorotrast is suggested to be the pathogenic pathway in the production of the Sanarelli-Shwartzman phenomenon after the preparative injection of thorotrast (1, 3).

We observed a typical consumption coagulopathy with fibrin thrombi in the glomerular capillaries after infusion of Lipofundin[R] and a sublethal injection of E. coli endotoxin. We interpreted these findings in the sense that reticulo-endothelial blockade induced by Lipofundin enhances the biological effects of endotoxin, since Lipofundin clearance, as India ink clearance, depends on the function of the reticulo-endothelial system (6).

Since phagocytosis of the reticulo-endothelial system can also be impaired by intravascular coagulation processes (5), the following studies on the influence of thorotrast and endotoxin on serum lipids were undertaken. Assuming that the phagocytic function of the reticulo-endothelial system might not only be impaired directly by thorotrast but also by the activation of intravascular coagulation processes and hyperlipoproteinemia, we tried to answer the question, whether the changes of serum lipids as were observed by us (4, 8, 10) after endotoxin are a result of the consumption coagulopathy. In this study we used the Sanarelli-Shwartzman phenomenon produced by thorotrast and endotoxin. If the changes in serum lipids after endotoxin are the result of a consumption coagulopathy the changes in serum lipids should be prevented by coumarins (9).

Material and Methods

Male and non-pregnant female rabbits weighing 1.5 to 2.4 kg were used throughout the study. Dry food and drinking water were given ad libitum. The endotoxin used was lipopolysaccharide B 055 : B 5 (Difco Laboratories, Detroit, USA). 20 µg of this endotoxin in 1 ml isotonic saline were injected per animal. The thorotrast was a highly dispersed solution of 24-26 % thorium dioxide in dextrine (Fellows Testagar, Detroit, USA), administered at a dosage of 3 ml/kg. The long-acting phenprocoumon (Marcumar[R], Hoffmann - La Roche, Grenzach) was used as coumarin. Blood samples were taken with PVC catheters introduced into the right ear veins of the rabbits, extending to the right atria. After centrifugation EDTA plasma was stored frozen at -22ºC until the lipid determinations were made.

Free fatty acids were determined according to the method of Dole and Meinertz (2), triglycerides according to the method of Schmidt and Dahl (11), esterified fatty acids according to the method of Schön and Zeller (12), and total cholesterol according to the method described by Watson (13). The thromboplastin time was determined with Behring reagents.

Statistical evaluation included analysis of variance with a partially hierarchic model, the F-test and Scheffé's test. After the last blood sample had been taken, all animals surviving the experiment were killed with an overdose of Nembutal[R]

Supported by Deutsche Forschungsgemeinschaft, Bad Godesberg

and dissected like the animals which had died. The organs were inspected macro-
and microscopically.

Results

Seven rabbits received a single i.v. injection of 3 ml thorotrast/kg body weight.
This injection caused an increase of plasma triglycerides, esterified fatty acids,
and total cholesterol. As compared to the initial values, the increase was significant
only in the case of cholesterol after 48 hours.

At necropsy the liver was moderately enlarged, and a few hemorrhagic pulmonary
infarctions and epicardial hemorrhages in the area of the left ventricle were found.

Six rabbits received a single i.v. injection of 20 μg endotoxin, which was followed by a significant increase of triglycerides and total cholesterol after 48 hours,
compared to the initial values. As after a single thorotrast injection, no significant
increase of free fatty acids and esterified fatty acids was observed. The animals
were unremarkable, both clinically and patho-anatomically.

Forty rabbits received consecutive injections of thorotrast and endotoxin. 3 ml
thorotrast/kg were given first, followed 4 hours later by 20 μg of E. coli endotoxin.
20 of the animals died within 24 hours. Blood samples after 48 hours could only be
taken from 14 animals. Five of these showed renal cortical necrosis macroscopically. Histologic examination revealed fibrin thrombi in the glomerular capillaries.
All animals had petechial adrenal hemorrhages and hemorrhagic pulmonary infarctions as well as multiple hemorrhages and infarctions of the spleen which was
hardened, enlarged and of a blue-black color. Enlargement of the right ventricle was
present in all cases, indicating acute right heart failure. In addition there were sub-
epicardial and thymic hemorrhages and in one case a subserous hemorrhage of the
colon was observed. Black-brown thorotrast deposits were found in the intima of
the aortic valve and the aortic arch.

Analysis of the serum lipids after the combined administration of thorotrast and
endotoxin revealed a significant elevation of free fatty acids, triglycerides, esterified fatty acids, and total cholesterol. In the case of free fatty acids, triglycerides
and esterified fatty acids, this elevation was most distinctly discernible after 24
hours. Total cholesterol reached its maximum after 48 hours. All increases were
significant at the $p < 0.01$ level.

Five days before the consecutive application of thorotrast and endotoxin, 45 rabbits received i.v. injections of phenprocoumon daily until a total dose of 70 mg was
reached. When thromboplastin time had decreased to 5 %, the animals were injected with thorotrast and endotoxin. None of the 19 animals who had received prophylactic phenprocoumon treatment had any macroscopic or microscopic necroses of
the renal cortex; no fibrin thrombi were present in the glomerular capillaries. The
spleens were also free of hemorrhages and infarctions. The mortality of the animals anticoagulated with phenprocoumon was markedly higher than those of the control animals; only 5 animals survived 48 hours after endotoxin administration. Most
of the rabbits had died of massive pericardial, pleural, or abdominal hemorrhages.

Despite pretreatment with phenprocoumon a significant increase of serum lipids
was noted after 24 hours, comparable to that of the untreated animals. Again, the
free fatty acids, triglycerides, esterified fatty acids and total cholesterol were involved. With the exception of the free fatty acids, the 48-hour values were even
higher than the 24-hour values in the animals anticoagulated with phenprocoumon
(Fig. 2 and Fig. 3).

Conclusions

The experiments described in this paper show:
1. Intravenous administration of thorotrast that is sufficient to prepare a rabbit for
the development of bilateral necrosis of the renal cortex, results in hyperlipoproteinemia within 24 to 48 hours. The statistically established parallel behaviour of

the serum lipids following a single dose of thorotrast or endotoxin is noteworthy (Fig. 1). This parallelism suggests a similar point of attack for thorotrast and endotoxin.

2. Following the combined intravenous injection of thorotrast and a non-lethal dose of endotoxin, spaced 4 hours apart, a marked and significant ($p < 0.01$) increase of free fatty acids, triglycerides, esterified fatty acids and total cholesterol occurs within 24 hours. The changes in serum lipid concentrations are similar to those produced by two intravenous injections of endotoxin, spaced 24 hours apart, i. e. they resemble those in the experimental model for inducing the generalized Shwartzman reaction.

3. Although anticoagulative pretreatment with phenprocoumon (Marcumar[R]) prevents bilateral necrosis of the renal cortex and thus the morphologic substrate of severe consumption coagulopathy following the combined application of thorotrast and endotoxin, it does not prevent the increase of free fatty acids, triglycerides, esterified fatty acids and cholesterol. The increase of plasma lipids after 48 hours is even more pronounced than after 24 hours. It must be concluded therefore that the serum lipid changes seen in experimental consumption coagulopathy produced by the application of thorotrast and endotoxin do not result from the activation of intravascular coagulation processes and from the development of bilateral necrosis of the renal cortex. If one assumes that thorotrast and endotoxin have only two features in common - the activation of intravascular clotting, and an impairment of reticulo-endothelial function - our results suggest that hyperlipoproteinemia after injection of thorotrast, endotoxin, alone or in combination, might be the result of an impairment of the reticulo-endothelial system.

References

1. Beeson, P. B. : Proc. Soc. exp. Biol. Med. 64 : 146 (1947).
2. Dole, V. P., Meinertz, H. : J. biol. Chem. 235 : 2595 (1960).
3. Good, R. A., Thomas, L. : J. exp. Med. 96 : 625 (1952).
4. Huth, K., Schoenborn, W., Börner, J. : Med. Ernähr. 8 : 146 (1967).
5. Lee, L., McCluskey, R. T. : J. exp. Med. 116 : 611 (1962).
6. Lemperle, G., Reichelt, M. : Med. Klin. 68 : 48 (1973).
7. McKay, D. G. : Fed. Proc. 22 : 1373 (1963).
8. Müller-Berghaus, G., Huth, K., Krecke, H.-J., Lasch, H. G. : Schweiz. med. Wschr. 94 : 1519 (1964).
9. Müller-Berghaus, G., Schneberger, R. : Brit. J. Haemat. 21 : 513 (1971).
10. Obst, R., Jannakopulos, E., Urbaschek, B., Huth, K., Müller-Berghaus, G. : Thromb. Diath. haemorrh. 26 : 474 (1971).
11. Schmidt, F. H., von Dahl, K. : Z. klin. Chem. klin. Biochem. 6 : 156 (1968).
12. Schön, H., Zeller, W. : Münch. med. Wschr. 104 : 2433 (1962).
13. Watson, D. : Clin. chim. Acta 5 : 637 (1960).

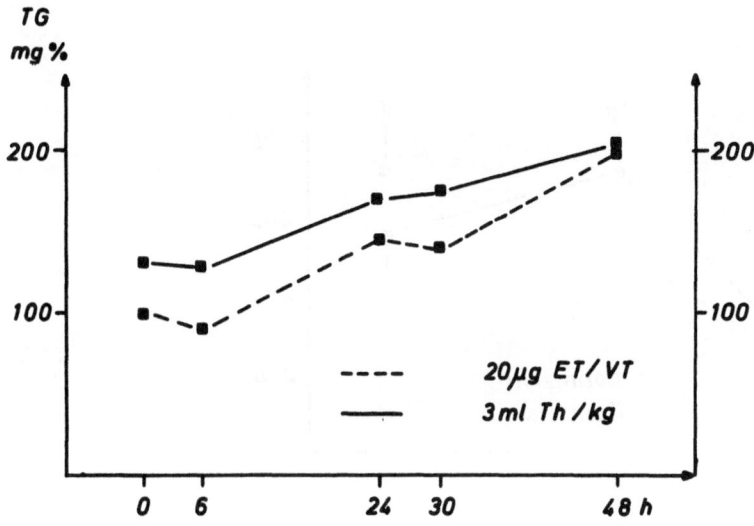

Fig. 1. Parallelism of increasing serum triglyceride concentration in the rabbit after 20 μg endotoxin per animal, resp. after 3 ml/kg thorotrast in animals not pretreated.

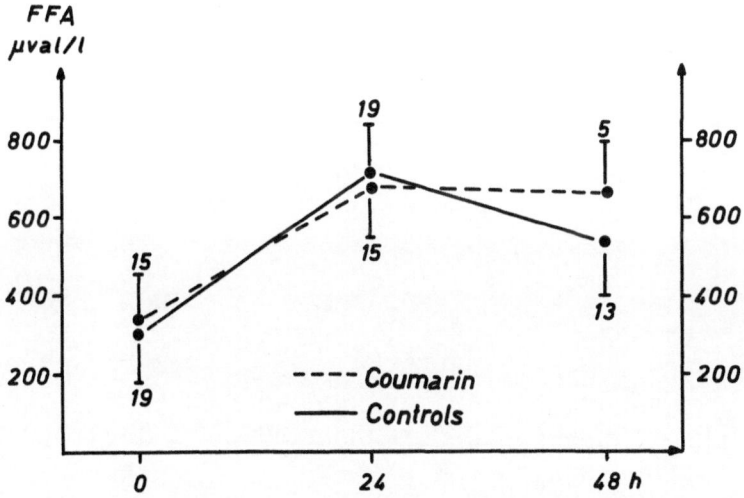

Fig. 2. Identical rise of plasma free fatty acid concentration in the rabbit after the combined administration of thorotrast and endotoxin. No difference between the animals pretreated with phenprocoumon and the untreated group.

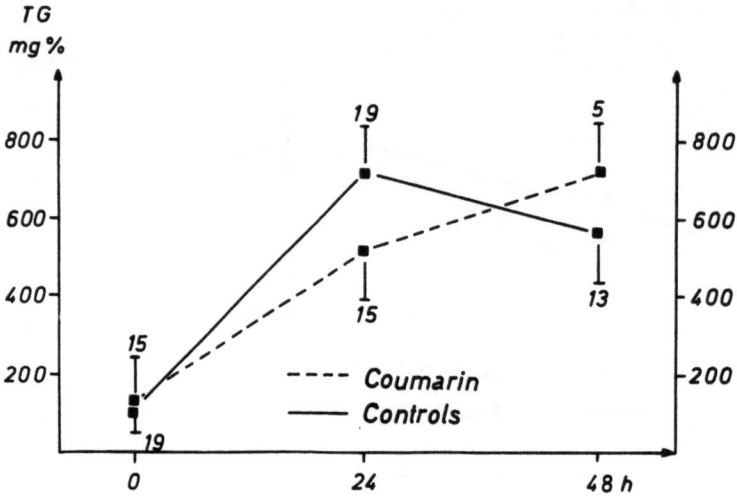

Fig. 3. Identical rise of plasma triglyceride concentration after the combined administration of thorotrast and endotoxin in anticoagulated and non-anticoagulated rabbits.

Changes of Lipoproteinlipase (LPL) after Intravenous Injection of Endotoxin

G. Oehler, R. Hassinger, F. W. Schmahl*, K. Huth*, and L. Róka

Plasma lipoproteins are assimilated by the peripheral tissues after hydrolytic splitting of the neutral fat portion has taken place through the action of lipoprotein-lipase (LPL) (1). The enzyme is supposed to be localized in the vascular walls. LPL activity is released into the plasma within few minutes after the intravenous injection of heparin or other polyanions.

Decreased LPL activity can result in hyperlipoproteinemia since the enzymatic disintegration of the blood lipids is impaired. Decreased LPL activity may be caused by a congenital enzyme defect (2), or it may be secondary to a primary disease (3, 4).

Parenteral endotoxin administration is known to result in hyperlipoproteinemia in the rabbit (5, 6). The question arose therefore whether decreased LPL activity contributes to the occurrence of hyperlipoproteinemia.

Methods

Rabbits of both sexes, weighing between 1.7 and 2.5 kg, were used. A PVC catheter was tied into the jugular vein of the animals for withdrawal of blood, and the injections. Since LPL activity depends on the kind and quantity of food intake, all animals were first exposed to a 36-hour preliminary period during which they were fasted for 12 hours. Subsequently the rabbits received Altromin standard food (Altrogge, Lage/Lippe) ad libitum for 12 hours. Thereafter they remained fasting until completion of the test.

For our studies Professor Urbaschek, Mannheim/Heidelberg, kindly provided us with endotoxin from E. coli 0 55 extracted according to the Boivin method (8). At the end of the preliminary period this endotoxin was injected into the ear vein of the rabbits at a dosage of 50/ug/kg body weight. Some animals received a second injection of the same dose of endotoxin 24 hours later.

Serum triglycerides and free glycerol were determined by the method of Eggstein and Kreutz (7). Serum cholesterol was measured according to Watson's method (8). Free fatty acid concentration was titrimetrically determined with the semi-automatic technique of Keul and co-workers (9). For the activation of LPL the animals were injected with 100 U heparin (Liquemin(R))/kg body weight. Five minutes later approximately 3 ml blood was withdrawn to which one drop Liquemin(R) was added to prevent coagulation. The plasma was separated by cold centrifugation at +2º C. LPL activity was measured according to Fredrickson's method (1) by which the amount of fatty acids released by the post-heparin plasma from a fat emulsion (Ediol(R)) is determined in vitro.

The results were statistically evaluated with Student's t-test and the partially hierarchic model of variance analysis.

Results

Fig. 1 shows a distinct rise of serum triglyceride concentration after the first endotoxin injection. The letter "s" indicates that the values of the endotoxin-treated animals at this time differ significantly ($p < 0.05$) from those of the controls in the Student's t-test.

From the 12th hour on after the first endotoxin injection serum cholesterol is sig-

* Supported by Deutsche Forschungsgemeinschaft

nificantly higher than in the control animals (Fig. 2). In these investigations no significant differences in serum free fatty acids and free glycerol could be observed between endotoxin-treated and control animals. LPL activity decreased markedly after endotoxin had been administered (Fig. 3).

Variance analysis of the results showed the time course of the LPL changes to be significantly different for a period of up to 30 hours between animals following endotoxin and the controls (probability of error 5 %). In Fig. 4, the measured mean values of LPL activity are plotted in per cent of the initial values to demonstrate that endotoxin causes a deviation from the uniform decline of enzyme activity observed in the control animals.

In order to further elucidate the time relation between injection of endotoxin and LPL decrease, we observed LPL activity in two-hour intervals for up to six hours after a single injection of endotoxin.

As can be seen in Fig. 5, LPL activity had already decreased by 38 % of the initial value when the first measurement was taken two hours after the injection. During the following four hours no substantial additional decrease occurred. Evaluation by variance analysis showed a significant difference in comparison to the control animals (probability of error 5 %).

Discussion

In previous studies with rabbits we have shown that the intravenous injection of endotoxin causes an increased release of free fatty acids into the blood (6, 10). The increase of peripheral lipolysis is probably not caused by direct action of endotoxin but by mediators, such as the catecholamines, which activate the hormone-sensitive tissue lipase.

The increased offer of free fatty acids enhances lipoprotein synthesis in the liver. This explains the hyperlipoproteinemia we observed after the administration of endotoxin. In our experiments the concentration of free fatty acids in the serum of the endotoxin-treated rabbits was not significantly higher than in the control animals. This may be explained by the fact that all animals had been fasted for a relatively long period of time before the blood samples were taken. As demonstrated by the high initial values of the free fatty acids, this fasting resulted in marked lipolysis which masked the mobilization of free fatty acids caused by endotoxin.

The results suggest that the occurrence of hyperlipoproteinemia after the administration of endotoxin is not only due to increased lipoprotein synthesis, but also to impaired degradation of blood lipids, since we found a distinct inhibition of LPL activity in the endotoxin-treated animals. This effect also is probably due to the influence of mediators. Catecholamines, for instance, are known for their ability to activate adipose tissue lipase and inhibit lipoproteinlipase at the same time (11).

The effect of these mediators would have to be at its maximum in the first two hours after a single endotoxin injection since we did not observe any substantial decrease of enzyme activity after this time.

In view of the short duration of impaired LPL activity, it cannot be excluded that endotoxin, rather than acting entirely or partly through mediators, directly inhibits LPL or blocks its release. This is supported by previous observations that endotoxin causes early lesions in the vascular walls (12, 15, 16) where LPL is probably concentrated (13).

Decreased LPL activity could therefore be the result of endothelial damage caused by endotoxin.

A certain discrepancy exists between our results and those of Schuler and co-workers (14) who found that post-alimentary lipemia in rats was reduced after administration of a polysaccharide preparation of gram-negative bacteria. Accordingly one would expect that endotoxin alone, like heparin, can activate LPL. So far, however, our orienting studies have shown that no lipolytic activity appears in the blood of rabbits if endotoxin alone is administered.

References

1. Fredrickson, D. S., Ono, K., Davis, L. : J. Lipid Res. 4 : 24 (1963).
2. Fredrickson, D. S., Levy, R. J. : Familial Hyperlipoproteinemia. In : Stauburg, J. B., Wyngaarden, J. B., Fredrickson, D. S. : The Metabolic Basis of Inherited Diseases (Mc Graw Hill, New York 1972).
3. Bagdade, J. D., Porte, D., Bierman, E. L. : New Engl. J. Med. 279 : 181 (1967).
4. Oehler, G., Huth, K., Schmahl, F. W., Róka, L. : Klin. Wschr. 51 : 350 (1973).
5. Le Quire, V. S., Hutcherson, J. D., Hamilton, R. L., Gray, M. E. : J. exp. Med. 110 : 293 (1959).
6. Huth, K., Karliczek, G. : Die Endotoxin-induzierte Hyperlipämie des Kaninchens. In : Bartelheimer, H., Heisig, N. : Aktuelle Gastroenterologie, p. 373 (Thieme-Verlag, Stuttgart 1968).
7. Eggstein, M., Kreutz, F. H. : Klin. Wschr. 44 : 262 (1966).
8. Watson, D. : Clin. Chim. Acta 5 : 637 (1960).
9. Keul, J., Linnet, N., Eschenbruch, E. : Z. klin. Chem. 6 : 394 (1968).
10. Huth, K., Müller-Berghaus, G., Krecke, H. J., Lasch, H. G. : Verh. dtsch. Ges. inn. Med. 70 : 437 (1964).
11. Wing, D. R., Salaman, M. R., Robinson, D. S. : Biochem. J. 99 : 648 (1966).
12. McGrath, J. M., Stewart, G. J. : J. exp. Med. 129 : 833 (1969).
13. Robinson, D. S. : Adv. Lipid Res. 1 : 133 (1963).
14. Schuler, W., Müller, G., Maier, F. : Schweiz. med. Wschr. 25 : 787 (1957).
15. Urbaschek, B. : Verh. dtsch. Ges. inn. Med. 72 : 752 (1966).
16. Urbaschek, B. : Zur Frage des Wirkungsmechanismus bakterieller Endotoxine und seiner Beeinflussung. Habilitationsschrift, Universität Heidelberg 1967.

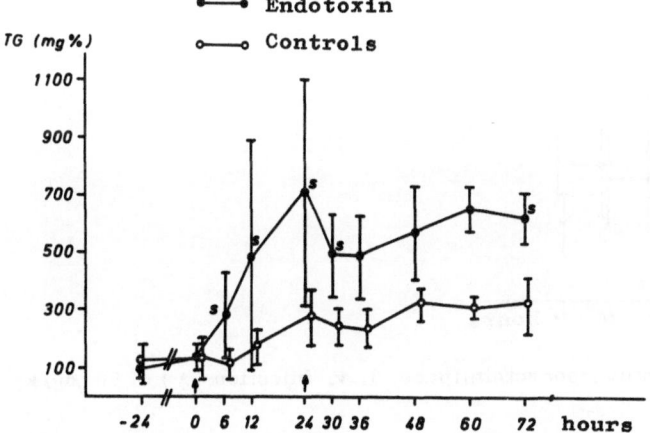

Fig. 1. Influence of endotoxin on serum triglycerides. I. v. injection (⬆) of 50 µg/kg endotoxin. (n = 8)

Cholesterol*(mg%)*

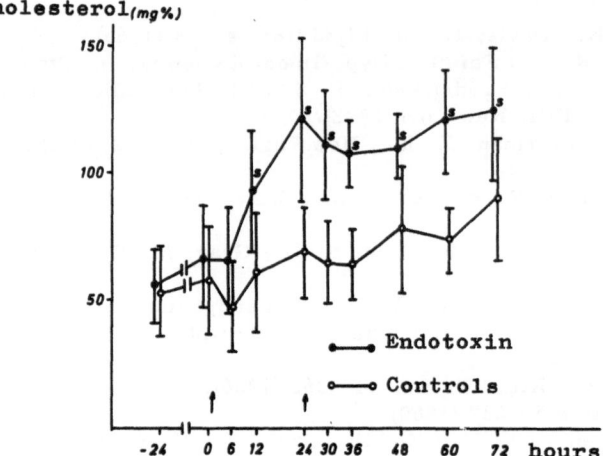

Fig. 2. Influence of endotoxin on serum cholesterol. I. v. injection (↟) of 50 /ug/kg endotoxin. (n = 8)

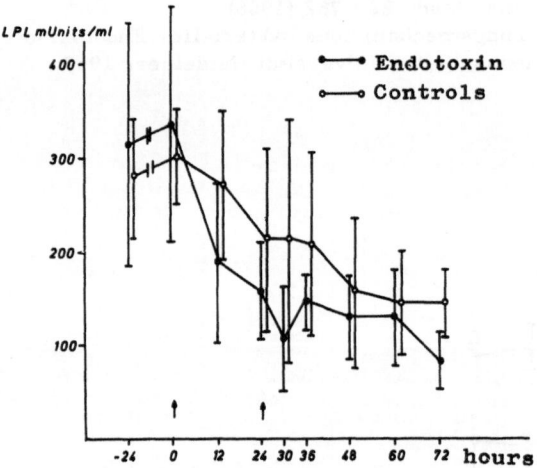

Fig. 3. Influence of endotoxin on lipoproteinlipase. I. v. injection (↟) of 50 /ug/kg endotoxin. (n = 8)

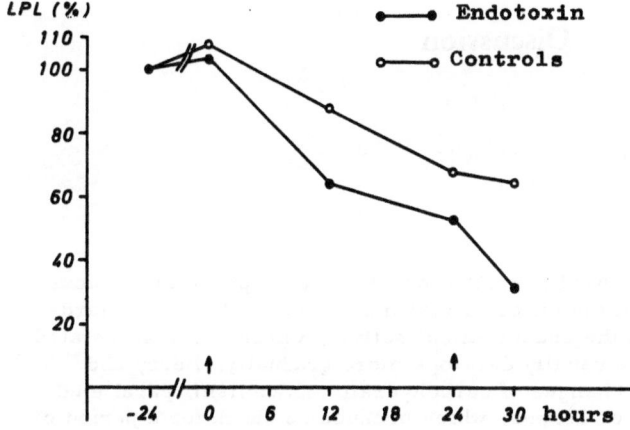

Fig. 4. Mean values of LPL activity in per cent of initial values. I. v. injection
($\uparrow$) of 50 μg/kg endotoxin. (n = 5)

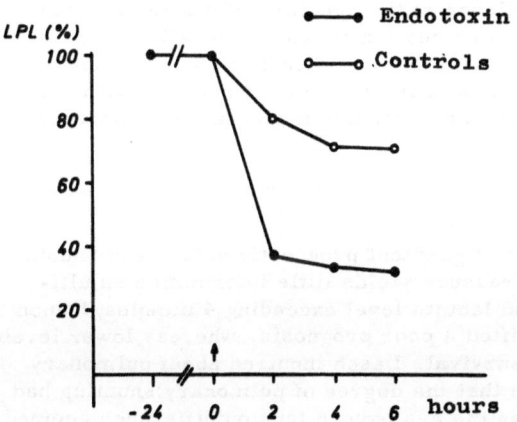

Fig. 5. Mean values of LPL activity in per cent of initial values. I. v. injection
($\uparrow$) of 50 μg/kg endotoxin. (n = 8)

Discussion

LUNDSGAARD-HANSEN

Berry, Clowes, McCallum and Neuhof pointed out that the response to endotoxin is very dose dependent and that the experimental model studied by Lundsgaard-Hansen is quite different from the usual clinical setting, where shock associated with gram-negative bacteremia usually develops more gradually. Berry and McCallum found early, subtle changes of carbohydrate metabolism, paralleled to some extent by the findings of Clowes, which produce no shock for a period of hours, but which presumably set the stage for a late and sudden circulatory break-down. Neuhof, too, infused a small dose of endotoxin slowly and observed a pronounced depression of oxygen uptake preceding the development of shock. Mela confirmed that the alterations of energy metabolism described by Lundsgaard-Hansen parallel the degree of shock. Greisman inquired about the situation in an equally severe hemorrhagic shock, and Lundsgaard-Hansen answered that in the absence of specific data, he could only suppose that it would be comparable to the endotoxic state. Beller reminded the audience of the species difference and Clowes recommended an experimental bacteremia or peritonitis as a clinically more relevant model. Lundsgaard-Hansen agreed with this and stressed the importance of defining precisely the relative proportions of living and dead bacteria within a given total number injected - a point which was probably inadequately considered in some recent publications.

SHUBIN

Graeff and Hey asked for the parameters of greatest prognostic value, and Shubin answered that whereas the mean blood pressure yields little information on ultimate survival or death, an arterial blood lactate level exceeding 4 mmoles/l upon admission to the shock unit usually signified a poor prognosis, whereas lower levels were associated with better chances of survival. Lasch inquired about pulmonary function studies, to which Shubin replied that the degree of pulmonary shunting had not been found very informative, whereas the a-a oxygen tension difference seemed to be more useful. Schmahl asked about the use of alpha receptor blockers, which Shubin had not found to have much influence on the various parameters measured by his group. Commenting on a remark by Shubin about the colloid osmotic pressure, Lundsgaard-Hansen confirmed that this was often very low in septic intensive care patients, and Shubin added that paradoxic values are sometimes obtained, such as a rising oncotic pressure in the face of developing pulmonary edema on a cardiac basis.

CLOWES

Lundsgaard-Hansen asked whether Clowes thought that the inhibition of oxygen consumption which he had demonstrated was due to a failure of delivery or utilization. Clowes held that both mechanisms were probably implied, besides a resistance to insulin with a deficient transport of glucose. Mela confirmed that at the mitochondrial level, oxygen delivery and utilization are both deficient at least in certain tissues, and asked Clowes whether he considered the deficiency of oxygen consumption to be local or general; the answer being that overall data only were available. Schmahl pointed to the difficulty of evaluating the in vivo tissue enzyme activities on the basis of in vitro measurements under standard optimum conditions and asked Clowes whether he had any data on the output of amino acids from skeletal muscle and its pattern in the septic state. Clowes replied that such studies were in progress, but that the available data referred only to the overall nitrogen balance and proteolysis.

BROBMANN

Schauer asked about Brobmann's view of dopamine, which was skeptical. Neuhof referred to studies by Hinshaw and inquired about any possible relationships between the diverging results obtained by Brobmann in the dog and in the monkey and the dose of endotoxin employed, suggesting that other dosage levels might have lessened the discrepancies. Brobmann had worked with the stated dose only. Clowes congratulated Brobmann on his presentation, which had reemphasized the fallacies inherent in the use of the canine model for endotoxin studies, and recommended the pig as a suitable experimental animal.

BERRY

Lundsgaard-Hansen asked Berry whether he had any data on a quantitative relationship between the enzymatic changes he had found and lethality, since this might provide information on their importance for the killing potential of endotoxin. Berry replied that it took a small amount of endotoxin only to produce the inhibition described, the main difference between various dosage levels being the duration of the effect, which if prolonged might presumably be fatal. In reply to a question by Mergenhagen, Berry recalled that the immunosuppressives he had used did not deplete the B cells completely. Chedid mentioned that neonatal thymectomy did not affect the behaviour of mice exposed to endotoxin and asked Berry whether the enzymatic changes he had found were confined to the liver or present in the brain, too. Berry's studies had, however, been done on liver only.

MELA

Berry asked whether the mitochondria were actually operating in vivo at the pH at which the lysosomal enzymes are active. Mela answered that a pH of 7.4 will suffice to produce some effects already and that damage is very severe at pH 6.5. Another important factor is the calcium concentration, the doubling of which inhibits mitochondrial respiration tremendously. Schmahl inquired about morphological changes of the mitochondria and cautioned against the use of an "average" tissue pO_2, since particularly under abnormal conditions, considerable gradients of oxygen tension may exist within a very limited cellular area. Mela stated that mitochondrial morphology was distinctly altered in her studies, that tissue hypoxia alone was insufficient to explain the changes observed, ischemia with local acidosis also being necessary, and that her group had initiated experiments considering the well-known, but difficult problem of tissue inhomogeneity with respect to pO_2. Clowes asked whether Mela's studies provided any information about the failure of substrate utilization which he had observed and which looked more like an uncoupling than an actual damage to the transport system. Mela replied that she had been unable to demonstrate any damage to the mitochondrial electron transport system, whereas substrate transport had not yet been specifically investigated. The cytochrome would respond to substrate but not to ADP. In reply to a final question of Lundsgaard-Hansen about possible relationships between the enhanced levels of free fatty acids known to be present in shock and mitochondrial damage, Mela thought that though this was theoretically possible, it was unknown whether the free fatty acid level within the cell would be sufficiently high to cause any damage.

NEKARDA

OEHLER

Berry asked for any relevant enzymatic changes in the livers of endotoxin-poisoned animals, but Oehler had not investigated this point. Springer asked Oehler to comment on earlier studies of Schuler concerning the relationship between endotoxins and lipemia. Oehler pointed to the species difference - rats versus rabbits - and to probable differences in the endotoxin preparations. In rabbits, no lipolytic enzymes could be demonstrated in the blood. Commenting on this question, Huth, K., remarked that he had observed a stimulation of lipolysis in rabbits treated with Westphal-Lüderitz as well as with Boivin preparations, which could be blocked with phenoxybenzamine. Like Schuler, he had also worked with pregnant rats, but in contrast to that author, he found an increased - not a decreased - hyperlipemia associated with a generalized Sanarelli-Shwartzman reaction following a single dose of endotoxin.

Summing up the presentations and discussions of the session on hemodynamics and metabolism, Lundsgaard-Hansen though that despite the problems still inherent in the use of various experimental models and animal species, it had served to bring out two important points quite clearly:

First, that endotoxins produce certain subtle and early alterations of carbohydrate metabolism, which though not detectable clinically, may play an important role in setting the stage for a later circulatory breakdown.

Second, that in the clinically manifest stage of circulatory alterations, the capillary exchange mechanisms, or the state of shock as such are probably the decisive factors for the breakdown of energy metabolism and thus for the killing effects of endotoxins.

Chapter VI

Pharmacology

Pharmacology

Pharmacological Aspects of Endotoxin Effects

H. Giertz

An outsider like myself is not able to give any news to the specialists attending this meeting but only to present you with my impressions from studying the literature. However, according to the conception of Dr. Urbaschek, representatives of pharmacology and pathology are to be involved in the discussion on endotoxemia and on pathophysiology of gram-negative infections. As a pharmacologist I am interested especially in the pharmacology of the mediators of endotoxin shock. There is no doubt that in shock some highly effective substances, for instance histamine, bradykinin or catecholamines, are released from stores or freshly produced from precursors as in the case of anaphylactic shock. Furthermore, certain chain reactions are activated of which only the complement activation should be considered.

Clear information about the importance of the mediators for the shock symptoms cannot be obtained from the literature. Only one point seems to be sure: If at all, most mediators play a role only in the early phase of shock. In this respect endotoxin shock is very similar to allergic shock where there is no doubt about the involvement of mediators in the early shock phase, whereas the importance of this involvement for the shock symptoms is obscure except in the case of only a few anaphylactic phenomena.

First, let us consider the role of histamine. There are great similarities between histamine and endotoxin shock (for references see Hinshaw, 9). It is most remarkable to observe that the difference of the shock phenomena between various animal species are the same in both shock types and also in anaphylaxis.

Besides the similarity of endotoxin shock to histamine shock a more important indication of the involvement of histamine is the histamine release in endotoxin shock which has been observed in all animal species. Dr. Schauer will give some references concerning this point.

It is an interesting phenomenon that in endotoxin shock of the dog the increase in the histamine level in plasma is accompanied by a decrease in histamine in the whole blood (11). In this respect endotoxin shock seems to differ from histamine shock and anaphylaxis. According to investigations in guinea pig anaphylaxis an increase in plasma histamine and a simultaneous decrease in blood histamine can be observed only when using brucellae as antigen, whereas in anaphylaxis induced by ovalbumin and by horse serum plasma histamine as well as blood histamine increase. The decrease in histamine in the whole blood seems to be typical in shock forms in which humoral factors are involved. Accordingly, this is also true for the Forssman shock of the guinea pig (for references see Giertz, 7).

An increase in plasma histamine and a simultaneous decrease in blood histamine can only be explained by the assumption that the liberation of histamine into the plasma runs parallel to the disappearance of histamine-containing blood cells, i.e. of platelets and of leukocytes. Indeed, as early as 1957 Weil and Spink (24) observed the disappearance of leukocytes and platelets in endotoxin shock of the dog.

The same authors pointed to another interesting fact, namely, that the decrease in systemic blood pressure induced by endotoxin lasts much longer than the transient increase in the blood pressure in the portal vein, which is typical in the dog, as has also been pointed out in this symposium by Brobmann. This increase is also typical of the action of histamine, so that from the short duration of this effect in endotoxin shock one may conclude that the involvement of histamine in this type of shock is very short, too. Indeed, the histamine release is very transient according to experiments of Vick and coworkers (23) which will be mentioned in Dr. Schauer's paper.

The final proof of the importance of histamine release for the shock symptoms would be an inhibitory effect by antihistamines. Data concerning such effects in endotoxin shock of dogs are contradictory. As in dog anaphylaxis, in most experiments antihistamine pretreatment was ineffective. The same is true for the pretreatment with histamine liberators which deplete the histamine stores. If any influence was observed at all, it merely concerned the early phase of shock, as was to be expected (10).

A new aspect of the participation of histamine in endotoxin shock was put forward by Schayer in 1960 (20) when he showed that in tissues of mice treated with endotoxin the histamine-producing capacity increases. Dr. Schauer will discuss this matter in detail. The histamine produced by this mechanism, the so-called induced histamine, seems to have special effects on the microcirculation which are said to be not inhibited by antihistamines.

The problem of the participation of histamine in endotoxin shock is all the more difficult because endotoxin is not only able to release histamine and to reinforce its production but it can also intensify its effects as has been observed using various histamine-sensitive preparations (22).

The endotoxin effects are not only similar to those of histamine but also to bradykinin. Erdös and Miwa (6) refer to some reasons for the assumption that kinins are involved in endotoxin shock of the dog as for instance: the appearance of a vasoactive material, possibly a kinin, in the circulation; a decrease in kininogen level in blood; the prevention of this drop by a kininogenase inhibitor.

The kinin production seems to be dependent on the animal species. Probably, it occurs by activation of the Hageman factor which can be also activated by contact with glass and in its turn activates a kininogenase. According to Erdös and Miwa (6) the dog has only one kininogenase which is not influenced by the Hageman factor, whereas rats, rabbits and humans have a second kininogenase which can be activated. Accordingly, Carretero and coworkers (5) showed that the decrease in blood pressure of the endotoxin-treated dog is not accompanied by a decrease in kininogen. On the other hand, kinin production by contact with glass is significantly inhibited in plasma of endotoxin-treated rabbits (6). However, a decrease in the Hageman factor rather than a decrease in kininogen accounts for this finding. The authors emphasize that the decrease in blood pressure in endotoxin shock cannot be produced by the production of bradykinin, because it cannot be prevented by carboxypeptidase in contrast to the bradykinin effect.

It is a matter of definition whether or not the catecholamines should be considered as mediators. Their pharmacological action is principally antagonistic to the effects of histamine and bradykinin. Their liberation in endotoxin shock has been proved more than once but differences between the species seem to exist. Brobmann and coworkers (3), as well as the paper of Brobmann read at this meeting have demonstrated that in dogs the rapid decrease in blood pressure induced by endotoxin is followed by a transient increase which does not occur in the monkey. Assuming that this increase is produced by catecholamines, this observation corresponds to Hinshaw's (9) statement that catecholamine liberation does not occur in the early phase of endotoxin shock of the monkey. In the dog as well in the cat an increase in the plasma catecholamine level is proved (8, 21). Besides the catecholamines another pressor substance, angiotensin, can be found in shock plasma (8).

The question whether the liberation of catecholamines has a good or harmful effect on the course of shock has not yet been answered. However, in rats a sublethal dose of endotoxin has lethal effects after adrenalectomy.

The original death rate could be reestablished by simultaneous treatment with adrenaline and corticosteroids. Adrenaline alone only had an incomplete effect (15).

Some catecholamine effects are intensified, others are inhibited by endotoxin. For instance, the activation of the adenylcyclase of mouse liver by adrenaline is reinforced by pretreatment with endotoxin (2). On the other hand, the positive inotropic effect of noradrenaline on the guinea pig atrium is inhibited after pretreatment with endotoxin (1). Further, it should be mentioned that in rats (18) as well as in dogs (19) tritium-labelled noradrenaline disappears from the circulation more rapidly in pretreated than in normal animals.

Whether other mediators than the above mentioned ones are involved in endotoxin shock is questionable. Recently, Kessler and coworkers (14) pointed to participation of prostaglandins which are no doubt involved in anaphylaxis.

The role of complement in endotoxin shock seems to be somewhat clearer than the participation of mediators and is further clarified by several papers at this meeting. In their excellent review of this field Mergenhagen and coworkers (16) state that C3 - C9 are especially involved, but also C1, C4, and C2 participate according to more recent investigations (13, 17). With respect to the importance of humoral factors for the disappearance of platelets induced by endotoxin - as already mentioned above - it is a very remarkable fact that this thrombocytopenia is abolished or inhibited in C4-deficient guinea pigs (13) or in C6-deficient rabbits or in rabbits in which C3 - C9 were inactivated by the cobra venom factor (4). A decrease in the death rate runs parallel to the inhibition of thrombocytopenia. However, also in the respect of complement participation in endotoxin shock some contradictions exist which as yet have not been cleared up. Johnson and Ward (12) observed not a decrease but an increase in the death rate after endotoxin application to C6 - deficient rabbits.

References

1. Bhagat, B., Cavanagh, D., Merrild, B.N., Rana, M.W., Rao, P.S.: Brit. J. Pharmacol. 39: 688 (1970).
2. Bitensky, M.W., Gorman, R.E., Thomas, L.: Proc. Soc. exp. Biol. Med. 138: 773 (1971).
3. Brobmann, G.F., Ulano, H.B., Hinshaw, L.B., Jacobson, E.D.: Amer. J. Physiol. 219: 1464 (1970).
4. Brown, D.L., Lachmann, P.J.: Int. Arch. Allergy 45: 193 (1973).
5. Carretero, O.A., Nasjletti, A., Fasciolo, J.C.: Experientia 26: 63 (1970).
6. Erdös, E.G., Miwa, I.: Fed. Proc. 27: 92 (1968).
7. Giertz, H.: Bildung und Freisetzung biologisch aktiver Substanzen unter besonderer Berücksichtigung des Histamins. In: Filipp, G.: Pathogenese und Therapie allergischer Reaktionen, p.424 (Ferdinand Enke Verlag, Stuttgart 1966).
8. Hall, R.C., Hodge, R.L.: J. Physiol. 213: 69 (1971).
9. Hinshaw, L.B.: Release of vasoactive agents and the vascular effects of endotoxin. In: Kadis, S., Weinbaum, G., Ajl, S.J.: Microbial toxins, Vol. 5, p.209 (Academic Press, New York and London 1971).
10. Hinshaw, L.B., Emerson, T.E., Iampietro, P.F., Brake, C.M.: Amer. J. Physiol. 203: 600 (1972).
11. Hinshaw, L.B., Vick, J.A., Carlson, C.H., Fan, Y.-L.: Proc. Soc. exp. Biol. Med. 104: 379 (1960).

12. Johnson, K. J., Ward, P. A. : J. Immunol. 106: 1125 (1971).
13. Kane, M. A., May, J. E., Frank, M. M. : J. clin. Invest. 52: 370 (1973).
14. Kessler, E., Hughes, R. C., Bennett, E. N., Nadela, S. M. : J. Lab. clin. Med. 81: 85 (1973).
15. Krawczak, J. J., Brodie, B. B. : Pharmacology 3: 65 (1970).
16. Mergenhagen, S. E., Snyderman, R., Gewurz, H., Shin, H. S. : Curr. Topics Microbiol. Immunol. 50: 37 (1969).
17. Philips, J. K., Snyderman, R., Mergenhagen, S. E. : J. Immunol. 109: 334 (1972).
18. Pohorecky, L. A., Wurtman, R. J., Taam, D., Fine, J. : Proc. Soc. exp. Biol. Med. 140: 739 (1972).
19. Rao, P. S., Bhagat, B. D., Cavanagh, D. : Proc. Soc. exp. Biol. Med. 141: 412 (1972).
20. Schayer, R. W. : Amer. J. Physiol. 198: 1187 (1960).
21. Spink, W. W., Reddin, J., Zak, S. J., Peterson, M., Starzecki, B., Seljeskog, E. : J. clin. Invest. 45: 78 (1966).
22. Urbaschek, B. : Addendum - The effects of endotoxins in the microcirculation. In: Kadis, S., Weinbaum, G., Ajl, S. J. : Microbial toxins, Vol. 5, p.261 (Academic Press, New York and London 1971).
23. Vick, J. A., Mehlman, B., Heiffer, M. H. : Proc. Soc. exp. Biol. Med. 137: 902 (1971).
24. Weil, M. H., Spink, W. W. : J. Lab. clin. Med. 50: 501 (1957).

Initial Liberation of Biogenic Amines and Effect of Further Mediators Following Application of Endotoxins

A. Schauer

In shock sensitive species shock symptoms develop very quickly after the application of endotoxins. This phenomenon has led to extensive comparisons of anaphylactic and endotoxin shocks (4). The anaphylactic shock clearly results from a massive liberation of histamine from tissue mast cells. According to Becker and Austen (3) this takes place under the influence of so-called homocytotropic antibodies after sensibilization of the mast cells, as shown in experiments with rats. In contrast to this relatively simple mechanism, the factors leading to endotoxin shock are many-faceted and much more complex.

To recognize the mediators contributing to the initiation of the endotoxin shock and to evaluate their importance these questions have to be answered: 1. What is the importance of biogenic amines, especially histamine? 2. Are the plasma kinins of any importance and what is their influence?

The question concerning the importance of catecholamines released in the course of circulatory recompensation efforts is also of interest for the present problems with respect to the initial phases of stasis in the microcirculation. I will not go into further detail on this issue in the present paper, as it represents a separate point for discussion.

The first problem can be subdivided into the following two questions:
a) that of initial liberation of histamine from its stores;
b) that of endogenous formation of histamine during the course of shock, which may then become directly, i. e. "intrinsically", effective in the region of its formation.

Concerning a): Initial liberation of histamine in endotoxin shock was long contested as a fact and with regard to its importance. To ascribe acutely released histamine as contributory to initiate progressive hypotension led to controversies. The main reason is the fact that histamine levels of blood plasma in a different species were measured by means of different methods and at different times in the course of shock. As histamine is highly diffusible and can be inactivated very quickly, this fact represents an essential item for explaining the discrepancies.

Considering these facts, any positive findings especially relating to the early phase of shock and gathered by expedient methods seem to be valuable. Weil and Spink (35), Hinshaw, Jordan and Vick (10) and Spink et al. (27) were the first to point out the increased histamine level of the blood in endotoxin shock. The study of Vick et al. (33) in 1971 is of particular importance, since it is selectively directed to the question concerning the role histamine plays at the beginning of endotoxin shock and its importance for a decrease in arterial blood pressure and as determinations of histamine were affected by the reliable fluorometric method according to Shore, Burkhalter and Cohn (26).

Intravenous injection of lethal doses of E. coli endotoxin resulted in a considerable decrease in blood pressure, as early as 30-60 seconds later. This was correlated with a strong increase in plasma histamine (Fig. 1). In addition, animals with low initial increases in histamine levels survived, whereas animals that died showed a large increase (Fig. 2). Urbaschek and co-workers (31) demonstrated the correctness of these conclusions by intravital microscopic demonstration of mast cell degranulation in rabbits, guinea-pigs and golden hamsters.

The exact mechanism of this degranulation is still completely unknown. In all species, including man, most of the body histamine is located in the mast cells. Therefore, during acute release, there will occur above all an emptying of these stores. Histamine liberation by homocytotropic antibodies has been demonstrated in rats (3).

In addition, a principle damaging the mast cells has been found in the lysosomes of granulocytes. The experiments of Cline et al. (6) gain special interest, as they demonstrate that endotoxin is taken up by granulocytes in the presence of complement, releasing kininase. However, it has not been established that these factors and processes are really significant. After absorption (so-called pexis) or disintegration of lipopolysaccharides, an exchange of the histamine located in the mast cell granules against strongly basic spermine and/or putrescine - these are found in endotoxin according to Nowotny (16) - by ion exchange must be considered according to the studies of Werle and Amann (36).

Concerning b): An inducible specific histidine decarboxylase (HDC) - probably located at or close to the capillary system - has been demonstrated by Schayer (23). We therefore carried out a systematic study concerning the influence of various endotoxins on this inducible enzyme (20-22).

Sprague-Dawley rats kept under standardized conditions and NMRI or LAF 1-mice were given E. coli endotoxin (E. coli 0111 : B4 lipopolysaccharide W of DIFCO Laboratories, USA) or endotoxin, which was made available to us by Dr. Metz, Max-von-Pettenkofer-Institut, Munich. In further experiments Salmonella anatum and Salmonella abortus equi endotoxins, which had also been provided by Dr. Metz, were used.

E. coli endotoxin was administered by intraperitoneal injection at a dose of 8 mg/kg to rats and at a dose of 10-13 mg/kg to mice. Rats were given Salmonella anatum endotoxin at a dose of 8 or 12 mg/kg, while Salmonella abortus equi endotoxin was administered at a dose of 15 mg/kg.

The animals were usually killed 6 hours after endotoxin administration. The organs were prepared under sterile conditions and homogenized with m/15 phosphate buffer at a pH value of 7.2. After centrifuging at 100 000 g, the supernatant was used as enzyme source for the measurements. Tetracycline was used to counteract any bacterial growth, possibly existing aminooxidase was inhibited by 3 x 10^{-4} M aminoguanidine.

Enzyme determinations were largely carried out by the isotopic dilution method indicated by Schayer (24) using 1.25 ng ring-labelled C^{14} histidine, or by a $C^{14}O_2$ method which had been developed jointly with Aures and Hakanson (1) and which corresponds in principle to the method of Aures and Clark (2).

Both methods showed inducible HDC-activity in lung tissue of mice and rats after endotoxin application (Fig. 3 and 4). The enzyme activity could be suppressed after two days of premedication with glucocorticoid hormone (soludecortin-H, 30-40 mg/kg/day). According to our in vitro inhibition tests, the effect of glucocorticoids is not competitive. Under the influence of growth hormone the enzyme activity was substantially increased after five days of premedication with 5.0 mg i.m. per day. In the case of mice, HDC-activity was higher; all animals died after endotoxin treatment. With rats, which are less sensitive to endotoxin than mice, not all animals died after endotoxin treatment.

The endotoxins of Salmonella anatum (8-12 mg/kg intraperitoneally) and Salmonella abortus equi (15 mg/kg intraperitoneally) also caused increases in HDC-activity in the lung tissue of rats (Fig. 5). With these endotoxins, the induction of histidine decarboxylase was even more pronounced than in the case of E. coli endotoxin. Under the influence of endotoxins, significantly higher HDC-activities were also found in kidney, liver and spleen. The activity in kidney and liver being very low and hardly measurable before endotoxin administration.

As HDC-activity can be demonstrated in numerous organs of the endotoxin-treated animals, the conceivable transfer of enzyme material from organs of high activity to those of lower activity had to be taken into account. This possibility was excluded, however, by repeated examinations of the sera, which revealed essentially no enzyme activity.

Further examinations concerning the increase and subsequent decay of HDC-activity with time in the lung tissue during the course of shock revealed maximum activity between the sixth and the twelfth hour after initiation of endotoxin shock. The subsequent decrease (8 hrs) in activity occurred at a still unchanged corticoid level (Fig. 6). The time curve shows that there is a distinct time-lag between the acute liberation from mast cells and the subsequent formation of "intrinsic" histamine. Concerning the action of histamine - liberated as well as newly formed - the potentiating effect of endotoxins on the pharmacological action of histamine is of special interest. This potentiating effect, reported by Urbaschek et al. (28) and later represented by Urbaschek and Versteyl (29, 30) amounts to up to 900 %.

The second question - are plasma kinins of any importance during endotoxin shock? - is answered more and more affirmatively as methods improve.

Among the group of researchers contesting the significance of kinins for progressive hypotension, Webster and Clark (34) did not find any lowering in the plasma kininogen level, and Erdös and Miwa (8), Carretero et al. (5) as well as Shah et al. (25) could not observe any definite increase in kinin. Decreases in kininogen were pointed out by Habermann (9), but he did not attach any significance to these findings, because serum proteins also decrease during shock. Consequently, it was argued that the kininogen/protein ratio would not change decisively. In contrast to this group there are other authors who have long drawn our attention to a contribution of the kinins in the course of endotoxin shock. In 1964, Kobold and co-workers (12) pointed to an enteral liberation of kinin under the influence of endotoxin. In 1968, Rothschild and Castania (18) thought the kinins to be responsible for early hypotension. Scharnagel et al. (19), Diniz et al. (7) as well as our laboratory demonstrated an unequivocal decrease in the kinin level of the blood.

The group of Kimball, Melmon and Wolff (11) recently found a direct increase of bradykinin in plasma under the influence of Salmonella abortus equi endotoxin (Fig. 7). Similar observations had already been made by Nies et al. (15) in rhesus monkeys, where, after the administration of E. coli and Pseudomonas pseudomallei endotoxins, a decrease in kininogen was observed. Here, the increased survival rates after protection by Trasylol observed by Meyer and Werle (14), by Massion and Erdös (13) as well as by our laboratory, should be briefly mentioned.

On the basis of all these findings, particularly in view of the elevated bradykinin levels in man found by Kimball et al. (11), the author attributes considerable importance to kinin formation as a contributory cause of the permeability increase and the development of hypotension. Further factors favoring progressive hypotension and increasing permeability are mainly 5-hydroxytryptamine liberated in the decomposition of thrombocytes and possibly also the "slow reacting substance" (17).

Due to their well-known pharmacological actions the neurohumoral liberation of acetylcholin (32) and possibly the activation of the renin-angiotensin system should rather be considered as factors counteracting endotoxin shock. The significance of these factors, however, cannot be clearly evaluated because of the complexity of the interactions involved.

Summary

Transient initial vasoconstriction is significant for the onset of stasis. During the first seconds to minutes at least, this could be a direct endotoxin effect. Later on, however, catecholamine action is mainly responsible. The presence of catecholamine is not required to bring about endotoxin shock, since during the first two hours no measurable increase of catecholamines has been found in the serum of primates (Hinshaw).

The most important substance known to induce progressive hypotension is histamine, which is either liberated from mast cells or formed by the inducible histidine decarboxylase. In addition to the normal histamine action on smooth muscle there is potentiation by endotoxin.

Now that methods for the direct measurement of increased body kinin levels are available, plasma kinins are rated much more important than at the time of indirect determination, when the decreases in kininogen levels had to be relied on to obtain estimates.

Histamine and the kinins are the main factors in the development of progressive hypotension and are responsible for the high increase of permeability.

References

1. Aures, D., Hakanson, R., Schauer, A.: Europ. J. Pharmacol. 3: 217 (1968).
2. Aures, D., Clark, W. G.: Ann. Biochem. 9: 35 (1964).
3. Becker, E. L., Austen, K. F.: J. exp. Med. 124: 379 (1966).
4. Braude, A. I.: Absorption, distribution, and elimination of endotoxins and their derivations. In: Landy, M., Braun, W.: Bacterial Endotoxins, p. 98 (Rutgers Univ. Press, New Brunswick, N. J. 1964).
5. Carretero, O. A., Nasjletti, A., Fasciolo, J. C.: Experientia 26: 63 (1970).
6. Cline, M. J., Melmon, K. L., Davis, W. C., Williams, H. E.: Brit. J. Haemat. 15: 539 (1968).
7. Diniz, C. R., Carvalho, I. F., Reis, M. L., Corrado, A. P.: 3rd Int. Congr. on vasoactive polypeptides, editor Rocha e Silva, Sao Paulo 1967.
8. Erdös, E. G., Miwa, I.: Fed. Proc. 27: 92 (1968).
9. Habermann, E.: Neue Aspekte der Trasyloltherapie, vol. 3, p. 35 (Schattauer, Stuttgart 1969).
10. Hinshaw, L. B., Jordan, M. M., Vick, J. A.: J. clin. Invest. 40: 1631 (1961).
11. Kimball, H. R., Melmon, K. L., Wolff, S. M.: Proc. Soc. exp. Biol. Med. 139: 1078 (1972).
12. Kobold, E. E., Lovell, R., Katz, W., Thal, A. P.: Surg. Gynec. Obstet. 118: 807 (1964).
13. Massion, W. H., Erdös, E. G.: J. Oklahoma State Med. Ass. 59: 467 (1966).
14. Meyer, A., Werle, E.: Europ. Pancreas Symp., Erlangen 1963 (Schattauer, Stuttgart 1964).
15. Nies, A. S., Forsyth, R. P., Williams, H. E., Melmon, K. L.: Circulation Res. 22: 155 (1968).
16. Nowotny, A.: Naturwissenschaften 58: 397 (1971).
17. Raskova, H., Vanecek, J.: Pharmacol. Rev. 16: 1 (1964).
18. Rothschild, A. M., Castania, A.: J. Pharm. Pharmacol. 20: 77 (1968).
19. Scharnagel, K., Greef, K., Luehn, R., Strohbach, H.: Arch. exp. Path. Pharmakol. 250: 176 (1965).
20. Schauer, A., Menzinger, I., Gielow, L.: Nature 212: 1249 (1966).
21. Schauer, A., Gielow, L., Calvoer, R.: Klin. Wschr. 45: 593 (1967).
22. Schauer, A., Gielow, L.: Verh. dtsch. Ges. Path. 51: 271 (1967).
23. Schayer, R. W.: Science 131: 226 (1960).
24. Schayer, R. W., Rothschild, Z., Bizony, P.: Amer. J. Physiol. 196: 295 (1959).
25. Shah, J. P., Shah, U. S., Appert, H. E., Howard, J. M.: J. Trauma 10: 255 (1970).

26. Shore, O.A., Burkhalter, A., Cohn, V.H.: J. Pharmacol. exp. Ther. 122: 182 (1959).
27. Spink, W.W., Davis, R.B., Potter, R., Chartrand, S.: J. clin. Invest. 43: 696 (1964).
28. Urbaschek, B., Kotowski, H., Bäurle, H.: Z. ImmunForsch. 122: 343 (1961).
29. Urbaschek, B., Versteyl, R.: Nature 207: 763 (1965).
30. Urbaschek, B., Versteyl, R., Götte, D.: Klin. Wschr. 43: 1012 (1965).
31. Urbaschek, B., Huth, K., Krecke, H.J.: Endotoxin nduced microcirculatory disturbances and interaction to various metabolic parameters including the coagulation system. Bibl. anat., vol. 10, p. 442 (Karger, Basel/New York 1969).
32. Vick, J.A., Hinshaw, L.B., Carlson, C.H., Fan, Y.L.: Proc. Soc. exp. Biol. Med. 104: 379 (1960).
33. Vick, J.A., Mehlman, Heiffer, M.H.: Proc. Soc. exp. Biol. Med. 137: 902 (1971).
34. Webster, M.E., Clark, W.R.: Amer. J. Physiol. 197: 406 (1959).
35. Weil, M.H., Spink, W.W.: J. Lab. clin. Med. 50: 501 (1957).
36. Werle, E., Amann, R.: Klin. Wschr. 34: 624 (1956).

HISTAMINE AND ENDOTOXIN SHOCK

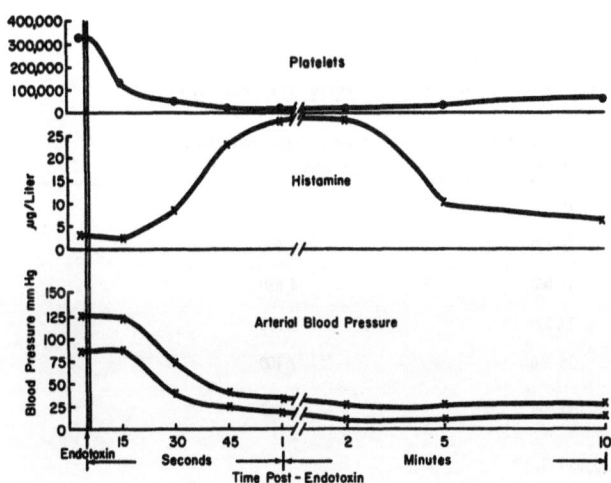

The effect of 1.0 mg/kg *E. coli* endotoxin on platelets, plasma histamine levels, and arterial blood pressure.

Fig. 1.

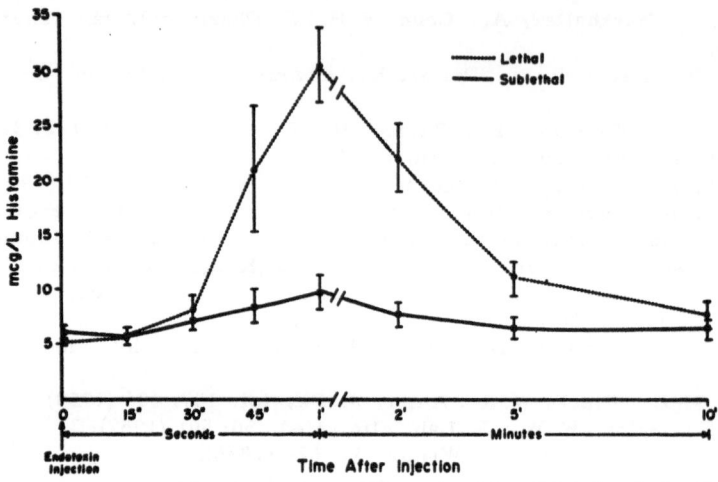

Changes in mean histamine levels ± SEM produced by all lethal and all sublethal doses of *E. coli* endotoxin.

Fig. 2.

Increase in histidine decarboxylase activity of mice lung in endotoxin shock; inhibitory effect of glucocorticoid pretreatment

Controls	Endotoxin treated animals	Glucocorticoid pretreatment and endotoxin application
0	15. 500	3. 700
0	18. 300	3. 000
0	17. 000	4. 680
0	17. 500	3. 350
	18. 580	7. 100
av. 0	17. 378	4. 365

Results in c. p. m. / min. / 100 mg BSH

Fig. 3.

Increase in histidine decarboxylase activity of rat lung following endotoxin application; inhibitory effect of glucocorticoid pretreatment

Controls	Endotoxin treated animals	Glucocorticoid pretreatment and endotoxin application
900	3. 900	500
0	8. 000	0
1. 100	3. 200	0
1. 500	1. 400	1. 300
0	3. 800	0
1. 400	2. 500	0
900	5. 500	
2. 000	3. 700	
4. 700	(3. 000 - 5. 000)	
av. 1. 388	4. 000	300

Results in c. p. m. / min. / 100 mg benzenesulfonylhistamine (BSH)

Fig. 4.

Increase in histidine decarboxylase activity of rat lung following Salmonella endotoxin dependent of dosage

Controls	S. anatum - Endotoxin 8 mg / kg i. p.
350	2. 300
1. 100	6. 400
3. 000	6. 600
800	4. 300
av. 1. 320	4. 900

Controls	S. anatum - Endotoxin 12 mg / kg i. p.
850	12. 000
1. 400	10. 000
3. 700	7. 800
600	9. 500
750	9. 100
av. 1. 460	9. 720

Controls	S. abortus equi - Endotoxin 15 mg / kg i. p.
950	11. 400
1. 100	6. 990
2. 000	7. 200
1. 400	0
av. 1. 360	6. 370

Results in c. p. m. / 100 mg BSH

Fig. 5.

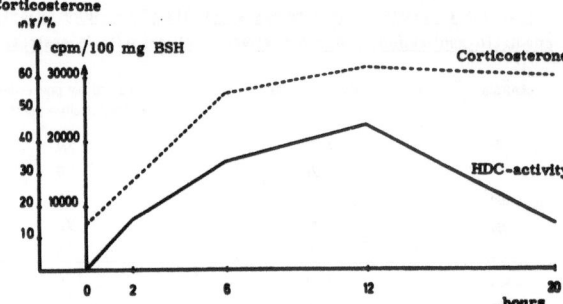

Temporal behaviour of histidine decarboxylase (HDC)
activity of mice lung during endotoxin shock and of
plasma corticosterone concentration in mice.

Fig. 6.

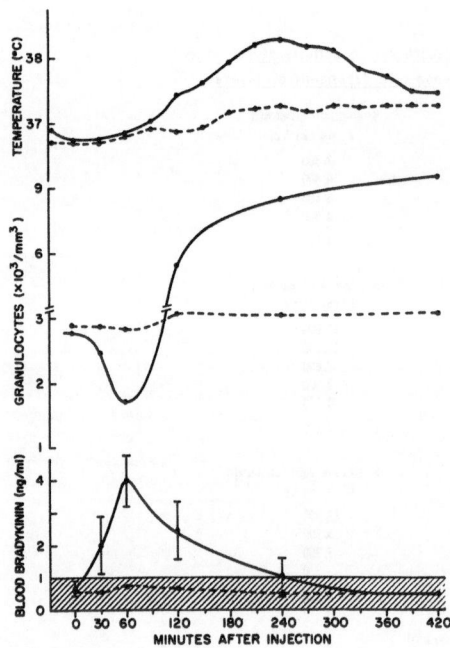

Fig. 7.

Some Aspects of the Effects of Endotoxin, Endotoxoid, and Biogenic Amines

B. Urbaschek and R. Urbaschek

Endotoxins certainly play a more important role in systemic diseases caused by gram-negative bacteria than for instance in gram-negative infections of the gastrointestinal tract, such as shigellosis. Considering the pathophysiology of infections with gram-negative bacteria endotoxins represent only one of the pathogenic principles of gram-negative bacteria.

It should be emphasized that the therapy of these infections causes great difficulties, since the administration of antisera, respectively hyperimmunoglobulin, does not seem to be successful. Single clinical observations by Barandun (2) using hyperimmunoglobulin and the findings of Braude and co-workers (8, 46, 56) and the results of McCabe (27, 28) give good evidence to further study these possibilities. Until now, however, no definite clinical study exists on the efficacy of immunoglobulin administered in gram-negative infections.

Besides the decrease in platelets and leukocytes, the disturbances in the microcirculation - the content of the vessels, the vessel wall and the perivascular spaces - are among the earliest detectable changes caused by endotoxins. The role of several mediators in the initial phase of endotoxemia have often been discussed. Thus, it is well-known that endotoxins cause the release of various vasoactive mediators; among others, these include histamine (18, 19, 55) and serotonin (9). With regard to these release reactions, endotoxins imitate the effect of antigen-antibody complexes. Depending on the serologic specificity of endotoxins preformed antibodies may play a role; however, antibodies are not necessary for the startermechanisms.

The release of histamine in anaphylactic shock is so rapid and violent that early counteractions, for instance of histaminase and the elimination of histamine via methylation in liver and kidney, cannot be effective to the same extend as in the case of endotoxemia. It was found that histaminase is released by heparin (42). Together with histamine this enzyme is released from the granules of degranulating perivascular mast cells which is one mechanism of histamine liberation in the initial stage of endotoxemia. In the course of endotoxemia, which is protracted in contrast to anaphylactic shock, the enzyme histaminase can be effective. A further source of histamine could be considered, namely that of "induced histamine" which is according to Schayer (39, 40) synthesized by an inducible form of histidine decarboxylase in or near the vascular endothelial cells, among others, following endotoxins or bacterial infection.

Our results may contribute to a better understanding of the pathophysiological efficacy of even small amounts of released biogenic amines, especially in the initial stage of endotoxin action. Endotoxins from different gram-negative bacteria or cold-heat-treated bacterial suspensions increase the effect of histamine on the isolated guinea pig uterus up to 900 % and the action of serotonin up to 400 % on the rat and mouse uterus (47, 48, 49). These findings correspond with the histamine-sensitizing effect of H. pertussis vaccine (37, 38, 41) and the increased sensitivity to serotonin (21, 35). In the microcirculation the combination of threshold doses of endotoxin and biogenic amines lead to all the disturbances produced by endotoxins including stasis and microbleedings. Each threshold dose by itself caused only a slowdown of blood flow. The pathogenetic significance of the circulus vitiosus concerning the endotoxin-caused release of biogenic amines and the sensitizing effect is obvious.

Regardless of the variability in mode of action of histamine and serotonin in different species and anatomical differences these mediators seem to be of special importance in the disturbances endotoxin causes in the capillary bed and in hemodynamic balance - especially in the initial stage. Endotoxin, histamine, and serotonin were found to induce similar disturbances in the capillary bed. The effect of endotoxin differed from that of biogenic amines only in the immediate initiation of the changes after administration of the latter.

In in vivo microscopic studies the capillary bed of the rabbit and guinea pig mesentery as well as the hamster cheek pouch were observed. The sequence of the severe disturbances in the microcirculation following endotoxin and biogenic amines have been described in detail (49, 53). The main changes are the following: Degranulation of perivascular mast cells, granulocytosis, walladherence of granulocytes, changes in plasticity and rouleaux-formation of erythrocytes, formation of mixed thrombi and microthrombi, swelling, respectively contraction and deviation of endothelial cells, increase in permeability of the vessel wall, and microbleeding. In the beginning formation of reversible aggregations of platelets occurs and later dose-dependent irreversible aggregates are formed, such as microthrombi with the participation of fibrin fibers, as could be verified by means of scanning electron microscopy (51, 53).

The decrease in platelet count precedes the formation of platelet aggregations observed in the microcirculation. The reversible aggregates of platelets do not initiate the plasma clotting system and seem to be without severe consequences. Thus, the release of serotonin, adrenaline, and adenine nucleotides from the platelets (7, 20, 24) and the slow demasking of platelet factor 3 only occurs at the time when irreversible platelet aggregations are formed. Besides the activation of the plasma clotting system via phospholipoproteins (22) the changes of the endothelial cells in the vitalmicroscopic studies reveal the early activation of the intrinsic system via factor XII. The deviation of endothelial cells enable for instance subendothelial collagen to contact the content of the vessels, so that factor XII can be activated. The various factors involved in the activation of the coagulation system by endotoxins have been described by Lüscher (24).

The described vitalmicroscopic observations concerning the damage of the endothelial cells after endotoxin are supported by studies of McGrath and Stewart (30) and Gaynor, Bouvier and Spaet (13) by means of electron microscopy. In electron microscopic studies Majno, Shea and Leventhal (25) have observed alterations of the endothelial cells such as swelling and contraction as consequence of histamine.

According to general pathological principles, different noxious substances lead to similar or identical reactions in the capillary bed in the sense of nonspecific acute inflammation (49, 53).

It has been shown here that endotoxins initiate several startermechanisms in the circulation and in the vessel walls. Pretreatment with detoxified endotoxins (endotoxoid) protects animals against lethal doses of endotoxins (49, 52, 53), and prevents microcirculatory changes which occur after endotoxin (49, 50). After pretreatment with Glyvenol[R] (Ciba, Basel), a glucofuranoside derivative, the described severe endotoxin-caused disturbances in the microcirculation also fail to appear within the first hour (49). However, Glyvenol[R] does not protect the animals from the lethal effects of endotoxins. These results indicate that besides the mechanisms discussed above changes in other parameters, especially metabolic parameters, for instance of the carbohydrate metabolism as has been demonstrated by Berry (4) might be of significance in the outcome of endotoxemia.

In cooperation with Lundsgaard-Hansen, Stork, and K. Huth several metabolic, hemodynamic and clotting parameters were measured in miniature pigs. In this context only a few results will be described. Endotoxin from E. coli 0111 and 055 were extracted according to Boivin's trichloroacetic acid method (5, 6). Together with M. Frodl detoxified endotoxins were prepared from these endotoxins with potassium methylate according to the method of Nowotny (36). Four groups of three pigs each were studied for four hours after beginning the infusion which lasted one hour. One group received endotoxin (250 /ug/kg), one group endotoxoid (150 /ug/kg), one group endotoxin (250 /ug/kg) after pretreatment with endotoxoid (150 /ug/kg), and the fourth group served as control.

Following intravenous infusion of endotoxin in the pig a glycolytic and gluconeo-genesis-inhibiting effect of endotoxin was observed, indicated by an increase in lactate and decrease in glucose. These in vivo results of the glycolytic and gluco-neogenesis-inhibiting effect correspond with findings of Berry, McCallum, and Schumer (3, 29, 43). It is a rather interesting fact that the carbohydrate metabo-lism is not altered after endotoxoid in pigs. Moreover, after pretreatment with endotoxoid the endotoxin-caused changes in lactate and glucose levels fail to appear.

The alkaline phosphatase is markedly increased after endotoxin. This increase seems to be due to the hepatotoxic action of endotoxins and the wall-adhering de-granulating leukocytes which are observable in the capillary bed. The endotoxin-caused increase is moderate after the pigs have been pretreated with endotoxoid.

The functional disturbances in mitochondria after endotoxin can be concluded from the finding that glutamate dehydrogenase markedly increases. Mela (31, 32) de-scribed disturbances in energy production of liver mitochondria during lethal endo-toxemia. After pretreatment with endotoxoid in our studies no increase in glutamate dehydrogenase occurred following endotoxin.

From the measured parameters during endotoxemia in pigs it cannot be con-cluded that there are endotoxin-specific alterations. It is rather presumable that interaction of the observed disturbances - such as hypoxia, decrease in arterial blood pressure and in pH, the changes in carbohydrate metabolism and in the coag-ulation system, changes which are expressed in the behavior of the microcircula-tion - plays a decisive role which can also be observed in shock states not due to endotoxin.

A rapid and sustained fall of complement levels following endotoxin administra-tion has been known for some time (16). Endotoxin effect on complement components is the result of alternate pathway activation (17, 34). Thus, activation of C3 pro-activator (C3PA) occurs with subsequent activation of C3 and the following compo-nents sparing the early components C1, C4, and C2 (10, 14, 33, 54). Besides induc-tion of hemolysis, bactericidal reaction and anaphylatoxin activity through C3a and C5a (23) chemotactic factors (44, 45) are liberated by endotoxins. Comparing nor-mal to C4-deficient guinea pig sera (11) it has been proposed by Frank, May, and Kane (12) that another pathway of complement activation through endotoxin utilizing the early components C1, C4, and C2 might be necessary for the well-known endo-toxin effects on the coagulation system. Using different detoxified lipopolysaccha-rides Gewurz et al. (15) could demonstrate the loss of the effect on C3 normally observed with the parent endotoxin.

In preliminary in vitro studies with Mauff (26) by means of immunofixation in agarosegel-electrophoresis (1), using human sera and specific antisera, activation of C3PA and subsequently C3 was observed with varying amounts of endotoxin. Con-siderably less activation of C3PA and C3 occurred with corresponding amounts of detoxified endotoxin (Fig.). The slight activating effect of detoxified endotoxin might be regarded either as the result of incomplete detoxification or is possibly due to activation via the early components. Addition of EDTA does not lead to C3PA or C3 activation by endotoxin or by detoxified endotoxin, presumably explained through absence of free Mg-(or Ca-)ions (17), inhibiting early components or bypass acti-vation.

The interesting difference in complement activation between endotoxin and de-toxified endotoxin might partially explain the fact that no microcirculatory disturb-ances occur after detoxified endotoxin, because some of the trigger substances, anaphylatoxin or chemotactic factors, are not generated. However, detoxified endo-toxin does not prevent in vitro the endotoxin effect on C3PA or C3 (Fig., pos. 8).

Whereas these results indicate that tolerance is not induced by inhibition of bypass activation through detoxified endotoxin, the causative mechanism remains to be found.

References

1. Alper, Ch. A., Johnson, A. M.: Vox Sang., Basel 17: 445 (1969).
2. Barandun, S.: Bibl. Haemat., Fasc. 17 (Karger, Basel/New York 1964).
3. Berry, L. J., Smythe, D. S.: Ann. N. Y. Acad. Sci. 88: 1278 (1960).
4. Berry, L. J.: Metabolic effects of bacterial endotoxins. In: Kadis, S., Weinbaum, G., Ajl, S. J.: Microbial toxins, vol. V., p. 165 (Academic Press, New York/London 1971).
5. Boivin, A., Mesrobeanu, I., Mesrobeanu, L.: C. R. Soc. Biol. 113: 490 (1933).
6. Boivin, A., Mesrobeanu, I., Mesrobeanu, L.: C. R. Soc. Biol. 114: 307 (1933).
7. Davey, M. G., Lüscher, E. F.: Biochim. biophys. Acta 165: 490 (1968).
8. Davis, Ch. E., Brown, K. R., Douglas, H., Tate, W. J., Braude, A. I.: J. Immunol. 102: 563 (1969).
9. Davis, R. B., Meeker, W. J., Jr., Bailey, W. L.: Proc. Soc. exp. Biol. Med. 108: 774 (1961).
10. Dierich, M., Bitter-Suermann, D., König, W., Hadding, U., Galanos, C., Rietschel, E. T.: Immunology 24: 721 (1973).
11. Frank, M. M., May, J., Gaither, T., Ellmann, L.: J. exp. Med. 134: 176 (1971).
12. Frank, M. M., May, J. E., Kane, M. A.: J. infect. Dis. 128: S176 (1973).
13. Gaynor, E., Bouvier, C., Spaet, Th. H.: Science 170: 986 (1970).
14. Gewurz, H., Shin, H. S., Mergenhagen, S. E.: J. exp. Med. 128: 1049 (1968).
15. Gewurz, H., Mergenhagen, S. E., Nowotny, A., Phillips, J. K.: J. Bact. 95: 397 (1968).
16. Gilbert, V. E., Braude, A. I.: J. exp. Med. 116: 477 (1962).
17. Götze, O., Müller-Eberhard, H. J.: J. exp. Med. 134: S90 (1971).
18. Hinshaw, L. B., Vick, J. A., Carlson, C. H., Fan, Y. L.: Proc. Soc. exp. Biol. Med. 104: 379 (1960).
19. Hinshaw, L. B., Jordan, M. M., Vick, J. A.: Amer. J. Physiol. 200: 987 (1961).
20. Holmsen, H., Day, H. J., Stormorken, H.: Scand. J. Haemat., Suppl. 8 (1969).
21. Kind, L. S.: Proc. Soc. exp. Biol. Med. 95: 200 (1957).
22. Lasch, H. G., Heene, D. L., Huth, K., Sandritter, W.: Amer. J. Cardiol. 20: 381 (1967).
23. Lichtenstein, L. M., Gewurz, H., Adkinson, N. F., Shin, H. S., Mergenhagen, S. E.: Immunology 16: 327 (1969).
24. Lüscher, E. F.: The activation of intravascular coagulation by endotoxin. In this volume.
25. Majno, G., Shea, S. M., Leventhal, M.: J. Cell Biol. 42: 647 (1969).
26. Mauff, R., Urbaschek, B.: In preparation.
27. McCabe, W. R.: J. Immunol. 108: 601 (1972).
28. McCabe, W. R., Kreger, B. E., Johns, M.: New Engl. J. Med. 287: 261 (1972).
29. McCallum, R. E., Berry, L. J.: Immunity 7: 642 (1973).
30. McGrath, J. M., Stewart, G. J.: J. exp. Med. 129: 833 (1969).
31. Mela, L., Bacalzo, L. V., Miller, L. D.: Amer. J. Physiol. 220: 571 (1971).
32. Mela, L., Miller, L. D., Nicholas, G. G.: Surgery 72: 102 (1972).
33. Mergenhagen, S. E., Snyderman, R., Gewurz, H., Shin, H. S.: Curr. top. Microbiol. Immunol. 50: 37 (1969).
34. Mergenhagen, S. E., Snyderman, R., Phillips, K.: J. infect. Dis. 128: S86 (1973).
35. Munoz, J.: Proc. Soc. exp. Biol. Med. 95: 328 (1957).
36. Nowotny, A.: Nature 197: 721 (1963).

37. Parfentjev, I.A., Goodline, M.A.: J. Pharmacol. exp. Ther. 92: 411 (1948).
38. Sanyal, R.K., West, B.B.: Int. Arch. Allergy 14: 241 (1959).
39. Schayer, R.W.: Amer. J. Physiol. 198: 1187 (1960).
40. Schayer, R.W.: Progr. Allergy 7: 187 (1963).
41. Schmutzler, W., Zschoch, H.: Naturwissenschaften 48: 134 (1961).
42. Schmutzler, W.H., Giertz, H., Hahn, F., Seseke, G.: Arch. exp. Path. Pharmakol. 250: 173 (1965).
43. Schumer, W.: Assessment of studies in the biochemistry and immunology of endotoxemia. In this volume.
44. Snyderman, R., Gewurz, H., Mergenhagen, S.E.: J. exp. Med. 128: 259 (1968).
45. Snyderman, R., Shin, H.S., Phillips, J.K., Gewurz, H., Mergenhagen, S.E.: J. Immunol. 103: 413 (1969).
46. Tate, W.J.,III., Douglas, H., Braude, A.I., Wells, W.W.: Ann. N.Y. Acad. Sci. 133: 746 (1966).
47. Urbaschek, B., Kotowski, H., Bäurle, H.: Z. ImmunForsch. 122: 343 (1961).
48. Urbaschek, B., Versteyl, R.: Nature 207: 763 (1965).
49. Urbaschek, B.: Zur Frage des Wirkungsmechanismus bakterieller Endotoxine und seiner Beeinflussung. Habilitationsschrift, Universität Heidelberg 1967.
50. Urbaschek, B., Branemark, P.-I., Nowotny, A.: Experientia 24: 170 (1968).
51. Urbaschek, B., Fritsch, H., Richter, I.E.: Klin. Wschr. 47: 1166 (1969).
52. Urbaschek, B., Nowotny, A.: Endotoxin tolerance induced by endotoxoid. In: Chedid, M.L.: La structure et les effects biologiques des produits bactériens provenant de germes gram-négatifs, p. 357 (Editions du Centre nationale de la recherche scientifique, Paris 1969).
53. Urbaschek, B.: The effects of endotoxins in the microcirculation. In: Kadis, S., Weinbaum, G., Ajl, S.J.: Microbial toxins, vol. V, p. 261 (Academic Press, New York/London 1971).
54. Wardlaw, A.C.: Endotoxin and complement substrate. In: Landy, M., Braun, W.: Bacterial endotoxins, p. 81 (Rutgers, New Brunswick, USA 1964).
55. Weil. M.H., Spink, W.W.: J. Lab. clin. Med. 50: 501 (1957).
56. Ziegler, E.J., Douglas, H., Sherman, J.E., Davis, Ch.E., Braude, A.I.: J. Immunol. 111: 433 (1973).

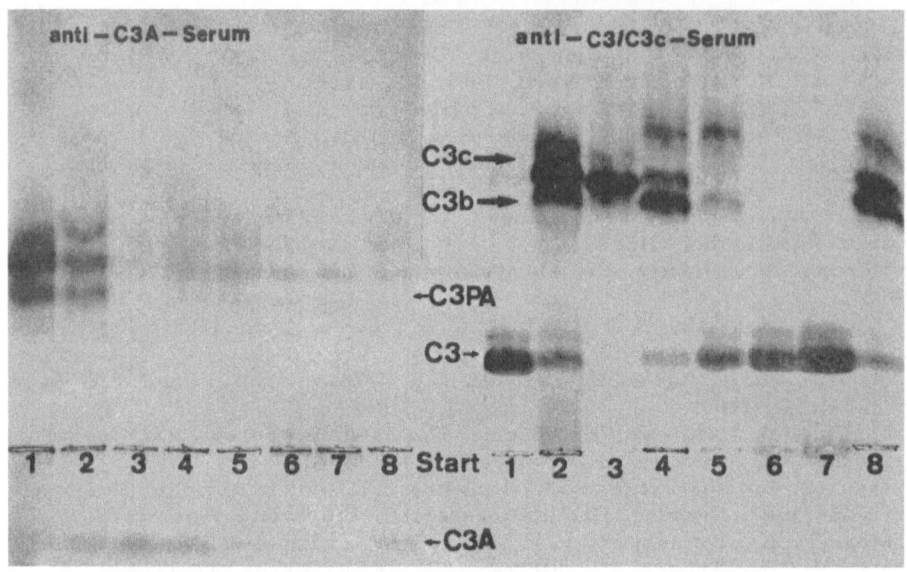

Fig. Immunofixation of agarosegel-electrophoresis with anti-C3-activator (C3A) serum and anti-C3/C3c serum. Incubation of fresh human serum (HS) with: cobra venom factor (CF), endotoxin and detoxified endotoxin.

1 = HS control
2 = aged HS (3 weeks, 4°C)
3 = HS + CF (2 hrs, 37°C)
4 = HS + endotoxin (10 mg/ml, 2 hrs, 37°C)
5 = HS + detoxified endotoxin (10 mg/ml, 2 hrs, 37°C)
6 = HS + EDTA + endotoxin
7 = HS + EDTA + detoxified endotoxin
8 = HS + detoxified endotoxin (10 mg/ml, 2 hrs, 37°C) + endotoxin (10 mg/ml, 2 hrs, 37°C)

HS of C3 phenotype S and C3-proactivator (C3PA) phenotype FS.

Biological Consequences of Endotoxin Interaction with Complement

A. L. Sandberg, R. P. Siraganian, and S. E. Mergenhagen

The interaction of endotoxin with the complement (C) system may account in part for some of the biological effects observed after endotoxin administration. Several years ago Gewurz, Shin and Mergenhagen (1) demonstrated the ability of endotoxin to preferentially consume the late C components (C3, C5, C6, C7, C8 and C9) with sparing of the early components (C1, C4 and C2). Additional recent studies have established that endotoxin is capable of activating an alternate C pathway which bypasses C1, C4 and C2 and enters the C sequence at the C3 stage (2, 3). However certain biological activities such as lysis of endotoxin coated erythrocytes appear to require the classical C pathway (4).

Since C activation at the surface of platelets results in the release of biologically active substances (reviewed in 5) these studies were designed to determine the role of the alternate C pathway in the in vitro release of histamine from rabbit platelets and to investigate the contribution of the platelets to the hypercoagulable state produced by endotoxin and other activators of the C system. We have previously shown that guinea pig gamma-1 immunoglobulins as well as the 5S pepsin derived $F(ab')_2$ fragments of guinea pig gamma-1 and gamma-2 antibodies activate only the alternate C pathway whereas intact 7S gamma-2 antibodies can activate the classical C sequence (6, 7, 8). $F(ab')_2$ fragments of rabbit immunoglobulin G also activate only the alternate C pathway as contrasted to the parent 7S antibodies which interact with the classical C pathway (9). It is therefore possible to distinguish platelet damage associated with the alternate pathway from that due to the classical pathway by the proper choice of activating agents.

Materials and Methods

C Activators. E. coli 026:B6 Lipopolysaccharide B was obtained from Difco Laboratories. Inulin purchased from C. P. Pfanstiehl Laboratories was dissolved in saline (4 mg/ml), heated for 30 min at 56° C, centrifuged at 1200 X g and the supernatant used. Guinea pig gamma-1 and gamma-2 antibodies were obtained by DEAE cellulose chromatography of serum from guinea pigs hyperimmunized with dinitrophenylated bovine gamma globulin (DNP-BGG) and characterized as previously described (10). 7S and 5S rabbit antibodies were prepared and characterized as in (11).

Platelet Reaction Mixtures. Platelets, plasma and complement activators were incubated as previously described (12). Platelets and plasma from both normal and genetically C6 deficient rabbits were utilized.

Histamine Analysis. Histamine in the supernatants of the reaction mixtures was determined by an automated spectrofluorimetric method (13).

Platelet Factor 3 Analysis. A modification of the Russell's viper venom assay of Hardisty and Hutton (14) was used to determine the time required for Russell's viper venom to induce clot formation in the presence of the reaction mixture supernatants, heparinized rabbit plasma and protamine sulfate.

Human C6. This reagent was obtained from Cordis Laboratories.

Results and Discussion

In order to establish a role for the alternate C pathway in the platelet release reaction gamma-1 and gamma-2 antibodies and specific antigen were incubated with rabbit platelets in the presence of normal rabbit plasma and the resultant histamine release determined. As shown in Table I the gamma-1 immunoglobulins, which activate only the alternate C pathway, were as effective in inducing histamine release as were the gamma-2 antibodies, which activate the classical C pathway. However a longer lag period was observed when histamine release occurred via the alternate C pathway. Entirely analogous results have been obtained when comparing other activators of the alternate C pathway with those of the classical C pathway. Thus the $F(ab')_2$ fragments of several rabbit immunoglobulin preparations, which activate only the alternate pathway, effectively initiate the platelet release reaction but require a longer lag period than do the intact parent antibodies, which activate the classical C pathway (11).

Endotoxin and immune aggregates are capable of enhancing blood coagulation (15, 16) as well as activating C suggesting an interrelationship between the two systems. Recently Zimmerman and Müller-Eberhard (17) demonstrated that inulin, an activator of the alternate C pathway, also enhances clotting of normal rabbit blood. Since both C pathways support the release of biologically active constituents from rabbit platelets further studies were carried out to investigate the role of the platelet as a possible point of juncture between the complement and coagulation systems. As shown in Table II activation of the classical C pathway by 7S immunoglobulins as well as activation of the alternate C pathway by 5S $F(ab')_2$ antibody preparations results in histamine release and the simultaneous enhanced release of Platelet Factor 3 of the intrinsic clotting pathway. In the absence of C activators the Russell's viper venom clotting time is 90 sec whereas the activation of C via either pathway decreases the clotting time by at least one half.

Zimmerman, Arroyave and Müller-Eberhard (18) have demonstrated a coagulation defect in genetically C6 deficient rabbits which could be corrected by the addition of C6. The possibility existed that the clotting abnormality in these rabbits could be explained by the need for an intact C system which upon activation would influence platelets. This hypothesis was examined by incubating platelets and plasma from normal and C6 deficient rabbits with activators of both C pathways and determining the resultant release of histamine and Platelet Factor 3. As shown in Table III endotoxin initiates platelet histamine release in the presence of normal rabbit plasma. However no histamine release occurs in the presence of C6 deficient plasma. C6 is also a requirement for the enhancement of Platelet Factor 3 release as indicated by a decrease in the Russell's viper venom clotting time in the reaction mixtures containing normal rabbit plasma but not in those containing the C6 deficient plasma. Other activators of both the alternate C pathway (inulin) and the classical C pathway (antibody-antigen aggregates) behave identically in that C6 is essential for histamine and enhanced Platelet Factor 3 release. Endotoxin appears to be capable of activating both C pathways and its mode of C interaction resulting in platelet damage has not yet been delineated.

The defect in the platelet release reaction in C6 deficient rabbits is inherent in the plasma rather than in the platelets (19). Platelets from C6 deficient as well as normal animals release biologically active substances in the presence of normal rabbit plasma. However platelets from both normal and C6 deficient rabbits are unaffected in the presence of the C6 deficient plasma.

The addition of purified C6 to the deficient plasma restores the ability of the plasma to support histamine and Platelet Factor 3 release as shown in Fig. 1. Whereas no histamine is released by inulin, an activator of the alternate C pathway, in C6 deficient plasma the addition of increasing concentrations of C6 restores this activity. The addition of C6 also allows the enhanced release of Platelet Factor 3 as demonstrated by the reduction of the Russell's viper venom clotting time.

A scheme portraying one point of interaction between the C and coagulation systems in the rabbit is shown in Fig. 2. Endotoxin as well as other activators of the alternate and classical C pathways interact with the C sequence including C6 to induce the release of Platelet Factor 3 which may then fulfill its role in the clotting sequence.

Summary

By virtue of their ability to activate the C sequence endotoxin and other agents initiate the release of vasoactive amines and the clotting factor, Platelet Factor 3, from rabbit platelets. Both the classical and alternate C pathways are effective in these release reactions and C6 is required regardless of the pathway traversed. The production of the hypercoagulable state observed after administration of endotoxin and other C activators may be attributed to the C dependent release of Platelet Factor 3.

References

1. Gewurz, H., Shin, H.S., Mergenhagen, S.E.: J. exp. Med. 128: 1049 (1968).
2. Marcus, R.L., Shin, H.S., Mayer, M.M.: Proc. nat. Acad. Sci., U.S.A. 68: 1351 (1971).
3. Frank, M.M., May, J., Gaither, T., Ellman, L.: J. exp. Med. 134: 176 (1971).
4. Phillips, J.K., Snyderman, R., Mergenhagen, S.E.: J. Immunol. 109: 334 (1972).
5. Osler, A.G., Siraganian, R.P.: Prog. Allergy 16: 450 (1972).
6. Sandberg, A.L., Osler, A.G., Shin, H.S., Oliveira, B.: J. Immunol. 104: 329 (1970).
7. Sandberg, A.L., Osler, A.G.: J. Immunol. 107: 1268 (1971).
8. Sandberg, A.L., Oliveira, B., Osler, A.G.: J. Immunol. 106: 282 (1971).
9. Reid, K.B.M.: Immunology 20: 649 (1971).
10. Oliveira, B., Osler, A.G., Siraganian, R.P., Sandberg, A.L.: J. Immunol. 104: 320 (1970).
11. Siraganian, R.P., Sandberg, A.L., Alexander, A., Osler, A.G.: J. Immunol. 110: 490 (1973).
12. Siraganian, R.P., Secchi, A.G., Osler, A.G.: J. Immunol. 101: 1130 (1968).
13. Siraganian, R.P.: Analyt.Biochem., in press.
14. Hardisty, R.M., Hutton, R.A.: Brit. J. Haemat. 12: 764 (1966).
15. McKay, D.G., Shapiro, S.S., Shanberge, J.: J. exp. Med. 107: 369 (1958).
16. Robbins, J., Stetson, C.A., Jr.: J. exp. Med. 109: 1 (1959).
17. Zimmerman, T.S., Müller-Eberhard, H.J.: J.exp. Med. 134: 1601 (1971).
18. Zimmerman, T.S., Arroyave, C.M, Müller-Eberhard, H.J.: J. exp. Med. 134: 1591 (1971).
19. Siraganian, R.P.: Nature (N.B.) 239: 208 (1972).

Table I.

Time Course of Histamine Release

Guinea Pig anti DNP-BGG[*]	Reaction Time	Histamine Release
	Min	%
γ-1	1	0
	10	0
	20	0
	45	74
γ-2	1	0
	10	37
	20	83
	45	86

*Antigen-DNP-BGG
 Antibody/Antigen ratio=25

Table II.

Platelet Factor 3 Release Due to C Activation

Antibody Preparation		Platelet Factor 3	Histamine Release
		Sec	%
Rabbit anti-human albumin	7S	43	62
	5S	40	53
Rabbit anti-ferritin	7S	39	61
	5S	33	81
Blanks		90	0

Table III.

Histamine and Platelet Factor 3 Release Reactions

Material Tested	mg/ml*	Histamine Release With		Platelet Factor 3 Release With	
		C6 Deficient plasma	Normal plasma	C6 Deficient plasma	Normal plasma
		%		Sec.	
None		0	0	105	120
Endotoxin	3.3	2	28	135	69
Inulin	0.67	0	81	120	44
Antigen-antibody aggregates	0.156	2	28	135	69
Antigen-antibody (soluble system)	0.026	1	79	117	42

*Final concentration in reaction mixtures

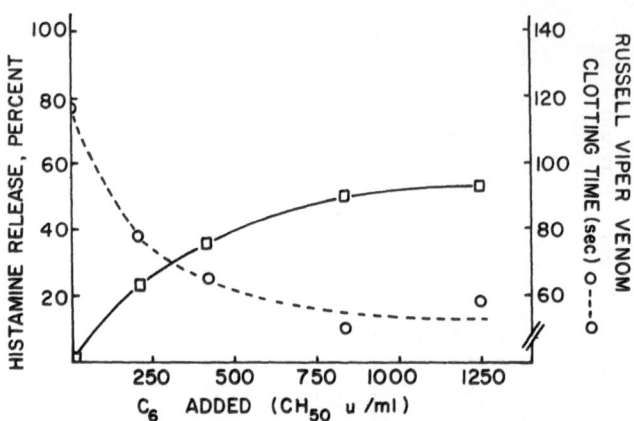

Fig. 1. Inulin induced histamine and Platelet Factor 3 release in C6 deficient rabbit plasma containing increasing concentrations of purified C6.

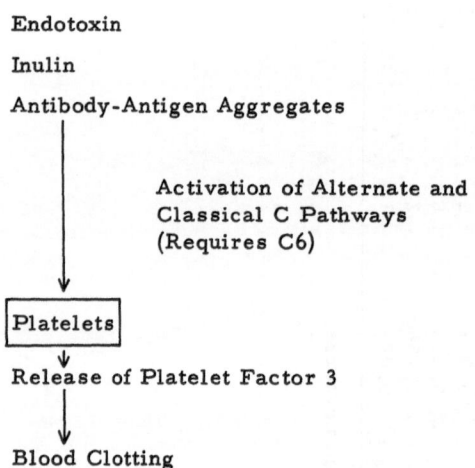

Fig. 2. Interaction of the complement and coagulation systems.

Influence of Lipid A and Various LPS-Preparations on the Third Component of Complement (C 3): Biological Significance of this Interaction

D. Bitter-Suermann and U. Hadding

Endotoxic lipopolysaccharides (LPS) are known to interact with the complement system. This interaction, leading to the activation of C3, takes place in normal as well as in C4-deficient guinea pig serum and thus represents an example for the alternate pathway of C activation starting with C3. This does not exclude the activation of C1, C4 and C2 provided a LPS-specific antibody is available.

For the experiments of this report three different LPS-preparations from Dr. Galanos (Freiburg) were used: LPS of S. minnesota smooth form (LPS); LPS of S. minnesota rough form, R 595(LPS-R); Lipid A coupled to BSA.

In dose experiments an increase of the total C3 turnover was found as function of LPS input, but when expressed as C3 consumption per μg LPS the relative turnover decreased. Lipid A was most efficient followed by LPS-R and LPS-S.

In various experiments the kinetics of the C3 turnover as induced by LPS and LPS-R have been studied, where marked differences occurred. While for LPS-S the C3 turnover at 37°C takes place during the first 12-15 minutes, LPS-R needs only 3-4 minutes (Fig. 1). During these experiments an observation was made, which we called "plateau formation" and which might be correlated with a detoxification mechanism: The amount of C3 consumption, i. e. the level of the plateau reached after the above mentioned intervals, is typical for given concentrations of LPS (intermediate plateau), while above a certain LPS-concentration additional LPS does not induce further C3 turnover (maximal plateau). The "maximal plateau" might be explained by limitation of the factors needed for the LPS dependent consumption of C3, whereas the reason for the "intermediate plateau" and thus for the termination of the action on C3 is not well understood. It is hypothesized that coating of the LPS molecule by serum proteins, e. g. albumin, inactivates the endotoxins with regard to their action on C3.

To support this assumption experimentally, we used ion deprived serum, in which LPS cannot initiate C3 turnover due to the missing Mg^{++}. This serum was not able to lyse EA unless Ca^{++} and Mg^{++} were added. LPS was incubated for various time intervals in such serum samples prior to the addition of Mg^{++}, which was followed by determination of the resulting C3 consumption. If Mg^{++} was not added to the mixture at the very beginning a reduced or even no C3 turnover was provoked during further 20 minutes of incubation at 37°C. The possibility of initiating C3 consumption and its degree as induced by addition of Mg^{++} to a preincubated mixture of LPS and ion free serum was a function of the time and temperature of preincubation. Short incubation or low temperature prior to the addition of Mg^{++} still allowed high C3 turnover. Higher temperatures ($24-37^{\circ}$C) led to fast "detoxification" of LPS. Another way of blocking LPS with respect to its potency of C3 activation was preincubation with albumin (1-4 %), while 1 % IgG, 0.05 % gelatin, or purified C3 (1 200 μg/ml), or purified C5 (350 μg/ml) have no effect.

Thus the termination of C3 consumption, i. e. the beginning of the plateau reflecting a first detoxification is probably due to the coating of LPS with serum proteins and not due to decay of the enzymatic activity of a LPS-cofactor complex since in the preincubation experiments such a complex had never existed. This interpretation is in accord with the immediate blocking effect of EDTA on the C3 turnover by LPS in serum. If there would exist a somewhat stable complex with enzymatic activity, it should still act after EDTA addition provided, of course, Mg^{++} is not needed for the cleaving of C3. Such an EDTA-insensitive C3 cleavage takes place for instance with the preformed C42 enzyme, or the cobra venom factor induced VF-C3PA-factor D-enzyme.

The biological significance of the LPS-complement interaction rests with the demonstration that this very fast reaction represents a full activation of the most important C3, which can be shown by the appearance of either hemolytically active C3b or anaphylatoxin activity due to C3a and C5a. The hemolytic efficiency of LPS induced C activation can be demonstrated either in an autologous system consisting of guinea pig E., g.p. serum and LPS coupled to the erythrocytes or in a heterologous system of sheep erythrocytes, g.p. serum and LPS. In this system LPS markedly interferes with a parallel lysis of E due to natural g.p. anti-sheep-erythrocytes antibodies and C142 (Fig. 2). This classical lysis is strongly inhibited and can be even better demonstrated with preformed EA, g.p. serum and LPS. This effect is due to high avidity of LPS to C1, as will be demonstrated by Dr. Hadding hereafter.

In conclusion, whereever endotoxic LPS in a living organism will get access to serum, within a few minutes there follows an activation of the complement system starting with C3 and thus the generation of biologically active split products. The deleterious process is soon limited probably by coating of the LPS with other serum proteins, but the early state of activation will be long enough to exert a pronounced effect on other systems, e.g. the clotting-system.

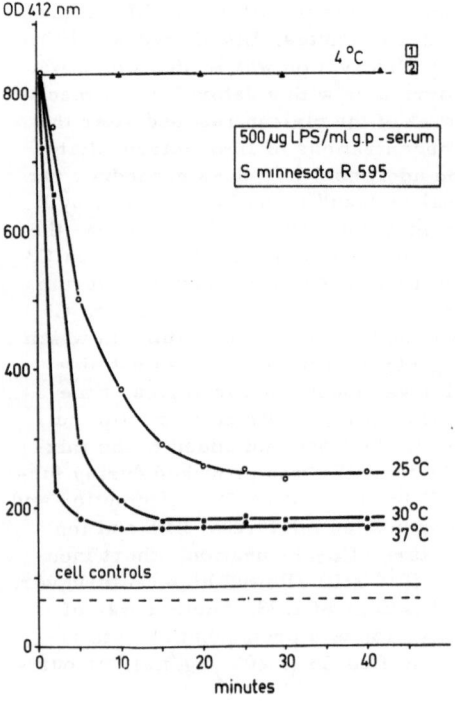

Fig. 1. Turnover of C3 in guinea pig serum by LPS-R as a function of time and temperature. Ordinate: Optical density values at 412 nm obtained from the hemolysis test. High values correspond to high amounts of C3, i.e. low consumption of C3.

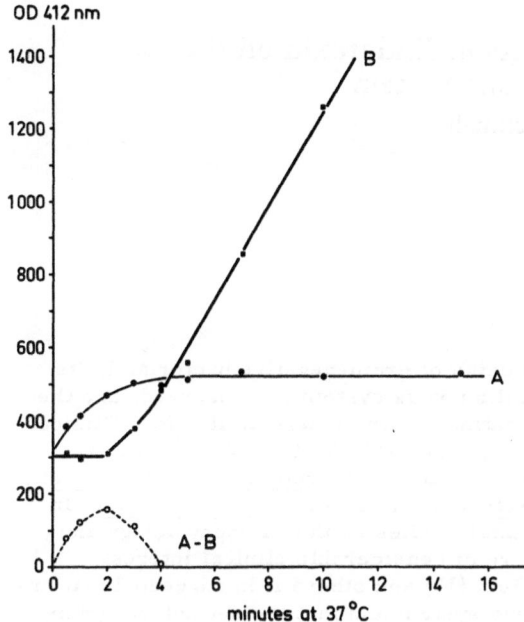

Fig. 2. Influence of LPS on the hemolysis of SRBC by guinea pig serum.
A: The following prewarmed reactants were mixed: 2 ml guinea pig serum, 2 ml
LPS-R (500 /ug/ml), 2 ml SRBC (5 x 10^8/ml). B: Serum, buffer, SRBC. At various
time intervals samples were taken and centrifuged. The degree of hemolysis was
determined by measuring the supernatant at 412 nm, while the cell sediment was
further incubated for 60 minutes at 37°C after resuspension in a solution containing
C5-C9 followed again by Hb-determination. The values shown in the figure represent the sum of both measurements corrected for the serum- and cell-control.

Some Aspects of the Effects of Endotoxin on the Central Nervous System

F. W. Schmahl

Septic shock is mainly caused by endotoxins of gram-negative bacteria. In its terminal stages disturbances of the central nervous system are observed, and the mortality rate is very high, even after treatment with modern antibiotics. Clinical reviews give a mortality rate of up to 80 % (1, 26, 28). Septic shock states are mainly caused by the following gram-negative bacteria: Proteus, Escherichia coli, Pseudomonas, Klebsiella, Salmonella, Neisseria, and Aerobacter aerogenes. In view of the high mortality rate, experimental studies on the pathophysiology and therapeutic management of septic shock are of considerable clinical interest.

Based on the findings of Penner and Klein (19) and others it is suggested that endotoxin acts directly on the central nervous system and that the secondary symptoms of endotoxin shock are entirely or partially triggered by this action (cf. 5, 12).

Penner and Klein, however, used the toxin of Shigella dysenteriae and studied its effect in dogs with crossed cerebral circulation, i. e. the brain of one dog received its blood supply from the circulation of a "donor" dog. Effects of the Shigella toxin were only seen in those animals whose brains were entered by the toxin through the crossed circulation. Aside from questions about the technical reliability with regard to separation of the crossed circulations in these experiments, the studies with Shigella toxin cannot be conclusive for "endotoxins of gram-negative bacteria".

In contrast to Escherichia coli and other gram-negative bacteria, Shigella dysenteriae produces an additional exotoxin having neurotoxic properties (cf. 4, 10, 27). In their studies Penner and Klein did not differentiate between the endotoxin and the neurotoxin of Shigella dysenteriae; accordingly, no conclusions can be drawn as to the endotoxins of other gram-negative bacteria.

This led us to re-investigate the effects of endotoxin on the central nervous system. Within the scope of this brief review only a few aspects of the effects of endotoxin on the central nervous system can be presented.

Considerable differences as to structure, vascularisation, blood flow, oxygen consumption and other metabolic parameters exist between the various parts of the brain. We started our experiments with investigating local tissue blood flow, oxygen supply and parameters of energy-supplying metabolism in tissue specimens of the cerebral cortex. Previous studies together with Lübbers and co-workers have shown that cortical oxygen consumption is almost five times as high as that of the white matter (15).

Our studies were performed in cats anesthetized with pentobarbital. Surgical approach, techniques for the registration of blood pressure and measurement of local cortical blood flow and cortical oxygen pressures, the freeze-stop technique for obtaining cortical tissue for biochemical analyses at the end of the experiment, and the methods for the excision and fixation of tissue for microscopic and electron microscopic studies have been described previously (2, 20, 22, 23). Sixty-five cats

Supported by Deutsche Forschungsgemeinschaft

were injected i. v. with 1 mg/kg endotoxin from Escherichia coli 055*.

Fig. 1 gives a characteristic example of mean arterial blood pressure, local cortical blood flow, and cortical oxygen pressure after intravenous injection of 1 mg/kg endotoxin. A transitory drop of mean arterial blood pressure is seen a few minutes after the injection, corresponding to the "early phase" of endotoxin shock. After the initial drop, blood pressure rises again to almost the initial value. This was observed in the majority of the animals. A gradual, progressive fall in blood pressure sets in approximately two hours after the injection, corresponding to the "late phase" of endotoxin shock.

Fig. 1 also illustrates that local cortical blood flow declines progressively in the late phase of endotoxin shock if mean arterial blood pressure has dropped below 70 mm Hg, the critical range for the autoregulation of cerebral blood flow. Cortical oxygen pressure, continuously recorded by means of a platinum multi-wire electrode at one point of the cortical surface, drops from an initial value of 40 mm Hg to 8 mm Hg in the experiment illustrated in Fig. 1.

More detailed information about the changes of oxygen supply in the cerebral cortex during endotoxin shock is obtained by measuring oxygen pressure fields, i. e. by measuring oxygen pressure at various points in a defined area of the cortex (gyrus suprasylvicus and marginalis) before the injection of endotoxin and at the end of the experiment.

Table I gives a characteristic example of measuring cortical oxygen pressure at 11 points of the gyrus suprasylvicus resp. marginalis with the platinum multi-wire electrode** before and 10 hours after the intravenous injection of 1 mg/kg endotoxin. The arithmetic mean of 11 measurements was 34. 8 mm Hg at the start of the experiment and 2. 6 mm Hg (7 measurements) at the end, 9 hours and 57 minutes after the injection of endotoxin.

In the experiment presented in Table I, mean arterial blood pressure dropped from an initial value of 102 mm Hg to 37 mm Hg at the end of the experiment.

In our example, measurements with the platinum electrode (which has 10 single wires) showed cortical oxygen pressures between 1 and 4 mm Hg at the end of the experiment. Keeping in mind that these pressures represent the mean value of local cortical PO_2 at the points of the cortex touched by the 10 wires of the electrode, it can be assumed that the critical mitochondrial oxygen pressure of approximately 1 mm Hg is not reached in some of the cortical cells (13, 14, 16, 20).

These considerably reduced cortical oxygen pressures are reflected in the results of the biochemical analyses which were performed immediately after the last readings of cortical oxygen pressures had been taken. The tissue specimens for these analyses were excised from the same area of the cerebral cortex, using the freeze-stop technique described previously. Table II shows the biochemical data in comparison to the metabolite analyses of 11 control animals which had been subjected to the same conditions of testing and anesthesia without receiving endotoxin. A decreased content of the energy-rich phosphates, creatinine phosphate (CrP) and ATP can be recognized as well as a decrease of the corresponding phosphorylation degrees of CrP/Cr and ATP/ADP. Also due to insufficient oxygen supply, glycolysis is considerably increased as evidenced by the considerably increased lactate levels and the lactate/pyruvate (Lac/Pyr) ratio, an indicator of the redox-state of the extramitochondrial nicotinamide adenine dinucleotide/reduced nicotinamide adenine dinucleotide (NADH/NAD) system.

---

* Endotoxin from Escherichia coli 055 had been extracted according to Boivin's method. The endotoxin was extracted and kindly provided by Professor Urbaschek, Institute for Hygiene and Medical Microbiology, Klinikum Mannheim, University of Heidelberg.

** 10 platinum wires, each wire having a diameter of 15 µg; for details see 20, 23.

Likewise, a markedly reduced phosphorylation degree of the CrP/Cr and ATP/ADP systems as well as a marked increase of the lactate levels and the redox quotient Lac/Pyr was observed in all of the animals whose mean arterial blood pressure had dropped below the critical range for the autoregulation of cerebral blood flow in the late phase of endotoxin shock, thus resulting in considerably decreased cortical blood flow and cortical oxygen pressures. For all parameters determined, the changes were significant in comparison to the control group (20, 21).

Sharp drop in cortical oxygen pressure in the late phase of endotoxin shock, leading to a break-down of the energetic potential of the cerebral cortex - as in the experiment shown in Table I - determines the fate of the animals. (Some of them died so quickly that the in vivo excision of tissue for biochemical analysis was impossible).

In contrast to these findings, no significant decreases of cortical CrP and ATP content and of the corresponding phosphorylation degrees of CrP/Cr and ATP/ADP nor significant increases of the Lac/Pyr quotient were seen in those animals the mean arterial blood pressure of which did not drop below 70 mm Hg (the critical range for the autoregulation of cerebral blood flow) within a period of up to 10 hours after the i. v. injection of 1 mg/kg endotoxin. (For details cf. 20, 21). Measurement of cortical oxygen pressures in these animals did not reveal any clear-cut shift of PO_2 distribution to lower values.

From these results we concluded that the break-down of the energetic potential of the cerebral cortex, which we observed in many of the animals in the terminal phase of endotoxin shock, is not due to the direct action of endotoxin on the cortical tissue but the consequence of insufficient blood flow and insufficient oxygen supply (20, 21).

These results are in agreement with microsopic and electron microscopic studies[*] on the cerebral cortex after the intravenous injection of endotoxin. During our experiments tissue specimens for these morphological studies were excised from the gyrus suprasylvicus resp. marginalis at various times after the injection.

Considering the high molecular weight of endotoxin[**] it probably does not cross the blood-brain-barrier. On the other hand, endotoxin is known to damage the endothelium or cause the separation of single endothelial cells in other areas of the vascular system (9, 17). Therefore it might well be that endotoxin damages the blood-brain-barrier and penetrates into the brain after this damage or destruction has occurred.

However, the morphological studies in our experiments did not give any evidence for a structural damage of the substrates of the blood-brain-barrier (endothelial cells, basal membranes, perivascular glia) by endotoxin. The endothelial cells merely showed so-called "reactive changes", in particular an increased number of organelles. These findings were interpreted as signs for an active response of the endothelial cells to the influence of endotoxin (20). Fig. 2, an electron micrograph, gives a typical example of the reactive endothelial changes found in cerebral vessels (precapillary from the cerebral cortex) after intravenous injection of endotoxin.

The reaction of the endothelial cells of the cerebral cortex, which differs from that in other areas of the vascular system, is understandable considering their morphological and functional characteristics. The endothelium of the brain is an integral component of the blood-brain-barrier (7, 11).

Thus the morphological findings give no evidence that the blood-brain-barrier is damaged or destroyed by endotoxin. This, however, would have to be the case for the direct action of endotoxin on the cortical tissue. Our studies are supported by the findings of Braude (3), who after intravenous injection of Cr^{51}-labelled endotoxin

---

[*] The morphological studies were performed by Professor W. Schlote, Dept. for Submicroscopic Pathology and Neuropathology, University of Tübingen.
[**] For this high-molecular lipopolysaccharide which tends to form aggregates, molecular weights of approx. 10^6 to 20×10^6 are mentioned (6, 18, 25).

did not observe any radioactivity in cerebral tissue.

The described studies of physiological and biochemical parameters as well as the morphological findings refer exclusively to the cerebral cortex. Accordingly, we cannot answer the question whether intravenously injected endotoxin triggers reactions by direct influence in areas of the hypothalamus, where the "blood-brain-barrier" has other characteristics of permeability and a different morphological structure.

Supplementary to the series of experiments described above, we started experimental studies on the effect of intravenously injected endotoxin on the energy-providing metabolism of the cerebral cortex in co-operation with Reinhard, Betz, and Schlote. Bypassing the blood-brain-barrier, 200 μg/kg endotoxin were injected into the cisterna cerebellomedullaris of cats, i. e. directly into the space of the cerebrospinal fluid. In this series CrP content and the phosphorylation degree of the CrP/Cr system already fell significantly when mean arterial blood pressure had not dropped below 70 mm Hg, the critical range for the autoregulation of cerebral blood flow, within 10 hours after the injection of endotoxin. Lac content and Lac/Pyr quotient were significantly elevated.

It is possible that endotoxin, if administered intracisternally, has direct noxious effects on the tissue of the cerebral cortex. Further experimental studies are in progress to elucidate this question, including measurements of cortical oxygen pressure distribution in correlation to biochemical analyses, and histologic and electron microscopic investigations.

Within the scope of these short and necessarily incomplete comments on endotoxin induced disturbances of the nervous system, it should be pointed out that the activation of the sympathetic nervous system is of great importance for the course of endotoxin shock. As in other forms of shock, increased activity of the sympathetic system also occurs in endotoxin shock, mainly via pressor receptors in the aortic arch and the carotid sinus as blood pressure drops. It should be kept in mind that the sympathetic system is represented in the brain by central nuclear areas in the hypothalamus, the medulla oblongata, and the cerebrum. Sympathetic activation causes the increased release of catecholamines at the peripheral nerve ends of the sympathetic system and from the adrenal medulla, which is innervated by the sympatics.

In the late phase of endotoxin shock as well as in shock states of other etiology, excessive activation of the sympathetico-adrenal system can have negative effects on the organism, since peripheral circulation, already insufficient, is further reduced. In experimental studies, phenoxybenzamine (Dibenzylin[R]), an alpha receptor blocker, has proved to be effective in lowering sympathetic tone during endotoxin shock (24). Hardaway and other authors have used this pharmacon therapeutically in patients with various kinds of shock, but only under strict consideration of the respective pathophysiological situation and not until a possible hypovolemia had been compensated (review cf. 8). Since peripheral circulation and oxygen uptake are of decisive importance in the prognosis of shock, further studies on the endotoxin induced activation of the sympathetic nervous system and its pathophysiological significance and treatment are urgently required.

References

1. Berk, J. L., Hagen, J. F., Dunn, J. M. : Surg. Gynec. Obstet. 130 : 1025 (1970).
2. Betz, E., Ingvar, D. H., Lassen, N. A., Schmahl, F. W. : Acta physiol. scand. 67 : 1 (1966).
3. Braude, A. I. : Absorption, distribution, and elimination of endotoxins and their derivatives. In: Landy, M., Braun, W. : Bacterial endotoxins, p. 98 (Rutgers, New Brunswick, N. J. 1964).

4. Davis, B. D., Dulbecco, R., Eisen, H. N., Ginsberg, H. S., Wood, W. B. : Microbiology. (Harper and Row, New York/Evanston/London 1969).

5. Göing, H : Arb. Paul-Ehrlich-Inst., Frankfurt 57 : 80 (1962).

6. Göing, H., Kaiser, P. : Ergebn. Mikrobiol. 39 : 243 (1966).

7. Hager, H. : Acta neuropath. 1 : 9 (1961).

8. Hardaway, R. M. : Syndromes of disseminated intravascular coagulation. With special reference to shock and hemorrhage. (Charles C. Thomas, Springfield, Ill. 1966).

9. Hoff, H. F., Gottlob, R., Blümel, G. : Naturwissenschaften 54 : 287 (1967).

10. Jawetz, E., Melnick, J. L., Adelberg, E. A. : Medizinische Mikrobiologie. (Springer, Berlin/Heidelberg/New York 1968).

11. Karnovsky, M. J : J. Cell Biol. 35 : 213 (1967).

12. Kroneberg, G., Sandritter, W. : Z. ges. exp. Med. 120 : 329 (1953).

13. Lübbers, D W. : The oxygen pressure field of the brain and its significance for the normal and critical oxygen supply of the brain. In : Lübbers, D. W., Luft, U. C., Thews, G., Witzleb, E. : Oxygen transport in blood and tissue, p. 124 (Thieme, Stuttgart 1968).

14. Lübbers, D. W. : Intercapillärer O_2-Transport und intracelluläre Sauerstoffkonzentration. In : Hess, B., Staudinger, Hj. : Biochemie des Sauerstoffs, p. 67 (Springer, Berlin/Heidelberg/New York 1968).

15. Lübbers, D. W., Ingvar, D., Betz, E., Fabel, H., Kessler, M., Schmahl, F. W. : Pflügers Arch. ges. Physiol. 281 : 58 (1964).

16. Lübbers, D. W., Kessler, M. : Oxygen supply and rate of tissue respiration. In : Lübbers, D. W., Luft, U. C., Thews, G., Witzleb, E. : Oxygen transport in blood and tissue, p. 90 (Thieme, Stuttgart 1968).

17. McGrath, J. M., Stewart, G. J. : J. exp. Med. 129 : 833 (1969).

18. Nowotny, A. : Bact. Rev. 33 : 72 (1952).

19. Penner, A., Klein, S. H. : J. exp. Med. 96 : 59 (1952).

20. Schmahl, F. W. : Regionale Durchblutung und Metabolite des energieliefernden Stoffwechsels in umschriebenen Gewebsarealen - insbesondere des Myokards und der Hirnrinde - unter verschiedenen pathophysiologischen Bedingungen. Habilitationsschrift, Gießen 1972.

21. Schmahl, F. W. : Effects of endotoxin shock on the oxygen supply and the levels of energy-rich phosphates of the cerebral cortex. In : Kessler, M., Bruley, D. F., Clark, L. C., Lübbers, D. W., Silver, I. A., Strauss, J. : Oxygen supply. Theoretical and practical aspects of oxygen supply and microcirculation of tissue, p. 256 (Urban und Schwarzenberg, München/Berlin/Wien 1973).

22. Schmahl, F. W., Betz, E., Talke, H., Hohorst, H. J. : Biochem. Z. 342 : 518 (1965).

23. Schmahl, F. W., Betz, E., Dettinger, E., Hohorst, H. J. : Pflügers Arch. ges. Physiol. 292 : 46 (1966).

24. Schmahl, F. W., Ohlemutz, A., Huth, K. : Verh. dtsch. Ges. inn. Med. 75 : 900 (1969).

25. Schramm, G., Westphal, O., Lüderitz, O. : Z. Naturforsch 7 b : 594 (1952).

26. Shubin, H. : Hemodynamic, respiratory and metabolic changes in bacterial shock. In this volume.

27. Weil, M. H., MacLean, L. D., Spink, W. W., Visscher, M. B. : J. Lab. clin. Med. 48 : 661 (1956).

28. Weil, M. H., Shubin, H., Biddle, M : Ann. intern. Med. 60 : 384 (1964).

Table I.

Measurement of Cortical Oxygen Pressures.

| Before Injection of Endotoxin | 9 hrs 52 min to 9 hrs 57 min after Endotoxin (1 mg/kg i.v.) |

	mm Hg		mm Hg
1.	36	1.	2
2.	49	2.	4
3.	20	3.	2
4.	33	4.	1
5.	27	5.	4
6.	24	6.	2
7.	17	7.	3
8.	35		
9.	47		
10.	54		
11.	41		

Mean Value 34.8 mm Hg   Mean Value 2.6 mm Hg

Table II.

Metabolite Contents in the Cerebral Cortex (10^{-6} Mol/g fresh weight) Tissue Excision 9 hrs 58 min after Endotoxin, 1 mg/kg i.v.

CrP	1.19	(3.90 ± 0.14)*
ATP	1.01	(1.72 ± 0.06)
Lac	9.7	(1.22 ± 0.10)
CrP/Cr	0.15	(0.54 ± 0.02)
ATP/ADP	2.3	(4.4 ± 0.2)
Lac/Pyr	47.5	(15.2 ± 1.1)

* The values in parentheses represent the metabolite analyses of 11 cats (control group) who did not receive endotoxin but were otherwise exposed to the same experimental conditions (Means and standard errors of means)

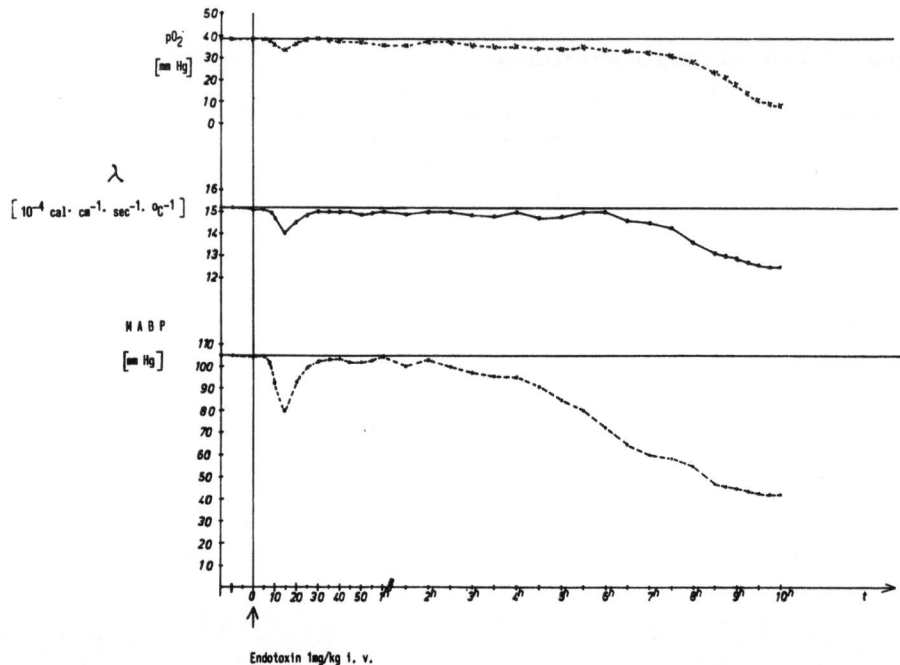

Fig. 1. Registration of mean arterial blood pressure, local cortical blood flow and cortical PO$_2$ in a cat after i.v. injection of 1 mg/kg endotoxin. Local cortical blood flow is measured in terms of heat clearance λ (2). Cortical PO$_2$ is measured by a 10-wire platinum electrode.

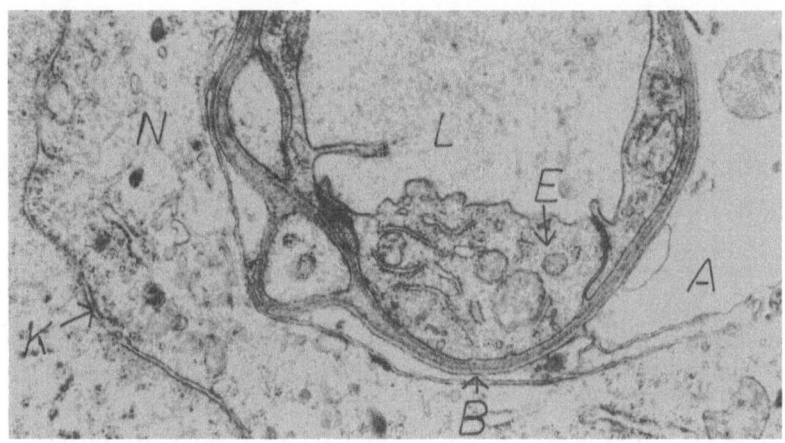

Fig. 2. Reactive endothelial changes in a cortical precapillary after endotoxin, 1 mg/kg i. v. intact endothelial plasma membrane. Increased granular endoplasmatic cisterns in the endothelial cytoplasm. Swollen perivascular astrocyte process. Nerve cell plasma unremarkable. Tissue excised 630 minutes after injection. Magnification 24,500 : 1. (Specimen from joint experiments with W. Schlote, figure from (2)).

A = perivascular astrocyte process   B = basal membrane
E = endothelial cell plasma   K = nuclear membrane
L = vascular lumen   N = plasma body of a nerve cell

Fixation by immersion 5 seconds after excision in 2 % glutaraldehyde (Polysciences, New York) in 0.2 m cacodylate buffer (2 hrs). Postfixation in 1 % O_sO_4 in the same buffer (1 hr). Embedding in araldite (Ciba).

Discussion

GIERTZ

SCHAUER

URBASCHEK

Müller-Berghaus: When did the first fibrin fibres appear? Perhaps, one should differentiate between platelet aggregation and fibrin formation.

Urbaschek: I agree. In the initial phase reversible platelet aggregates are formed. As soon as irreversible platelet aggregations are observed, dose-dependent fibrin formation occurs, approximately 50 minutes after i. v. endotoxin injection. I should add that endotoxoid induces tolerance to endotoxin in miniature pigs as well as in guinea pigs. We studied 25 miniature pigs. Together with Huth, Frankfurt, we examined the blood coagulation of these pigs.

Neuhof: Dr. Urbaschek, you stated that vasoactive substances are released in the early phase of endotoxic shock. What impresses me the most is that the dose of endotoxin appears to influence the onset of hemodynamic changes as well as the time when vasoactive substances are released. We worked with the endotoxin which you provided, the LD_{50} being 100 μg/kg. When one injects this amounts intravenously the main reaction, as I have shown, occurs after 16 ± 3 minutes; that is, degranulation of mast cells. Hillebrand, using the same intravenous dose in rabbits, observed hemodynamic effects in the same time period. Measuring the pulmonary pressure, which appears to be the most sensitive parameter, we found after 2 to 5 minutes hemodynamic changes in approximately 50% of the animals. When one increases the dose, the main reaction occurs at 1 to 3 minutes rather than after 16 minutes. One seldom sees a second reaction after 20 minutes. This observation relates to the mode by which these vasoactive substances are liberated. With higher doses, degranulation of the mast cells occurs after 1 to 3 minutes. The hemodynamic reaction does not simply appear to shift from one point in time to another. Rather, one observes either a very early or late reaction or both.

Mergenhagen: Does endotoxin per se cause contraction of smooth muscle?

Urbaschek: There are some reports in the literature that endotoxin per se causes contraction of smooth muscles. In all of our experiments we did not note contraction of smooth muscle per se. We also could not detect histamine release from the isolated smooth muscle, measuring the histamine content of the bath fluid after addition of endotoxin.

Mergenhagen: This may be so, because all of these effects in vitro, such as the contraction of the isolated smooth muscle of the guinea pig ileum with endotoxin, require serum. The components that are contributed in the serum are the complement, the anaphylactic toxin fragment, and C5 or C3. The degranulation of mast cells in vitro with the release of heparin and histamine depends on activation of the complement system by endotoxin. So, all these effects that are mediated and observed on mast cells and smooth muscles and, as Dr. Sandberg will show on platelets, can be explained on the basis of activated complement.

Wissler: I would like first to comment on Dr. Mergenhagen's remark regarding anaphylatoxin action. Fresh, washed thrombocytes are not influenced by contact with serum activated by either endotoxin or yeast cell walls. Moreover, thrombocytes are not affected by anaphylatoxin in the crystallized state. In our opinion alteration of thrombocytes in vivo might be due to the change in the membranes of the vesicles and not through a mediator action. There is another possible explanation: It is probable that the change in the vesicles may be due to hydrophobic action and thus the thrombocytes may aggregate, but anaphylatoxin per se does not affect thrombocytes. Another point: When you initiate shock with anaphylatoxin in the guinea pig, there is no thrombocytopenia, but with cruder preparations this does occur. We still do not know whether this latter effect depends on a series of mediators. In contrast, in the in vitro system with washed thrombocytes, they are not affected. Probably it is only a physical effect on the membranes of the vesicles and not activation by a mediator.

Bitter-Suermann: I would like to comment on Dr. Wissler's remarks. It has never been postulated that C3a or C5a have a direct effect on thrombocytes. The effect of complement and activation of complement on that cell-type is either possible via C3b and the immuneadherence receptor of thrombocytes. This cannot be tested with your activated serum, because in this case it will be decayed. The other possibility is via the activation of C6 and the possible interaction of the 6th component with platelets. Those are the two possibilities of interaction, which, I believe, should be discussed, but not the interaction of C3a and C5a with platelets.

Wissler: Certainly, I hope I was not misunderstood. I do not believe that anybody has stated this. Since anaphylatoxin is very active - for instance it liberates histamine - one might be led falsely to the idea that it also affects thrombocytes. This is not the case for anaphylatoxin per se. It was only an understatement of the effect of anaphylatoxin.

Koslowski: I would like to ask Dr. Schauer about the effects of kinin. It has been postulated that the kininogen reservoir in the organism is not sufficient to produce kinins in amounts necessary for eliciting systemic effects. Does this theory still prevail?

Schauer: No, but the perfusion and the formation is decisive. Most of the measurements have been done in a very short time and there do not exist long-lasting measurements, as far as I know. The capacity in itself would not alone be adequate in order that it be really decisive.

Beller: What is the significance of the two-phase platelet aggregation in regard to shock? At least, we can say that platelet aggregation is presumably of very minor consequence for the development of the Shwartzman phenomenon.

Urbaschek: In regard to the significance of the platelets it should be differentiated between the early drop in platelets and the later formation of real microthrombi. The aggregation of platelets is biphasic: First reversible and only later dose-dependent irreversible platelet aggregations are formed, then the release reactions from the damaged and aggregated platelets begin with all the known consequences. Years ago, Hardaway has indicated that appearance and extend of microthrombi are indicators of the outcome of endotoxemia. Dr. Lüscher, who is an expert in this area, will deal with this tomorrow.

SANDBERG

Alexander: I would like to know in this very beautifully done study, whether you have gone on and looked at possible deficiencies of C8 or C9, since these seem to be more membrane-active materials.

Sandberg: There are no known genetic deficiencies of the later complement components. However, we feel that the reaction that we are looking at is due to platelet lysis, and our evidence for this is a release of lactic dehydrogenase which occurs simultaneously with histamine or serotonin or factor 3 release. Most lytic reactions probably require all nine components, but we have not done this work.

McCabe: I would like to offer some clinical support for these pretty demonstrations of the role of complement in infections with gram-negative organisms. We have previously studied C3 levels in patients with bacteremia with gram-negative bacilli. We found no decrease in the ordinary patient, but a significant decrease in patients who subsequently developed shock or died. These studies have been extended, using the same material, by Dr. Peter Schur and in Dr. Austin's laboratory. The interesting facet is that there is no decrease in the components 1, 4 and 2. This is primarily involvement of the alternate system in these patients despite the presence of antibodies.

Mergenhagen: Dr. Leddy at the University of Rochester has a patient who is completely deficient in the 6th component of complement and, according to a recent report given at the Complement Workshop, this patient shows no abnormal coagulation in his activated plasma. Am I correct?

Sandberg: That is true and I think we have to keep in mind the species we are discussing. I think there might be extreme differences in the platelets among the different species.

Springer: In the alternate pathway of complement do you need component C3?

Sandberg: Yes.

Springer: Does the alternate pathway of complement work without any fragment of any antibody whatsoever? That means without any gammaglobulin or its fragments? You mentioned at the beginning some fragments, can you do away with these?

Sandberg: This is a question which we are currently investigating. The role of antibody in activating the alternate pathway really has not been clearly resolved. I, at this point, would tend to say, it very well may be involved in some of our studies involving specificity in activation of serum that has been totally depleted of immunoglobulin by insoluble immunoabsorbants and then restored with antibody preparations. The question, however, is not answered yet.

Springer: There was a long conflict about properdin a decade ago. What is the difference between aggregated IgM and properdin and how exactly do you define properdin?

Sandberg: Properdin is defined as an euglobulin which carries none of the antigenic determinants of G, M, A, or D. It has a molecular weight of about 220,000. There are different magnesium- or ion-requirements for activation of properdin than for the classical complement pathway.

Bitter-Suermann: One example that antibodies do not participate in the alternate pathway is the cobra venom factor-induced C3 activation which is most clearly defined, because it is a shortcut of the whole alternate pathway and only C3 proactivator and factor D are needed, to have an active enzyme which cleaves C3 without addition of any other component. This is a substantial part of the alternate pathway and no properdin is needed for it.

Sandberg: However, that requires an additional protein for expression of lysis. I think there may be great differences in the two pathways.

Bitter-Suermann: But not for the activation of C3. There you have no additional component.

Sandberg: No, except that it is not properdin-requiring and I think it is a different pathway with different entry into the complement system.

Bitter-Suermann: Yes, but it is a full activation of C3 with all consequences.

Springer: That indicates that properdin is an enzyme. This enzyme function can be taken up instead of properdin by the enzyme which you mentioned from the cobra venom. Is this correct?

Bitter-Suermann: Yes, that is correct. Can cobra venom factor induce the same effects as endotoxin in your system, Dr. Sandberg?

Sandberg: We have not been successful in getting platelet lysis by cobra venom factor. There have been reports from Pickering's laboratory that cobra venom factor will aggregate platelets. We have not observed this.

Bitter-Suermann: Is the lysis of platelets in your system mediated by a method like the reactive lysis of Lachman also, that is, lysis mediated by the C5, 6, and 7? What do you think about this mechanism of lysis?

Sandberg: I do not know what the mechanism is at this point. It is a lytic reaction and it involves the factor at least through C6. Beyond that I cannot go.

Beller: Can you explain the observation of Nagayama and co-workers and myself a couple of years ago, namely, differences in platelet factor 3 and serotonin release with different endotoxins? We worked with relatively pure substances which were provided by Westphal and by Nowotny.

Sandberg: No, I do not. We have just used the one endotoxin preparation discussed here, which, if looking at the data, was not as active as some of our immune aggregates. I do not know whether there are differences between the endotoxins and their ability to be entrapped in the matrix surrounding the platelet and hence differing in their ability to activate complement at the platelet membrane.

Müller-Berghaus: How do you explain the difference between your results and the results of Müller-Eberhard? He observed, as far as I know, release of platelet factor 3 and serotonin from thrombocytes of rabbits with C6 deficiency.

Sandberg: To my knowledge, he did not look at factor 3. He has postulated that factor VII and VIII become entrapped on the surface. You get a change in the membrane surface with deletion of at least two proteins from the platelet membrane.

BITTER-SUERMANN

Hadding: I have a comment on the influence of LPS on activated C1. In contrast to the effect of LPS on C3 in serum, the activated first component is strongly inhibited by various LPS-preparations. As was shown by our colleague Dr. Loos, LPS interacts with the subcomponent C1q and this interferes with the binding capacity of macromolecular C1.

Wissler: Dr. Bitter-Suermann, how active was your purified C3?

Bitter-Suermann: In the first slide you have seen the C3 consumption expressed in terms of site-forming units, and in this system we had 1×10^{11} site-forming units of C3 and the amounts of the latter were diminished by over 95%.

SCHMAHL

Drommer, Hannover/Germany: We have also studied the central nervous system of the cat and I can confirm your findings on capillary endothelium. In pigs we also did not find defects with electronmicroscopy but anabolic processes indicating the activation of endothelial cells. Furthermore, we observed damage of the endothelial cells of arterioles including edema and destruction of the media. We were able to prove that high molecular ferridin was capable of passing through the damaged endothelial cells.

Schmahl: We have compared our findings with yours. How is the E. coli toxin you used defined?

Drommer: It is a mixture of endotoxin and neurotoxin which was proved by Schimmelpfennig in Göttingen. The neurotoxin is the component responsible for the prolongation of shock, which we intended to induce.

Schmahl: With regard to the capillaries your results agree with ours. The discrepancies in the results observed in the arterioles might be explained by the different preparations, since we used endotoxin, while you used a preparation with a neurotoxic component.

Schauer: You referred to the dehiscence of the endothelial cells and you did not mention the ganglion cells. This is a problem which is difficult for neuropathologists to evaluate. Do studies exist on the depletion of serotonin in the brain, since there are relatively large amounts of serotonin present in certain areas of the brain? Following intravenous injection of endotoxin we observed swelling of the ganglion cells and also of the neuron after prolonged manifestation of hypoxia together with a marked decrease in pO2. After a long period of time, during which the pO2 remained constant, a decrease occurred. The evaluation of the morphology following the intercisternal administration has to be worked out especially by means of electronmicroscopy.

Raynaud: Who has prepared the E. coli endotoxin and do you have any idea of the mode of the preparation?

Schmahl: Dr. Urbaschek has prepared the endotoxin.

Urbaschek: The endotoxin was prepared according to Boivin's trichloroacetic acid method.

Chapter VII

Blood Coagulation

Consumption Coagulopathy in the Course of Endotoxinemia

G. Müller-Berghaus and H. G. Lasch

The exposure of the circulating blood to bacterial endotoxin, be it in a patient or in an animal, results in characteristic changes in the coagulation and the fibrinolytic enzyme systems. The reaction of the body to endotoxin is similar in its appearance in different species but may considerably differ in its quantitative response and sequence of events.

In the acute reaction of the coagulation system to endotoxin, a decrease in platelet counts (30) and impaired functioning of these blood cells are observed (16, 22). All the plasmatic coagulation factors as well as antithrombin III may be reduced in their activities (17, 23, 26, 30). Similar changes are observed in the fibrinolytic enzyme system, as decreases in plasminogen levels and plasminogen activator activities (11, 31, 42, 43). Activation products of intravascular coagulation can be demonstrated with specific tests, e. g. soluble fibrin (fibrin monomer complexes) and fibrinopeptide A (8, 13, 40, 41). The split products produced by fibrinolysis activation are a characteristic finding in patients with endotoxinemia or in animals after intravenous injection of endotoxin (1, 32). These changes in the hemostatic system are correlated with a hemorrhagic diathesis. Autopsy may, but does not necessarily have to, reveal occlusions of small vessels with fibrin-rich microclots as well as hemorrhages and ischemic necroses (18, 29).

The occurrence of microclots on the one hand and the observation of a hemorrhagic diathesis on the other hand are commonly considered diametrically opposed phenomena. However, a hemorrhagic diathesis may be the paradoxical result of generalized or locally excessive intravascular blood coagulation. The deduction that a hemorrhagic diathesis may be the consequence of intravascular coagulation explains the morphological alterations and the altered coagulation data observed in patients with endotoxinemia. For this event of intravascular coagulation with consecutive hemorrhagic diathesis, Selye (45) coined the term "thrombohemorrhagic phenomenon" in order to describe the simultaneousness of coagulation and hemorrhage. McKay (29) and Hardaway (12) introduced the term "disseminated intravascular coagulation". Lasch et al. (23) described the same process as "Verbrauchskoagulopathie" (consumption coagulopathy) since the breakdown of the hemostatic mechanism in the course of this phenomenon is caused by the in vivo-consumption of blood coagulation factors and platelets.

Definition

Consumption coagulopathy begins with the entry of procoagulant material or activity into the circulating blood; progresses to the stage of free thrombin activity within the circulation, causing fibrin monomers and platelet aggregation; is associated with activation of the fibrinolysis system causing fibrin and/or fibrinogen split products to develop; and is not terminated until the hemostatic mechanism and the vasomotor apparatus have returned to normal. Consumption coagulopathy may or may not cause occlusions of arterioles, capillaries and venules of various organs by fibrin-rich microclots. Consequently, the coagulation process may or may not impair the functioning of these organs.

Already the description of the dynamic course of consumption coagulopathy points to the complexity of mechanisms decisive in initiating intravascular coagula-

Supported by the Deutsche Forschungsgemeinschaft, Bad Godesberg, Germany

tion. Two questions have to be answered in order to understand the mechanisms involved in consumption coagulopathy induced by endotoxinemia: 1. How is the coagulation system triggered by endotoxin? 2. Which additional mechanisms are necessary to determine the extent of fibrin formation and the localization of microclots into certain capillary areas?

Trigger Mechanisms of Intravascular Coagulation Induced by Endotoxin

Direct activation of the coagulation system by endotoxin does not play any major role, since amounts of endotoxin highly active in vivo only cause minor changes in the coagulation system in vitro. More likely, indirect activation of the coagulation system is mediated by one or several additional factors. Several mediators have been experimentally investigated: 1. Platelets (4, 27, 33) 2. Leukocytes (14, 15, 25, 39) 3. Endothelial cells (6, 7, 28, 36) 4. Complement system (2, 5, 47) 5. Hageman factor (26, 38). An evaluation of these mediators will be presented in the following papers.

Mechanisms for Localizing Fibrin in Endotoxinemia

Two steps are necessary for the generation of capillary microclots (34): First, the coagulation system has to be activated to form fibrin monomer which will remain in solution by forming soluble complexes. These complexes can be removed from the circulation without harm to the healthy body. Secondly, soluble fibrin may be precipitated, thus forming microclots. This second phase has to be assumed to explain the local occurrence of microclots. The generation of soluble fibrin ubiquitously in the circulation on the one hand and the absence of primarily parietal fibrin thrombi on the other hand suggest the generalized activation of intravascular coagulation in the circulation. Polymerization and/or precipitation of soluble fibrin may occur if the concentration of fibrin monomers exceeds a critical concentration. The infusion of fibrin monomer into heparinized rabbits, i. e. with inhibited thrombin formation, can generate microclots in the glomerular capillaries of the kidneys (37). Polymerization and/or precipitation of soluble fibrin can also be induced by endotoxin itself, in which case anticoagulation with heparin does not prevent the process of precipitation (34, 35): Fibrinemia is induced in rabbits by infusion of ancrod, the procoagulant fraction of the Malayan pit viper venom. Infusion of ancrod decreases the fibrinogen levels below 50 mg/100 ml, but microclots do not occur. If in addition to ancrod infusion a sublethal dose of endotoxin is intravenously administered, microclots generate in the glomerular capillaries as they are typical of the generalized Shwartzman reaction. Since heparin treatment does not prevent the occurrence of microclots after ancrod and endotoxin administration, endotoxin has to cause precipitation of soluble fibrin independent of the coagulation system. This effect of endotoxin is different from its procoagulant activity.

The second phase in the course of consumption coagulopathy induced by endotoxin may very well fail to appear, and especially in the patient, since the occurrence of microclots in patients is a relatively rare event in comparison to cases with diagnosed consumption coagulopathy (19).

The Coagulation Balance and the Shift of the Dynamic Equilibrium Toward Consumption Coagulopathy

The occurrence of microclots in the peripheral circulation is dependent on control mechanisms of coagulation apart from the extent of activation of intravascular coagulation. These control mechanisms are: 1. Inhibitors of coagulation as antithrombins (10, 20, 46); 2. Clearance function of the reticulo-endothelial system (9, 24); 3. Fibrinolytic enzyme system (3, 11, 44); and 4. The circulation which is decisive in the transport of activated coagulation factors to the clearing organs (18, 21). The well-functioning of the control mechanisms is important for the patient's survival, whereas treatment of the hemorrhagic diathesis itself is generally quite successful.

This short summary of events in endotoxin-induced consumption coagulopathy demonstrates the complexity of reactions leading to the accelerated turnover of coagulation components. The following papers by clinicians as well as by investigators will present and elucidate the specific problems in greater detail.

References

1. Beller, F. K., Theiss, W. : Thromb. Diath. haemorrh. 29 : 363 (1973).
2. Brown, D. L., Lachmann, P. J. : Int. Arch. Allergy 45 : 193 (1973).
3. Condie, R. M., Hong, C. Y., Good, R. A. : J. Lab. clin. Med. 50 : 803 (1957).
4. Evensen, S. A., Jeremic, M. : Brit. J. Haemat. 19 : 33 (1970).
5. Fong, J. S. C., Good, R. A. : J. exp. Med. 134 : 642 (1971).
6. Gaynor, E. : Blood 41 : 797 (1973).
7. Gaynor, E., Bouvier, C., Spaet, T. H. : Science 170 : 986 (1970).
8. Godal, H. C., Abildgaard, U., Kierulf, P. : Coagulation 3 : 343 (1971).
9. Good, R. A., Thomas, L. : J. exp. Med. 96 : 625 (1952).
10. Good, R. A., Thomas, L. : J. exp. Med. 97 : 871 (1953).
11. Graeff, H., Mitchell, P. S., Beller, F. K. : Lab. Invest. 19 : 169 (1968).
12. Hardaway, R. M. : Syndromes of disseminated intravascular coagulation. With special reference to shock and hemorrhage. (Thomas, Springfield 1966).
13. Heene, D. L., Matthias, F. R. : Thrombos. Res. 2 : 137 (1973).
14. Horn, R. G. : J. infect. Dis. Suppl. 128 : 134 (1973).
15. Horn, R. G., Collins, R. D. : Lab. Invest. 18 : 101 (1968).
16. Horowitz, H. I., Des Prez, R. M., Hook, E. W. : J. exp. Med. 116 : 619 (1962).
17. Kleinmaier, H., Goergen, K., Lasch, H. G., Krecke, H.-J., Bohle, A. : Z. ges. exp. Med. 132 : 275 (1959).
18. Lasch, H. G., Heene, D. L., Huth, K., Sandritter, W. : Amer. J. Cardiol. 20 : 381 (1967).
19. Lasch, H. G., Huth, K., Heene, D. L., Müller-Berghaus, G., Hörder, M.-H., Janzarik, H., Mittermayer, C., Sandritter, W. : Dtsch. med. Wschr. 96 : 715 (1971).
20. Lasch, H. G., Krecke, H.-J., Rodriguez-Erdmann, F., Sessner, H. H., Schütterle, G. : Folia haemat. (N. F.) 6 : 325 (1961).
21. Lasch, H. G., Mechelke, K., Nusser, E., Sessner, H. H. : Z. ges. exp. Med. 129 : 484 (1958).
22. Lasch, H. G., Rodriguez-Erdmann, F., Krecke, H.-J. : Verh. dtsch. Ges. inn. Med. 66 : 993 (1960).
23. Lasch, H. G., Rodriguez-Erdmann, F., Schimpf, K. : Klin. Wschr. 39 : 645 (1961).
24. Lee, L. : J. exp. Med. 115 : 1065 (1962).
25. Lerner, R. G., Goldstein, R., Cummings, G. : Proc. Soc. exp. Biol. Med. 138 : 145 (1971).
26. Lerner, R. G., Rapaport, S. I., Spitzer, J. M. : Thromb. Diath. haemorrh. 20 : 430 (1968).
27. Margaretten, W., McKay, D. G. : J. exp. Med. 129 : 585 (1969).
28. McGrath, J. M., Stewart, G. J. : J. exp. Med. 129 : 833 (1969).
29. McKay, D. G. : Disseminated Intravascular Coagulation. An Intermediary Mechanism of Disease. (Hoeber, Harper and Row, New York 1964).
30. McKay, D. G., Shapiro, S. S. : J. exp. Med. 107 : 353 (1958).
31. Merskey, C., Johnson, A. J., Pert, J. H., Wohl, H. : Blood 24 : 701 (1964).
32. Merskey, C., Kleiner, G. J., Johnson, A. J. : Blood 28 : 1 (1966).
33. Müller-Berghaus, G., Goldfinger, D., Margaretten, W., McKay, D. G. :

Thromb. Diath. haemorrh. 18 : 726 (1967).

34. Müller-Berghaus, G., Hocke, M. : Thrombos. Res. 1 : 541 (1972).
35. Müller-Berghaus, G., Hocke, M. : Brit. J. Haemat. 25 : 111 (1973).
36. Müller-Berghaus, G., Lasch, H. G. : Thromb. Diath. haemorrh. 9 : 335 (1963).
37. Müller-Berghaus, G., Róka, L., Lasch, H. G. : Thromb. Diath. haemorrh. 29 : 375 (1973).
38. Müller-Berghaus, G., Schneberger, R. : Brit. J. Haemat. 21 : 513 (1971).
39. Niemetz, J., Fani, K. : Nature New Biol. 232 : 247 (1971).
40. Niewiarowski, S., Gurewich, V. : J. Lab. clin. Med. 77 : 665 (1971).
41. Nossel, H. L., Canfield, R. E., Butler, V. P., Jr. : Plasma fibrinopeptide A concentration as an index of intravascular coagulation. IVth Internat. Congr. Thrombos. Haemostas., Abstracts, p. 237, Vienna 1973.
42. Phillips, L. L. : Clin. Obstet. Gynec. 7 : 325 (1964).
43. Phillips, L. L., Margaretten, W., McKay, D. G. : Amer. J. Obstet. Gynec. 100 : 319 (1968).
44. Scott, G. B. D., Blaszczynski, N. : Arch. Path. 80 : 70 (1965).
45. Selye, H. : Thrombohemorrhagic Phenomena (Thomas, Springfield 1966).
46. Shapiro, S. S., McKay, D. G. : J. exp. Med. 107 : 377 (1958).
47. Zimmermann, T. S., Müller-Eberhard, H. J. : J. exp. Med. 134 : 1601 (1971).

The Activation of Intravascular Coagulation by Endotoxin

E. F. Lüscher

The intravascular activation of the blood clotting system is one among the many consequences of the presence of endotoxin in the blood stream. It is a very serious consequence indeed, since fibrin deposition in the microcirculation of vital organs is often responsible for the lethal outcome of endotoxin intoxication. This is most pronounced in the generalized Shwartzman reaction, a very complex phenomenon, which has recently been reviewed again by McKay (28).

This article will deal primarily with the role of the cellular compartment in the events which finally lead to thrombin formation. This seems permissible, since the next articles in this series will deal with the trigger mechanism of endotoxin in a more general way (cf. 14, 30) and because blood platelets, leukocytes, endothelial cells, and the cells of the reticulo-endothelial system play an undisputed role in this activation process.

1. Blood platelets

The activation of intravascular coagulation is linked to a more or less drastic reduction in the number of circulating platelets. Since thrombin is a powerful inducer of platelet alterations which culminate in their aggregation, in the release of a variety of substances from their storage organelles (17), and in the activation of their contractile system (22), this is not unexpected : altered and aggregated platelets are removed from the circulation by the reticulo-endothelial system, or kept back in the capillary system, particularly of the spleen, liver, and lung. Looked at this way, the platelets are the target of the end-product of the prothrombin activation sequence, and their number in circulation therefore, like fibrin formation, an indicator for intravascular thrombin formation.

Thrombin may be formed by two pathways. The so-called "extrinsic" system is based on the availability of tissue thromboplastin, a potent lipoprotein proactivator present in almost all cells. It is of great importance that tissue thromboplastin is liberated upon the disruption of cells ; intact cells, even in the course of blood coagulation, remain inert. Tissue thromboplastin is a direct activator (in conjunction with factor VII) of the prothrombin converting enzyme, factor X.

The second, "intrinsic" pathway of thrombin formation starts with the activation of factor XII or Hageman factor ; it then includes the formation of two phospholipid-enzyme complexes, the first one involved in factor X-activation, the second in prothrombin-activation. Since all clotting factors are present in adequate amounts in the plasma, the only limiting factors are the activator of factor XII, and the phospholipid procoagulant. The latter, the essential carrier material for topochemical enzyme reactions, is provided mainly by the blood platelets ; it has been termed platelet factor 3 (PF3). As outlined above, all cells possess latent phospholipid procoagulants : what makes the platelet unique, is the fact that PF 3 is made available upon a wide variety of external stimuli. Since the procoagulant phospholipids of the platelet are normal membrane constituents, it must be assumed that their "unmasking" is linked to a dramatic reorganization of the plasma membrane. Zwaal et al. (51) have recently shown that in the erythrocyte, phosphatidyl ethanolamine and phosphatidyl serine, both active procoagulants, are found almost exclusively on the inner side of the membrane. The making-available of PF 3 therefore could consist in the inside-out-transfer of these membrane constituents. Such an "activation" of the platelet is not exclusively achieved by thrombin (and certain other proteases), but also by quite unrelated materials, such as collagen, antigen-antibody complexes or aggregated IgG, ADP (in the presence of fibrinogen and Ca^{2+}-ions), adrenaline, serotonin, basic polymers, and, finally, by the contact with platelets which have undergone a primary membrane alteration (cf. 24). Since several of these inducers, e. g. ADP, serotonin, and adrenaline, are released by the platelet upon a primary stimulus, a self-sustained chain reaction is to be expected, in the course of which more and more platelets, which enter into close contact with each other will undergo release and will make PF 3 available. The latter will in turn participate in intrinsic thrombin formation, thus generating again a powerful inducer of platelet alterations.

a) Endotoxin as a direct inducer of platelet alterations

It is well established that in vitro endotoxin will directly interact with rabbit platelets, whereby a plasmatic cofactor as well as small amounts of Ca^{2+}-ions are required (3, 4, 5, 6, 19, 34, 48). Such platelets will undergo all the alterations described above : They will show a release reaction and PF 3 availability (6, 18, 19, 34). There is a distinct lag phase in the platelet response, as compared with the other inducing agents (3, 5). Obviously, the active inducer of platelet alterations is first formed in a time-consuming reaction. Since no complement fixation accompanies the endotoxin-platelet reaction, it is unlikely that the classical complement (C) pathway is involved (4); on the other hand, it is of interest that the activation product is removed from plasma (where it seems to persist up to 24 hours; cf. 48) by adsorption on zymosan (4).

Thus, for the rabbit, a plasma-mediated, direct effect on platelets, which could account for the subsequent activation of the intrinsic clotting system, is established. It should be kept in mind, though, that this involves, first the persistence of the activated platelets in circulation, and, second, the simultaneous activation of the contact factors (XII and XI). In this context, the observation by Walsh (47) that platelets altered by ADP or collagen are capable of activating the factors XII and XI, or directly factor XI, respectively, is of particular interest. It implies that a platelet triggered into activity by any external agent will be self-sustained in the activation of the intrinsic clotting system; it also corroborates the observations on factor XII consumption in the generalized Shwartzman reaction and similar conditions (31, 32). In line with this possibility is the observation by Margaretten and McKay (26), according to which a second injection of endotoxin will not induce intravascular coagulation in thrombopenic rabbits. On the other hand, it is noteworthy that the substitution of the second endotoxin application by phospholipid or PF 3 does not lead to the development of the generalized Shwartzman reaction (33). This might mean that these products cannot replace the altered platelets in the activation of the contact factors (cf. 47). Endotoxin seems to adhere to platelets (16, 41), even when this is not followed by secondary reactions, such as in platelets from dogs or guinea pigs, or from rats (38). It is thus obvious that the rabbit platelet with its pronounced susceptibility to endotoxin, even under in vitro conditions, is a very special case.

b) Indirect effects of endotoxin on platelets

In vitro, human platelets are inert towards endotoxin (29), although a slow release of serotonin, without concomitant aggregation and PF 3 - availability have been described (34). We have recently investigated the effect of immune complexes on human platelets suspended in plasma and found that this reaction proceeds in two phases. The first one is a very fast reaction, during which some release (of ^{14}C-serotonin) occurs; it is followed, after a lag phase of several minutes, by a second, more complete release reaction (35). Further studies showed that the time-consuming, second phase reaction was dependent on the elaboration of a product which was also obtained with zymosan, and the formation of which was prevented by pretreatment with cobra venom factor. Together with other evidence, this strongly suggested the participation of the alternate (or shunt-) pathway of C 3 - activation (36). Similar conclusions were reached independently by Zucker and Grant (50). Since endotoxin activates the C-system through the alternate (C 3) pathway (49), it seemed worthwhile to repeat these experiments with E. coli lipopolysaccharide. Although perfectly active in C 3 - activation, this material proved completely inert towards platelets. It is of interest that the active zymosan complex, in addition to the components of the C 3 - shunt, must also contain fibrinogen and an ill defined IgG - component in order to induce platelet alterations. It might be that the inactivity of endotoxin towards platelets, which it shares with inulin and aggregated IgA (36), is explained in terms of its incapacity to take up these two plasma components. Obviously the reaction of the platelet with the active zymosan complex is based on a multi-receptor system on the platelet surface, which is not activated by the endotoxin - C 3 - activator complex. In this context it is of interest that the electrophoretic mobility of human platelets, contrary to rabbit platelets, is not influenced by endotoxin (13).

Since under in vivo conditions endotoxin induces in man platelet alterations which are as pronounced as those in the rabbit, it must be concluded that these alterations result from indirect effects, i. e. that the human platelet is in fact the secondary target of other, toxin-induced factors.

2. Other cells affected by endotoxin

The effect of endotoxin on leukocytes has been discussed in chapter 4 of this book. It is of interest that mononuclear leukocytes contain a factor capable of causing the release reaction in platelets (15); furthermore leukocytes are a source of tissue thromboplastin (37).

Another cell which seems to be primarily affected by endotoxin is the endothelial cell. This is of considerable interest for several reasons: Endothelial cells are contractile (25); in culture they show morphological alterations to external stimuli, among others to thrombin (2), and thus, to a certain extent resemble platelets in their reactivity. Endotoxin in vivo has a rather dramatic effect on endothelial cells (27, 45, 46): They suffer severe damage and show gross morphological changes, partly due to contraction (27), culminating in their release from the vascular wall (11) and their appearance in the circulating blood (10). Under these conditions the now denuded endothelium will exert a profound effect on the clotting system: platelets will adhere to endothelial cells (46) and to the connective tissue in gaps in the endothelial layer (20, 40, 42, 43, 44) and undergo alterations; furthermore, collagen of the subendothelial layer is a known activator of the contact factors and finally, tissue thromboplastin may become available. Thus, the disturbance of the vascular endothelium may, by different pathways, contribute to the intravascular activation of the clotting system. Finally, it is established that endothelial cells are the seat of fibrinolytic activity (cf. 29); it has in fact been described that their disturbance by endotoxin depleates this activity (21) and that this is a most important contributory factor in the persistence of local fibrin deposits, e. g. in the glomeruli (1).

3. Conclusions

Of the many contributing factors which join together in the activation of intravascular coagulation, only very few have been discussed here. The role of the platelets as the centers of intrinsic thrombin formation has been particularly stressed, although it is realized that this view is not undisputed (cf. 7, 8, 9, 39). Although an amazing number of factors involved in the development of intravascular coagulation has already been recognized (see schemes in 23 and 28), there is little doubt that the subtle interplay between the components of this most complex system remains decisive for the final result. Furthermore, it should not be overlooked that local phenomena, which escape the observation of changes in the circulating blood, may be of considerable importance.

Finally, great emphasis must be placed on the fact that a discussion such as the one presented here, which is limited to the activation processes, is apt to give a distorted picture. The interplay between activation of coagulation, fibrinolysis, platelet aggregation and release reaction must be seen against the background of the availability of inhibitors, of the hemodynamic conditions, and of the activity of the reticulo-endothelial system, to mention but a few other contributory factors. In view of this complexity it is not astonishing that divergent views are to be found now, and will probably continue to exist for some time.

References

1. Bergstein, J. M., Michael, A. F.,jr. : Thromb. Diath. haemorrh. 29 : 27 (1973).
2. Booyse, F. M., Shepro, D., Rosenthal, M., McDonald, R. I. : Properties of cultured endothelial cells. Ser. Haemat. 6 : in press.
3. Des Prez, R. M. : J. exp. Med. 120 : 305 (1964).
4. Des Prez, R. M. : J. Immunol. 99 : 966 (1967).
5. Des Prez, R. M., Bryant, R. E. : J. exp. Med. 124 : 971 (1966).
6. Des Prez, R. M., Horowitz, H. I., Hook, E. W. : J. exp. Med. 114 : 857 (1961).
7. Evensen, S. A., Jeremic, M. : Thromb. Diath. haemorrh. 19 : 556 (1968).
8. Evensen, S. A., Jeremic, M. : Brit. J. Haemat. 19 : 33 (1970).
9. Evensen, S. A., Jeremic, M. : Scand. J. Haemat. 7 : 413 (1970).
10. Gaynor, E., Bouvier, C. A., Spaet, T. H. : Clin. Res. 16 : 535 (1968).
11. Gaynor, E., Bouvier, C. A., Spaet, T. H. : Science 170 : 986 (1970).
12. Giacomelli, F., Wiener, J., Spiro, D. : J. Cell Biol. 45 : 188 (1970).
13. Gröttum, K. A., Hjort, P. F., Jeremic, M. : Thromb. Diath. haemorrh. 22 : 192 (1969).
14. Heene, D. : Blood coagulation mechanism and endotoxins : hemostatic defect in septic shock. In this volume.
15. Henson, P. M. : J. exp. Med. 131 : 287 (1970).
16. Herring, W. B., Herion, J. C., Walker, R. I., Palmer, J. G. : J. clin. Invest. 42 : 79 (1963).
17. Holmsen, H., Day, H. J., Stormorken, H. : Scand. J. Haemat. Suppl. 2 (1969).
18. Horowitz, H. I. : Blood 25 : 600 (1965).
19. Horowitz, H. I., Des Prez, R. M., Hook, E. W. : J. exp. Med. 116 : 629 (1962).
20. Hugues, J., Mahieu, P. : Thromb. Diath haemorrh. 24 : 395 (1970).
21. Lipinski, B., Worowski, K., Jeljaszewicz, J., Niewiarowski, S., Rejniak, L. : Thromb. Diath. haemorrh. 20 : 285 (1968).
22. Lüscher, E. F., Bettex-Galland, M. : Path. Biol. 20 : Suppl. 89 (1972).
23. Lüscher, E. F., Pfueller, S. L. : Disseminated intravascular coagulation (From an immunological viewpoint). In : F. Duckert : Immunological Mechanisms in Blood Coagulation, Thrombosis and Haemostasis, p. 129 (Schattauer, Stuttgart 1971).
24. Lüscher, E. F., Pfueller, S. L., Massini, P. : Ser. Haemat. 6 : 382 (1973).
25. Majno, G., Shea, S. M., Leventhal, M. : J. Cell Biol. 42 : 647 (1969).
26. Margaretten, W., McKay, D. G. : J. exp. Med. 129 : 585 (1969).
27. McGrath, J. M., Stewart, G. J. : J. exp. Med. 129 : 833 (1969).
28. McKay, D. G. : Thromb. Diath. haemorrh. 29 : 11 (1973).
29. Mueller-Eckhardt, Ch., Lüscher, E. F. : Thromb. Diath. haemorrh. 20 : 336 (1968).
30. Müller-Berghaus, G. : Trigger mechanism of endotoxin-induced intravascular coagulation. In this volume.
31. Müller-Berghaus, G., Lasch, H. G. : Thromb. Diath. haemorrh. 26 : 58 (1971).
32. Müller-Berghaus, G., Schneeberger, R. : Brit. J. Haemat. 21 : 513 (1971).

33. Müller-Berghaus, G., Goldfinger, D., Margaretten, W., McKay, D. G. : Thromb. Diath. haemorrh. 18 : 726 (1967).
34. Nagayama, M., Zucker, M. B., Beller, F. K. : Thromb. Diath. haemorrh. 26 : 467 (1971).
35. Pfueller, S. L., Lüscher, E. F. : Studies on the mechanisms of the human platelet release reaction induced by immunological stimuli. I. Complement-dependent and complement-independent reactions. : J. Immunol., in press.
36. Pfueller, S. L., Lüscher, E. F. : Studies on the mechanisms of the human platelet release reaction induced by immunological stimuli.II. The effects of zymosan. : J. Immunol., in press.
37. Rapaport, S. I., Hjort, P. F. : Thromb. Diath. haemorrh 17 : 222 (1967).
38. Roy, A. J., Djerassi, I., Neitlich, H., Farber, S. : Amer. J. Physiol. 203 : 296 (1962).
39. Simmons, J., Rapaport, S. I., Hjort, P. F. : Proc. Soc. exp. Biol. Med. 124 : 742 (1967).
40. Spaet, T. H., Stemerman, M. B. : Ann. N. Y. Acad. Sci. 201 : 13 (1972).
41. Spielvogel, A. R. : J. exp. Med. 126 : 235 (1967).
42. Stemerman, M. B., Baumgartner, H. R., Spaet, T. H. : Lab. Invest. 24 : 179 (1971).
43. Trancer, J. P., Baumgartner, H. R. : Nature 216 : 1126 (1967).
44. T'sao, C. H., Glagov, S. : Brit. J. exp. Path. 51 : 423 (1970).
45. Urbaschek, B. : Verh. dtsch. Ges. inn. Med. 72 : 752 (1967).
46. Urbaschek, B. : Zur Frage des Wirkungsmechanismus bakterieller Endotoxine und seiner Beeinflussung. Habilitationsschrift, Universität Heidelberg (1967).
47. Wals, 'P. N. : Platelet coagulant activities : Evidence for multiple, different function of platelets in intrinsic coagulation. Ser. Haemat. 6 : (1973) in press.
48. Yamazaki, H., Shinamoto, T., Murase, H., Shinamoto, T. : Blood 30 : 792 (1967).
49. Zimmermann, T. S., Müller-Eberhard, H. J. : J. exp. Med. 134 : 1601 (1971).
50. Zucker, M. B., Grant, R. A. : Aggregation and release reaction induced in human platelets by zymosan. : J. Immunol., in press.
51. Zwaal, R. A., Roelofsen, B., Colley, C. M. : Biochim. biophys. Acta 300 : 159 (1973).

Trigger Mechanism of Endotoxin-Induced Intravascular Coagulation

G. Müller-Berghaus

According to the classical theory of blood coagulation, the in vitro activation of coagulation in the intrinsic pathway starts with the activation of factor XII (Hageman factor) and in the extrinsic pathway with release of tissue thromboplastin. In both pathways, platelets with their surface and their constituents play an important role as accelerators of coagulation activation. In this short review, I would like to concentrate on the significance of Hageman factor activation (intrinsic pathway), of tissue thromboplastin from leukocytes and/or endothelial cells (extrinsic pathway) and of platelets (extrinsic and intrinsic systems).

Activation of the Contact System of Blood Coagulation

In man, the occurrence of consumption coagulopathy is accompanied by a decrease in Hageman factor activity (2, 19). In the same manner, Hageman factor activity drops pronouncedly when endotoxin is injected into rabbits (17, 28). The significance of Hageman factor in triggering intravascular coagulation is questioned, however, since the inhibition of Hageman factor activation by lysozyme does not prevent the activation of intravascular coagulation (28). Furthermore, anticoagulation with warfarin or phenprocoumon eliminates the decrease in Hageman factor activity after endotoxin administration, suggesting that Hageman factor activation represents the result but not the cause of intravascular coagulation (17, 28). These experiments exclude the possibility that injury to endothelial cells and exposure of the underlying collagen might activate Hageman factor since anticoagulation with warfarin prevents the decrease in Hageman factor activity but does not inhibit endothelial damage (6). These findings suggest that Hageman factor is not essential in triggering intravascular coagulation by endotoxin.

Release of Procoagulant Material from Granulocytes

In man as well as in different animal species the granulocytes in the circulating blood appear to be the principal target cell of endotoxin. Intravenous injection of small amounts of endotoxin leads to a decrease in the number of circulating granulocytes, followed by rebound granulocytosis. Thomas and Good (34) were the first to demonstrate the significance of granulocytes in the pathogenesis of endotoxin-induced intravascular coagulation. They showed that endotoxin does not initiate disseminated intravascular coagulation when animals are rendered granulocytopenic by treatment with nitrogen mustard. The first evidence supporting the role of granulocytes in triggering intravascular coagulation was recently presented by Horn and Collins (11). When granulocytopenic rabbits were transfused with a large number of granulocytes (1.5 x 10^9 cells), the subsequent endotoxin injection initiated intravascular coagulation to the same extent as in control rabbits with previously normal leukocyte counts. In this experiment model intravenous granulocyte transfusion could be substituted by i n t r a - a o r t i c infusion of isolated leukocyte granules. If these animals were injected with endotoxin immediately after infusion of leukocyte granules, disseminated intravascular coagulation occurred, demonstrating the necessity of granules for the initiation of intravascular coagulation by endotoxin.

Experiments reported in this review were supported by the Deutsche Forschungsgemeinschaft, Bad Godesberg, Germany.

Most interestingly, the i n t r a v e n o u s infusion of leukocyte granules concomitantly with endotoxin did not induce disseminated intravascular coagulation (11).

Niemetz and Fani (29) confirmed the experiments of Horn and Collins and extended the knowledge of the role of leukocytes in triggering intravascular coagulation by
endotoxin. Lerner et al. (16) found that leukocytes incubated in vitro in plasma or
serum develop a procoagulant activity like tissue thromboplastin. Niemetz and Fani
(29) demonstrated that leukocytes from endotoxin-injected rabbits exhibit a tenfold
increase in tissue factor activity in comparison to leukocytes from control rabbits.
The infusion of these "activated" leukocytes from endotoxin-treated rabbits into normal rabbits caused activation of intravascular coagulation, as judged by clot formation in the lungs, whereas infusion of normal leukocytes did not initiate coagulation.

Thomas et al. (35) suggested a second effect of leukocytes in promoting blood coagulation. The authors proposed a hypothesis that endotoxin causes the release of
highly charged polymers in precipitating endotoxin-mediated damaged fibrinogen.
In the terms used today this would mean precipitation of soluble fibrin (soluble fibrin monomer complexes) by cellular basic proteins or lysosomal protein fractions
of granulocytes. Under in vitro conditions, these highly charged macromolecules
can indeed precipitate soluble fibrin (8, 14). Under in vivo conditions, however, the
endotoxin-induced precipitation of soluble fibrin occurs in heparin-treated as well
as in granulocytopenic rabbits (4, 26). Thus, granulocytes are potentially involved
in endotoxin-induced activation of extrinsic intravascular coagulation, but do not
play a significant role in precipitating soluble fibrin.

Release of Procoagulant Material from Platelets

The incubation of platelet-rich rabbit plasma with endotoxin results in platelet
aggregation and release of serotonin and platelet factor 3 into the plasma (3, 12).
Likewise, human platelets can be aggregated by bacterial endotoxin (30). Stetson
(33) demonstrated that endotoxin causes aggregation of platelets in vivo. The ensuing thrombocytopenia is a well-known finding after intravenous injection of endotoxin into rabbits (13, 22, 31). The injected endotoxin is fixed onto platelets within
minutes (9). In vivo, platelet damage becomes manifest as platelets remaining in
the circulating blood have lost most of their total platelet factor 3 activity (15). This
activity seems to be released to the surrounding plasma (12). Platelet factor 3 released by endotoxin is thought to be the most potent agent of platelets accelerating
blood coagulation. However, infusion of purified platelet factor 3 or phospholipids
representing up to 420 % of the platelet factor 3 activity normally available in a rabbit or a monkey failed to initiate intravascular coagulation to a degree causing microclot formation (21, 24). These experiments and the data of Epstein and Quick
(5) demonstrate the relatively weak effect of platelet factor 3 in vivo. In the same
manner, the reduction of platelet decrease by treating the animals with inhibitors
of platelet aggregation does not reduce the occurrence of disseminated intravascular
coagulation (32). Yet platelets seem to be essential in endotoxin-induced disseminated intravascular coagulation since thrombocytopenia prevents the occurrence of
microclot formation (18).

Still the intriguing question remains which role is played by the platelets in endotoxin-induced intravascular coagulation. Since the significance of platelets in triggering intravascular coagulation can not be proven without doubt, platelets may only
represent an essential accelerator but not mediator in triggering intravascular coagulation as indicated in Fig. 1.

The Role of Endothelium in Endotoxin-Induced Activation of Intravascular Coagu-
lation

In 1934, Apitz (1) believed derangement of the endothelial wall to be the primary
cause for the changes observed after endotoxin injection into rabbits. For more
than 30 years, however, it could not be clarified whether damage to the endothelium

associated with endotoxinemia is a secondary phenomenon after deposition of plate-
let-leukocyte-fibrin clots or a primary event prior to deposition of blood elements.
Functional changes of the endothelium were observed in aortas of animals injected
with endotoxin. Already fifteen to thirty minutes after endotoxin injection these vessel segments develop a higher procoagulant activity than the aortas of control rabbits (27). These effects of the endothelium may be due to release of procoagulant
material from the endothelial cells and/or exposure of the underlying basement
membrane with activation of the contact factors of blood coagulation or aggregation
of platelets and leukocytes.

Hoff et al. (10) described very early occurring ultrastructural changes in arteries after endotoxin injection. These authors found electron dense particles residing in intra-endothelial vacuoles possibly resembling the injected endotoxin. Morphological studies by McGrath and Stewart (20) demonstrated severely damaged endothelial cells already one hour after endotoxin administration, i.e. at a time when functional changes of the vessel wall were observed in a coagulation test system (27).

The involvement of vascular lesions in triggering intravascular coagulation by
endotoxin is supported by the findings of Gaynor et al. (7) who demonstrated circulating endothelial cells released from the vessel wall by the action of endotoxin.
This effect of endotoxin is observed already 5 minutes after endotoxin injection and
cannot be prevented by prior anticoagulation, indicating that endothelial damage is
a primary event and independent of coagulation. Most interestingly, nitrogen mu-
stard-induced neutropenia does not protect rabbits from endotoxin-induced endothelial injury (6). Since neutropenia prevents activation of intravascular coagulation,
endotoxin-induced endothelial damage does not seem to represent the mediator for
triggering intravascular coagulation, but may be an accelerator favoring coagulation of an already activated coagulation system.

Summary and Conclusions

The experimental data for the participation of granulocytes, platelets, contact
factors of blood coagulation and endothelial cells in triggering endotoxin-induced intravascular coagulation are reviewed. There is substantial evidence suggesting
that several essential factors are involved in intravascular coagulation (theory of
pluricausality 23). However, granulocytes rich in tissue factor activity seem to be
the main activator of endotoxin-induced intravascular coagulation. These granulocytes initiate intravascular coagulation by the extrinsic pathway of blood coagulation
(Fig. 1). Platelet factors, especially platelet factor 3, are weak procoagulant material and can function as accelerators only, but platelet factor 3 by itself cannot initiate intravascular coagulation. The sequence for the activation of intravascular coagulation by endotoxin is outlined in Fig. 1, in which the role of endothelial cells is
left out. It may very well turn out that endothelial cells develop an accelerator function, but activator quality has been excluded. More likely, the role of endotoxin-induced endothelial damage subsists in presenting a surface for the deposition and precipitation of soluble fibrin, since endotoxin mediates precipitation or polymerization of soluble fibrin in animals with inhibited coagulation system as well as in neu-
tropenic rabbits (4, 25).

References

1. Apitz, K. : Virchows Arch. path. Anat. 293 : 1 (1934).
2. Colman, R. W., Robboy, S. J., Minna, J. D. : Amer. J. Med. 52 : 679 (1972).
3. Des Prez, R. M., Horowitz, H. I., Hook, E. W. : J. exp. Med. 114 : 857
   (1961).
4. Eckhardt, T., Müller-Berghaus, G. : Verh. dtsch. Ges. inn. Med. 80 : (1974).
5. Epstein, E., Quick, A. J. : Proc. Soc. exp. Biol. Med. 83 : 453 (1953).
6. Gaynor, E. : Blood 41 : 797 (1973).

7. Gaynor, E., Bouvier, C., Spaet, T. H. : Science 170 : 986 (1970).
8. Hawiger, J., Collins, R. D., Horn, R. G. : Proc. Soc. exp. Biol. Med. 131 : 349 (1969).
9. Herring, W. B., Herion, J. C., Walker, R. I., Palmer, J. G. : J. clin. Invest. 42 : 79 (1963).
10. Hoff, H. E., Gottlob, R., Blümel, G. : Naturwissenschaften 54 : 287 (1967).
11. Horn, R. G., Collins, R. D. : Lab. Invest. 18 : 101 (1968).
12. Horowitz, H. I., Des Prez, R. M., Hook, E. W. : J. exp. Med. 116 : 619 (1962).
13. Kleinmaier, H., Goergen, K., Lasch, H. G., Krecke, H.-J., Bohle, A. : Z. ges. exp. Med. 132 : 275 (1959).
14. Kopéc, M., Wegrzynowicz, Z., Latallo, Z. S. : Thromb. Diath. haemorrh. Suppl. 40 : 253 (1970).
15. Lasch, H. G., Rodriguez-Erdmann, F., Krecke, H.-J. : Verh. dtsch. Ges. inn. Med. 66 : 993 (1960).
16. Lerner, R. G., Goldstein, R., Cummings, G. : Proc. Soc. exp. Biol. Med. 138 : 145 (1971).
17. Lerner, R. G., Rapaport, S. I., Spitzer, J. M. : Thromb. Diath. haemorrh. 20 : 430 (1968).
18. Margaretten, W., McKay, D. G. : J. exp. Med. 129 : 585 (1969).
19. Mason, J. W., Colman, R. W. : Thromb. Diath. haemorrh. 26 : 325 (1971).
20. McGrath, J. M., Stewart, G. J. : J. exp. Med. 129 : 833 (1969).
21. McKay, D. G., Müller-Berghaus, G., Cruse, V. : Amer. J. Path. 54 : 393 (1969).
22. McKay, D. G., Shapiro, S. S. : J. exp. Med. 107 : 353 (1958).
23. Müller-Berghaus, G. : Thromb. Diath. haemorrh., Suppl. 36 : 45 (1969).
24. Müller-Berghaus, G., Goldfinger, D., Margaretten, W., McKay, D. G. : Thromb. Diath. haemorrh. 18 : 726 (1967).
25. Müller-Berghaus, G., Hocke, M. : Thrombos. Res. 1 : 541 (1972).
26. Müller-Berghaus, G., Hocke, M. : Brit. J. Haemat. 25 : 111 (1973).
27. Müller-Berghaus, G., Lasch, H. G. : Thromb. Diath. haemorrh. 9 : 335 (1963).
28. Müller-Berghaus, G., Schneberger, R. : Brit. J. Haemat. 21 : 513 (1971).
29. Niemetz, J., Fani, K. : Nature New Biol. 232 : 247 (1971).
30. Ream, V. J., Deykin, D., Gurewich, V., Wessler, S. : J. Lab. clin. Med. 66 : 245 (1965).
31. Shimamoto, T., Yamazaki, H., Sagawa, N., Iwahara, S., Konishi, T., Maezawa, H. : Proc. Japan Acad. 34 : 450 (1958).
32. Slabber, C. F., Theiss, W., Beller, F. K. : J. Med. 3 : 341 (1972).
33. Stetson, C. A. : J. exp. Med. 93 : 489 (1951).
34. Thomas, L., Good, R. A. : J. exp. Med. 96 : 605 (1952).
35. Thomas, L., Smith, R. T., v. Korff, R. : J. exp. Med. 102 : 263 (1955).

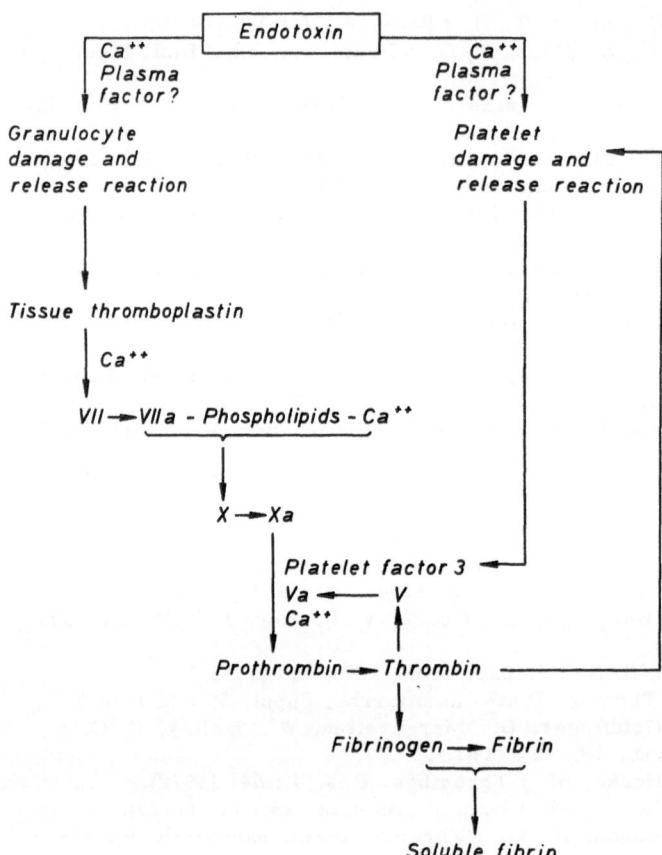

Fig. 1.   Activation of intravascular coagulation by endotoxin. Granulocyte damage represents the trigger whereas platelet factor 3 constitutes an accelerator of intravascular coagulation only.

Blood Coagulation Mechanism and Endotoxins:
Hemostatic Defect in Septic Shock

D. L. Heene

The effect of endotoxins and bacterial toxins on the blood clotting system has been clearly demonstrated by means of the Sanarelli-Shwartzman phenomenon in the animal experiment (1, 2). The involvement of diffuse intravascular coagulation as the essential pathophysiological event is of decisive implication for the pathomorphological picture as well as for the course of the syndrome.

In human pathology there are several clinical equivalents, mainly related to septic conditions, such as bilateral cortical necrosis of the kidneys following septic abortion, which have to be understood as a localizing phenomenon of consumption-coagulopathy (3, 4). The role of endotoxins as trigger mechanism of consumption-coagulopathy is still under experimental investigation. Besides intravascular liberation of procoagulant activity following endotoxin-induced damage of the vascular endothelium and platelets, one of the main mechanisms for induction of consumption-coagulopathy and its perpetuation seems to be the inhibition of physiological compensating mechanisms acting against procoagulant stimulation of the coagulation system (5, 6, 7). Decrease of the inhibitor potential, of secondary fibrinolysis and of the clearence capacity of the reticulo-endothelial system promote the development of diffuse intravascular coagulation and microthrombosis within the vascular periphery. Furthermore, endotoxins may provoke the generation of localized or generalized impairment of microcirculation by alteration of hemodynamic factors, which itself may be considered as a clot-promoting event by means of stasis (8, 9). Diffuse intravascular coagulation thus may be regarded either as consequence or cause of shock state.

With this pathophysiological background in mind it is of interest to evaluate whether in clinical cases with septic conditions, especially in septic shock, there is evidence of typical changes within the coagulation system which can prove the development of consumption-coagulopathy or diffuse intravascular coagulation, respectively. In order to evaluate such analytical criteria, coagulation analyses of 25 patients with septic shock were studied retrospectively. In these patients coagulation studies were performed at the beginning of the clinical symptoms of shock and continued at regular intervals. These cases were observed at the intensive care units of the surgical, neurosurgical and medical departments of the University hospital in Giessen during the past 3 years.

Results

At the time of the first coagulation profile all patients revealed clinical evidence of shock and acute renal failure. The basic conditions are summarized in Table I. So far as bacteriological identification was performed in most of the cases infection was due to gram-negative bacteria. One patient with a lung abscess suffered from septicemia caused by Staphylococcus aureus. The mortality was 80 %. Five patients revealed clinical evidence of hemorrhagic diathesis, which was demonstrated analytically to be caused by secondary hyperfibrinolysis in two cases. Severe bleeding tendencies were observed also in four patients with bone marrow insufficiency, mainly caused by secondary thrombocytopenia.

The analytical data are shown as mean values with standard deviation in Table II and Fig. 1 as compared to normal values. The following parameters were analyzed: platelet count, fibrinogen concentration, maximal amplitude of the thrombelastogram, streptokinase resistance, partial thromboplastin time (PTT), thromboplastin time, factor V and X activity, thrombin time, reptilase time, staphylococcal clumping test

(SCT) and ethanol test. All results showed significant differences from normal values ($p < 0.001$). Critical evaluation of these results raises the question whether the laboratory diagnosis of consumption-coagulopathy may be definitely established by means of these analytical data. Thrombocytopenia, in the first place, seems to be an essential feature. With regard to the low platelet count one would expect a further decrease of the maximal amplitude of the thrombelastogram. However the latter is also dependend on the fibrinogen concentration of plasma which was found to be elevated in most of the cases. Hyperfibrinogenemia is not considered as an analytical criterion of consumption-coagulopathy (10, 11). High fibrinogen levels are usually found in acute or chronic inflammatory processes and are interpreted to be the result of a nonspecific increase of the alpha-2-glycoproteins. Hyperfibrinogenemia is obligatory in septic conditions and its presence does not exclude the involvement of diffuse intravascular coagulation. By means of the analytical data presented in this study it is difficult to prove the evidence of an increased intravascular fibrinogen turn-over. The ethanol test, sensitive to circulating fibrinmonomer indicating indirectly the intravascular action of thrombin, was positive in 96 % of the cases, however, it should be considered that in the presence of high fibrinogen levels above 400 mg % the number of false-positive tests increases (12).

The involvement of secondary fibrinolysis is clearly demonstrated by means of assays indicating the presence of fibrinogen degradation products such as thrombin time, reptilase time and staphylococcus clumping test. Only in 2 patients was there analytical evidence of hyperfibrinolysis with hypofibrinogenemia of less than 200 mg % and extreme prolongation of thrombin time as well as of reptilase time. The other cases showed only minor involvement of the fibrinolytic system. The prolongation of the reptilase time may be at least partially explained by the presence of fibrinmonomer in plasma and with regard to hyperfibrinogenemia (13).

The remaining parameters such as partial thromboplastin time, factor V and factor X activity, reveal the actual decrease of the hemostatic potential and thus the degree of the hemostatic defect. In almost all cases the streptokinase resistance was definitely increased. With regard to methodological aspects this may be due to the high fibrinogen levels in the samples assayed, however, it can also be interpreted as indicator of an excessive increase of kinase inhibitor activity. By means of these studies it is not possible to decide whether the increase in streptokinase resistance was induced by the influence of a previous streptococcal infection. This phenomenon requires further detailed investigation both from the bacteriological side and from fibrinolytic parameters.

Discussion

The data presented in this study demonstrate that the main features of the hemostatic defect encountered in septic conditions associated with shock are thrombocytopenia, increased fibrinogen levels and decrease of plasmatic coagulation factors. By critical interpretation of the whole analytical profile the diagnosis of consumption-coagulopathy is not definitely established, especially the analysis of the fibrinogen derivatives does not give satisfactory information with regard to its dependence upon the fibrinogen concentration of plasma. However it can be pointed out that by means of follow-up analysis the progressive development of the hemostatic defect and the consumption of the hemostatic potential can be detected during the early phase of septic shock, thus giving essential information on the possible involvement of diffuse intravascular coagulation in vivo. The alterations of coagulation parameters demonstrated in this study can be considered as essential criteria of the coagulation defect in septic shock and, once encountered in septic conditions, should be interpreted as analytical features of consumption-coagulopathy (14).

With regard to the therapeutical consequences anticoagulant therapy with heparin (100 - 400 u/kg body weight/24 hrs) should be recommended in cases with evidence of septic shock. However, heparin therapy is only of prophylactic value, since microthrombosis of the vascular periphery once established during the development of septic shock may not be reversed by means of anticoagulant measures. In consideration of the highly increased streptokinase resistance effective fibrinolytic therapy with streptokinase can not be achieved.

References

1. Kleinmaier, H., Goergen, K., Lasch, H. G., Krecke, H. J., Böhle, A. : Z. ges. exp. Med. 132 : 275 (1959).
2. Rodriguez-Erdmann, F., Krecke, H. J., Lasch, H. G., Bohle, A. : Z. ges. exp. Med. 134 : 109 (1960).
3. Lasch, H. G., Krecke, H. J., Rodriguez-Erdmann, F., Sessner, H. H., Schütterle, G. : Fol. haemat. 6 : 325 (1961).
4. Lasch, H. G., Heene, D. L., Huth, K., Sandritter; W. : Amer. J. Cardiol. 20 : 381 (1967).
5. Müller-Berghaus, G. : Thromb. Diath. haemorrh. Suppl. 50 : 5 (1971).
6. Beller, F. K., Graeff, H., Gorstein, F. : Amer. J. Obstet. Gynec. 103 : 544 (1969).
7. Beller, F. K., Douglas, G. W. : Obstet. Gynec. 41 : 521 (1973).
8. Lasch, H. G. : Med. Welt 31 : 1780 (1967).
9. Siegenthaler, W., Lüthy, R., Vetter, H., Siegenthaler, G. : Schweiz. med. Wschr. 102 : 593 (1972).
10. Heene, D. L. : Thromb. Diath. haemorrh. Suppl. 50 : 33 (1971).
11. Merskey, C., Johnson, A. J., Kleiner, G. J., Wohl, H. : Brit. J. Haemat. 13 : 528 (1967).
12. Schwabe, G., Heene, D. L., Krause, W. H. : Verh. dtsch. Ges. inn. Med. 78 : 719 (1972).
13. Heene, D. L., Lasch, H. G. : Internist 14 : 154 (1973).
14. Heene, D. L., Lasch, H. G. : Thromb. Diath. haemorrh. Suppl. 46 : 139 (1971).

Table I.

Coagulation Studies in Patients with Septic Shock and Renal Failure: Clinical Evaluation

Pathogenesis	No. of Patients	Exitus	Hemorrh. diathesis	Pres. Cause of Hemorrh. diathesis
Menengitis	3	2	1	
Lung Abscess	1	1		
Cholangitis	2	1	1	Hyperfibrinolysis
Pyonephrosis	3	2		
Leukemia	2	2	(2)	Thrombocytopenia
Agranulocytosis	2	2	(2)	Thrombocytopenia
Septic Abortion	2	1	1	Hyperfibrinolysis
Post Partal	2	2	1	
Peritonitis	8	7	1	
Total	25	20	5	
%	100	80	20	

Table II.

Coagulation Studies in Patients with Septic Shock (n = 25)

		$\overline{X}$	S. D.	(Normal Values)	
Platelets	$x\ 10^3/mm^3$	52.4	27.6	184.3	$\pm$ 46.5
Fibrinogen	mg/100 ml	418.9	270.8	278.7	$\pm$ 80.2
TEG : m_e	mm	38.2	18.4	53.2	$\pm$ 4.1
Sk-Resist.	10^6 u	1.14	0.3	0.198	$\pm$ 0.088
PTT	sec	65.6	19.7	43.7	$\pm$ 6.9
Quick-Test	%	38.4	23.3	93.0	$\pm$ 12.9
Factor V	%	45.3	16.3	92.5	$\pm$ 17.2
Factor X	%	44.0	21.6	96.2	$\pm$ 7.4
Thrombin time	sec	27.6	6.7	19.8	$\pm$ 3.1
Reptilase time	sec	33.8	10.4	21.5	$\pm$ 2.1
Staph. clump. test	$\log_2$	5.0	1.5	3	$\pm$ 1
Ethanol-Test	% / n	96.0		1	

for all abnormal values $p < 0.001$

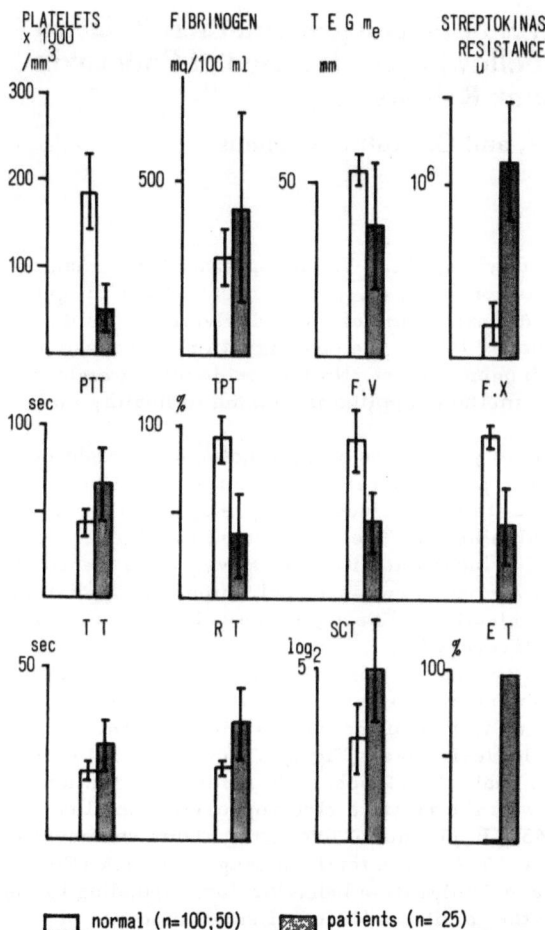

Fig. 1. Coagulation parameters in patients with septic shock (n = 25). Platelets, fibrinogen, TEG m_e : maximal amplitude of thrombelastogram, streptokinase resistance, PTT : partial thromboplastin time, TPT : thromboplastin time F. V : factor V, F. X : factor X, TT : thrombin time, RT : reptilase time, SCT : staphylococcal clumping test, ET : ethanol test. Normal values : white columns, patients values : grey columns.

High Molecular Weight Derivatives of Fibrinogen and their Relation to the Total Fibrinogen Content Following the Infusion of Endotoxin in Pregnant Rabbits

R. von Hugo, H. Graeff, and G. Müller-Berghaus

High molecular weight derivatives of fibrinogen can be demonstrated in human cases of disseminated intravascular coagulation (DIC) (1, 2) in concentrations of up to 15 % of the total fibrinogen content (3). At least some of these derivatives could also be shown in uncomplicated human pregnancy (4) but the concentrations observed amounted only to 2 - 3 % of the total fibrinogen content. No high molecular weight derivatives could be demonstrated with the methods applied in plasma of healthy male blood donors.

Distribution pattern and physicochemical properties of high molecular weight derivatives of fibrinogen were investigated in pregnant rabbits after the induction of DIC by endotoxin infusion. Fibrinogen and its derivatives were precipitated from plasma samples by addition of 2.5 M ß-alanine (5). The redissolved precipitate was submitted to agarose gel filtration, and fibrinogen and its derivatives were identified in the eluted fractions by PAA gel electrophoresis and intra gel immunoprecipitation. Fibrinogen related antigen material was identified directly in the gel by intra gel immunoprecipitation using a monovalent antiserum (2).

The amount of high molecular weight derivatives was estimated by planimetric evaluation of the elution pattern and was expressed in per cent of the total fibrinogen content. In virgin rabbits approximately up to 5 % of the total fibrinogen content in plasma consisted of high molecular weight derivatives (Fig.1). They were eluted in a small shoulder prior to the fibrinogen peak. Electrophoretic studies in fractions of the shoulder revealed a protein band with the relative electrophoretic mobility of 0.28×10^{-5} cm^2 / V sec (fibrinogen 0.43). In plasma of pregnant rabbits the concentration of the 0.28 derivative amounted to 15 % of the total fibrinogen content (Fig 2).

A component with an elution pattern and a migration behavior corresponding to the 0.28 derivative was also observed after the in vitro addition of small amounts of thrombin to plasma (6). The in vivo occurrence of the 0.28 derivative may, therefore, also result from the action of thrombin on fibrinogen. The elution position after agarose gel filtration and the migration behavior in PAA gels are in agreement with the assumption that the 0.28 derivative is a dimer, possibly formed by fibrinogen and a fibrin monomer, similar to the one described by Shainoff and Page (7) as cryoprofibrin. It is conceivable that the increased amount of the 0.28 derivative during pregnancy indicates hypercoagulability and enhanced susceptibility to DIC.

Following 6 hours of endotoxin infusion (25 µg/kg/hr) the number of platelets declined significantly to mean values of 60,000/m^3, consumption coagulopathy was indicated by a low fibrinogen content, and plasma hemoglobinemia was accompanied by the occurrence of distorted red cells. All animals developed complete renal shutdown. Intravascular coagulation was autoptically verified by extensive glomerular microthrombosis. Blood samples were taken at the end of the infusion period. According to the elution pattern of agarose gel filtration high molecular weight derivatives made up to 40 % of the total fibrinogen content (Fig. 3). Yet they did not consist of the 0.28 derivative only. Other derivatives with a molecular weight higher than that of the 0.28 derivative were eluted at the elution position of the void volume. This pattern of two or more high molecular weight derivatives of fibrinogen with a shift to the higher molecular weight products seems to be pathognomonic of severe forms of DIC. There is some indication that the amount and pattern of high molecular weight derivatives of fibrinogen might enable a detailed description of DIC in the experimental animal.

Summary

Distribution pattern and physicochemical properties of high molecular weight derivatives of fibrinogen were investigated in pregnant rabbits after the infusion of endotoxin. In plasma from pregnant rabbits high molecular weight derivatives amounted to approximately 15 % of the total fibrinogen content, whereas virgin rabbits revealed up to 5 % only. The derivatives observed during pregnancy consisted predominantly of a thrombin mediated derivative with an electrophoretic mobility of 0.28×10^{-5} cm^2/V x sec. High concentrations of this derivative might be indicative for the so-called state of hypercoagulability during pregnancy. Following the induction of DIC by endotoxin infusion, 40 % of the total fibrinogen content consisted of high molecular weight derivatives with a shift to higher molecular weight products.

References

1. Vermylen, J., Donati, M. B., Verstraete, M. : The identification of fibrinogen derivatives in plasma and serum by agarose gel filtration. In : Verstraete, M., Vermylen, J., Donati, M. B. : Fibrinogen degradation products, p. 219 Scand. J. Haematol., Suppl. 13 (1971).
2. Graeff, H., von Hugo, R. : Thromb. Diath. haemorrh. 27 : 610 (1972).
3. Graeff, H., von Hugo, R. : Fibrinogen derivatives with a higher molecular weight than that of fibrinogen in a case of premature separation of the placenta. Submitted for publication.
4. Graeff, H., von Hugo, R., Hafter, R. : In vivo formation of soluble fibrin monomer complexes in human plasma. Thrombosis Res., in press.
5. Straughn, W. III, Wagner, R. H. : Thromb. Diath. haemorrh. 16 : 198 (1966).
6. von Hugo, R., Graeff, H. : Thrombosis Res. 3 : 183 (1973).
7. Shainoff, J. R., Page, J. H. : J. exp. Med. 116 : 687 (1962).

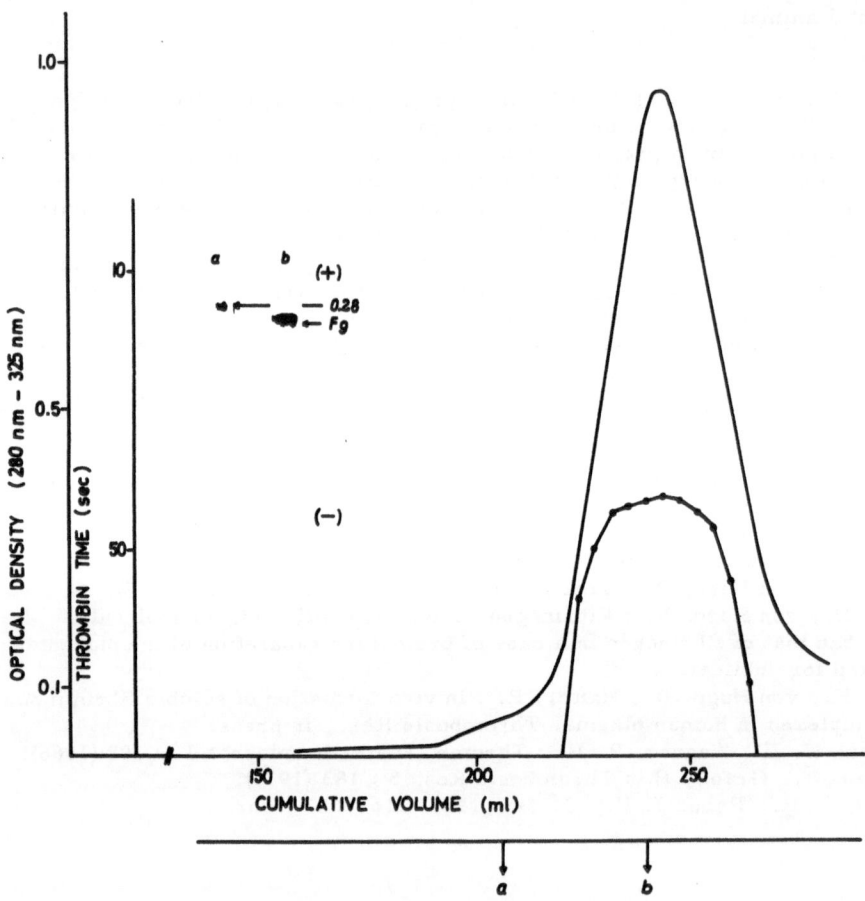

Fig. 1. Elution pattern following agarose gel filtration of ß-alanine precipitated plasma from virgin rabbits. PAA gels on the left represent fractions eluted at a) and b). Thrombin times in the eluted fractions (─•─•─•─).

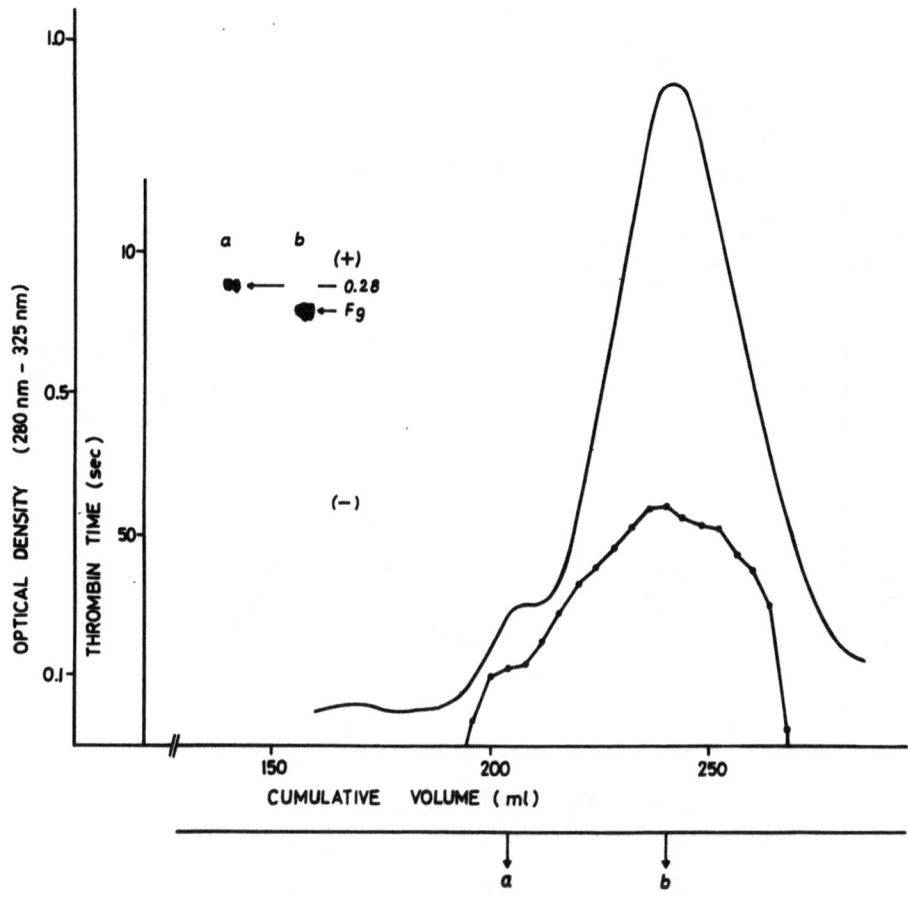

Fig. 2. Elution pattern following agarose gel filtration of ß-alanine precipitated plasma from pregnant rabbits. PAA gels on the left represent fractions eluted at a) and b). Thrombin times in the eluted fractions (—•—•—•—).

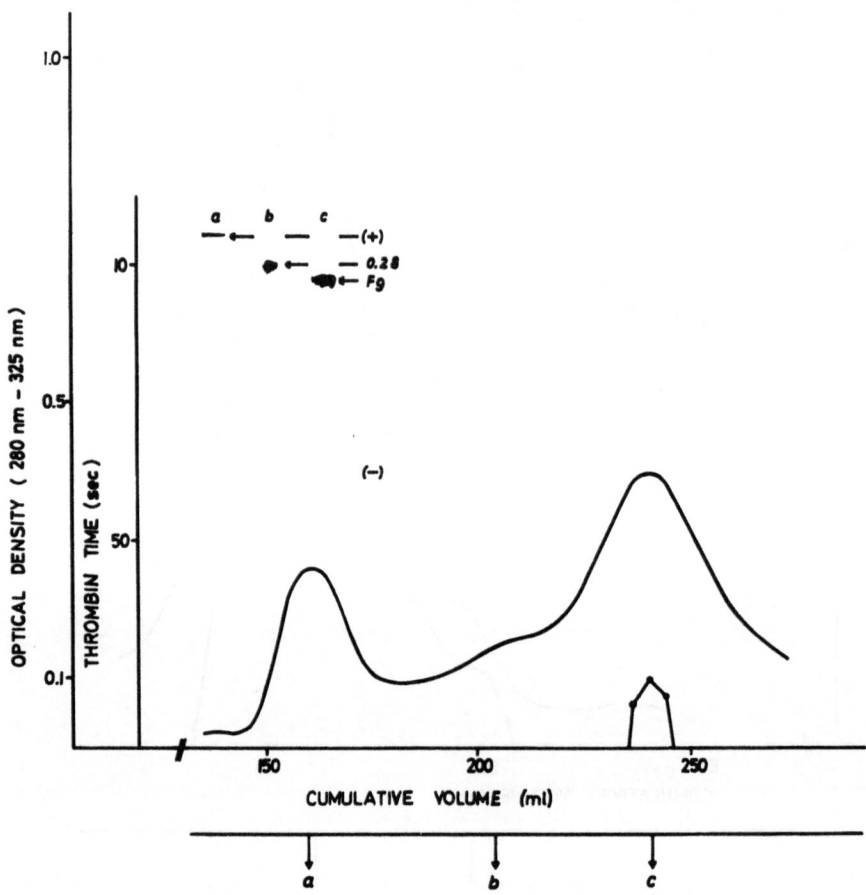

Fig. 3. Elution pattern following agarose gel filtration of ß-alanine precipitated plasma from pregnant rabbits after 6 hours of endotoxin infusion. PAA gels on the left represent fractions eluted at a), b) and c). Thrombin times in the eluted fractions (—•—•—•—).

Endotoxin Effects on Kidney Morphology and Function

F. K. Beller

Glomerular capillary fibrin deposition is considered as the hallmark of the generalized Shwartzman reaction (GSR) (24, 40). Apitz in his original publication (1) emphasized renal cortical necrosis and coagulation defects as criteria for the GSR but did not recognize fibrin deposition as the leading criterion.

The various experimental animal models were summarized up to 1970 by Beller (3). Many modifications were published since that time. It seems reasonable to question the validity of the term GSR for various mechanisms in a variety of species. Especially the transfer of animal data to human disease seems to be open to question. The introduction of a variety of new terms like thrombohemorrhagic phenomenon (34) to name only one seems to indicate the insecurity to use the GSR as standard terminology for the morphologic endpoint in human disease. In addition, terms like consumptive coagulopathy and disseminated intravascular coagulation, are used frequently synonymously without clear definition, especially so far as human disease is concerned.

The following presentation is related to renal effects of endotoxin. Our group has evaluated in 1967 a modification of the original model (5, 9). The "lag phase" between the two intravenous i n j e c t i o n s of endotoxin was eliminated by a continuous intravenous endotoxin i n f u s i o n. The decrease in coagulation factors, especially platelet count, plasma count, plasma fibrinogen, factor V and VIII was correlated with glomerular fibrin deposition and renal function (9, 18, 19). The data obtained provided evidence that the amount of deposited fibrin expressed by glomerular involvement is correlated with the increasing activation of the coagulation system. However, this working hypothesis had to be abandoned when the infusion model was used in the pregnant and non-pregnant rat (33). The virgin rat was considered to be endotoxin-resistant since neither one nor more spaced endotoxin injections produce glomerular fibrin deposition. It was frequently overlooked that even one injection of endotoxin results in a decrease in platelet count, plasma fibrinogen, factor V and VIII all nearly identical to the coagulation deficiency in the rabbit (Fig. 1). The virgin rat therefore reacts to endotoxin with glomerular fibrin deposition or what was termed in our laboratory disseminated intravascular coagulation (DIC). Both consumptive coagulopathy and DIC as well can be induced in the pregnant rat by one injection of endotoxin or in the virgin rat by a continuous infusion (33). The rat is therefore reacting species specifically to endotoxin and presents a suitable animal for demonstrating the difference between consumptive coagulopathy and disseminated intravascular coagulation (DIC).

It became obvious later that the infusion medium is of significance to explain the difference in results between injection and infusion of endotoxin. Glomerular fibrin deposition can be induced in the virgin rat after a single injection of endotoxin followed by a subsequent infusion of physiologic saline. There is a close correlation between the dose of endotoxin injected and the amount of saline infused subsequently (36) (Fig. 2). Replacement of saline by glucose following the injection does not result in glomerular fibrin deposition. The pathologic mechanism of saline is not yet clarified. The infusion of hypertonic saline without the preceding endotoxin injection does not change the coagulation system (unpublished results).

Injection or infusion of endotoxin induces a decrease in platelet number in a variety of species, e.g. the rabbit, the rat, the hamster and the rhesus-monkey. There is ample evidence to assume that endotoxin reacts with platelets by a cytotoxin action

since the effect can be produced <u>in vitro</u> also (15). The platelet drop can also be e-
licited by the injection or infusion of a modified endotoxin (endotoxoid) which does
not produce glomerular fibrin deposition, either in pregnant or in virgin animals
(10, 17, 39). The decrease in platelet number cannot be prevented by heparin or streptokinase or by aspirin (2) but by pyrimidine derivatives (35).Pyrimidine derivatives,
however, do not prevent glomerular fibrin despite normal platelet count (35).

One of the prerequisites for the development of glomerular fibrin deposition is
the inhibition of activator activity of the fibrinolytic enzyme system. Species with
low renal activator activity are more susceptible to DIC as for instance the hamster,
the rat and the rabbit than species with higher concentration such as the dog (16).
Pregnant individuals are more susceptible than non-pregnant animals (8, 16). Also
in human pregnancy activator activity was found decreased as compared to non-pregnant individuals (16). However, despite the rather low activity the rabbit is able to
lyse glomerular fibrin deposition (9). Lee demonstrated in the rabbit that infusion
of thrombin does result in glomerular fibrin deposition only if the animal was prepared by epsilon-amino-caproic acid (EACA). Similar data were obtained in pregnant rats (23). The inhibition reaction seems to be universal since it was shown that
EACA could be substituted by nearly all known proteolytic inhibitors, including Trasylol (12). The pregnant rat is more susceptible to endotoxin measured by fibrin deposition than the virgin rat in all dose ranges employed (8) (Fig. 3). Virgin rats prepared with estrogens and progesterone were more susceptible to endotoxin (8, 38).
The significant hormone to produce this effect was shown to be estrogen (38) (Fig. 4).
Graeff et al. (18) demonstrated that rabbit kidneys filled with glomerular fibrin deposition lose their activator activity. These data were recently confirmed using a similar histochemical technique (13). Fibrinolytic inhibition seems, however, only one
prerequisite for the development of glomerular fibrin deposition.

The higher susceptibility of pregnant rats to glomerular fibrin deposition is transferable to virgin rats by exchange transfusion (37). The transferable material is as
yet unknown. As a working hypothesis the transferable factor may be high molecular
fibrin derivatives. The production of high molecular fibrin derivatives by Acron resulted in glomerular fibrin deposition if the animal was prepared by fibrinolytic inhibition or an endotoxin injection (26, 27). Similar results were produced by an infusion of purified high molecular fibrinogen derivatives (18). It is somewhat puzzling
that glomerular fibrin deposition was not preventable by heparin.

It is unknown whether high molecular fibrin derivatives, which are not identifiable
by staining for fibrin, are present as an early effect of endotoxin (14). We are per-
forming experiments at present with the fibrinmonomer staining by Bleyl et al.(14)
and immunofluorescence, to clarify this point. Accumulation of high molecular fibrin
derivatives in the microcirculatory bed may provide an explanation for early partial
renal necrosis as observed in rabbits (9) and human beings in septic shock (unpublished data).

Kidney function was correlated with glomerular fibrin deposition by Graeff et al.
(18, 19) in the rabbit. Renal shutdown was preceded by a short lasting polyuric phase.
The kidney is unable to concentrate and urokinase excretion decreases. The subsequent oliguric phase develops in close correlation with glomerular fibrin deposition.
These effects are not present in animals which have a consumptive coagulopathy as
induced by an endotoxin injection or an infusion of endotoxoid (17). There is only a
gradual difference between pregnant or non-pregnant animals.

The pathophysiologic difference between consumptive coagulopathy and DIC indicates the significance for test systems to identify fibrin deposits in the microcirculation. Coagulation tests failed to provide this information. However, the increase
in plasma hemoglobin correlates with glomerular fibrin deposition (6, 32, 33) (Fig. 5).
Fibrin(ogen) breakdown products correlate only in extremely high concentrations (6)
(Fig. 6a and 6b). Urokinase excretion decreases in correlation with renal shutdown
(20, 39).

Increase in plasma hemoglobin and decrease in urokinase excretion to identify fibrin deposition in the microcirculation were shown to be a valuable test system also in human septic shock.

It is yet unknown whether endotoxin prepares the vasculature of the glomerular capillaries by mechanisms in addition to its known effects as a trigger for coagulation activity. Recent data by Müller-Berghaus et al. indicate a special action on the kidney (28). Even more important for the evaluation of treatment schedules seems to be the question whether fibrin is formed in the microcirculation or may be the result of formation in the general circulation, thus depositing fibrin as an embolic phenomenon.

References

1. Apitz, K.: J. Immunol. 29: 255 (1935).
2. Beller, F.K.: Z. Rechtsmed. 70: 55 (1972).
3. Beller, F.K.: Experimental animal models for the production of disseminated intravascular coagulation. In: Bang, N., Beller, F.K., Deutsch, E., Mammen, E.: Thrombosis and Bleeding Disorders. (Thieme, Stuttgart; Academic Press, New York 1971).
4. Beller, F.K.: Thromb. Diath. haemorrh., Suppl. 34: 125 (1964).
5. Beller, F.K., Graeff, H.: Nature 215: 295 (1967).
6. Beller, F.K., Theiss, W.: Thromb. Diath. haemorrh. 29: 363 (1973).
7. Beller, F.K., Douglas, G.W.: Obstet. Gynec. 41 : 521 (1973).
8. Beller, F.K., Schoendorf, T.: Gynec. Invest. 3 : 176 (1972).
9. Beller, F.K., Graeff, H., Gorstein, F.: Amer. J. Obstet. Gynec. 103 : 544 (1969).
10. Beller, F.K., Douglas, G.W., Graeff, H.: Zur Pathologie des Endotoxinschocks. In: Navratil, E., Heiss, H. Vortr. Wiss. 5. Akad. Tag dtschspr. Profess. Geb. Gyn. Graz 1968, p. 110 (Thieme, Stuttgart 1971).
11. Beller, F.K., Douglas, G.W., Morris, R.H., Johnson, A.J.: Amer. J. Obstet. Gynec. 101 : 587 (1968).
12. Beller, F.K., Mitchell, P.S., Gorstein, F.: Thromb. Diath. haemorrh. 18: 427 (1967).
13. Bergstein, J.M., Michael, A.F.: Thromb. Diath. haemorrh. 29: 27 (1973).
14. Bleyl, J., Sebening, H., Kuhn, W.: Thromb. Diath. haemorrh. 22: 68 (1969).
15. Des Prez, R.M., Horowitz, H.I., Hook, E.W.: J. exp. Med. 114: 357 (1961).
16. Epstein, M.D., Beller, F.K., Douglas, G.W.: Obstet. Gynec. 32 : 494 (1968).
17. Graeff, H., Schuetz, H.J., v. Hugo, R., Kuhn, W., Immich, H., Bleyl, J., Beller, F.K.: The effect of endotoxin and "endotoxoid" infusion in pregnant rabbits. (In press 1974).
18. Graeff, H., Mitchell, P.S., Beller, F.K.: Lab. Invest. 19:169 (1968).
19. Graeff, H., Beller, F.K.: Thromb. Diath. haemorrh. 20: 420 (1968).
20. Graeff, H., Belzer, M., Immich, H., Kuhn, W.: Int. J. Gynec. Obstet. 9 : 141 (1971).
21. Good, R.A., Thomas, L.: J. exp. Med. 97: 871 (1953).
22. Lee, L.: J. exp. Med. 115: 1065 (1962).
23. Margaretten, W., Zonker, H.O., McKay, D.G.: Lab. Invest. 13: 552 (1964).
24. McKay, D.G.: Disseminated intravascular coagulation. (Heeber Med. Div. Harper and Row, New York 1964).
25. McKay, D.G., Latour, J.G., Lopez, A.M.: Thromb. Diath. haemorrh. 26: 71 (1971).
26. Müller-Berghaus, G., Mann, B.: Thromb. Res. 2: 305 (1973).
27. Müller-Berghaus, G., Hocke, M.: Brit. J. Haemat. 25: 111 (1973).
28. Müller-Berghaus, G., Róka, L., Lasch, H.G.: Thromb. Diath. haemorrh. 29: 375 (1973).

29. Müller-Berghaus, G., Hocke, M.: Thromb. Res. 1: 541 (1972).
30. Müller-Berghaus, G., Schneberger, R.: Brit. J. Haemat. 21: 513 (1971).
31. Müller-Berghaus, G., McKay, D.G.: Lab. Invest. 17: 276 (1967).
32. Regoeczi, E., Rubenberg, M.L., Brain, M.C.: Lancet I: 601 (1967).
33. Schoendorf, T., Rosenberg, M., Beller, F.K.: Amer. J. Path. 65: 51 (1971).
34. Selye, H.: Thrombohaemorrhagic Phenomena. (Thomas, Springfield, Ill. 1966).
35. Slabber, C.F., Theiss, W., Beller, F.K.: J. Med. 3: 341 (1972).
36. Theiss, W., Beller, F.K.: J. Lab. clin. Med. 81: 473 (1973).
37. Theiss, W., Beller, F.K.: Surg. Gynec. Obstet. 135 : 713 (1972).
38. Theiss, W., Beller, F.K.: Amer. J. Obstet. Gynec. 115 : 775 (1973).
39. Theiss, W., Graeff, H. et al.: Thromb. Diath. haemorrh. 23: 368 (1970).
40. Thomas, L.: Annu. Rev. Physiol. 16: 467 (1954).

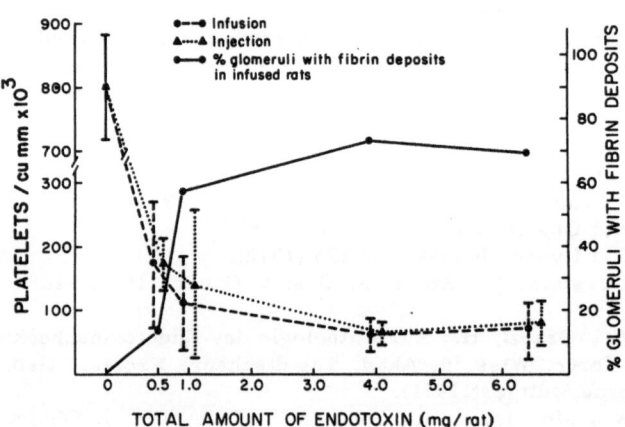

Fig. 1.   Platelet number after a single injection or a continuous infusion of endotoxin. Glomerular fibrin deposition is present only after the infusion (from Schoendorf et al.,33).

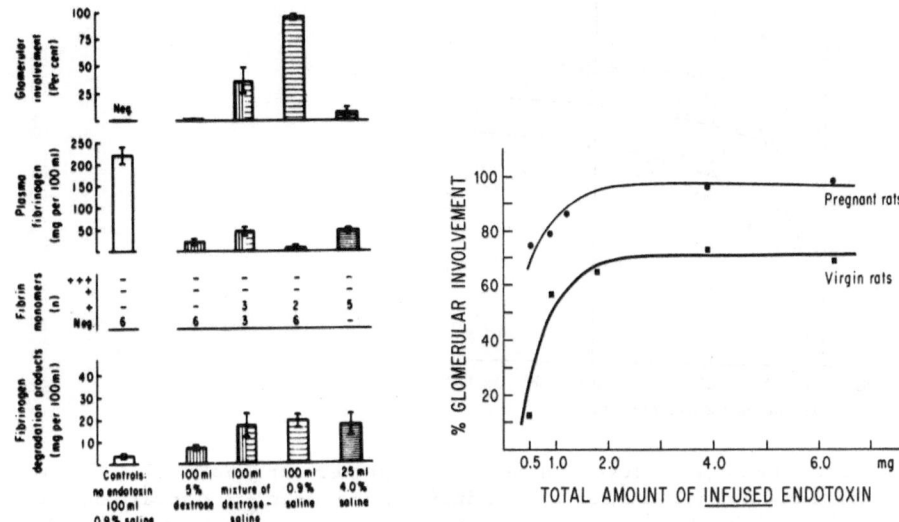

Fig. 2.   Fig. 3.

Fig. 2.   Glomerular fibrin deposition after a single injection of endotoxin followed by an infusion of saline (from Theiss and Beller, 38).

Fig. 3.   Dose response curve to infused endotoxin after 6 hours measured by glomerular fibrin deposition in virgin and pregnant rats (from Beller and Schoendorf, 8).

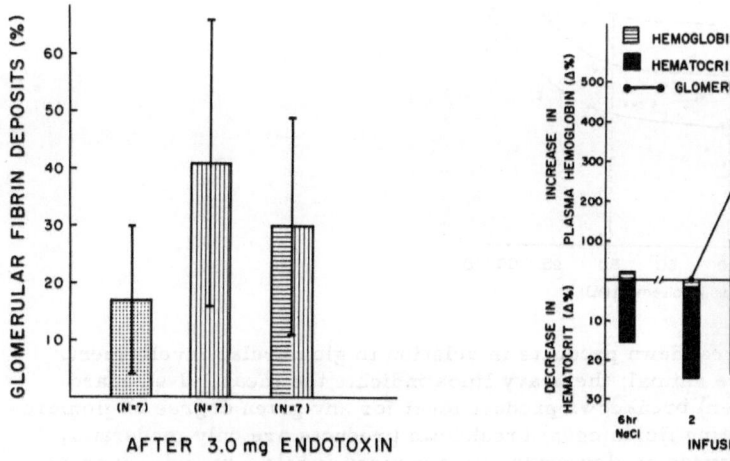

Fig. 4.   Fig. 5.

Fig. 4.   Glomerular fibrin deposition after infusion of endotoxin. First column: control, second column after pretreatment with stilbestrol, third column after pretreatment with stilbestrol and gestagen (from Theiss and Beller, 38).

Fig. 5.   Hematocrit and plasma hemoglobin changes (◁ %) after infusion of endotoxin correlated with glomerular fibrin deposition (from Schoendorf et al., 33).

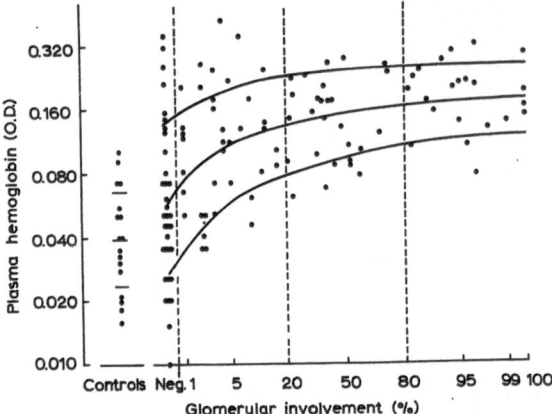

Fig. 6a.   Plasma hemoglobin levels in relation to glomerular involvement. Each point represents one animal; the heavy lines indicate the mean ± 1 standard deviation of the plasma hemoglobin level for any given degree of glomerular involvement. While endotoxin treated animals without histological evidence of DIC reveal only minor increase of plasma hemoglobin levels as compared with untreated controls, the occurrence of even moderate glomerular involvement results in a considerable increase in plasma hemoglobin levels.

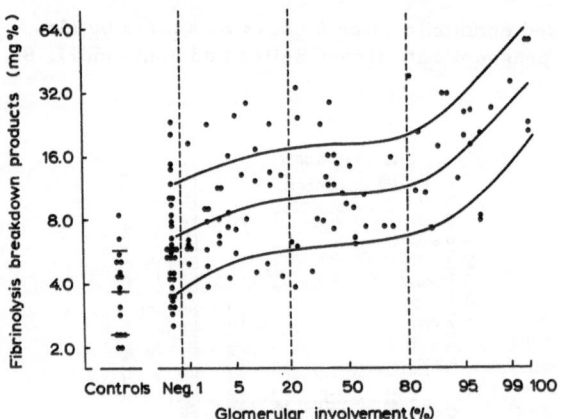

Fig. 6b.   Fibrin(ogen)breakdown products in relation to glomerular involvement. Each point represents one animal; the heavy lines indicate the mean ± 1 standard deviation of the fibrin(ogen) breakdown product level for any given degree of glomerular involvement. Circulating fibrin(ogen) breakdown products are only moderately increased with lesser degrees of glomerular involvement, while a massive increase is observed in severe DIC when glomerular involvement exceeds 80% (from Beller and Theiss, 6).

Pathomorphology of Intravascular Coagulation Following Endotoxin Administration

U. Bleyl

A pathognomonic morphologic equivalent of endotoxic shock is presently unknown. For clinical pathologists the pathomorphology of endotoxic shock essentially is a pathomorphology of shock in general, that means the histomorphologic equivalent of a polyetiologic, acute hemodynamic insufficiency with generalized disturbances of the microcirculation and of blood distribution, resulting in decreased oxygen supply in the peripheral circulation and in metabolic acidosis (16, 20, 45).

Nevertheless a series of typical, regularly detectable pathomorphologic findings is to be observed during infections with gram-negative bacteria and endotoxemia. Regardless of the underlying disease and the topographical anatomy of bacterial infection, these findings can be considered as characteristic, morphologic consequences of endotoxemia and as endotoxin-caused acute hemorrhagic insufficiency. Even though the initial pathophysiologic phenomena, which occur in the course of endotoxemia and result in endotoxic shock, are not entirely clarified, comparative human pathologic and experimental investigations without doubt have shown that the pathomorphology of endotoxic shock is determined by the direct influence of endotoxins on the release of biogenic amines and on the interrelationship between the peripheral microcirculation with its regulatory mechanisms and the hemostatic system.

Therefore, it was my particular concern to demonstrate the morphologic equivalents of endotoxin-caused generalized disturbances in the microcirculation and in hemostasis. The histomorphologic results will give the opportunity to discuss, while considering the characteristic functional organic alterations, the significance of the release of biogenic amines in connection with the pathophysiology and pathomorphology of endotoxic shock.

The earliest morphologic manifestation of endotoxin-caused disturbances of microcirculation and hemostasis are the frequently pronounced aggregations of platelets in the capillary bed, as demonstrated in vitalmicroscopic studies by Urbaschek (53, 54). These disturbances parallel the clinical determination of a suddenly occurring thrombocytopenia (Fig. 1). Initially these aggregates are observed predominantly on typical sites of disturbances of the microcirculation and blood distribution and in the subsequent veins of the peripheral circulation. Soon thereafter, they can be seen in the pulmonary circulation where they are filtered as microemboli of aggregated platelets from the venous blood. Via a viscous metamorphosis with subsequent thrombocytolysis, the dose-dependent irreversible aggregation of platelets in the capillary bed and the connected venous regions induce the release of vasoactive substances such as catecholamines, serotonin, and histamine, as well as the release of thromboplastic and heparin-neutralizing activities. The thrombocytogenic substances, serotonin and histamine, released in the periphery in parallel to the aggregation of platelets and emboli, and both catecholamines are able to induce a marked vasoconstriction in the pulmonary circulation which can be accentuated by any kind of hypoxemia and metabolic acidosis of the blood. Thus, in the initial stage of endotoxic shock the lung already represents the reacting organ to catecholamines released in the peripheral circulation as well as to emboli of aggregated platelets.

The causative initiators of platelet aggregations are presently unknown. A direct effect of endotoxins on platelets does not seem to exist since in vitro studies, in which platelet-rich plasma was incubated with endotoxin, aggregates of platelets were not inducible. At present especially endotoxin-induced endothelial lesions with desquamation of endothelial cells (12) are considered as a cause of platelet aggregation. Alternatively endotoxin-caused release of intramural procoagulative activities from the vascular wall (12, 15, 48), as well as, endotoxin-induced aggregation of leukocytes with the formation of so-called granulocytic plugs (55, 56), have been discussed as possible stimuli of platelet aggregation.

In view of the twofold irritation of the pulmonary circulation by thrombocytogenic vasoconstriction and by the occurrence of platelet emboli, it is not surprising that hemodynamics of endotoxic shock in the initial stage is characterized by a dose-dependent decrease in blood pressure as consequence of increased capillary resistance in the pulmonary circulation. Experimental studies, however, revealed that this initial stage of endotoxic shock is frequently reversible. Through the release of cate-cholamines this initial stage turns into a "compensatory stage" in which the mean arterial blood pressure returns to normal values. In general only after an interval without symptoms or with minor clinical symptoms the typical clinical manifestation of endotoxic shock develops with generalized disturbances of the microcirculation, generalized turnover of fibrinogen and of other plasma coagulation factors, and with the formation of soluble, intravascular, intermediary products of fibrinogen-fibrin-transformation.

The formation of so-called fibrin monomers is the earliest morphologic manifestation of the disturbances of microcirculation and hemostasis in the third stage of endotoxic shock. Fibrin monomers represent the intermediary products of thrombin-induced proteolytic cleavage of two low-molecular fibrinopeptides, fibrinopeptide A and fibrinopeptide B, each from the terminal site of the fibrinogen molecule. Therefore, fibrin monomers are smaller in size than the parent fibrinogen molecule. Within the circulation fibrin monomers form higher molecular complexes with the parent fibrinogen molecule (so-called fibrinogen-fibrin-monomer-complexes) which can be dissociated in the presence of 5 molar urea. These complexes remain soluble as long as they do not polymerize under the influence of thrombin.

Circulating fibrin monomers can be detected histomorphologically in the capillary bed if their local concentration is high enough. Under these circumstances highly diluted alcohol is capable of artificially aggregating fibrin monomers and of polymerizing them (5, 6). This fact is applied clinically in the so-called alcohol-test for fibrin monomers (17). Gels formed in vitro from aggregated and polymerized fibrinogen-fibrin-monomer-complexes differ from fibrin aggregated and polymerized in vivo when compared by means of electron microscopy and light microscopy (22). In histologic slides peculiar long-stretched, bundel-shaped, mycelium-like precipitates are observed which are frequently seen in close association with platelet aggregates (Fig. 2). These can be demonstrated by means of PAS-reactivity, as well as with conventional staining methods for fibrin (6). Comparative clinical and patho-anatomical studies show that circulating fibrinogen-fibrin-monomer-complexes which are clinical and morphological signs of a preceding activation of coagulation accompanied by increased turnover and consumption of coagulation factors ("consumption coagulopathy") (23, 24, 25), can be detected (Fig. 3) long before the formation of highly polymerized intravascularly precipitating fibrin derivatives.

The circulating fibrin monomers are eliminated by cells of the reticulo-endothelial system. Similar to other corpuscular elements fibrin monomers are adsorbed in a biphasic process onto the surface of reticulo-endothelial cells (Fig. 2). In a second phase they are taken up into the cell and intracellular proteolysis begins. During intracellular digestion the reticulo-endothelial cells lose their ability to adsorb fibrin monomers from the blood, referred to as "blockade of the reticulo-endothelial system". Besides and apparently independent from intracellular fibrinolysis by the RES fibrin monomers underlie humoral fibrinolysis through the plasminogen-plasmin-system and its activators.

High molecular intermediary (fragment X, fragment Y) and low molecular (plasmin-resistant) terminal (fragment D, fragment E) fibrin split products ("fibrin degradation products") are formed. These products are characterized by a lacking (fragment Y) or a highly reduced (fragment X) coagulability, by a marked antithrombin activity (fragment Y) and especially by a polymerization-inhibiting character (anti-polymerase-activity) (28, 29, 30, 31, 32). Apparently easier than fibrinogen molecules these fibrin degradation products form soluble complexes (so-called fibrin-monomer-fibrin-degradation-complexes). In these complexes they inhibit further polymerization of fibrin monomers (1, 26). Moreover, fibrin degradation products are incorporated into soluble and circulating highly polymerized intermediary fibrin and thereby block further coagulability and polymerization of fibrin polymers (21) by the formation of abnormal and unstable fibrin polymers (so-called defect polymers).

Intermediary and terminal fibrin degradation products cannot be detected histomorphologically. However, abnormal and unstable "defect polymers", resulting from incorporation of fibrin degradation products (FDP) into highly polymerized intermediaries of fibrinogen-fibrin-transformation, are detectable in the capillary bed (Fig. 4-6). The morphological equivalent of these "defect polymers" are 3 to 30 μ large, hyaline globules or so-called Siegmund-Schindler globules (Fig. 6). Comparative immunohistochemical and electron microscopic studies showed that these microthrombi, demonstrable with the PAS-reaction and the tryptophan procedure, essentially consist of fibrinogen- or fibrin derivatives. However, they only occasionally show periodicity of 180-210 Å (46, 47) so characteristic of the ultrastructure of fibrin. As a rule these globules electron microscopically appear as densely packed, mosaic-like filaments without typical transverse striation. The filaments also can be penetrated with or surrounded by disseminated formations of fibers with typical transverse striation (Fig. 4 and 5). Histologically, mycelium-like filaments are observed especially on the surface, but also in the center, of the hyaline precipitates. Obviously these filaments are again equivalents of incorporated intermediaries of fibrinogen-fibrin-transformation (Fig. 4). In view of the pathophysiologic inhomogeneity of the defect polymers which are formed in vivo and precipitate in the capillary bed of the organs, the principle heterogeneity of the light- and electron microscopic finding on the globules is not surprising.

Under the condition of a functioning humoral fibrinolysis, an ubiquitous activation of coagulation and an increased turnover of coagulation factors (consumption coagulopathy), as a rule lead to the formation of fibrinogen-fibrin-monomer-complexes in the capillary bed and to the formation of FDP-containing globules, but not to the formation of highly polymerized microthrombi. The decisive initiator of activation of fibrinolysis in endotoxic shock is the peripheral hypoxemia which parallels the generalized microcirculatory disturbances. According to studies of Bergstein and Michael (3) this hypoxia leads to liberation of endothelial plasminogen activators. The process coincides with consumption of plasminogen. Polytopic highly polymerized intravascular fibrin thrombi, the typical equivalent of generalized intravascular activation of the coagulation system in endotoxic shock, in principle are formed under conditions when:
1. the localized or generalized humoral fibrinolysis is inhibited;
2. the localized or generalized humoral fibrinolysis is exhausted by an excessive supply of substrates;
3. a fibrinolytic potential, which per se is quantitatively and qualitatively sufficient, cannot be effective in the course of shock-specific generalized or localized disturbances of the microcirculation and blood distribution.

Thus, the precipitation of highly polymerized intravascular microthrombi in each form of shock represents a problem of balance between the intensity of local or generalized activation of coagulation and the intensity of local or generalized fibrinolytic clearance activity. The balance is essentially influenced by shock-specific vasoactive factors.

A pronounced but not complete inhibition of the humoral fibrinolytic potential is especially manifested during pregnancy (41). Gravidity predisposes to precipitation of polytopic highly polymerized fibrin thrombi. The inhibition of fibrinolysis prevents the resolution of highly polymerized, intravascularly precipitated microthrombi (2, 18, 19, 52) and this favors the formation of multilocular organic necrosis (33). It is not surprising that endotoxic shock in septic abortion and in chorionamnionitis via this inhibition of fibrinolysis, almost regularly leads to precipitation of fibrin-rich intravascular microthrombi in glomerular capillaries of the kidney with reactive impairment of the glomerular filtration rate, clinically detectable oligoanuria and manifestations of a bilateral renal cortical necrosis (23, 34, 35, 40, 49, 57). In 1941 Duff and More (14) reported that two-thirds of all cases with bilateral renal cortical necrosis occur in pregnant women.

Histochemical studies by our group using the method of fibrinolysis-autography according to Todd showed that in states other than pregnancy endotoxic shock also coincides with an almost regular decrease in endothelial plasminogen-activator activity in different human organs. These results were predicted for some time based on experimental studies in rabbits following Shwartzman-equivalent injections or infusion of endotoxin (3, 18). It has not been clarified whether the assumption of endothelial plasminogen-activator activity is to be evaluated as a consequence of a consumption of this activity or as a consequence of the disturbed cellular function during shock-induced disturbances of the microcirculation. Independent from this pathogenic alternative, the decrease of endothelial plasminogen-activator activity favors, also in states other than pregnancy, the formation of highly polymerized fibrin-like intravascular microthrombi.

The histomorphologic picture of highly polymerized fibrin thrombi after localized or generalized inhibition or exhaustion of fibrinolysis is characterized by inhomogenous, partly massive, partly fine-fibered, intensively PAS-positive precipitates (Fig. 7). After treatment of the native specimens with fluorescein-labelled anti-human-fibrinogen sera these precipitates show intense specific secondary fluorescence (Fig. 8). They adhere to the capillary wall, at first in a loose connection and later they are broadly attached to the wall (Fig. 9). Beginning endothelial activation can be observed at necropsy after a surviving time longer than 24 hours. By electron microscopy these highly polymerized precipitates of fibrin appear as bundles of fibers of varying thickness, which partly have a constant axial periodicity of 180-220 $\overset{\circ}{A}$ and in part consist of longitudinal fibrillae without demonstrable periodicity. Sometimes aggregates of platelet residues with complete loss of granules or intact platelets with partial loss of granules are observed between the bundles of fibers. At the same site, precipitates of plasma are regularly seen resulting from occlusions of gamma globulins and lipoproteins in fibers (42) as detected by immunofluorescence.

With the exception of endotoxic shock during pregnancy the site of predilection for the intravascular precipitates of highly polymerized microthrombi are primarily the pulmonary capillaries, but also the glomerular capillaries of the kidney and the sinusoids of the organs of the RES. Only occasionally PAS-positive microthrombi are observed in the sinusoids of the adrenals, in the pituitary gland, as well as in the capillaries of the tunica mucosa and submucosa of the gastrointestinal tract. The topography of disseminated microthrombi obviously is essentially influenced by the variability of the sympathetic-adrenergic activity of the organs and by the intensity of "shock-specific" vasomotion (36). Studies by Müller-Berghaus and McKay (37, 38) demonstrated that stimulation of alpha-adrenergic receptor sites of the kidney leads to precipitation of fibrin in the glomeruli of pregnant rats. Blocking these receptors with dibenzamine or dibenzyline prevents the formation of microthrombi. By sympathetic denervation of the kidney Palmerio et al. (39) were able to suppress the formation of bilateral cortical necrosis following two Shwartzman-equivalent injections of endotoxin. Also adrenalectomy prevents the manifestation of fibrin-rich intravascular microthrombi in the circulation of the kidney after two injections of endotoxin. This is due to the decrease in catecholamines responsible for the stimulation of alpha-adrenergic receptor sites and the decrease in sensitivity of the striated musculature of the vessels to catecholamines when glucocorticoids are diminished. According to Latour (27) the injection of glucocorticoids leads to an increased reactivity of the arteriolar musculature to catecholamines.

The frequent occurrence of disseminated microthrombi in the pulmonary circulation also results from the special topography of the vessels of the lung and from the filtration function of the pulmonary circulation for the corpuscular aggregates formed in the peripheral circulation. Outside of the pulmonary circulation the intensity of aggregations is favored by the fact that two-thirds of the total blood volume circulates in the venous system where the velocity of the blood stream is markedly slower than in the arterial regions.

In recent years experiments have revealed that besides the disseminated micro-thrombosis a further characteristic equivalent of generalized intravascular activation of coagulation, the well-known pulmonary hyaline membranes (4, 7, 9, 10), can be manifested in the lung as a consequence of this special topographic situation (Fig. 10). Pulmonary hyaline membranes frequently are massive and homogenously condensed intra-alveolar precipitates of fibrin and other plasma proteins (Fig. 11). Between these precipitates residues of membrane profiles are visible in varying abundance as well as cytoplasmatic vesicles, osmophilic lamellar corpuscles, fragments of mitochondria, nuclear fragments and residues of erythrocytes. In electron microscopic studies the extravascular fibrin derivatives appear as a dense network of varying, broad, branched precipitates which show, like fibrin, a regular periodic transverse striation with a periodic length of 210 Å. The fibers of fibrin are adjacent to the alveolar epithelium or to the denuded alveolar walls. Only occasionally parts of a second fine membrane can be observed in the region between the precipitates of fibrin and the alveolar wall. These fine membranes correspond to the residues of the surfactant activity necessary for the function of the lung.

We performed systematic pathological and experimental studies on shock-lungs, lungs with respiratory distress syndrome, lungs with so-called uremic pneumonitis, and on rabbit lungs with experimentally induced long-lasting oxygen intoxication. These studies revealed that the formation of pulmonary hyaline membranes in these etiologically heterogeneous diseases and pathogenic, intermediary mechanisms constantly coincides with a generalized intravascular coagulation (4, 7, 8, 9, 10, 11, 13). Shock-induced endothelial lesions, uremic capillaropathy and oxygen-caused disturbances in permeability of the capillaries of the lung favor the extravasation of soluble intravascular fibrinogen-fibrin-intermediaries and their extravascular polymerization as pulmonary hyaline membranes. Like disseminated microthrombi, hyaline membranes are characteristic signs of intravascular activation of coagulation as well as of increased turnover of intravascular factors and of the formation of circulating intermediaries of fibrinogen-fibrin-transformation. The significance of extravascular polymerizing fibrinogen derivatives for the inhibition of the surfactant activity on the alveoles (so-called anti-atelectasis-factors, "surfactant", 43, 44, 50, 51) and for the pathogenesis of the so-called shock-lung in this context has been described in detail (10).

Like disseminated intravascular microthrombi, pulmonary hyaline membranes cannot be considered as a pathognomonic morphologic equivalent of endotoxic shock. The occurrence of hyaline membranes is merely the morphologic equivalent of the disturbances of hemostasis characteristic of endotoxic shock. As extravascular precipitates of fibrin the hyaline membranes, however, can demonstrate endotoxin-caused intravascular disturbances of hemostasis, even in the case when the intravascular precipitates are destroyed by the shock-induced fibrinolytic activity. Taylor and Abrams (50, 51) demonstrated that extravascular fibrinolysis can be inhibited by the intra-alveolar surfactant activity. Resolution of hyaline membranes obviously

only results when the so-called shock-lung becomes a medium for bacterial superinfection and when a lobular bronchopneumonia (Fig. 12) with diffuse granulocytic
infiltration develops. In case the resolution of the membranes fails to appear their
organization follows (Fig. 13) with formation of intra-alveolar and septal fibrosis.
The patients die with the clinical picture of a rapidly progressive pulmonary insufficiency even when endotoxic shock seems to be under control.

References

1. Alkjaersig, N., Fletcher, A. P., Sherry, S.: J. clin. Invest. 41: 917 (1962).
2. Beller, F. K., Mitchell, P. S., Gorstein, F.: Thromb. Diath. haemorrh. 17: 427 (1967).
3. Bergstein, J. M., Michael, A. F.: Thromb. Diath. haemorrh. 29: 27 (1973).
4. Bleyl, U.: Verh. dtsch. Ges. Path. 55: 39 (1971).
5. Bleyl, U., Sebening, H., Kuhn, W.: Thromb. Diath. haemorrh. 22: 68 (1969).
6. Bleyl, U., Kuhn, W., Graeff, H.: Thromb. Diath. haemorrh. 22: 87 (1969).
7. Bleyl, U., Büsing, C. M., Krempien, B.: Virchows Arch. path. Anat. Abt. A 348: 187 (1969).
8. Bleyl, U., Büsing, C. M.: Klin. Wschr. 48: 13 (1970).
9. Bleyl, U., Heilmann, K., Adler, D.: Klin. Wschr. 49: 71 (1971).
10. Bleyl, U., Büsing, C. M.: Pathogenese pulmonaler hyaliner Membranen. In: Wiemers, H.: Lungenveränderungen bei Langzeitbeatmung, p. 19 (Thieme, Stuttgart 1973).
11. Bleyl, U., Werner, Ch., Büsing, C. M.: Klin. Wschr. 52 (1974). In press.
12. Bouvier, C. A., Gaynor, E., Cibtron, J. R., Bernhardt, B., Spaet, T. H.: Thromb. Diath. haemorrh., Süppl. 40: 163 (1970).
13. Büsing, C. M., Bleyl, U.: Virchows Arch. path. Anat. 363: 113 (1974).
14. Duff, G. L., More, R. H.: Amer. J. med. Sci. 201: 428 (1941).
15. Gaynor, E., Bouvier, C. A., Spaet, T. H.: Science 170: 986 (1970).
16. Gelin, L. E.: Flüssigkeitsersatz im Schock. In: Bock, K. D.: Schock, Pathogenese und Therapie, p. 372 (Springer, Berlin/Göttingen/Heidelberg 1962).
17. Godal, H. C., Abildgaard, U.: Scand. J. Haemat. 3: 342 (1966).
18. Graeff, H., Mitchell, P. S., Beller, F. K.: Lab. Invest. 19: 169 (1968).
19. Graeff, H., Schütz, H. J., von Hugo, R., Kuhn, W., Immich, H., Bleyl, U., Beller, F. K.: Arch. Gynäk. 216: 205 (1974).
20. Hardaway, R. M.: Syndromes of disseminated intravascular coagulation (Thomas, Springfield, Ill. 1966).
21. Hirsh, J., Fletcher, A. P., Sherry, S.: Amer. J. Physiol. 209: 415 (1965).
22. Kuhn, W., Bröcker, G., Graeff, H., Bleyl, U., Frost, H.: Klin. Wschr. 49: 106 (1971).
23. Lasch, H. G., Krecke, H. J., Rodriguez-Erdmann, F., Sessner, H. H., Schütterle, G.: Folia haemat., N. F. 6: 325 (1961).
24. Lasch, H. G., Heene, D. L., Huth, K., Sandritter, W.: Amer. J. Cardiol. 20: 381 (1967).
25. Lasch, H. G., Huth, K., Heene, D. L., Müller-Berghaus, G., Hörder, M. H., Janzarik, J., Mittermayer, C., Sandritter, W.: Dtsch. med. Wschr. 96: 715 (1971).
26. Latallo, Z. S., Fletcher, A. P., Alkjaersig, N., Sherry, S.: Amer. J. Physiol. 202: 681 (1962).
27. Latour, J. G., Prejean, J. B., Margaretten, W.: Amer. J. Path. 65: 189 (1971).
28. Marder, V. J.: Immunologic structure of fibrinogen and its plasmin degradation products: Theoretical and clinical considerations. In: Laki, K.: Fibrinogen, p. 339 (New York 1968).
29. Marder, V. J.: Thromb. Diath. haemorrh., Suppl. 39: 187 (1970).

30. Marder, V. J., Shulman, N. R., Carroll, W. R.: Trans. Ass. amer. Physicians 80: 156 (1967).
31. Marder, V. J., Shulman, N. R., Carroll, W. R.: J. biol. Chem. 244: 2111 (1969).
32. Marder, V. J., Shulman, N. R.: J. biol. Chem. 244: 2120 (1969).
33. Margaretten, W., Zunker, H. O., McKay, D. G.: Lab. Invest. 13: 552 (1964).
34. McKay, D. G., Shapiro, S. S.: J. exp. Med. 107: 353 (1958).
35. McKay, D. G., Shapiro, S. S.: J. exp. Med. 107: 369 (1958).
36. Messmer, K., Brendel, W.: Med. Welt 22: 1159 (1971).
37. Müller-Berghaus, G., McKay, D. G.: Lab. Invest. 17: 276 (1967).
38. Müller-Berghaus, G., Davidson, E., McKay, D. G.: Obstet. Gynec. 30: 774 (1967).
39. Palmerio, C., Ming, S. C., Frank, E., Fine, J.: J. exp. Med. 115: 609 (1962).
40. Pfau, P., Lasch, H. G., Günther, G.: Gynaecologia 150: 17 (1960).
41. Phillips, L. L.: Modifications to the coagulation mechanism during pregnancy. In: Corey, H. M.: Modern trends in human reproduction physiology, vol. I, p. 190 (Butterworth, Washington 1963).
42. Regoeczi, E.: J. Haematol. 14: 279 (1968).
43. Said, S. I., Davis, R. K., Davis, W. M., Banerjee, C. M., El-Gohary, M.: Fed. Proc. 22: 339 (1963).
44. Said, S. I., Avery, M. E., Davis, R. K., Banerjee, C. M., El-Gohary, M.: J. clin. Invest. 44: 458 (1965).
45. Schneider, M.: Zur Pathophysiologie des Schocks. In: Horatz, K., Frey, M. R.: Schock und Plasmaexpander, p. 1 (Springer, Berlin/Göttingen/Heidelberg 1964).
46. Skjorten, F.: Acta path. microbiol. scand. 73: 489 (1968).
47. Skjorten, F.: Acta path. microbiol. scand. 76: 361 (1969).
48. Spaet, T. H., Gaynor, E., Bouvier, C.: Thromb. Diath. haemorrh., Suppl. 45: 157 (1971).
49. Studdiford, W., Douglas, C. W.: Amer. J. Obstet. Gynec. 83: 1229 (1962).
50. Taylor, F. B., Jr., Abrams, M. E.: Physiologist 7: 269 (1964).
51. Taylor, F. B., Jr., Abrams, M. E.: Amer. J. Med. 40: 346 (1966).
52. Theiss, W., Graeff, H., Bleyl, U., Immich, H., Kuhn, W.: Thromb. Diath. haemorrh. 23: 369 (1970).
53. Urbaschek, B.: Zur Frage des Wirkungsmechanismus bakterieller Endotoxine und seiner Beeinflussung. Habilitationsschrift, Universität Heidelberg 1967.
54. Urbaschek, B.: The effects of endotoxins in the microcirculation. In: Kadis, S., Weinbaum, G., Ajl, S. J.: Microbial toxins, vol. V, p. 261 (Academic Press, New York/London 1971).
55. Wilson, J. W., Ratliff, N. B., Hackel, D. B.: Amer. J. Path. 58: 337 (1970).
56. Wilson, J. W., Ratliff, N. B., Young, W. G., Hackel, D. B., Mikat, E.: Changes in the morphology of leukocytes trapped in pulmonary circulation during hemorrhagic shock. In: Ditzel, J., Lewis, D. H.: Microcirculatory approaches to current therapeutic problems, p. 41 (Karger, Basel 1971).
57. Zander, J.: Septischer Abort und bakterieller Schock (Springer, Heidelberg/Berlin/New York 1968).

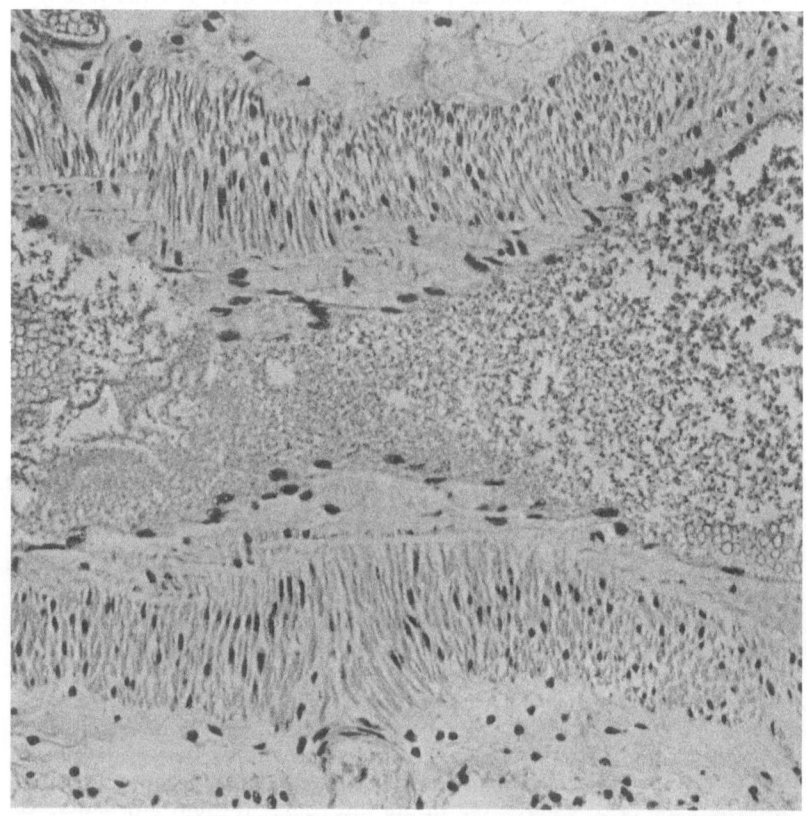

Fig. 1. Intravascular aggregates of platelets in a pulmonary artery after the second injection of endotoxin in the rabbit.

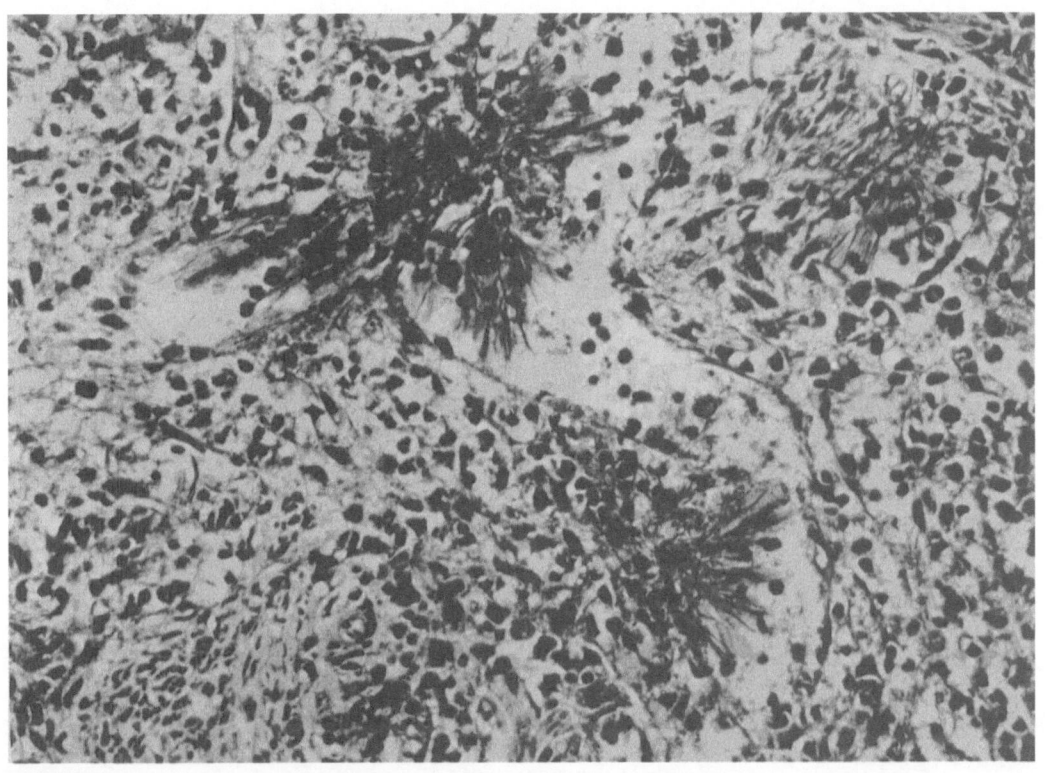

Fig. 2.   Pericellular, tuft-like, mycelium-like precipitates of <u>in vitro</u> aggregated fibrin monomers near the endothelium of the sinus and the reticulum cells of a rabbit spleen after thrombin infusion and ethanol fixation.

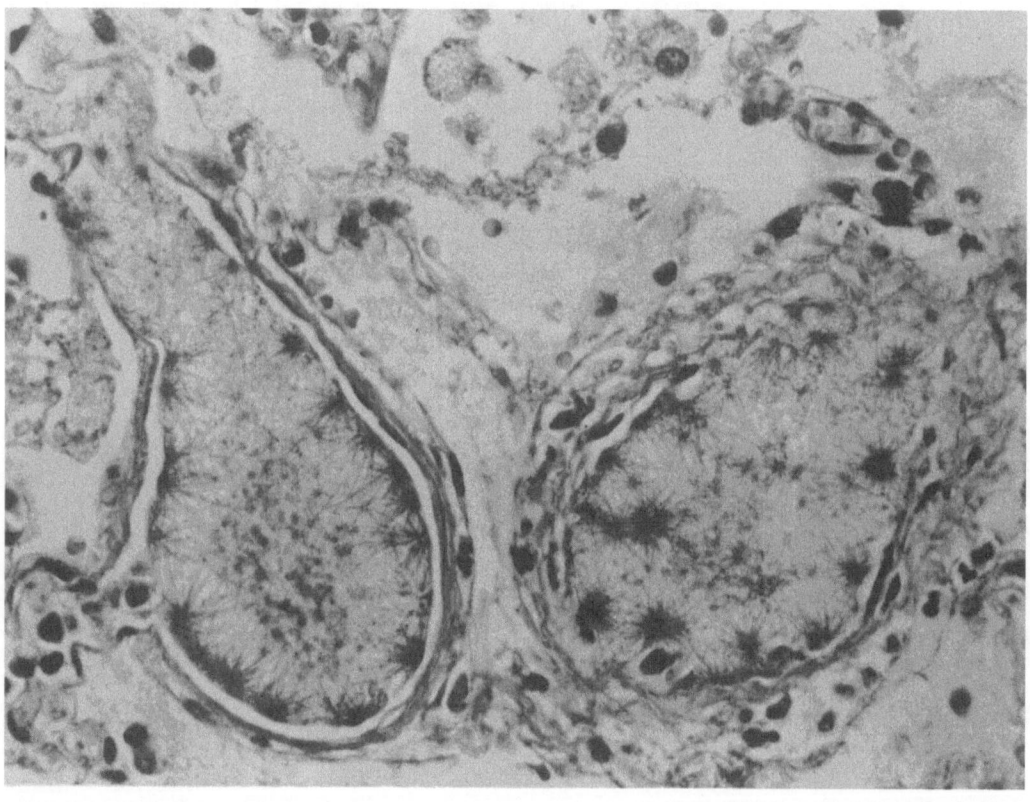

Fig. 3.   Tuft-like, mycelium-like precipitates of fibrinogen-fibrin-intermediates
in the pulmonary veins.

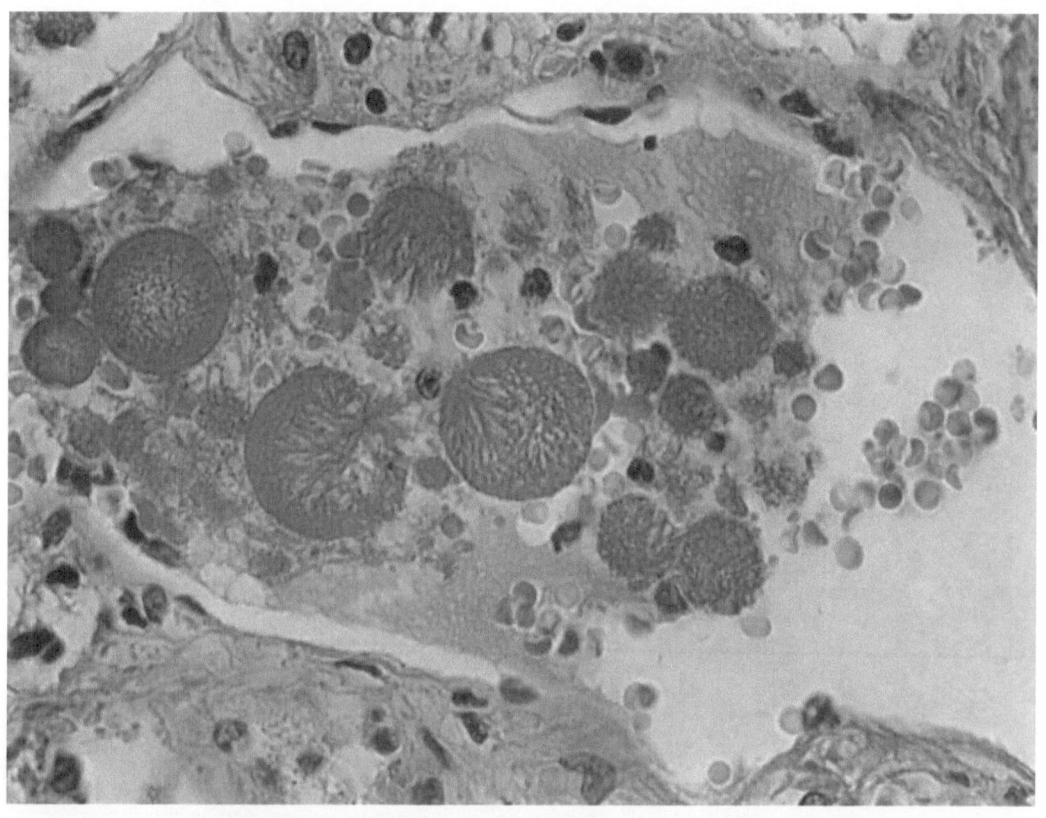

Fig. 4. Globules, formed by tuft-like, mycelium-like intermediates of fibrinogen-fibrin-transformation in a human pituitary gland after endotoxic shock.

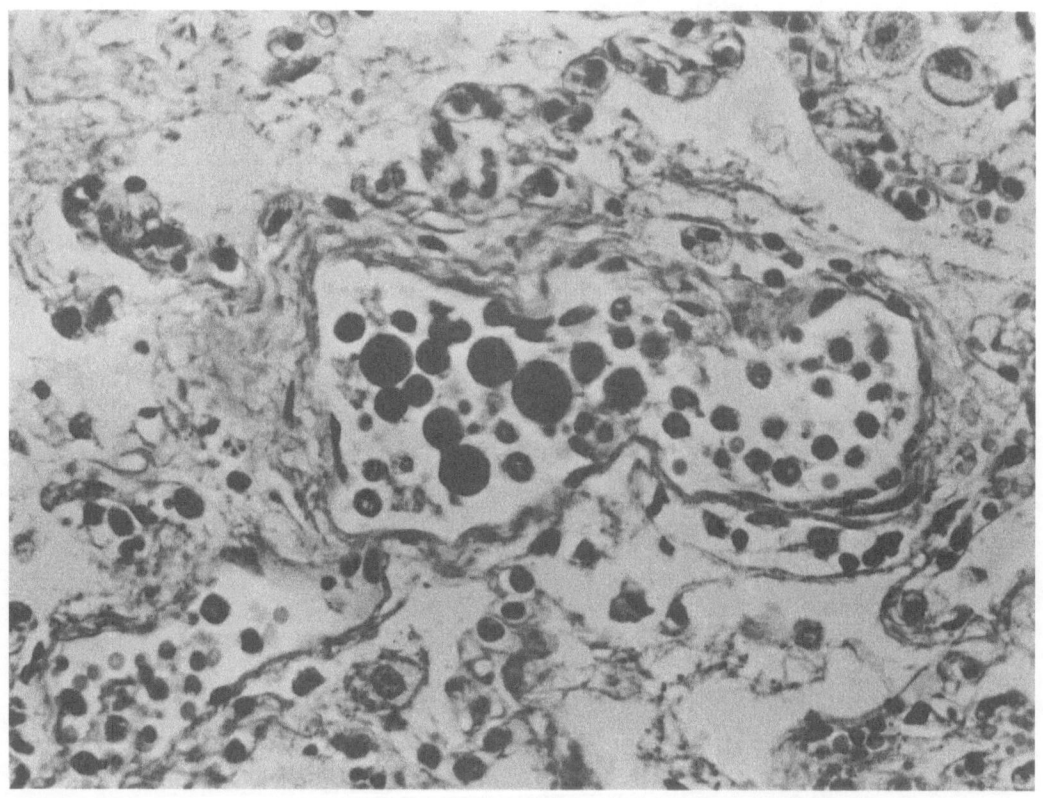

Fig. 5. Globules surrounded by tuft-like, mycelium-like precipitates of only partially polymerized fibrinogen-fibrin-intermediates in a human lung after endotoxic shock.

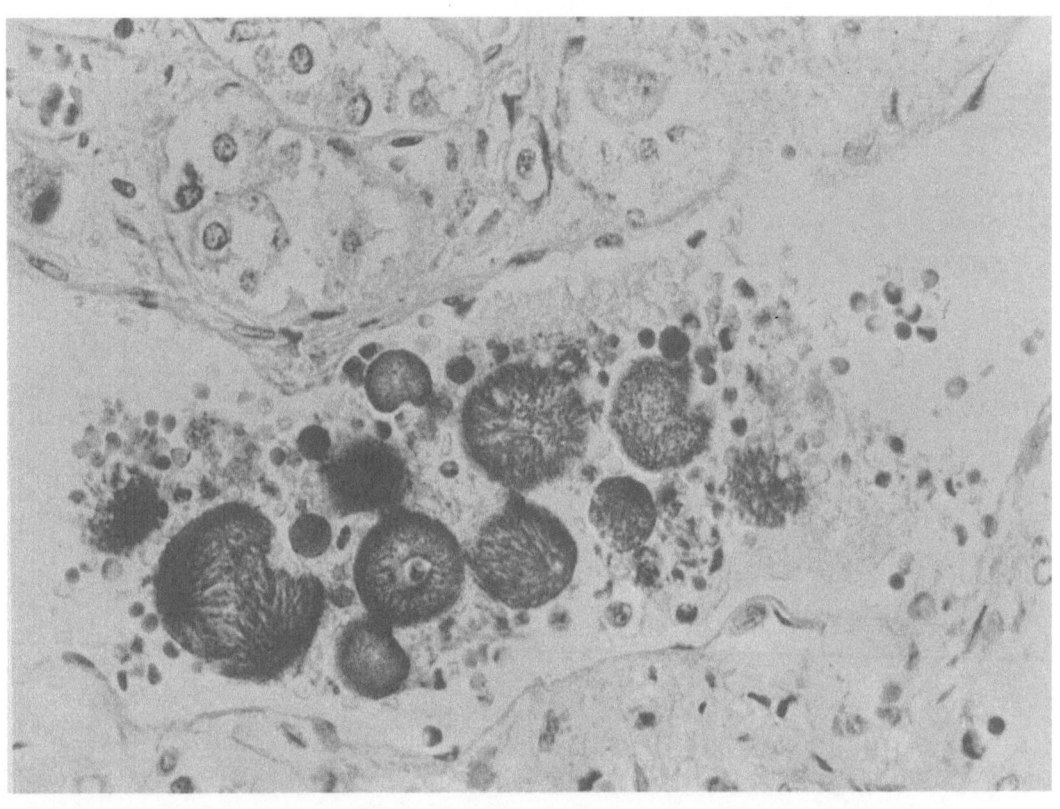

Fig. 6. Typical globules resulting from so-called defective polymers formed as complexes of fibrin monomers and fibrin degradation products.

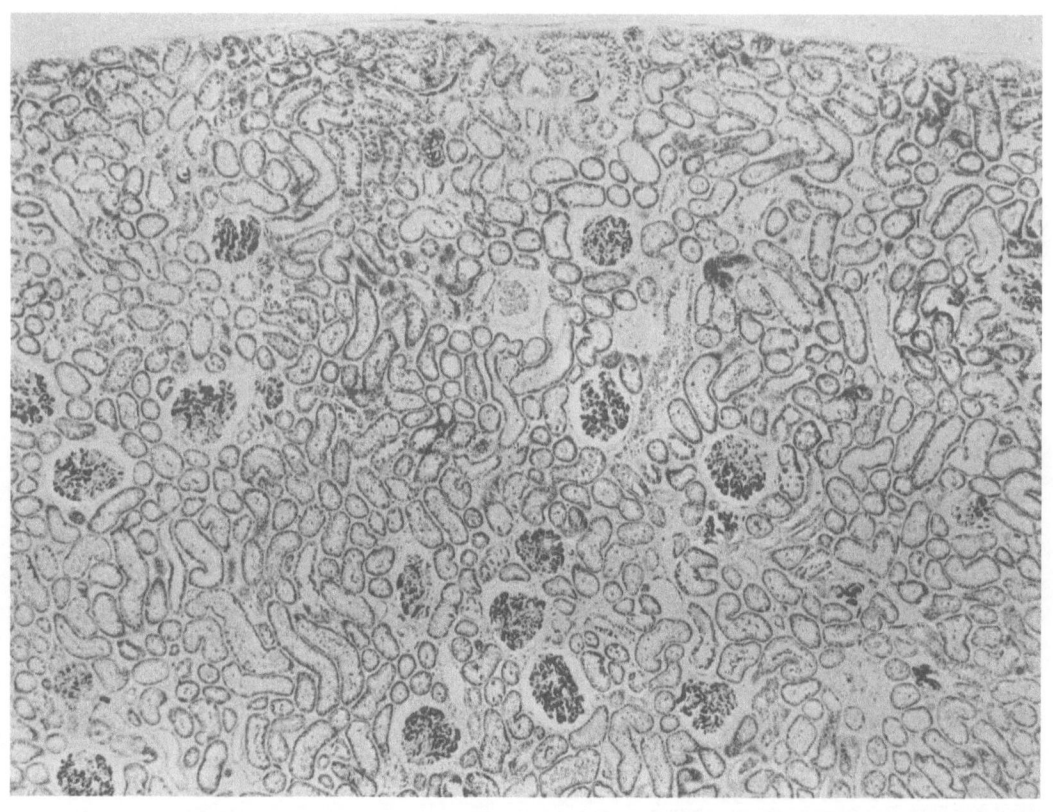

Fig. 7.   Disseminated intravascular microthrombi in the capillaries of rabbit glomeruli after two injections of endotoxin.

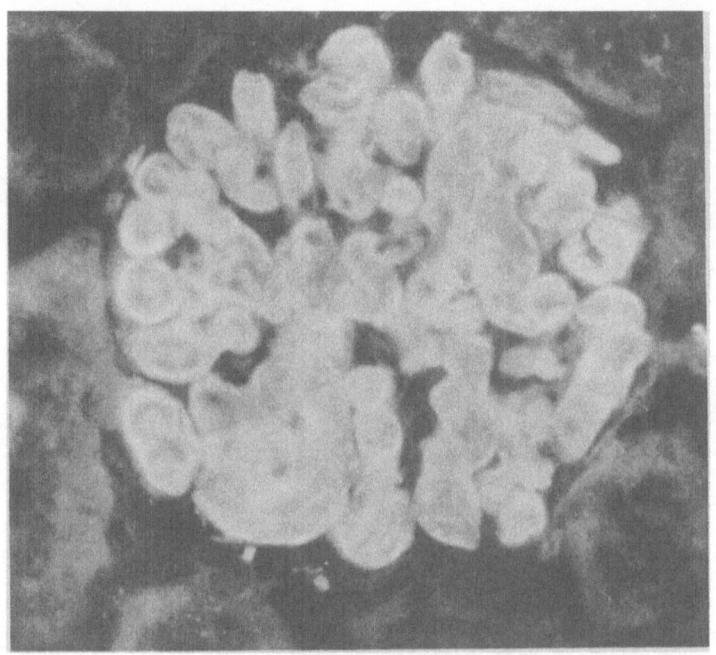

Fig. 8.   Specific fluorescence of highly polymerized fibrin derivatives in a rabbit kidney after treatment with anti-human fibrinogen serum.

2

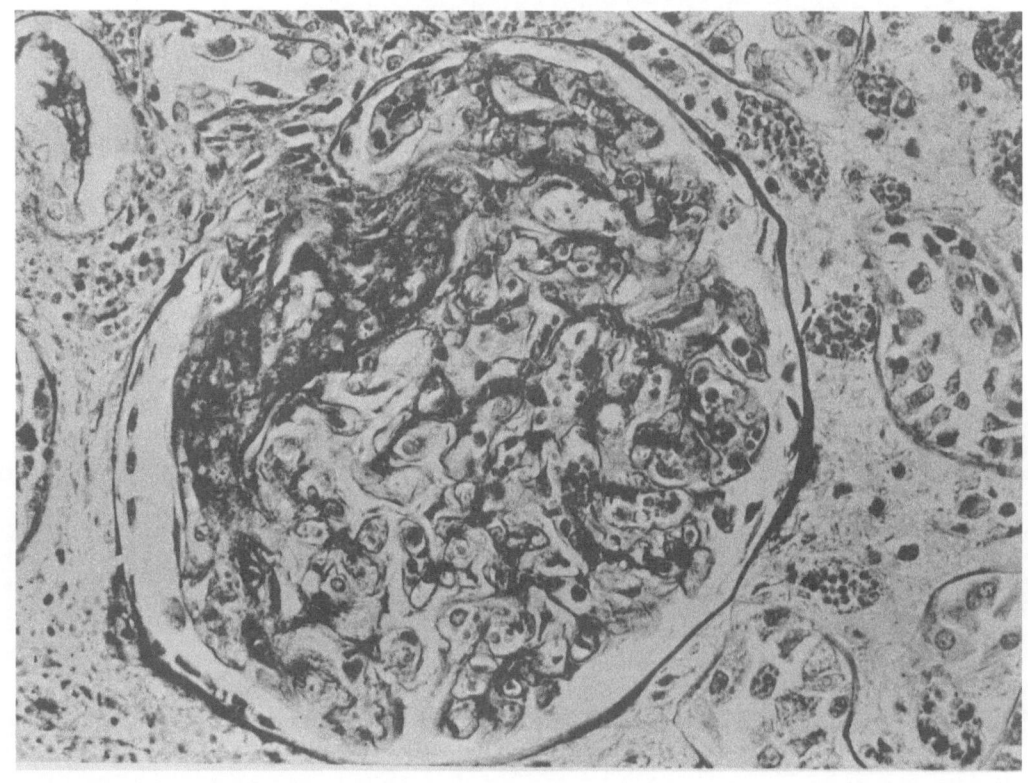

Fig. 9.   Highly polymerized intravascular fibrin derivatives deposited in the glomerular capillaries of a human kidney after endotoxic shock.

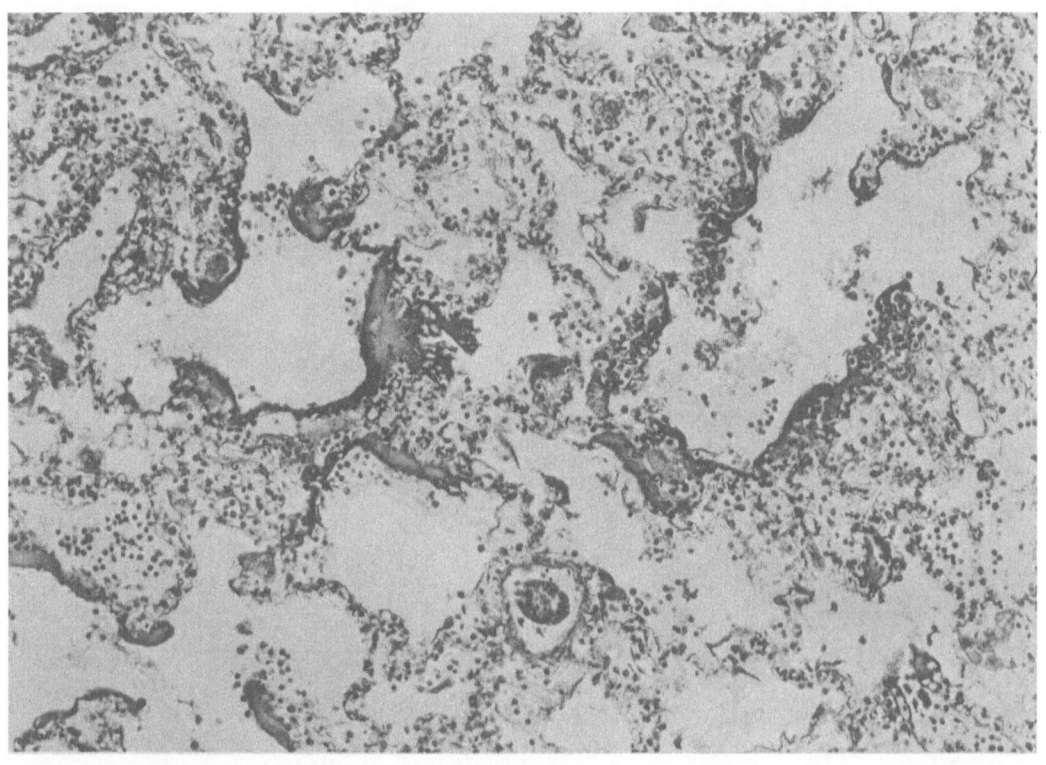

Fig. 10.   Typical pulmonary hyaline membranes and fibrin-rich microthrombi in the pulmonary microcirculation after endotoxic shock in man.

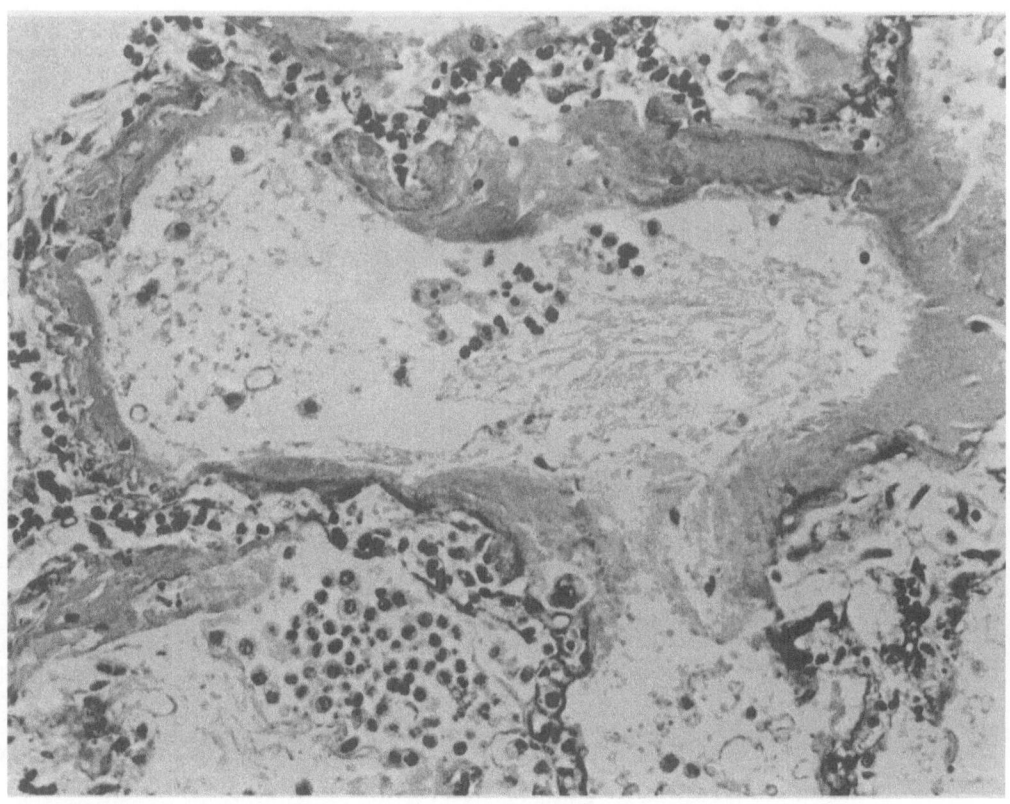

Fig. 11. Highly condensed fibrin-rich pulmonary hyaline membranes in the ductuli and sacculi alveolares of the lung after endotoxic shock in man.

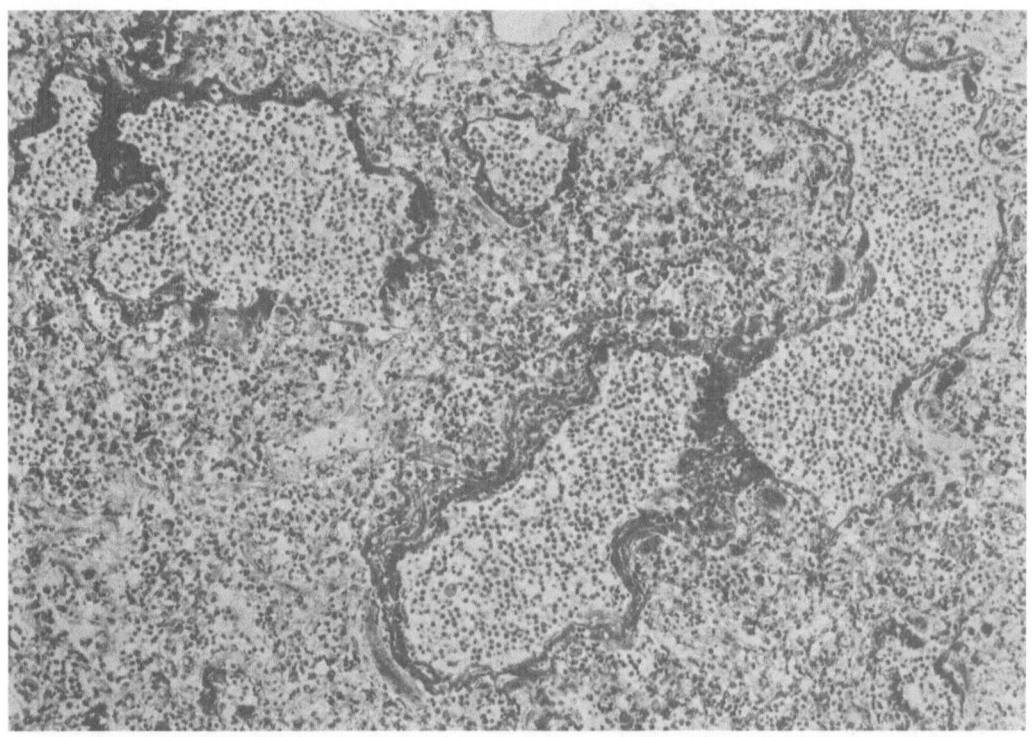

Fig. 12. Diffuse leukocytic infiltration of the lung with beginning extravaxcular fibrinolysis of hyaline membranes after endotoxic shock in man. This is manifested as so-called shock-lung and secondary pneumonia.

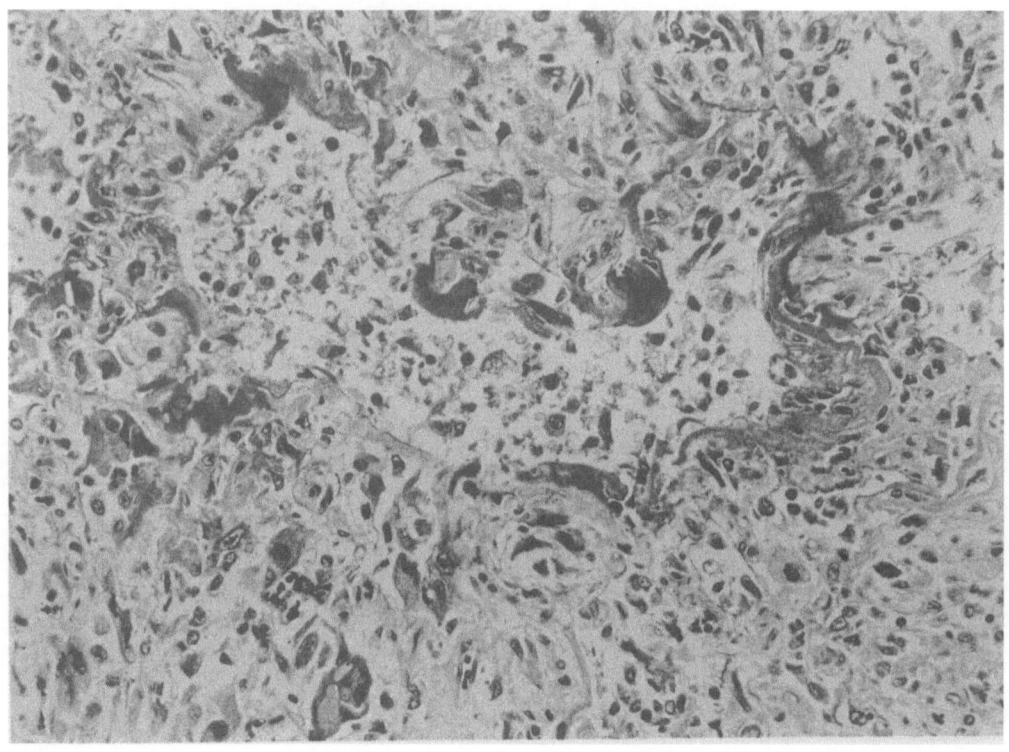

Fig. 13.   Beginning organization of pulmonary hyaline membranes after prolonged
endotoxic shock and manifestation of a so-called shock-lung in man.

Discussion

MÜLLER-BERGHAUS

LÜSCHER

Lundsgaard-Hansen: Aspirin is a known inhibitor of platelet functions and according to recent reports it also has a certain protective effect against endotoxin. Is it conceivable that there exists a connection between these two effects?

Lüscher: Aspirin inhibits the release reaction and the secondary aggregation of platelets and therefore should also interfere with their participation in the activation of the intrinsic clotting system. To my knowledge Dr. Beller has been working on the interrelationship of endotoxin and blood coagulation. Perhaps he would like to comment on this?

Beller: Dr. Slabber in my laboratory has used antiaggregants of the pyrimidine-class in endotoxin-treated rabbits. He succeeded in preventing completely the normally occurring drop in the number of circulating platelets. Nevertheless, after 10 hours of endotoxin infusion, such animals, inspite of a constantly high platelet count, developed fibrin deposits in their glomerular capillaries. This observation is in agreement with the findings of McKay and Müller-Berghaus, who demonstrated that it is possible to produce a generalized Shwartzman reaction in thrombocytopenic animals. All this speaks against the assumption that platelets are the essential trigger for intravascular coagulation.

Nowotny: I was interested in your description of the phenomenon that cephalin-type phospholipids may appear on the surface of "activated" platelets, where they participate in the activation of blood coagulation. This is another indication for the liquid state of the cell membrane, and parallels perhaps the "capping phenomenon" observed in immunocytes, which also must be explained in terms of the mobility of receptor sites within the membrane, where they can move and flow, and even submerge and emerge.

Lüscher: In this context it is of interest that the binding capacity for concanavalin A of platelets is increased to about twofold after treatment with thrombin. This again indicates that rather dramatic structural alterations occur in the membrane as the result of a rather well defined primary stimulus.

Beller: You have pointed to the rather striking species differences in the reactivity of platelets towards endotoxin. Why then is the final outcome the same, e. g. in man and rabbits? Both show activation of the clotting system and a low platelet count.

Lüscher: Let us assume, endotoxin acts in man primarily on endothelial cells and on leukocytes. In both cases an activation of the extrinsic clotting system will ensue, and the formed thrombin will then affect the platelets. In the rabbit, the platelet will be the prime target of endotoxin. In both cases, the platelets, indirectly or directly, will be affected.

Clowes: Are prostaglandins not also found among the different substances released from the platelets?

Lüscher: The substances mentioned here were those released from specific storage organelles in a very fast reaction. This process is quite different from the leakage of components of the cytoplasm through an injured membrane. Prostaglandins are partly synthesized by the platelet, and would rather be expected among the cytoplasmic constituents. This is the case for PGE_2 which potentiates platelet aggregation; on the other hand, PGE_1 which is an extremely potent inhibitor of aggregation, is not a normal platelet constituent.

Clowes: There is a very practical aspect to my question. To my knowledge at least two groups have shown that platelets may accumulate in the lung, where they undergo the release reaction. Serotonin, adenine nucleotides, and the prostaglandins are all vasoconstrictors in the lung and vasodilators in the periphery. This in part may account for some of the hemodynamic and metabolic changes that we were speaking of yesterday and is one of the features which might be blocked if there were proper methods of preventing platelet aggregation and release.

Lüscher: It is difficult to tell how important the contribution of the prostaglandins in such a mechanism would be. Since relatively large amounts of other highly vasoactive materials are released, this possible consequence of the platelet release reaction nevertheless seems most interesting indeed.

Lundsgaard-Hansen: Cortical steroids, particularly methyl-prednisolone in high dosage have been used extensively in the treatment of septic shock. Their effect seems to be beneficial, and to consist primarily in the prevention of the so-called shock-lung by inhibiting the aggregation of platelets and leukocytes and by interfering with the adhesion of blood cells to damaged endothelium. Is anything known which would corroborate this clinical observation, which is supported to some extent by light-microscopic findings?

Lüscher: In vitro, cortical steroids have no striking effect on blood platelets. In view of the fact that in man, platelets react indirectly to endotoxin, many other ways of interfering with platelet activation are possible.

Beller: I think, Dr. Lundsgaard is referring to the work of Cavanagh in St. Louis who produced a movie from which it can be seen that the endotoxin-induced constriction of renal arteries is released by the administration of a large dose of corticosteroids. Obviously, all attempts at explaining this action of the steroids have failed so far.

MÜLLER-BERGHAUS

Urbaschek: What action does histamine and on the other hand 5-hydroxytryptamine have on platelets? What response do human platelets show to both amines? To what extend are such mediators responsible for platelet changes in human gram-negative septicemia?

Lüscher: Human platelets react, as has already been mentioned, to 5-hydroxytryptamine, adrenaline and noradrenaline, and very slightly to histamine. The role of active biogenic amines in the induction and potentiation of platelet changes may in no way be underestimated.

Drommer, Hannover/Germany: Dr. Müller-Berghaus, do you have an explanation for the fact that endotoxin-caused endothelial damage varies from organ to organ and species to species? In certain organs no particular changes are seen, whereas in the central nervous system an activation of the endothelial cells appears to have been observed.

Müller-Berghaus: These studies with which we are concerned here, have been

405

conducted for practicle reasons on the aorta. In other vascular regions, it is often difficult to differentiate between primary and for example secondary changes caused by fibrin formation.

HEENE

Hennemann: If I have understood you correctly, you propose the use of heparin in every case of sepsis. In our series of septic cases, only 50% experienced shock and only 20% had a manifest hemorrhagic diathesis. In spite of this, is it justified then to use heparin in all cases of sepsis?

Heene: The early phase of sepsis is usually characterized by hypercoagulability and hyperfibrinogenemia. These need not be clinically apparent. As a result we feel prophylactically administered heparin has a place in therapy here.

Hennemann: Would you include cases of sepsis in the course of leukemia which are, as we know, often accompanied by disturbances of the coagulation system as for example thrombocytopenia.

Heene: In such cases heparin should only be used after predetermining clotting parameters. Do positive values for instance of the fibrin derivatives result undoubtedly a consumption coagulopathy exists and heparin should be given.

Lechner, Wien/Austria: What meaning do you attach to the decrease in prothrombin-complex factors and factor V resulting from liver damage during sepsis?

Heene: It is not easy to answer this question. Severe disturbances of the liver function should be expressed in a decrease in fibrinogen, which is not the case in septicemia. In order to clarify this problem one would have to do turnover rate studies.

von HUGO

Lüscher: Some years ago, Shainoff and Page described a so-called cryoprofibrin which consists of a complex containing molecules from which only one peptide A is removed. Do you think that you are dealing with the same or a comparable material?

von Hugo: Our 0.28 derivative probably is a dimer formed from fibrin monomer and fibrinogen. We do not know whether it contains material from which only one or both peptides are removed.

Lüscher: The "cryoprofibrin" is said to precipitate with heparin and in the cold. Is this also true for your material?

von Hugo: We only used precipitation in the cold and have no experience with heparin.

Graeff: It should be mentioned that the appearance of the 0.28 derivative mostly is correlated with a positive ethanol-gelation test.

BELLER

Hennemann: You considered various causes of anuria in your hypothetic case and mentioned in this context hypovolemia, tubulus- and cortical necrosis. Is tubulus necrosis not the direct consequence of hypovolemia?

Beller: Accepted.

Braude: What is the concentration of sodium chloride used to produce the fibrin deposits in the glomeruli after administration of endotoxin?

Beller: Isotonic

Braude: Are such deposits also produced in the classical way, i.e. by giving a second dose of endotoxin?

Beller: In the rat the generalized Shwartzman reaction cannot be produced by repeated injections of endotoxin.

Braude: What do you think is the mechanism by which the isotonic saline provokes this deposition of fibrin?

Beller: I have no idea at the present time. It might be that saline modifies or potentiates an activator system which in turn is stimulated by endotoxin. In saline abortion the activators may be derived from the disintegrating placenta; but at the present time this is a speculation.

Braude: In the virgin rat with glomerular fibrin deposits, is there also grossly visible cortical necrosis?

Müller-Berghaus: In rats nephro-cortical necrosis is infrequently seen macroscopically; in fact we found only ca. 5 in 1,000 rats studied. In virgin rats a Shartzman phenomen, here defined as fibrin exudate in the kidney, can be elicited by inhibiting fibrinolysis. In this case a strong reaction results with one dose of endotoxin.

Springer: Are consumption coagulopathies not also observed in certain cases of diabetes? What is the mechanism behind this, and how could it possibly relate to the electrolyte potentiation?

Fischer, Wien/Austria: Coagulation studies in patients with coma diabeticum reveal indeed the presence of consumption coagulopathy. The causes are acidosis and disturbed fat - free fatty acid relationships, which in turn trigger coagulation.

Clowes: In connection with the effects of saline on the kidney it strikes me that this is parallel to the kind of damage which can be inflicted on the lung by the use of cristalloid solutions following a primary injury brought about by a variety of agents such as materials derived from platelets or other activators. By the use of mannitol or of albumin instead of electrolytes these lesions are a good deal less often observed. I can cite you one example: In heat stroke, in the course of which changes as those discussed here are often observed, the incidence of renal failure used to be very high. Since the advent of mannitol, it is virtually down to zero. In patients with severe infections we routinely use mannitol at the outset of a shock state or whenever circulation is reduced and it is my belief that the incidence of severe renal failure has been greatly lowered. It could be that Dr. Beller's animals and patients are showing a severe oedema in the glomerular area, which results in impaired kidney function and tubular injury. I really think that we see much less of this cortical necrosis than has been reported here - the chief problem is much more lower nephron nephrosis.

Beller: I agree that we should systematically try other solutions, and mannitol certainly will be on our list. I must say, though, that I am not so optimistic; even those patients who have no obvious cortical necrosis will show, about three months after the septic shock, impaired kidney function.

Clowes: This very severe situation which results from the combination of parturition and endotoxin shock is perhaps different. I wanted to comment only on the effect of endotoxin in producing DIC and to point to the possible importance of changes in capillary permeability and of secondary oedema in both, the kidney parenchyma and the alveoli.

Beller: All our experiments were done in animals at term; in early pregnancy the situation is quite different. The typical hypersensitivity of pregnancy, most likely also in humans, is found only at term; an early abortion relates much more to the non-pregnant than to the "pregnant" state.

Lüscher: Dr. Beller, you have done some work on the importance of local fibrinolytic activity. Polish workers have made such studies even earlier, and recently Bergstein and Michaels have shown convincingly that the glomeruli belong to the organs with a low fibrinolytic activity. Furthermore, it is known that this activity of the endothelial cells can be reduced, even in vitro, by a variety of plasma-mediated stimuli, perhaps by thrombin and most likely also by osmotic shock. This would then mean that certain organs are apt to develop capillary fibrin deposits not because of the presence of particularly efficient activators of coagulation, but rather due to the absence of a fibrin-removing system.

Beller: This certainly is an interesting concept. We have recently repeated the experiment which shows that the activator content in a kidney loaded with fibrin is indeed very low. A co-worker in my laboratory has spent over a year on a comparative study on the fibrinolytic activator in the kidneys of various species. It became quite obvious that those species, like the rabbit or the rat, in which it is easy to induce the Shwartzman reaction, have a considerably lower activator activity than e.g. the dog or the guinea pig. Furthermore, the same difference is found between pregnant and non-pregnant animals. It remains to be seen in what the specific effect of endotoxin consists.

BLEYL

Steinbereithner, Wien/Austria: In our group Lempert and Müller have studied the surfactant activity in amniotic fluid during repeated amniocenthesis. In normal pregnancy this activity is markedly increased from the 23rd week on. In risk pregnancies, however, the surfactant activity remains low even at 30 to 36 weeks of gestation. This would mean that at an early point in time the possibility of hypoxia and acidosis must be taken into account.

Bleyl: In so-called ideopathic respiratory distress syndrome the intrauterine surfactant activity plays a decisive role. The formation of hyaline membranes have primarily nothing to do with surfactant activity, however, intrauterine asphyxia leads to an abrupt decrease in surfactant activity and in fact is due to the appearance of extravasated fibrinogen. Under these circumstances then the surfactant activity is influenced by hyaline membrane formation, but in principle they should be differentiated from one another. Hyaline membranes have not yet been observed in utero.

Steinbereithner: With respect to the often discussed respirator lung, it would be interesting to know how many of the adult cases shown had been on a respirator.

Bleyl: About 80%. In rabbits using 100% O_2, hyaline membrane formation can be induced along with generalized activation of the clotting system. Fibrinmonomer complexes and fibrinogen split products are observed. In accord with these results are observations made 20 years ago that pure O_2 can cause platelet aggregation.

Steffen, Wien/Austria: From your results it can be concluded that the typical appearance of the Shwartzman phenomen is observed when the RES is blocked. Endotoxin is rapidly taken up by the Kupffer cells and this could explain such blocking in the Sanarelli-Shwartzman phenomenon. How did you block the RES?

Bleyl: Partly, these were Sanarelli-Shwartzman experiments, the RES was partly blocked with macromolecules.

Schultz, Hannover/Germany: In the continuous thrombin infusion you described, did you find as for example in the kidney secondary changes resulting from extravasation of fibrinmonomers or polymerates? With spontaneous or experimental erysipeloid in pig gnotobiotes we found that intravascular coagulopathy is seen regularly with a concomitant extravasation. With the aid of immunofluorescence, C3, fibrinogen and certain globulins are detected. These findings are less pronounced in the lung than in the kidney. Moreover, we regularly observed a marked tendency toward edema in the periarticular spaces as well as in the joints. Experimental erysipeloid in the rat revealed these changes also in the aorta. We observed plaques in the aorta which lead to focal aortic fibrosis from the 21st day on. I realize that these experiments differ from yours to a great extend, however, the advantage could be a possible study of the secondary consequences.

Bleyl: In the Sanarelli-Shwartzman phenomenon one observes frequently for example in the lymphatic of the renal calyces more or less fibrin-rich coazervates which precipitate in the extravascular spaces. Also globules, as have been shown here, have been observed. The fate of these substances is uncertain since we have not conducted long time studies. In lung alveoli, fibrinolysis caused by granulocytes occur. The "surfactant" is able to block completely all other fibrinolytic activities in the alveoli. The precipitates persist and are not dissolved by direct fibrinolysis. Many animals died in these experiments. Only 7 out of 18 rabbits survived the continuous thrombin infusion longer than 36 hours. A subtotal defibrination with multiple microthrombi resulted and the animals died in a diffuse shock state.

Marget: Neonatologists often overlook sepsis because they interpret this clinical picture as RDS. How did you exclude infection in both of your patient groups?

Bleyl: We could only partially exclude infection in the perinatal period because we did not have the placentae. As we know very well placental infection plays an essential role. We limited our cases to those which were characterized by the ideopathic respiratory distress syndrome. The bacteriological status was not evaluated.

Deutsch, Wien/Austria: Are the globules, which you described, present intra vitam, or are they formed as a result of fixation?

Bleyl: These globules are formed intra vitam. They are frequently found in fibrin-rich precipitates which form as a result of blocked RES or inhibition of fibrinolysis. In especially great numbers they appear in the capillary bed, when for example urokinase following fibrin is infused. However, one must work rapidly in order to prevent post mortem fibrinolysis from breaking down all material present.

Huth, K.: Did you note pulmonary hyaline membranes in the rabbit also associated with the classical Sanarelli-Shwartzman phenomenon?

Bleyl: Only in some cases. After the second injection defibrination is so pronounced, that evidently there are not enough circulating fibrinmonomers present which could extravasate and polymerize in the extravascular spaces. After the first injection no such changes are observed.

Chapter VIII

Clinical Aspects Including Shock

Clinical Features in Shock Associated with Gram-Negative Bacteremia

H. Shubin, M. Weil, and H. Nishijima

Introduction

Gram-negative enteric bacilli account for the majority of cases of bacterial shock. The organisms most commonly implicated include Escherichia coli, Klebsiella-Enterobacter, Proteus, Pseudomonas, Bacteroides, and Salmonella. In some instances, gram-positive cocci may produce infections which account for bacterial shock. Systemic infection due to Neisseria, Clostridia, and infections by organisms other than bacteria, including viruses, rickettsiae, and fungi also occasionally may be complicated by shock.

Bacteremia is infrequently complicated by shock prior to the age of 40 except in infants in the neo-natal period and in women during pregnancy. When patients with septic abortion are excluded, bacterial shock is approximately twice as common in men as in women. This reflects the greater incidence of urinary tract infection in older men. Blood dyscrasias, diabetes, and chronic liver disease predispose to bacteremia and shock (1). In the majority of patients, bacteremia and shock are ushered in by some type of manipulation. Surgery and instrumentation of the genitourinary tract are the most common frequent causes. Other triggering events include surgery of the gastrointestinal tract and manipulation of infected wounds especially after severe burns. Patients who have prolonged treatment with corticosteroids (1, 2, 3), immunosuppressive drugs, radiotherapy, or protracted treatment with antimetabolites are particularly susceptible.

Mechanisms. The specific organism responsible for infection in a given patient is important not only with respect to antibiotic treatment but also for understanding the mechanism of shock and its management (4). When shock is related to infection by gram-positive bacteria, rickettsiae and viruses, there is often an increase in vascular permeability, a loss of intravascular fluid and vasodilation involving the arterial resistance vessels, all of which may contribute to the development of hypotension and shock. When shock occurs as a complication of gram-negative enteric bacteremia, the patient initially may present with warm extremities. Arterial vasodilation usually occurs during the early pyrogenic phase of the infection together with hyperthermia, hyperventilation often with respiratory alkalosis, and an increase in pulse pressure and cardiac output (5, 6, 7). Only subsequently are the classical clinical manifestation of shock noted, including cold and moist extremities, arterial vasoconstriction with a reduced pulse pressure and oliguria. The cardiac output is diminished and peripheral arterial resistance usually is elevated. A considerable portion of the total volume appears to be pooled in the venous capacitance bed resulting in a reduction in the effective circulating blood volume and hence a reduction in the volume of blood returned to the heart and lowered cardiac output. Perfusion failure is manifested clinically by decreased mental alertness, confusion and stupor due to decreased cerebral blood flow, oliguria due to decreased renal blood flow, and cool skin due to a reduction in skin blood flow. Lactic acid accumulates in tissues

The research programs of the Shock Research Unit are supported by grants from the John A. Hartford Foundation, Inc., New York, and by the United States Public Health Service research grants HE 05570 and GM 16462 from the National Health Institute and grant HS 00238 from the National Center for Health Services Research and Development.

and fluid spaces because the blood flow to muscle and viscera is inadequate to provide for aerobic metabolism (3, 4). Intravascular coagulation has been implicated in the shock syndrome. Whether this represents a specific reaction to the polysaccharide or a secondary effect of the stagnation which results when the blood flow is markedly curtailed, as yet, has not been clarified.

Clinical Features

Gram-negative enteric bacteremia typically is manifest by a shaking chill followed by a temperature spike. This often occurs between two and 24 hours following genitourinary surgery or manipulation of the genitourinary tract. An alteration in the patient's behavior may be an early manifestation of the reduced arterial pressure and flow. Although the skin may be warm and dry initially, as the hemodynamic defect progresses the skin becomes clammy. Many patients present with gastrointestinal signs including vomiting and diarrhea, often with green and sometimes bloody stools.

A paradoxical finding in the early period of shock is the reduction in the white blood cell count, particularly of the polymorphonuclear cells. Subsequently, leukocytosis is noted. Increases in serum glutamic oxalic transaminase, serum glutamic pyruvic transaminase, and lactic dehydrogenase are commonly observed as a consequence of cellular injury due to perfusion failure. Hyperglycemia is often noted and this reflects secretion of increased amounts of catecholamines from the adrenal medulla. Serum amylase may be considerably elevated in the absence of pancreatic injury and this has been attributed to decreased renal clearance of amylase. Electrocardiographic abnormalities, particularly ST segment and T wave changes are noted in the majority of patients (1). These changes do not reflect a cardiac basis for the shock state, but are a secondary manifestation of the reduced perfusion (8).

Rationale of Treatment

Infection. Appropriate antibiotic treatment may triple the chances for survival (1). If the physician awaits the results of the antibiotic sensitivity studies before starting antibiotic therapy, the potential benefit from such therapy may be lost. If infection in the urinary tract, a wound, or other site has been previously recognized and bacteria cultured at an earlier time, the bacteremia may be reasonably attributed to the same organisms. If in vitro antibiotic sensitivity studies on these bacteria are available, antibiotics may be chosen accordingly (3). Bacterial populations and antibiotic sensitivities vary from one hospital to another and change over time within the same hospital. Antibiotic sensitivity tests are useful not only in the individual patient but also in the hospital at large. The availability of current antibiotic sensitivity tests within the hospital may be useful in selecting the appropriate antibiotic, before the results of the patient's own blood cultures are available. Prior to administration of antibiotic to patients, three blood samples for culture are taken in sequence with a single vein puncture and the samples are labeled one, two and three. If bacteria are cultured from the first but not the other samples, the first sample is more likely to contain a contaminant. Antibiotic treatment is then started immediately, based on information recurrent antibiotic sensitivities of gram-negative organisms.

Corticosteroids. The administration of corticosteroids in dosages which are many times greater than those required for adrenal replacement, may lead to a reduction in systemic reactions to endotoxin and controlled fever. Such pharmacologic doses of corticosteroids also may tend to prevent nonspecific injury to cell structures. Corticosteroids also increase cardiac output and decrease peripheral arterial resistance. These effects are opposite to the hemodynamic defects that are observed in patients with bacterial shock and in some animals following injection of bacterial endotoxin. In our own experience, survival has been significantly greater in patients who received large doses of synthetic corticosteroid drugs (1, 9). Similar findings were noted by Christy (10), Melnick (11), and Motsay (12). Current high dosage corticosteroid therapy, however, is based on extrapolations from experimental studies and retrospective analyses of survival in groups of patients who received such therapy or in whom it was withheld. Since controlled studies are unavailable, the therapeutic clinical efficacy of corticosteroid analogs in bacterial shock is not yet proven.

Blood Volume. Large quantities of fluid may be sequestered at a site of inflammation or lost because of fever, vomiting, or diarrhea. Since perfusion failure is intensified when intravascular volume is depleted, replacement of fluids may be a key therapeutic feature. Repetitive measurement of central venous pressure (CVP) or of pulmonary artery wedge pressure (PAWP) or pulmonary artery diastolic pressure (PADP) are helpful for evaluating the capability of the heart to pump the volume of fluid which is infused (13, 14). Additional guidelines to the patient's response to fluid infusion are provided by the arterial pressure, state of the sensorium, urine output, measurements of arterial blood PO_2, PCO_2, pH and plasma volume and red cell mass. Even in patients in whom the plasma volume is not reduced, infusion of fluids usually increases cardiac output as long as the PAWP or CVP do not indicate overload. A 5 % solution of Normal Human Albumin is the preferred volume expander. Whole blood should not be administered unless clearly needed. Isotonic sodium chloride with 5 % Normal Human Serum Albumin, USP or Plasma Protein Fraction, USP provides a useful combination. The macromolecular dextrans (molecular weight 70,000 to 80,000) gelatin solutions, and hydroxyethyl starches (polymers of sorghum corn starch) have been employed as volume expanders. However, we do not favor their routine use because of adverse reactions, including an increased risk in blood clotting defects, interference with subsequent crossmatching of blood and hypersensitivity reaction.

Acid-base Status. Acidemia frequently develops in patients with bacterial shock. The rationale for administering sodium bicarbonate in an attempt to restore the pH to a more normal value is open to question. The development of acidemia is a consequence of the compromised tissue perfusion. The acidemia is best corrected by improving tissue perfusion. In patients who have had a cardiac arrest, sodium bicarbonate may be given in amounts ranging from 40 to 160 meq. of sodium, but the effects on blood pH and osmolality should be monitored (15).

Anticoagulants. There is justification for the administration of heparin to patients who have gross bleeding due to consumption coagulopathy, presumably due to diffuse intravascular coagulation. However, this treatment is required in only the exceptional patient. In patients with meningococcemia, however, it is a relatively more common complication (16).

Vasoactive Drugs. There is widespread agreement that neither levarterenol (Levophed) nor metaraminol (Aramine) should be used for routine treatment. To the contrary, there is evidence that these vasopressor drugs increase peripheral vasoconstriction and further reduce effective blood flow. In those patients in whom there is concurrent myocardial failure due to coronary insufficiency, there may be an indication for a temporary increase in the perfusion pressure and coronary blood flow. However, their value even for this purpose remains controversial.

When peripheral vasoconstriction persists despite the use of volume expanders, isoproterenol (Isuprel), a beta-adrenergic stimulant, or alpha-adrenergic blocking agents such as phenoxybenzamine (Dibenzyline) or phentolamine (Regitine), which dilate arterial vessels, may be considered. Phenoxybenzamine is not currently available as an approved drug. Isoproterenol, which is a cardiac stimulant, may induce tachycardia and arrhythmias. Phentolamine and isoproterenol may reduce blood pressure even further. However, when perfusion failure is progressive, the use of these agents on a temporary basis may be warranted. We prefer phentolamine since isoproterenol acts not only as a cardiac but also as a more general metabolic stimulant.

Isoproterenol increases the tissue oxygen demands and thus the need for increased blood flow at a time when blood flow is already markedly reduced. When congestive heart failure is present, the administration of digoxin (Lanoxin), intravenously, is indicated.

Treatment

Prompt and continuing close attention to the patient's respiratory status is often the key to survival (17). With the increasing effectiveness of the treatment for shock, patients often subsequently succumb to respiratory failure, the so-called "shock lung" syndrome. Oxygen should be used judiciously with the concentration of inspired oxygen being regulated in response to the arterial PO_2. Measurements of arterial PCO_2, likewise are taken at frequent intervals to gauge the adequacy of the ventilatory exchange. When indicated, endotracheal intubation and mechanical ventilation should be promptly instituted.

Antibiotics. When the results of antibiotic sensitivity studies are not yet available, gentamycin (Garamycin) is the drug of choice in our unit. Gentamycin is administered intramuscularly in amounts of 1.5 mg/kg every eight hours. Based on in vitro studies, it has been shown that it is effective against more than 90 % of common strains of gram-negative enteric organisms, with the exception of Proteus, to which in vitro sensitivity approximates 75 %. This bactericidal drug is nephrotoxic and has both vestibular and ototoxicity, especially when renal function is impaired. Gentamycin is also very effective against Staphylococcus aureus and thus affords a special advantage when gram-negative enteric and Staphylococcus infections are both present. Kanamycin (Kantrex) is the drug of second choice. The loading dose of 7.5 mg/kg is followed every six hours by a dose of 2.5 mg/kg administered by deep intramuscular injection. When the urine output is less than 800 ml per day, as it is in many patients who have been in shock, the dose of both gentamycin and kanamycin usually needs to be reduced. Guidelines for the use of these and other commonly used antibiotics are summarized in the Table (5, 18).

When gentamycin is administered, no additional agent may be needed. However, when kanamycin is used, it may be supplemented with chloramphenicol. This is administered in amounts of 1.0 g diluted in 50 ml of saline and is injected into a central vein over a period of 3 to 15 minutes. Subsequent doses of 1.0 g each are given every six hours. As an alternative to chloramphenicol, cephalothin (Keflin) may be given intravenously in doses of 1.5 g every four hours. If the infection is known to be due to penicillin sensitive Proteus species or to a concommitant infection with gram-positive organisms, ampicillin may be used in amounts ranging from 6 to 12 g daily. Penicillin in a dose of 12.5 g (20 million units) also may be administered by the intravenous route. When there is acute renal failure, the sodium salt is preferred over potassium penicillin since this contains 1.67 mEq of potassium per million units of penicillin. In infections due to Pseudomonas, carbenicillin is likely to be effective when infused in amounts of 20 to 40 g per day. Klebsiella-Enterobacter infections which are resistant to kanamycin and chloramphenicol may be treated with colistimethate (Coly-Mycin) or polymixin B(Aerosporin) intramuscularly in the doses shown in the Table. However, gentamycin is usually very effective. Tetracycline drugs are bacteriostatic, and their use is limited to those instances in which there is demonstrable superiority over bactericidal agents on the basis of the in vitro sensitivity studies.

Administration of antibiotics for periods of one or two days prior to instrumentation or surgery has proven relatively ineffective in preventing bacteremia and shock in patients who have low grade infections of the urinary tract. We prefer that antibiotics be withheld for a period of one week prior to operation. If this is not feasible, soluble chemotherapeutic drugs, such as sulfonamide or nitrofuran may be used before surgery but antibiotics should be withheld until there is a specific indication.

Corticosteroids. Synthetic cortisone analogs are preferred since experimental animal studies reveal lesser toxicity when these steroids are administered in massive doses. Dexamethasone phosphate (Decadron) is administered intravenously in an initial dose of 40 mg and this is followed at intervals of four to six hours with doses of 20 mg each. Only three or four injections are needed in most patients. Methylprednisolone succinate(Solu-Medrol) may be selected instead in an initial dose of 200 mg followed by doses of 100 mg at four to six hour intervals. Even larger doses of up to 6 mg/kg of dexamethasone and 30 mg/kg of methylprednisolone are recommended by Lillehei (19). We are not aware of objective data, however, which document enhanced therapeutic efficacy with these larger doses. Once shock has been reversed the corticosteroid treatment is stopped abruptly without weaning. Suppression of adrenal cortical function is not a problem in patients in whom corticosteroid treatment is confined to a one or two day period. Upper gastrointestinal hemorrhage is a specific and not uncommon complication of corticosteroid therapy and such treatment should be immediately stopped if there is evidence of upper gastrointestinal tract bleeding.

Fluid Challenge. To assess volume deficits or pump failure, aliquots of between 100 and 200 ml of crystalloid fluid (isotonic saline or dextrose solution) or colloid (albumin 5 % solution) are administered. PAWP or alternatively PADP pressure measurements are obtained during a 10 minute observation period. Following this, fluid is administered intravenously. When the PAWP or PADP is less than 15 cm H_2O (11 torr) the fluid is administered at a rate of 20 ml/min for a period of 10 minutes. If the PAWP or PADP increases by more than 7 cm H_2O (5 torr) above the initial pressure, the fluid infusion is stopped. If the PAWP or PADP does not exceed the control pressure by more than 3 cm H_2O (2 torr) at the end of 10 minutes, or if it decreases to within that range over a second 10 minute observation period, an additional aliquot of 200 ml of fluid is administered over 10 minutes. If the PAWP or PADP increases by 3 to 7 cm H_2O (2-5 torr) above the initial pressure and continues to exceed the control value by more than 3 cm H_2O (2 torr) after the second 10 minute observation period, the subsequent aliquot of fluid is reduced to 10 ml over the next 10 minutes. The process is then repeated. Several liters of fluid may be required to restore effective circulation. When the PAWP and PADP are not available the CVP may be used instead. If the CVP increases by more than 5 cm H_2O (4 torr) above the initial pressure, the infusion is discontinued. If the CVP does not exceed the control pressure by more than 2 cm H_2O (1.5 torr) at the end of 10 minutes, or if it decreases to within that range over a second 10 minute observation period, an additional aliquot of 200 ml of fluid is administered over 10 minutes. When the CVP increases by 2 to 5 cm H_2O (1.5-4 torr) above the initial pressure and continues to exceed the control pressure by more than 2 cm H_2O (1.5 torr) after the second 10 minute observation period, the subsequent aliquot of fluid is reduced to 100 ml over the next 10 minutes. The process is then repeated.

Vasoactive Drugs. The vasopressor drugs should be used rarely. When there is evidence of acute coronary insufficiency, they may be administered in doses which raise blood pressure just enough to maintain urine flow, preserve mentation and reduce electrocardiographic evidence of myocardial ischemia. The target blood pressure level is usually 20 to 30 mm less than the "normal" systolic pressure. Metaraminol (Aramine) is the preferred pressor agent. Since excessive vasoconstriction may further reduce effective blood flow and increase the workload on the heart, careful and continuous titration of the dosage is necessary to avoid these potentially disasterous effects. Metaraminol in a dose of 200 to 500 mg is added to one liter of physiological salt solution or 5 % dextrose. It is administered by continuous intravenous infusion. Metaraminol is preferred since it does not cause local ischemic tissue injury if the needle or catheter is dislodged from the vein. Levarterenol may be used instead, in amounts of 16 mg per liter of fluid, providing 10 mg of phentolamine mesylate (Regitine) is added to each liter of fluid.

In those patients in whom other treatment is ineffective, isoproterenol (Isuprel) may be considered. Administration of isoproterenol should not be undertaken unless volume repletion has been accomplished, otherwise profound hypotension may ensue. An increase in cardiac output is often noted during infusion of between 0.20 and 25 μg of isoproterenol per minute. This may be achieved, however, at the price of a greatly increased need for oxygen by the myocardium. We prefer phentolamine for such patients after the volume has been expanded and particularly in those instances in which clinical signs of severe vasoconstriction are not improved by more routine treatment. Phentolamine may be infused in amounts ranging from 0.1 to 2 mg per minute over a period of 20 to 30 minutes. With the infusion of phentolamine, there may be a reduction in blood pressure. This may be promptly reversed by adminis-tration of additional quantities of fluid, preferably equal parts of 5 % Human Serum Albumin and physiological salt solution.

Other Measures. Hyperthermia may be controlled with a cooling blanket. How-ever, there is no evidence that depression of temperature to subnormal levels is de-sirable. Chilling should be avoided, since it increases the oxygen requirements. For treatment of intravascular coagulation, 5000 international units of heparin are in-jected intravenously every four hours, guided by periodic measurements of blood clotting. Surgical treatment may be indicated in some patients for control of the bac-terial infection. Abcesses should be promptly drained and grossly infected tissues removed. Hysterectomy may be life-saving in patients with shock complicating sep-tic abortion. Surgical drainage of a large intraabdominal abscess may be necessary if antibiotics are to control the infection.

Summary

Bacteremia due to gram-negative enteric organisms is responsible for the major-ity of instances of shock complicating bacterial infection. Treatment of the infection and maintenance of adequate blood volume are the primary initial considerations in treatment. Administration of three or four doses of corticosteroids over a period of 24 hours is recommended for routine treatment. Metaraminol and levarterenol are rarely indicated. The use of isoproterenol or phentolamine is limited to those pa-tients who do not respond to volume repletion. Diffuse intravascular coagulation complicated by bleeding diathesis may be an indication for anticoagulation with he-parin. After successful management of the hemodynamic defects, some patients may develop pulmonary failure which as yet is poorly understood. Intensive respiratory management often is required for a successful outcome.

References

1. Weil, M.H., Shubin, H., Biddle, M.: Ann. intern. Med. 60: 384 (1964).
2. Du Point, H.L., Spink, W.W.: Medicine 48: 307 (1969).
3. Weil, M.H., Shubin, H.: Diagnosis and treatment of shock. (Baltimore: Williams & Wilkins Co., 1967).
4. Kwaan, H.M., Weil, M.H.: Surg. Gynec. Obstet. 128: 37 (1969).
5. Simmonds, D.H., Nicoloff, J., Guze, L.B.: JAMA 174: 2196 (1960).
6. Blain, C.M., Anderson, T.O., Pietras, R.J., Gunnar, R.M.: Arch. intern. Med. 126: 260 (1970).
7. Blair, E.: Amer. J. Surg. 119: 433 (1970).
8. Barnett, J.A., Sanford, J.P.: JAMA 209: 1514 (1969).
9. Nishijima, H., Weil, M.H., Shubin, H., Cavanilles, J.M.: Hemodynamic and metabolic studies on shock associated with gram-negative bacteremia. Medicine. In press.
10. Christy, J.H.: Amer. J. Med. 50: 77 (1971).

11. Melnick, I., Litvak, A.: J. Urol. 96: 257 (1966).
12. Motsay, G. J., Dietzman, R. H., Ersek, R. A., Lillihei, R. C.: Surgery 67: 577 (1970).
13. Weil, M. H., Shubin, H., Rosoff, L.: JAMA 192: 668 (1965).
14. Forrester, J. S., Diamond, G. A., Swan, H. J. C.: JAMA 222: 59 (1972).
15. Mattar, J. A., Weil, M. H., Shubin, H., Stein, L.: Amer. J. Cardiol. 29: 279 (1972).
16. Evans, R. W., Glick, B., Kimball, F., Lobelle, M.: Amer. J. Med. 46: 910 (1969).
17. Weil, M. H., Shubin, H.: JAMA 207: 337 (1969).
18. Kunin, C. M.: Ann. intern. Med. 67: 151 (1967).
19. Motsay, G. J., Dietzman, R. H., Schultz, L. S., Romero, L. H., Lillehei, R. C.: Effects of massive doses of corticosteroids in experimental and clinical gram-negative septic shock. In: Forscher, B. K., Lillehei, R. C., Stubbs, S. S.: Shock in Low- and High-Flow States, p.303 (Excerpta Medica, Amsterdam 1972).

Table

Antibiotic Doses with Adjustments for Renal Failure*

Antibiotic	Initial Dose	Subsequent single dose, interval, and route	Adjusted interval between doses	
			Oliguria**	Azotemia***
Gentamycin sulfate	1.5 mg/kg	1.5 mg/kg, q 8 h, I.M.	3-4 days	2 days
Kanamycin sulfate	7.5 mg/kg	2.5 mg/kg, q 6 h, I.M.	3-4 days	2 days
Colistimethate sodium	5 mg/kg	2.5 mg/kg, q 12 h, I.M.	3-4 days	2 days
Polymyxin B sulfate	2.5 mg/kg	0.6 mg/kg, q 6 h, I.M.	3-4 days	2 days
Streptomycin	15 mg/kg	7.5 mg/kg, q 6 h, I.M.	3-4 days	2 days
Chloramphenicol	1.0 g	0.5 g †, q 4 h, I.V.	6 hours	6 hours
Ampicillin	2.0 g	1.0 g, q 4 h, I.V.	12 hours	6 hours
Cephalothin sodium	3.0 g	1.5 g, q 4 h, I.V.	24 hours	12 hours
Tetracycline	1.0 g	0.25 g, q 4 h, I.V.	3-4 days	2 days

*Modified from Barnett and Sanford (5); **Creatinine clearance ≦10 ml/min; ***Creatinine clearance >10 ml/min; †reduce dose in patients with severe liver disease; I.M.=deep intramuscular injection; I.V.=slow intravenous injection of infusion over 3 to 15 minutes.

The Pulmonary Response to Sepsis and Endotoxin: Clinical and Experimental Observations

G. H. A. Clowes, Jr.

In recent years, it has become apparent that the seriously infected patient is more than usually prone to pulmonary complications, which are all too often fatal. Burke et al. (11) recognized the importance of respiratory support in peritonitis to relieve excessive work of breathing and severe hypoxemia. Clowes et al. (13, 14) described a sequence of events in septic patients which is reproducible in laboratory animals rendered septic by the induction of peritonitis. An interstitial pneumonitis (Phase I) is succeeded, if unchecked, by bronchopneumonia (Phase II).

The importance of the lung lesions associated with trauma and particularly with sepsis was emphasized by the high incidence of pneumonia and respiratory failure in the wounded military personnel in the Vietnam war (20). As a measure of the magnitude of the problem in civilian practice, a survey of 8, 000 consecutive admissions to two general services* disclosed that overall hospital mortality was 2.1%; respiratory insufficiency was responsible or contributed to death in 57% of those who died. This value rose to 70% when sepsis was present. A more recent study by Drs. Thomas O'Donnell and Erwin Hirsch, of injured patients who underwent partial hepatectomy for ruptured liver, is illustrative. All were in hypovolemic shock and received approximately the same amount of transfused blood. All exhibited a moderate degree of hypoxemia and shunting in the immediate postoperative period. However, severe pulmonary dysfunction occurred between six and ten days later, only in those patients who became seriously infected. Although the mortality associated with respiratory failure was 50% in the septic group, there were no postoperative deaths among the uninfected patients.

From these clinical and experimental observations, one is led to the conclusion that sepsis plays the single greatest role in the induction of progressive pulmonary lesions. Since cultures from peritonitis and major wound infections usually contain coliform and other enteric organisms, the question immediately arises as to what the role of endotoxin may be in producing the lung lesions. Following a brief description of the syndrome and the pathological sequence of events in the lung, it is the purpose of this paper to review the clinical and experimental evidence relating to endotoxin and other circulating factors of significance in the induction of pneumonitis followed by bronchopneumonia in the septic patient (13, 35).

Pulmonary Morphological Abnormalities: Assessment of the lung at autopsy is difficult because of the usual prolonged course and agonal changes in patients who die of sepsis. From the work of Proctor et al. (44) and others (7, 8), it is apparent that functional abnormalities of gas exchange and compliance are occurring at a time when the lungs appear to be normal by ordinary criteria or by X-ray. The lesion responsible might be referred to as the early pulmonary lesion of sepsis. At thoracotomy for drainage of mediastinal sepsis or for other reasons early in the septic course, the lungs appear grossly normal, except for punctate hemorrhagic areas diffusely scattered on the visceral pleura. Serous pleural fluid, at times con-

---

* Medical College Hospital, Charleston, S. C. and Harvard Surgical Service, Boston City Hospital, Boston, Ma.

The experimental and clinical results from our clinic reported in this paper were derived from research supported by Research Contract # DA-49-193-MD-2860 of the U.S. Army Research and Development Command.

taining small quantities of hemoglobin, may be present. At this stage, culture of
such fluid usually is negative.

Biopsies of apparently uninvolved lung in the e a r l y   s t a g e   are characterized by: 1. interstitial septal edema, 2. intravascular congestion, 3. infiltration
of leukocytes which are predominantly large mononuclear cells or lymphocytes, and
4. diffuse focal alveolar collapse (13), referred to by Border (8) as "focal atelectasis". In certain specimens, proteinacious material may be observed within some
of the open alveoli. The so-called alveolar pseudomembrane described in certain
cases may be partially dried or insoluble intra alveolar protein (37). It is not always present, and its significance is not clear. Some peribronchial and peritubular
interstitial hemorrhage may be present.

When the lung lesion progresses to the l a t e   s t a g e   of bronchopneumonia,
it becomes edematous. At autopsy, the weight is increased by 50 to 100% above normal. It is red, resembling liver tissue, and inflates with difficulty. Histologically,
there is much interstitial edema containing fibrin.Proteinacious fluid is present in
the alveoli, many of which are collapsed. There is massive infiltration of septal
and peribronchial tissue by polymorphonuclear leukocytes which are also present in
large numbers in airways and alveoli. Certain areas appear to be totally consoli-
dated by inflammatory reaction. Intra-alveolar and interstitial accumulations of red
cells are common.

Clinical Pattern: Moderate hypoxemia frequently occurs within 12 to 24 hours
after the onset of fulminating sepsis. While breathing air, the arterial oxygen tension (PaO_2) may fall to the range of 50 to 60 mm Hg despite the presence of subnormal arterial carbon dioxide tension ($PaCO_2$) of 25 to 35 mm Hg. At this stage, examination of the chest may disclose only coarse breath sounds. The lung fields by
X-ray may be clear or, at most, exhibit a slight degree of ground glass appearance.
No doubt, this phenomenon is due to the fact that the focal alveolar collapse found
in the early stage (Phase I) of the pulmonary lesions is not sufficiently confluent to
produce X-ray density. Subsequently, it is not possible to halt the progression of
the pulmonary lesion prior to the development of bronchopneumonia, coarse rhonchi,
decreased breath sounds, and other evidence of confluent pneumonia or atelectasis
may be obtained by physical examination. X-ray then shows a mottled appearance
or the coarse density characteristic of bronchopneumonia and edema.

As the pulmonary lesion progresses, pulmonary vascular resistance increases,
as shown by the data in Table I. The increase of pulmonary arterial (PA) pressure
contrasts with the normal behavior of the pulmonary circulation in which recruit-
ment of capillaries prevents an elevation of PA pressure as the cardiac output rises.
Not infrequently, the elevated pulmonary arterial pressure accompanied by the high
circulatory demand of sepsis leads to right heart failure. Venous pressure under
these circumstances may rise to values of 25 cm H_2O or higher. At the same time,
measurements of pulmonary occluded pressure with the Swan-Ganz catheter may
show normal values for the left arterial pressure. Thus, it is demonstrated that
there is resistance to blood flow in the lungs which becomes particularly pronounced
in the late stage of bronchopneumonia.

Pulmonary Function: The respiratory alkalosis and hypoxemia typical of the
e a r l y   i n t e r s t i t i a l   p n e u m o n i t i s   are evidence of both excess ventilation and the presence of a pulmonary "shunt". The latter, defined as a reduction
of the ventilation-perfusion ratio (57), is the result of venous blood passing through
the capillaries of unventilated alveoli. The exact measurement of shunt requires
knowledge of mixed venous and arterial oxygen content as well as cardiac output (6).
However, a reasonable measurement may be made by determining the alveolar-arterial gradient of oxygen tension (PaO_2 - PaO_2) (37). Although the shunt may be detected when the patient is breathing air, as shown in Fig. 1, it is more easily quanti-
fied when the respiratory mixture is raised to 100% for a period of twenty minutes
or more to allow equilibration. Making allowance for vapor pressure of 50 mm Hg

at body temperature, the alveolar oxygen tension is then 700 mm Hg. PaO_2 is normally between 450 and 500 mm Hg. An A-a gradient of greater magnitude is evidence of a shunt or venous admixture. Whereas, the normal degree of shunting ranges from 3 to 6%, values up to 20% are found during the early phase with clinically clear lungs. As bronchopneumonia develops, shunts of 30% or more are not uncommon when PaO_2 falls to 60 mm Hg on 40% O_2 respiratory mixture and to 120 mm Hg when breathing 100% O_2.

At low cardiac outputs, when the lung lesion limits oxygenating capacity, increased oxygen extraction and low mixed venous pO_2 values increase the A-a pO_2 gradient. Thus, both the degree of shunt and the cardiac output are of importance in determining the PaO_2 (42).

As the lung lesion progresses, Proctor et al. (44) observed in wounded septic military personnel that the increase of shunting which rose from 10 to 30% during 5 days was accompanied by a reduction of compliance which elevated the work of breathing. Powers and his associates (42) found a similar relationship regarding the functional residual capacity (FRC). The highest shunts were accompanied by a reduction of FRC from 2 liters to less than 500 ml.

It becomes evident, then, that as the work of ventilation increases, the circulatory requirement to perfuse the muscles of respiration also rises (41). The circulatory stress is further augmented by the arterial oxyhemoglobin desaturation which parallels both the reduction of compliance and FRC. Thus, a respirator, at times supplemented by positive end expiratory pressure (PEEP) (2), when properly employed with a cuffed endotracheal tube, may accomplish three beneficial function:
1. Collapsed alveoli may be re-expanded to reduce shunting and increase PaO_2.
2. The work of ventilation may be eliminated. 3. Improvement of the first two functions often reduces the circulatory demand and the requirement for cardiac output which usually decreases. However, it is important to remember that PEEP in excess of 15 cm H_2O may result in reduced cardiac output because of impedance of venous return and an increase of pulmonary capillary resistance in overdistended alveoli (43). Therefore, it is essential to monitor circulatory function carefully when PEEP is employed. An excessive reduction of cardiac output is deleterious from the standpoint of metabolic homeostasis in all parts of the body.

Experimental Evidence

The progressive lung lesion observed in septic man can be reproduced experimentally in dogs, rats, pigs and primates (14, 19, 24). Within 10 to 12 hours following the induction of bacterial peritonitis, an elevation of pulmonary arterial pressure and a moderate degree of arterial hypoxemia become evident (14). The initial lesion found in pulmonary biopsies is septal edema, associated with infiltration of large mononuclear leukocytes and lymphocytes. As in other inflammatory responses, the intercellular spaces of the capillary endothelium may open to 40A to permit the escape of large protein molecules and cells into the interstitial space (33). Electron microscopic studies not only demonstrate severe swelling and disruption of capillary endothelium (15, 58), but also demonstrate the passage of fibrin, as observed by Teplitz (55).

Elevation of pulmonary vascular resistance may rise more than 100% above the normal value. The mechanism is not clearly demonstrated. It appears that some type of venoconstriction may occur in response to certain possible circulating agents such as peptides activated by endotoxin or by lysozomal enzymes at the site of injury or infection. Venoconstriction may be related to intimal swelling (59). It is not mediated by smpathetic nerves (26). Thus, the increased capillary pressure and the augmented capillary permeability probably both contribute to the apparent cellular-vascular congestion and to the interstitial edema.

Within 12 to 18 hours, diffuse collapse of alveoli can be shown in the larger experimental animals to be accompanied by evidence of "shunting". It is at this stage of the early lesion that the X-ray of the lung fields is clear or reveals only modest evidence of pulmonary edema. The lung lesion may then progress to the late stage in which polymorphonuclear leukocytes stick to the vascular membrane and subsequently infiltrate into the alveolar septa and airways (13, 14, 59). Erythrocytes also appear in the alveoli accompanied by the proteinacious material. The picture subsequently resembles the classic bronchopneumonia (36).

Metabolism and Surfactant Synthesis: The presence of diffuse alveolar collapse observed histologically early in the progressive development of pneumonitis, suggests a failure of the pulmonary surfactant system. Surfactant is a complex, principally of lipid and protein. L-dipalmitoy phosphatidyl choline accounts for 70 to 80% of the phospholipid present in airway washings and appears to be responsible for the low surface tension characteristics which prevent the water-lined alveoli from collapsing (1, 22, 30). In addition, the surfactant system also plays an important role in preventing bacterial invasion of pulmonary parenchyma, not only by bacteriostatic qualities, but also by promoting alveolar bronchial cleansing through ciliary action (30, 39, 52). Failure in this regard may account for the progression from the early stage of pneumonitis to confluent bronchopneumonia and atelectasis.

Evidence points to the large alveolar cells (Type II) as the site of surfactant synthesis. These cells have a relatively high metabolic rate (10). Decay of surface activity (as measured by physical methods) occurs after pulmonary vascular occlusion or loss of normal ventilation (22). To study the synthesis and secretion of surfactant in sepsis, our group measured the rate of incorporation of 32 P into the total phospholipids, and phosphatidyl choline in both airway washings and in lung tissue washed clear of blood. Comparing fed, starved, and rats rendered septic by cecal ligation, it became evident that both starvation and non-thoracic sepsis interfere with synthesis, storage and secretion of surfactant phospholipids (50). Evidence further suggests that transudation of plasma constituents into the septae and alveoli during sepsis further inactivates existing surfactant (54).

Circulating Agents Capable of Injuring the Lung in Sepsis: Aerobic and anaerobic cultures in more than 60% of the biopsies from the lungs in experimental animals at this early stage of pneumonitis fail to demonstrate the presence of culturable organisms (14). As occurs in the patients, the early Stage I pneumonitis in animals either resolves, or it progresses on to bronchopneumonia. Thus, it appears that a factor other than pulmonary bacterial invasion (hematogenous pneumonia) is responsible for the early post traumatic and septic lung lesion. Farrington et al. (21), working in our laboratory, employed a perfused canine pulmonary lobe in situ to demonstrate that whole blood from shock or septic dogs produced, within half an hour, the complete histological pattern of the early pneumonitis. At the same time, the perfusion pressures increased on the average 71 ± 12%, indicating a rise in pulmonary vascular resistance. Flow pressure curves confirmed this finding. All of this occurred while the non-perfused remainder of the dog's normal lung, through which his own blood was circulating, remained normal. Endotoxin added to the perfusing blood produces an identical result, and Harrison et al. (24) found that the result in terms of elevated resistance and edema formation was even more dramatic when live E. coli organisms were added to the perfusate.

There seems to be little doubt that proteases are at work in the induction of these effects. Voss et al. (56), also in our laboratory, showed that the administration of trasylol (a protease inhibitor), 50,000 units to a septic dog, two hours before drawing blood for lung perfusion, almost entirely prevented development of the lung lesion in the perfused lobe. This therapy also prevented the elevation of pulmonary resistance by the blood from the septic animal.

In subsequent experiments, it has been observed that the non-protein fraction of plasma from septic patients and animals prepared by cold vacuum filtration, which contains substances with molecular weights from 1,000 to 10,000 when added to normal perfusing blood is capable of inducing the same pulmonary changes (12, 56). This

suggests a number of possibilities as being responsible for induction of the lesion:
1. the presence of bacterial endotoxin working directly upon the capillary membrane
of the lung, 2. the presence of thrombin secondary to the induction of diffuse intravascular coagulation (DIC), 3. the presence of fibrinopeptides produced by the activation of plasmin, concomitant with DIC, 4. aggregation of platelets and release of
serotonin, histamine, ADP, or prostaglandin, 5. activation of the kinin system and
6. the possibility that other vasoactive peptides are generated from other protein
by release of proteolytic lysozomal enzymes, particularly from macrophages.

To assess the possible role of each of these factors, we have undertaken a series
of experiments, at times reproducing the results of previous workers in this field.
Cuevas et al. (18) observed that, whereas endotoxin infusion caused rabbits to develop the characteristic lung lesion, animals made resistant to endotoxin failed to
develop pneumonitis. In order to separate the possible action of endotoxin directly
upon the pulmonary capillary membrane from its secondary effects of inducing DIC
by activation of the clotting mechanism, a series of experiments have been carried
out, employing the isolated perfused rabbit lung. Whereas, DIFCO endotoxin in small
doses (0. 05 - 0. 5 mg), when added to the perfusing rabbit blood, produced a rise in
perfusion pressure within a short time, reaching a level of 200% by 3 hours, the
addition of three times the usual heparin dose to the perfusing blood virtually oblit-
erated the response. Kux et al. (32) found that when endotoxin was added to blood
rendered non-coagulable with acid citrate, the perfused pulmonary lobe failed to
show an increase of resistance. Thus, the role of the clotting system, and possibly
calcium, in the action of endotoxin is indicated. Furthermore, when plasma or albumin in saline were substituted for the whole blood perfusate, little or no reaction
occurred. Under these conditions, during satisfactory experiments, no pulmonary
edema took place. These findings strongly suggest that endotoxin does not directly
affect the lung, confirming the earlier results of Kuida and Hinshaw (25, 31). When
the buffy coat containing the white cells and possibly the platelets was added to the
plasma perfusion, the introduction of endotoxin induced the same effect as with whole
blood. Thus, it would appear that endotoxin injures the lung through the medium of
the white cells, or platelets, either by destruction of the former or by aggregation
of the latter (32, 45). This supports the view that the active substance must be produced secondarily by endotoxin to cause the lung lesion and the pulmonary vasoconstriction. However, it remains to determine definitely the interacting roles of the
platelets, the polymorphonuclear leukocytes and the globulin constituents of the whole
plasma responsible for clotting.

DIC occurs in many clinical situations associated with the lung lesion under consideration, including endotoxemia (7, 23). The presence of thrombin may be the ultimate pathway (23). Olsson et al. (38) have found that infusion of thrombin experimentally produces peripheral vasodilation, and at the same time, an increase of
pulmonary vascular resistance. Both fibrin formation and platelet aggregation are
induced by thrombin infusion. The roles of intravascular obstruction by micro emboli and thrombosis in the elevated vascular resistance is doubtful. The reaction
has been found to occur in animals defibrinated with reptilase, and can also be rapidly reversed by the administration of heparin. Both findings argue against obstruction
of the pulmonary vessel and favor a reaction of pulmonary vascular constriction.

Platelet aggregation is recorded frequently in septic shock and other states involving DIC. With radioactive, chromium-marked platelets, Bergentz, Lewis and
Ljungqvist (5) demonstrated platelet accumulation in the lung as the blood platelet
count falls under these conditions. ADP and serotonin, both pulmonary vascular vasoconstrictors (9, 47) are released from aggregated platelets. Both also produce
peripheral dilation. Acetylsalicylic acid (ASA) does not inhibit the aggregation of
platelets as demonstrated by the presence of thrombocytopenia, but it does not prevent the pulmonary hypertension and airway constriction after the infusion of thrombin into dogs (45). Thus, one is led to the conclusion that most of the increase of

pulmonary vascular resistance is related to the release of vasoactive substances from platelets. However, inhibition of serotonin by methylsergide also fails to block the reaction. Another possible series of substances from platelets may be the prostaglandins, $E_2 F_2a$ (53). Their release can be blocked by ASA. These agents have been found to be pulmonary vasoconstrictors, as well as bronchoconstrictors in the dog (51).

Although Bayley (4) found that infusion of fibrinopeptides increased pulmonary vascular resistance, Olsson (38) observed that defibrinated dogs responded to thrombin infusion by pulmonary hypertension in the absence of evident clotting. Thus, the role of fibrinopeptides remains in doubt.

Activation of the kinin system through Hageman Factor (XII) (46) may play a role in the pulmonary reaction. It has been clearly demonstrated that the infusion of bradykinin into the pulmonary artery causes an increase of the pulmonary vascular resistance (29). Our earlier work with the isolated perfused dog lung gave the same response, and, in addition, produced a morphological change in the lung, characterized principally by interstitial edema. Evidence from experiments in primates carried out by Mason (34), indicate that endotoxin infusion causes a reduction of pre-kallikrein. Attar et al. (3) have shown that bradykinin, activated in significant quantities in shocked and septic patients, plays a significant part in plasma loss. Measurements made by ourselves in cooperation with Drs. Richard Talamo and Robert Coleman of the Massachusetts General Hospital in seriously ill septic patients, demonstrated a significant reduction of pre-kallikrein, which was most marked in those patients who were hypotensive at the time of the observations (17). In the same group, bradykinin was elevated above 3 ng per ml in 60% of the patients who were hypotensive but without liver disease, and in 66% of the patients who were hypotensive with liver disease. This is in contrast to an elevation of 28% and 12% respectively in the patients without hypotension. However, no exact correlation between the plasma bradykinin concentration and the extent of pulmonary shunting could be made.

The possibility that any of these agents of small molecular weight circulating in the non-protein fraction of the plasma are capable of injuring the lung directly is suggested by recent experiments employing the in vitro rabbit lung perfusion. If the perfusate is albumin (5% in Ringer lactate solution) and if the blood has been carefully washed out of the lung, the addition of endotoxin produces virtually no edema or vasoactivity. This is an effect previously observed by Kuida et al. (31) and Hinshaw et al. (25) by using a perfusate of gelatin. On the other hand, in 6 to 10 of our experiments, addition of the "active fraction" (molecular weight 1,000 to 10,000) derived from the plasma of septic dogs, caused an increase of perfusion pressure ranging from 18 to 48% which occurred within 15 minutes. A moderate degree of edema became apparent in seven. In no experiment in which the "active fraction" was derived from plasma of a normal animal, did these reactions occur.

In 7 of the 10 fractions from the plasma of the septic animals, endotoxin was demonstrated by Dr. Pedro Cuevas and his colleagues in Dr. Jacob Fine's laboratory, using the limulus lysate assay (48). Values ranged as high as 2.5 /ug/ml. Yet this can be discounted, since endotoxin apparently does not react with the lung when albumin or gelatin solution is used for the perfusion.

Conclusions

Thus, the exact nature of the substance or substances present in the venous blood after injury, shock or sepsis, which cause edema and pulmonary vasoconstriction, remains unknown. Endotoxemia in the presence of whole blood causes these same pulmonary changes (28). Whether endotoxin activates the complement to produce the injury to the lung through the mediation of leukocytes (16, 27), or whether lysozomal enzymes are released, it is almost certain that other substances, including vasoactive peptides are present which directly cause the early pulmonary abnormalities. The fact that experimentally, trasylol is capable of blocking the development and transfer of the active agent to the perfused lung preparation from a septic animal argues strongly that peptidases are involved in its production. The conclusion, then, must be reached that somewhere in the chain of inducing this active factor in the plasma, whether derived from platelets, clotting activity, or other sources, peptidases and peptides in abnormal concentrations must be involved. Furthermore, it is evident that the elevation of pulmonary vascular resistance which is found in septic patients (13) probably plays a role in shock by preventing the cardiovascular system from satisfying the high circulatory demand which is essential for recovery (49).

References

1. Abrams, M.E.: J. appl. Physiol. 21: 718 (1966).
2. Ashbaugh, D.G., Petty, T.L., Bigelow, D.B., Harris, T.M.: J. thorac. cardiovasc. Surg. 57: 31 (1969).
3. Attar, S.M.A., Tingey, H.B., McLaughlin, J.S., Cowley, R.A.: Surg. Forum 18: 46 (1967).
4. Bayley, M.B., Clements, J.A., Osbahr, A.J.: Circulat. Res. 21: 469 (1967).
5. Bergentz, S.E., Lewis, D.H., Ljungvist, U.: Trapping of platelets in the lung after experimental injury. Microcirculatory approaches to current therapeutic problems, p.35 (Karger, Basel 1971).
6. Berggren, S.M.: The oxygen deficit of arterial blood caused by non-ventilating parts of the lung. Acta physiol. scand. Suppl. 11 (1942).
7. Blaisdell, F.W., Lim, R.C., Stallone, R.J.: Surg. Gynec. Obstet. 130: 15 (1970).
8. Border, J., Gallo, E.: J. Trauma 6: 176 (1966).
9. Borst, H.G., Berglund, E., McGregor, M.: J. clin. Invest. 36: 669 (1957).
10. Buckingham, S., Heinemann, H., Sommers, S.C., McNary, W.E.: Amer. J. Path. 48: 1027 (1966).
11. Burke, J.F., Pontoppidan, H., Welch, C.E.: Ann. Surg. 158: 581 (1963).
12. Clowes, G.H.A., Jr.: The pulmonary response to circulating agents in post traumatic and septic states. In: Haberland, G.L., Lewis, D.H.: New Aspects of Trasylol Therapy, p.72 (Schattauer, Stuttgart/New York 1973).
13. Clowes, G.H.A., Jr., Farrington, G.H., Zuschneid, W., Cossetts, G.R., Saravis, C.A.: Ann. Surg. 171:663 (1970).
14. Clowes, G.H.A., Jr., Zuschneid, W., Turner, M., Blackburn, G.L., Rubin, J., Toala, P., Green, G.: Ann. Surg. 167: 630 (1968).
15. Coalson, J.J., Guenter, C.A., Hinshaw, L.B.: J. exp. mol. Path. 12: 84 (1970).
16. Cochrane, C.G., Aiken, B.: J. exp. Med. 124: 433 (1966).
17. Colman, R.W., O'Donnell, T.F., Talamo, R.C., Clowes, G.H.A., Jr.: Clin. Res. 21: 596 (1973).
18. Cuevas, P., de la Maza, M., Gilbert, J., Fine, J.: Arch. Surg. 104: 319 (1972).
19. DePalma, R., Coil, J., Davis, J., Holden, W.: Surgery 62: 505 (1967).
20. Eiseman, B., Ashbaugh, D.G., Eds.: Pulmonary effects of non-thoracic trauma. J. Trauma 8 (Williams and Wilkens Company, Baltimore 1968).
21. Farrington, G.H., Saravis, C.A., Cossette, G.R., Clowes, G.H.A., Jr.: Surgery 68: 136 (1970).
22. Finley, T.N., Tooley, W.G., Swenson, E.W.: Amer. Rev. resp. Dis. 89: 372 (1964).
23. Hardaway, R.M.: Syndromes of Disseminated Intravascular Coagulation (Charles C. Thomas, Springfield, Illinois 1966).
24. Harrison, L.H.J., Hinshaw, L.B., Coalson, J.J., Greenfield, L.J.: J. thorac. cardiovasc. Surg. 61: 795 (1971).
25. Hinshaw, L.B., Kuida, H., Gilbert, R.P., Visscher, M.B.: Amer. J. Phyiol. 191: 293 (1957).
26. Ingram, R.H., Szidas, J.P., Skalak, R., Fishman, W.P.: Circulat. Res. 22: 801 (1968).

27. Janoff, A., Zelig, J.D.: Science 161: 702 (1968).
28. Kass, E.H., Wolff, S.M., Eds.: Bacterial Lipopolysaccharides: Chemistry, Biology and Clinical Significance of Endotoxins. J. infect. Dis. 128 (1973).
29. Kellermeyer, R.W., Graham, R.C.: New Engl. J. Med. 279: 754, 802, 859, 866 (1968).
30. Klaus, M.H., Clements, J.A., Havel, R.J.: Proc. nat. Acad. Sci. 47: 1858 (1961).
31. Kuida, H., Hinshaw, L.B., Gilbert, R.P., Visscher, M.B.: Amer. J. Physiol. 192: 335 (1965).
32. Kux, M., Coalson, J., Massion, W.H., Guenter, C.A.: Ann. Surg. 175: 26 (1972).
33. Majno, G., Palade, G.E.: J. biophys. biochem. Cytol. 11: 57 (1961).
34. Mason, J.W., Kleeberg, V.R., Dolan, P., Colman, R.W.: Amer. int. Med. 73: 545 (1970).
35. McLean, A.P.H., Duff, J.H., MacLean, L.P.: J. Trauma 8: 891 (1968).
36. Moon, V.H.: Amer. J. Pathol. 24: 235 (1948).
37. Moore, F.D., Lyons, J.H., Jr., Pierce, E.C., Jr., Morgan, A.P., Drinker, P.A., MacArthur, J.P., Dammin, G.J.: Post-Traumatic Pulmonary Insufficiency (Saunders, Philadelphia 1969).
38. Olsson, P., Radegran, K., Taylor, G.A.: Cardiovasc. Res. 4: 443 (1970).
39. Pattle, R.E.: Physiol. Rev. 45: 48 (1965).
40. Peltier, L.F.: Surgery 40: 665 (1956).
41. Peters, R.M.: Pulmonary mechanics in septic shock. In: Hershey, S.G., Del Guercio, L.R.M., McConn, R.: Septic Shock in Man.(Little, Brown, Boston

42. Powers, S.R., Burge, R., Leather, R., Monaco, V., Newell, J.: J. Trauma 12: 1 (1972).
43. Powers, S.R., Jr., Mannal, R., Neclerio, M., English, M., Marr, C., et al.: Ann. Surg. 178: 265 (1973).
44. Proctor, H.J., Ballantine, T.V.N., Broussard, N.D.: Ann. Surg. 172: 2 (1970).
45. Radegran, K., McAslan, C.: Acta anaesth. scand. 16: 76 (1972).
46. Ratnoff, O.D.: Adv. Immun. 10: 145 (1969).
47. Reeves, J.T., Jockl, P., Merida, J., Leathers, J.E.: J. appl. Physiol. 22: 475 (1967).
48. Reinhold, R.B., Fine, J.: Proc. Soc. exp. Biol. Med. 137: 334 (1971).
49. Rubin, J.W., Clowes, G.H.A., Jr.: Surg. Clin. N. Amer. 49: 489 (1969).
50. Rubin, J.W., Clowes, G.H.A., Jr., MacNicol, M.F., Gavin, J.B.: Amer. J. Surg. 123: 461 (1972).
51. Said, S.: Some r 1971). ory effects of prostaglandins E_2 and F_2a.In: Prostaglandin Symposium of the Worcester Foundation for Experimental Biology, p. 267 (Interscience, New York 1967).
52. Scarpelli, E.M.: The Surfactant System of the Lung. (Lea and Tebiger, Philadelphia 1968).
53. Smith, J.B., Willis, A.L.: Nature New Biol. 231: 235 (1971).
54. Taylor, F.B., Jr., Abrams, M.E.: Physiologist 7: 269 (1964).
55. Teplitz, C.: J. Trauma 8: 700 (1968).
56. Voss, H.J., MacNicol, M.F., Saravis, C.A., Altug, K., Clowes, G.H.A., Jr.: Surg. Forum 22: 27 (1971).
57. West, J.B.: Ventilation/Blood Flow and Gas Exchange. (Blackwell Scientific Oxford 1965).
58. Wilson, J.W.: Surg. Gynec. Obstet. 134: 675 (1972).
59. Wilson, J.W., Ratliff, N.B., Mikat, E.,et al.: Chest 59: 36S+ (1971).

Table I.

Clinical observations on the relationship of hemodynamics, blood gases, and acid-base balance during period of maximal response. Value ± Standard Deviation (Number of Observations).

	Survivors		Deaths	
	Uneventful convalescence	Septic	Uneventful convalescence	Septic
Number of pts. hemodynamic	30	22	0	16
Cardiac index (1/m²/min)	3.1 ± 0.1 (60)	4.2 ± 1.2 (22)	– –	4.9 ± 2.0 (15)
Cen. ven. pres. (cmH$_2$O)	4 ± 2 (60)	9 ± 4 (24)	– –	12 ± 5 (16)
Pulm. art. Systolic pres. (mmHg)	22 ± 5 (12)	36 ± 7 (8)	– –	38 ± 10 (7)
Art. mean bld. pres. (mmHg)	90 ± 14 (60)	91 ± 12 (24)	– –	88 ± 19 (16)
Art. bld. gas (breathing air) PaO$_2$ (mmHg)	85 ± 6 (45)	65 ± 12 (20)	– –	53 ± 14 (16)
PaCO$_2$ (mmHg)	38 ± 5 (45)	31 ± 6 (24)	– –	26 ± 9 (16)
Arterial Blood pH	7.38 ± .04 (45)	7.48 ± .05 (24)	– –	7.44 ± 10 (16)
Buffer base (mval/l)	+1 ± 2 (45)	-2 ± 3 (24)	– –	-7 ± 4 (16)
Excess lactate mM	0.5 ± 0.2 (31)	1.4 ± 0.7 (20)	– –	2.5 ± 1.8 (14)

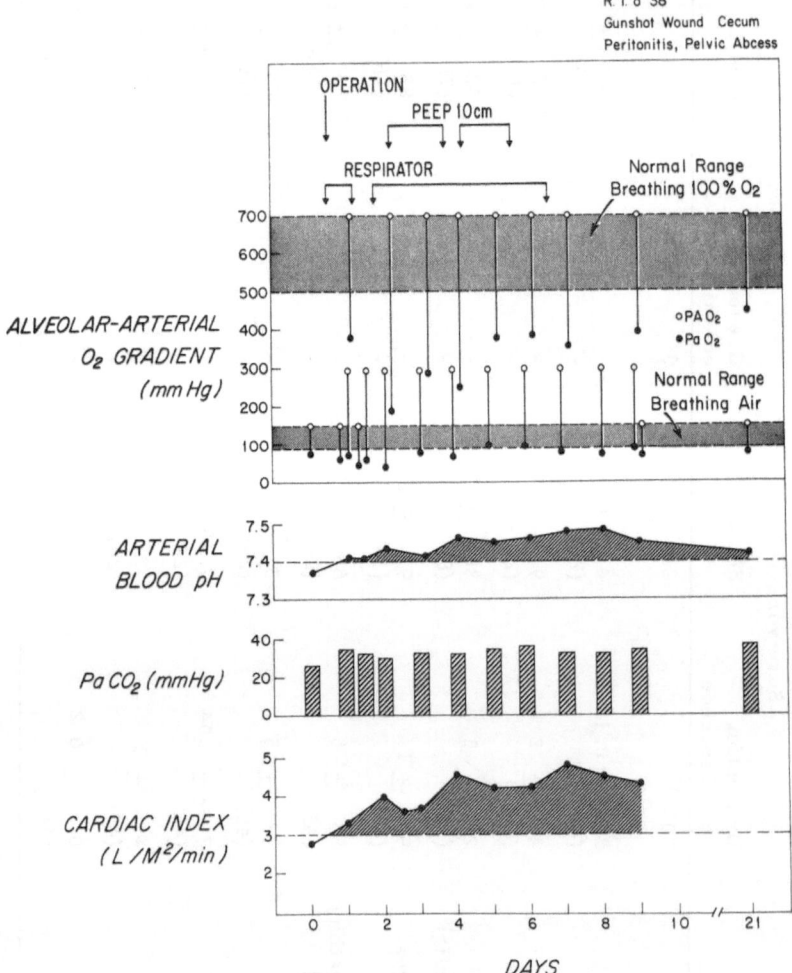

Fig. 1. The course of a patient with generalized peritonitis secondary to a neglected cecal wound. Please note the progressive increase of pulmonary shunting as demonstrated by the increased alveolar-arterial oxygen tension gradient at FIO_2 of 1.0, 0.4, 0.2. Note the improvement with the use of post end expiratory pressure (PEEP). The high cardiac index is also typical of septic patients who do well and recover.

On the Escape of Endotoxin from the Intestine

H. Gans and K. Matsumoto

We became interested in endotoxin as a result of our work on liver failure. We noted during experimentally induced acute hepatic insufficiency in the dog a number of changes (1, 2) which could be attributed to the development of gram-negative sepsis of enteric origin (3). Subsequent reports, notably those by Bjørneboe (4) and by Triger (5), indicate that in patients with liver cirrhosis E. coli antibody titers are markedly increased. This suggests that during hepatic insufficiency in man enteric organisms or their products are also able to escape from the intestine (6).

How is this possible, since as Dr. Brobmann again reiterated yesterday, very few investigators have been able to confirm unequivocally the presence of such an escape mechanism?

To explain the observed changes one can postulate that either enteric organisms and their endotoxins continuously escape from the intestine in considerable quantities, but that these materials are no longer cleared by the liver during hepatic insufficiency, or, that during this condition the efficiency of the mechanisms that normally retain gram-negative organisms and their products within the intestinal lumen, mechanisms which Dr. Cottier summarized for us so admirably day before yesterday (7), is somehow interfered with. For example, it is conceivable that intestinal permeability is increased as a result of portal hypertension and that routes other than the portal vein are followed in the development of sepsis. The fundamental difference between these 2 mechanisms is that in the first instance we are dealing with normal intestinal function and with a primary defect of the liver - while in the second circumstance the primary defect resides in the intestine.

If we scrutinize the first proposition, it is quite clear that it, in its turn, can be broken down further into 2 basic questions.

1. is endotoxin absorbed from the normal intestine into the portal vein in amounts sufficiently large to cause clinical manifestations? and

2. if the material enters the portal vein is the liver able to remove it? The latter question was of some significance to us since we found that the clearance of Cr^{51} labeled endotoxin from blood in hepatectomized dogs occurred at the same rate as in the intact dog, which made us wonder whether the liver was indeed important in this regard (8). I do not wish to go into this question at this time. Suffice to say that we showed that the liver is quite capable of clearing and detoxifying endotoxin and to refer here to a recently published report on this subject from our laboratory (9).

So let us move on then to the next question which is, is endotoxin normally absorbed into the portal vein in quantities sufficiently large to cause problems, if the liver would fail to clear and detoxify it?

To study this problem we made use of rats with a Thiry Vella fistula, which is an isolated, bypassed bile-free ileal segment, prepared as illustrated in Fig. 1. After opening the abdomen in the midline the vascular arcade immediately proximal to the ileocecal valve was divided together with the bowel and the ileum was measured along the mesenteric border in a proximal direction for the required distances. The bowel and mesentery were then divided proximally at the measured point. Intestinal continuity was re-established by approximating the two ends of the ileum with a continuous 7-0 suture, while the two ends of the isolated ileal loop were ex-

Study supported by NIH-GRSG 5 SOI RR-05396-II and NIH 5 ROI-HOL 14267.

teriorized as proximal and distal ileostomies through separate stab wounds in the left flank.

One week later, after the animals had fully recovered they were lightly anesthetized with ether following which pursestring sutures were placed around both ileostomies. Then a soft rubber catheter was threaded into the loop through the proximal ileostomy through which we injected a mixture of Cr^{51} labeled (1/2 mg) and unlabeled endotoxin (4 1/2 mg) (Difco Laboratories), and an indicator dye, methylene blue (0.1ml-0.25 %). This dye is readily absorbed from the intestine and rapidly excreted in the urine, staining it blue. Both agents were suspended in 3ml of isotonic saline, distilled water or hypertonic (50 %) glucose in water. Then the animals were given lead acetate intravenously (3 1/2 mg/100 g body weight), which has been shown to sensitize the rat to endotoxin more than a thousand fold, presumably through its effect on the reticuloendothelial system (9).

By frequently changing filterpaper placed underneath the wirebottom cages in which these animals were housed were we able to follow the dye excretion in the urine and to collect sequential urine samples for determination of radioactivity excreted in the urine. Also, the dye allowed us to detect possible spillage from either ileostomy. If this occurred to any significant degree the animal was discarded from the study. Data summarized in Table I show how sensitive a Pb acetate treated rat is for intravenous endotoxin; with as little as 1 /ug of endotoxin/ 100 g body weight the mortality rate, irrespective as to whether the endotoxin is administered over a protracted period or as a bolus, ranges from 65 - 80 %.

In our first experiment we made 25 cm T-V loops in 2 groups of rats, the first had undergone end-to-side porta-caval shunt operation, allowing the portal vein blood to bypass the liver, while animals of the second group had undergone only a sham operation consisting of cross-clamping the portal vein and inferior vena cava for 20 minutes, which is the time required to complete a porta-caval shunt in the rat.

In both groups of animals the administration of lead acetate alone was well tolerated, none of the animals died as a result (see Table II). Also, if lead acetate was injected 4 hours after endotoxin was placed in the bypassed ileal segment every animal of either group survived. However, if lead acetate was injected at the s a m e time the endotoxin was placed in the loop 1 out of 10 sham-operated rats and 4 out of 10 porta-caval shunt rats succumbed.

The finding that porta-caval shunt rats tolerated lead acetate alone or lead acetate and endotoxin 4 hours apart would suggest that neither the porta-caval shunt procedure nor the lead acetate alone can be held responsible for the observed increase in mortality in the porta-caval shunt group. Instead, it would suggest that the simultaneous administration of these 2 agents in porta-caval shunt rats resulted in animals that were more susceptable to endotoxin than were the sham-operated rats. Because the porta-caval shunt permits portal vein blood to bypass the liver, the increase in mortality may be due to a diminished clearance by the liver of portal vein blood and the endotoxin it contains. This would suggest that at least part of the endotoxin absorbed from the isolated loop is transported by way of the portal vein. Certainly the amount of endotoxin absorbed is minute, considerably less than 1 /ug/100 g body weight. If it were as much as 1 /ug, we could expect that at least twice the number of porta-caval shunt rats would have died.

In our next experiment we lengthened the intestinal loop from 25 cm to 40 cm. As a result (see Table III) the mortality rose from 10 % to 36 % indicating that the larger the surface area, the more endotoxin is absorbed. It should be pointed out, here, that none of these animals had a porta-caval shunt.

This observation poses an interesting problem. It suggests that, although the amount of endotoxin absorbed is minute, not enough is cleared by the liver to save every animal. We have to conclude, therefore, that some of the endotoxin is able to bypass the liver. The only other route available for its escape is via the lymphatics.

If part of the endotoxin, indeed, escapes by way of the lymphatics, our first proposition, which states that in hepatic insufficiency, failure of hepatic clearance is responsible for possible endotoxemia of enteric origin, becomes untenable. Since it seems that a considerable part of the endotoxin that escapes from the intestine bypasses the liver, it is obvious that the role of the liver is not a primary but rather a secondary one.

This is an important distinction. It is indeed unfortunate that limitation of time does not allow me to elaborate on this point.

We are left then with the 2nd proposition, namely, that if E. coli sepsis and endotoxemia are frequent complications of liver insufficiency this is probably not so much because of failure of hepatic clearance but rather because the simultaneous involvement of the intestine allows for the escape of more endotoxin.

This proposition remains to be investigated. On reviewing the literature we found that presently only few conditions are known to be associated with increased absorption of intestinal contents. These include total body irradiation (10), graft versus host reaction (11) and possibly treatment with certain chemotherapeutic agents.

The mechanism we choose to induce an increased absorption from the intestine utilized the effect of osmotic shock on the intestine. We resorted to it because of its ready accessibility, the ease with which we were able to control it and the uniformity of results it elicited. We found that the mortality rate rose steeply when hypo- or hyper-osmolar solutions were used to resuspend the endotoxin prior to its installation in a 40 cm T-V loop of a lead acetate sensitized rat (see Table IV). As is shown in Table V this increased mortality was not caused by a nonspecific, dehydrating effect of the hypertonic glucose, rather that it represented a highly s p e c i f i c   reaction.

These animals, at autopsy, exhibited a number of very marked changes of which the most unexpected one was the presence of radioactivity in the peritoneal cavity. Hence under these circumstances, labeled endotoxin had escaped from the gut lumen into the free abdominal cavity.

These changes lent themselves to a further, semi-quantitative analysis. As can be seen from Table VI, at autopsy none of the animals that received endotoxin in isotonic saline had radioactivity in the abdominal cavity. In contrast, in 9/10 rats that received endotoxin in 50 % glucose in water was free radioactivity found to be present in the peritoneal cavity at the time of death, approximately 5-6 hours after the beginning of the experiment. This group of animals had been sensitized with lead acetate. If no lead acetate was given only 5/10 animals had radioactivity in the peritoneal cavity at the time of sacrifice, 6 hours after the start of the experiment. Also, the relative amount of radioactivity in the first group of rats that died as a result of endotoxemia was much larger than in the latter group that was sacrificed, suggesting that either lead acetate itself or the conditions that preceeded death from endotoxemia, such as shock, were associated with an increased permeability of the gut wall, thus promoting the transfer of endotoxin from the gut lumen to the peritoneal cavity.

Daniele et al. previously showed that if E. coli endotoxin is placed in the peritoneal cavity, part of it remains there while some leaves, predominently via the thoracic ducts (12). This then would indicate that, if intestinal permeability increases, significant quantities of endotoxin (and possibly other intestinal contents) are able to leave the intestine, not by way of the portal vein but by way of the peritoneal cavity, to be taken up by the abdominal lymphatics and transported further via the thoracic ducts, suggesting that also under those conditions not much endotoxin escapes directly into the portal vein circuit.

This study, which is still in progress, has raised a number of interesting problems. There is, for instance, the possibility that lead acetate - as I pointed out - may affect intestinal permeability. Hence we are currently repeating a number of the described studies in adrenalectomized rats. Also, in order to further demonstrate the role of the lymphatics, we intend to repeat some of these studies in rats with a thoracic duct fistula. An important question to be answered relates to the problems that led us to perform these studies in the first place, namely - what is the cause of the sepsis of intestinal origin observed during hepatic insufficiency? We believe that we now have some of the models that will allow us to study this particular problem under controlled conditions in the laboratory.

References

1. Mori, K., Quinlan, R., Richter, D., Kaster, R., Tan, B., Gans, H.: Surg. Gyn. Obst. 131: 919 (1970).
2. Tan, B.H., Mori, K., Richter, D., Quinlan, R., Gans, H.: Surg. Gyn. Obst. 132: 263 (1971).
3. Gans, H., Mori, K., Lindsey, E., Kaster, B., Richter, D., Quinlan, R., Dineen, P., Tan, B.H.: Surg. Gyn. Obst. 132: 783 (1971).
4. Bjørneboe, M., Prytz, H., Ørskov, F.: Lancet I: 58 (1972).
5. Triger, D.R., Alp, M.H., Wright, R.: Lancet I: 60 (1972).
6. Gans, H., Matsumoto, K., Mori, K.: Lancet I: 1181 (1972).
7. See Also: Rhodes, R.S., Karnovsky, M.J.: Loss of macromolecular barrier function associated with surgical trauma to the intestine. Lab. Invest. 25: 220 (1971).
8. Gans, H. et al.: Unpublished observation, 1970.
9. Mori, K., Matsumoto, K., Gans, H.: Ann. Surg. 177: 159 (1973).
10. Smith, W.W., Alderman, I.M., Schneider, C., Cornfield, J.: Proc. Soc. exp. Biol. Med. 113: 778 (1963).
11. Julita, J.W.: J. infect. Dis. 128: S99 and Keats, D.: idem. S101 (1973).
12. Daniele, R., Singh, H., Appert, H., Pairent, F.W., Howard, J.M.: Surg. 67: 484 (1970).

Table I.

Mode of Injection	Dose of Endotoxin	No. of Animals Studied	Mortality
Rapid Injection			
Endotoxin	1 μg/100 g	10	0/10
Endotoxin	3 mg/100 g	10	9/10
Lead Acetate (5mg/rat)	0	5	0/5
Lead Acetate (5mg/rat) + Endotoxin	1 μg/100 g	12	8/12
Slow Infusion*			
Endotoxin	1 μg/100 g	5	0/5
Endotoxin	3 mg/100 g	5	4/5
Lead Acetate (5 mg/rat)	0	5	0/5
Lead Acetate (5 mg/rat) + Endotoxin	1 μg/100 g	18	13/18

*Over 1 hour period (with a Harvard pump).

Table II.

Mortality Following the Installation of Endotoxin

in 25cm. Long Rat T-V Fistulae

Sham Operated Group:	Porta-Caval Shunt Group:

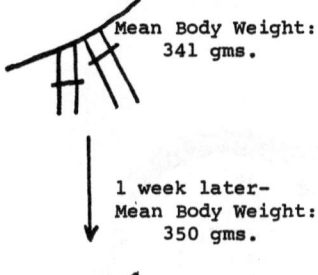

Mean Body Weight:
341 gms.

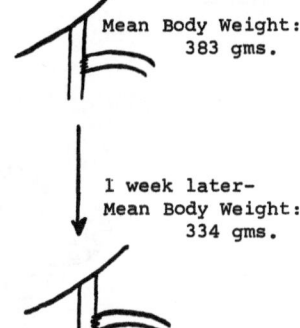

Mean Body Weight:
383 gms.

1 week later–
Mean Body Weight:
350 gms.

I week later–
Mean Body Weight:
334 gms.

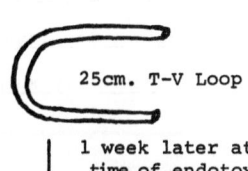

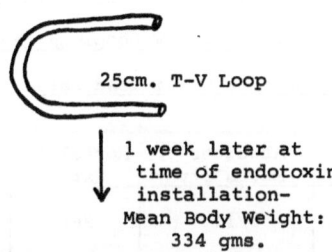

25cm. T-V Loop

25cm. T-V Loop

1 week later at
time of endotoxin
installation–
Mean Body Weight:
334 gms.

1 week later at
time of endotoxin
installation–
Mean Body Weight:
314 gms.

Mortality Rate

Pb acetate i.v. alone (3.5mg/100gm body weight)	0/10	0/10
Pb acetate i.v. + Endotoxin (in loop) 4 hours apart	0/10	0/10
Pb acetate i.v. + Endotoxin (in loop) simultaneously	1/10	4/10

434

Table III.

Effect of the Length of the Bypassed Ileal Segment on the Amount of Endotoxin Absorbed (as Expressed in the Mortality Rate of Lead Sensitized Rats).

	25 cm T-V loop	40 cm T-V loop
Mortality	$^1/10$ (10 %)	$^5/14$ (36 %)

Table IV.

Effect of the Suspension Medium Used to Reconstitute the Endotoxin

(after its Installation into 40cm. T-V Loops) on the

Mortality Rate of Lead Acetate Sensitized Rats

Endotoxin suspended in	Absolute Mortality	Mortality Rate
isotonic solution	5/14	36%
hypotonic solution	12/19	64%
hypertonic solution	13/14	93%

Table V.

Presence or Absence of Radioactivity in Abdominal Cavity at Time of Sacrifice or Death of the Rat

	Presence of Cr 51	Relative Activity
isotonic solution	0/10	
hypertonic solution E + E * and Pb acetate	9/10	++++
hypertonic solution E + E * (no Pb acetate)	5/10	++

Table VI.

Effect of 50 % Dextrose (in the Presence or Absence of Endotoxin and Lead Acetate) on the Mortality Rate of Rats with a 40 cm T-V Loop

Pb acetate i. v.	endotoxin in T-V loop	absolute mortality	mortality rate
-	+	1/20	5 %
+	+	19/20	95 %
+	-	9/10	90 %

Construction of a Thiry-Vella Fistula

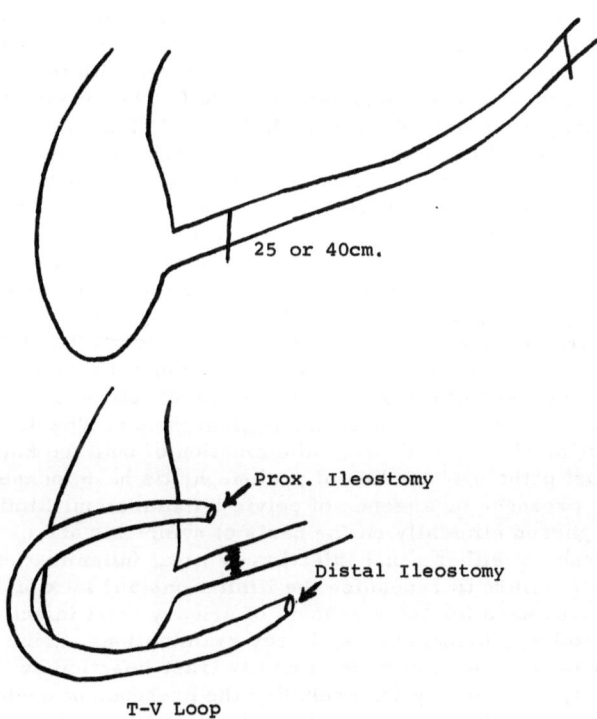

25 or 40cm.

Prox. Ileostomy

Distal Ileostomy

T-V Loop

Fig. 1.

Pyelonephritis: Pathogenesis and Unsolved Problems
W. R. McCabe

Ten to fifteen years ago, little difficulty would have been encountered in describing the pathogenesis of pyelonephritis. At that time, the recent renaissance of interest in pyelonephritis had convinced most physicians that pyelonephritis was extremely prevalent, was the most common cause of renal insufficiency, and that persistent bacteriuria and chronic pyelonephritis were practically synonymous. Subsequent studies have suggested that our original estimates of the prevalence of pyelonephritis were excessive and that persistent bacteriuria cannot be equated with chronic pyelonephritis. Indeed, most evidence indicates that in adults, without urinary tract obstruction, progression of bacteriuria or acute pyelonephritis to renal insufficiency is either infrequent or only occurs over a very protracted period of time. The relative infrequency of renal insufficiency resulting from urinary tract infection becomes quite apparent when the prevalence of bacteriuria is compared with the number of annual fatalities from renal disease, as shown in Table I. This presents estimates, derived by Kass, of the prevalence of bacteriuria in the United States in 1960 (1), with reported deaths from renal disease, Nephritis and Nephrosis and Infections of the Kidney (2). Thus, even if the assumption is made that all reported deaths from renal disease are actually due to pyelonephritis, only 0.5 % per year or 5 % per 10 years of bacteriurics develop fatal renal insufficiency. Similar re-examinations of many of our concepts of pyelonephritis have resulted in several major controversies concerning basic aspects of this disease. These have even extended to questioning whether chronic pyelonephritis has a bacterial etiology (3).

Much of this apparent controversy is attributable to two basic problems. The first problem is largely semantic and reflects the use of imprecise and inexact terms to describe infections and pathologic changes of the urinary tract. An excellent example of this problem may be seen in the use of the term, pyelonephritis. The derivation of the word, pyelonephritis, clearly indicates inflammation of both the kidney and the renal pelvis; yet, most pathologic studies of pyelonephritis have focused on renal changes and ignored the presence or absence of pelvic inflammation. Similarly, pyelonephritis is often diagnosed clinically on the basis of symptoms and urine culture results without evidence of either renal infection or renal inflammation. The second problem relates to our failure to recognize the limitations and lack of specificity of the various procedures used for the diagnosis of urinary tract infections. Although clinical symptomatology, urinalyses, X-ray examinations, urine cultures, and renal biopsy are invaluable for indicating urinary tract infection or renal involvement, none is entirely satisfactory for excluding the presence of pyelonephritis. As a result, the clinician is often uncertain about the actual type of urinary tract disease which is present and confused about the most appropriate type of therapy. This uncertainty is often reflected by over-zealous treatment of asymptomatic bacteriuria and inadequate treatment and follow-up of serious renal infections.

Bacteriuria

The concept of bacteriuria is one of the areas in which misconception and confusion has been greatest. Two English urologists, Barrington and Wright, were

These studies were supported by research grants (AI 09584 and AI 11116) and a training grant (5 T01-AI213) from the National Institute of Allergy and Infectious Diseases, National Institute of Health, U.S. Public Health Service.

actually the first to demonstrate that urinary tract infections of sufficient significance to induce bacteremia following urethral manipulation were uniformly associated with bacterial counts of 10^6/ml of urine (4). However, it was not until Kass popularized quantitative bacterial cultures of the urine that this technique gained wide acceptance (1). Kass was quite explicit that bacteriuria in excess of 10^5 per ml of urine was primarily of value in distinguishing contamination from bacteriuria resulting from urinary tract infection. He clearly specified that, although pyelonephritis was usually associated with significant bacteriuria, the presence of significant bacteriuria was not indicative of pyelonephritis (5, 6). Despite such admonitions, there was an almost uniform tendency to equate chronic pyelonephritis with chronic bacteriuria. Subsequent studies, previously mentioned, have indicated that progression of bacteriuria to renal insufficiency is relatively infrequent in the adult without obstructive lesions (7-10). This could reflect the failure of either qualitative or quantitative urine cultures to identify whether infection originates from the lower or upper urinary tract and most instances of bacteriuria representing lower tract infection. The development of the bladder wash-out and urethral catheterization technique by Stamey et al. has provided a method for localization of the site of urinary tract infection (11). A modification of this technique by Fairley et al. also allows identification of upper urinary tract infection but is more adaptable for routine clinical use (12). Studies utilizing these techniques have provided information concerning the relative frequency of lower and upper tract infections in various groups of bacteriurics (10-13). Fig. 1 presents the results of localization of the site of infection in three groups of bacteriurics as modified from the reports of Fairley et al. (13). As may be seen, bacteriuria originated from the upper urinary tract in over 50 % of each of the three groups. Renal localization of bacteriuria might be anticipated among patients with recurrent or acute symptomatic infections, but the finding of upper tract localization in more than one-half of unselected pregnant women with asymptomatic bacteriuria is somewhat surprising. If upper tract bacteriuria is as frequent in other patients with asymptomatic bacteriuria as in asymptomatic bacteriuria of pregnancy, the relative rarity of significant renal injury cannot be attributed to the infrequency of renal infection. Since upper tract infection appears to occur as often as lower tract infection, renal infection also appears to produce renal insufficiency only infrequently. These observations clearly illustrate the necessity of identifying those factors which lead to the development of renal damage from urinary tract infection and those patients who are destined to develop renal damage. It must be emphasized, however, that recognition of the infrequency of the development of renal injury in bacteriurics should not be construed to indicate that this condition does not require adequate therapy and follow-up. Prevention of even a few cases of renal insufficiency provides an adequate indication for treatment. In addition, prevention of the morbidity of acute symptomatic urinary infections and the increased risk of premature delivery resulting from episodes of acute pyelonephritis during pregnancy provide other cogent indications for the eradication of bacteriuria.

In contrast to the relative infrequency of the development of significant decreases in renal function or hypertension during prospective studies of prolonged bacteriuria, average blood pressure and urea nitrogen levels are usually greater in bacteriurics than in controls when large groups of patients are initially evaluated (7, 8). Although this could reflect the result of prolonged prior urinary tract infection, it could also represent an increased proclivity for the acquisition of bacteriuria in patients with hypertension or other forms of renal disease. There is one specific type of renal functional impairment, loss of maximal concentrating ability, which does appear to be definitely associated with renal bacteriuria, however. Several studies have demonstrated that renal bacteriuria is almost uniformly associated with impairment of maximal concentrating ability of the kidney and that this is corrected by eradication of the infection (14, 15).

Despite deficiencies in knowledge concerning some aspects of the course of pyelonephritis, considerable information has been acquired concerning the acquisition and pathogenesis of urinary tract infections through careful clinical and experimental studies. The aerobic gram-negative bacilli present in the fecal flora have uniformly predominated as etiologic agents in all studies of asymptomatic bacteriuria and acute and chronic pyelonephritis. However, the extreme rarity of anaerobes, numerically the most important constituents of the fecal flora, as a cause of urinary tract infection has not been entirely explained. E. coli has uniformly been the most frequent etiologic agent in all types of urinary tract infection although its relative prevalence decreases among patients with obstruction and recurrent infections. The demonstration that E. coli of the same serotype as those found in the urine, are usually present in the feces provides strong support that these infections usually derive from bacteria originally present in the fecal flora (16, 17). More recent studies by Stamey et al. have indicated that colonization of the vaginal vestibule with large numbers of gram-negative bacilli precedes bladder infection (18). The suggestion, based on clinical observations, that sexual activity in the female predisposes to the transport of bacteria up the urethra and into the bladder has also received support from experimental studies by Bran, Levison, and Kaye (19). These investigators demonstrated that manual urethral massage, simulating the effects of coitus, served to introduce urethral bacteria into the bladder (19). Once bacteria have been introduced into the bladder, those mechanisms which determine whether the bacteria will be eradicated or whether they will persist, multiply, and cause infection have not been clearly defined. Several host factors, mechanical bladder emptying of urine containing bacteria (20), antibacterial activity of the urine (21), and antibacterial activity of the bladder mucosa (22), have been identified which inhibit the development of bladder infection. Less is known concerning factors which favor bacterial persistence and multiplication and the subsequent development of infection. There is evidence, however, that once infection has developed, such patients have a greatly enhanced susceptibility to the acquisition of new urinary tract infections suggesting some impairment of bladder defense mechanisms (9, 23). In addition, the demonstration by Stamey et al. that colonization of the vaginal vestibule routinely precedes acquisition of bacteriuria also suggests a defect in those host mechanisms responsible for inhibiting vaginal colonization with gram-negative bacilli (18).

The mechanism of extension of bladder infection to the kidney has also been a subject of considerable investigation and some degree of controversy. It is generally accepted that most instances of pyelonephritis caused by gram-negative bacilli result from the ascending rather than the hematogenous route. For a time, ureteral reflux was considered a primary cause of renal infection which uniformly required surgical correction. However, this concept failed to recognize the necessity for preceding bladder bacteriuria to provide infected urine for reflux to the kidney. It has been subsequently demonstrated that endotoxin and killed gram-negative bacilli were capable of halting ureteral peristalsis (24) and that the eradication of infection often resulted in the disappearance of reflux (9). Other workers have also shown that, although surgical correction of reflux aborts symptomatic episodes of infection, it does not prevent recurrence of bacteriuria (25). These findings have tended to diminish the importance of reflux as a primary pathogenetic mechanism in pyelonephritis. Since it has been shown that motile bacteria can ascend against a moving column of urine and that inert particles, analogous to non-motile bacteria, theoretically can work their way upstream against a pulsatile flow (26), reflux does not appear to be a necessary precursor for renal infection.

A number of host and bacterial factors have been demonstrated which predispose to the persistence and multiplication of bacteria in the kidney. K antigens of E. coli, which inhibit phagocytosis, are the best documented of the bacterial virulence factors predisposing to renal involvement. Two separate studies have demonstrated that strains of E. coli containing large quantities of K antigen produce a higher proportion of renal infections than bladder infections while strains containing lesser amounts of K antigen tend to produce only bladder infections (27, 28). Urease production has also been postulated to be associated with nephrophathogenicity (29).

In addition, certain characteristics unique to the kidney also enhance susceptibility to infection. The milieu of the renal medulla, the major site of bacterial infection in pyelonephritis, is inimical to many of the host defense mechanisms of major importance in protection against gram-negative bacilli. Beeson and Rowley's investigations demonstrated that homogenates of kidney blocked the bactericidal effect of normal serum against gram-negative bacilli through the inactivation of C4 by ammonia generated in the renal medulla (29). Rocha and Fekety also demonstrated that mobilization of polymorphonuclear leukocytes in the medulla in response to a variety of injurious stimuli is considerably delayed in comparison to mobilization into the renal cortex (30). Further evidence of the inimical environment of the renal medulla on defense mechanisms was observed in Chernew and Braude's demonstration of depression of phagocytosis of bacteria by polymorphonuclear leukocytes in solute concentrations analogous to those seen under various conditions in the renal medulla (31). Thus, available evidence indicates that bactericidal activity of serum against gram-negative bacilli and both emigration and phagocytic capacity of leukocytes are all decreased by the milieu of the renal medulla. These demonstrations of decreased resistance to infection of the renal medulla cannot help but make us wonder why extensive renal damage does not result more often from renal bacteriuria. Very little is known about the factors which prevent such an occurrence.

Another of the peculiar paradoxes of pyelonephritis is the discrepancy between the nature of this disease in man and experimental animals. It has been extremely difficult to produce chronic pyelonephritis with persistent renal infection, except with enterococci or when obstructive lesions are present, in experimental animals. In general, bacteria disappear from the kidney concomitant with the appearance of antibody against the infecting organism in animals (32). In contrast, persistence of bacteriuria is usual in human infections despite the presence of high titers of type-specific antibody (33, 34).

The demonstration of progression of renal insufficiency in the absence of bacteriuria and persistence of bacterial antigen in the kidney in experimental and human pyelonephritis after disappearance of viable bacteria from the kidney and urine (35-38), suggests that immunologic mechanisms might contribute to the renal injury in chronic pyelonephritis. Preliminary studies by Kalmanson et al. using parabiotic animals have provided some support for this concept (39).

Pathologic Features of Chronic Pyelonephritis

As mentioned previously, the histologic changes observed in chronic pyelonephritis involve both the pelvo-calyceal system and the renal interstitium. The interstitial renal changes, round cell infiltration, tubular dilatation and atrophy, colloid casts, with relative sparing of the glomeruli, characteristic of pyelonephritis are not specific, however, and may be observed to result from a variety of other causes which are listed in Table II. The similarities of the renal lesions observed in these conditions was recognized earlier by our European colleagues than in the United States (40, 41). They pioneered in the use of chronic interstitial nephritis as an all-inclusive term for lesions with similar histologic features and identified specific etiologies whenever possible. This long list emphasizes why identification of bacterial infection as a cause of interstitial nephritis may be difficult. However, closer appraisal of the list provides an adequate explanation for emphasizing the importance of inflammatory changes in the renal pelvis. Of the 19 possible causes of interstitial inflammatory changes listed, only three, bacterial infections, obstruction, and analgesic abuse, have been demonstrated to be associated with pelvo-calyceal changes.

As a result of the increasing recognition that interstitial nephritis may have multiple causes other than bacterial renal infection, the frequency of post-mortem diagnosis of chronic pyelonephritis has decreased substantially. Fig. 2 depicts the frequency with which chronic pyelonephritis was diagnosed at post-mortem examination of male and female patients at Yale University from 1957 through 1964 as reported by Freedman (42). During this period, the frequency of diagnosis of pyelonephritis has decreased by 3 to 8-fold. This represents a considerably lower frequency of chronic pyelonephritis than was reported in similar autopsy surveys 10 to 15 years ago (43, 44).

The increasing reluctance to diagnose chronic pyelonephritis on a histologic basis may have become excessive, however, and resulted in an underestimate of the frequency of this disease. Some workers now consider such a diagnosis untenable unless a history of symptomatic renal infection can be elicited and concomitant bacteriuria is demonstrated (45). This completely ignores the repeated demonstrations that renal bacteriuria may be asymptomatic or that the symptoms of acute pyelonephritis may subside spontaneously despite persistence of bacteriuria (23). This attitude also fails to recognize that renal infection may produce renal scarring which will persist despite disappearance of bateriuria. We have attempted to circumvent these problems by developing a technique which would allow identification of prior bacterial renal infection after the disappearance of bacteriuria and viable bacteria from the kidney. These studies were based on earlier demonstrations of prolonged persistence of bacterial antigen in the kidney after disappearance of viable bacteria (35, 36) and utilized antiserum to an antigen, Common Enterobacterial Antigen, shared by most gram-negative bacilli for immunofluorescent staining. Studies in experimental animals demonstrated that the persistence of Common Enterobacterial Antigen in the kidney paralleled that of O-antigen (37). Subsequent studies showed that this technique uniformly allowed identification of bacterial antigen in all cases of acute pyelonephritis studied which were caused by Common Antigen containing bacteria (38). Fig. 3 demonstrates immunofluorescent staining of a specimen obtained at nephrectomy from a patient with acute pyelonephritis. Bacteria may be seen within the lumen and within epithelial cells of tubules. Bacterial antigen (white) may be seen within the interstitial tissues within macrophages, and in tubular epithelial cells. We have utilized this technique to study 7 patients whose kidneys have shown histologic changes consistent with pyelonephritis but who did not have bacteriuria at the time of study and who had no history of symptomatic urinary tract infection. Residual bacterial antigen was detected in the kidneys of 6 of the 7 patients suggesting that prior unrecognized renal infection had occurred. These results suggest that this technique appears to offer promise for the identification of prior renal infection in instances of interstitial nephritis of inapparent etiology. It should be emphasized, however, that the high frequency of detection of bacterial antigen in patients with "abacterial" pyelonephritis is not considered to reflect a bacterial etiology for a similar proportion of all patients with interstitial nephritis of unknown etiology. These results are reflective only of the group of patients studied. Determination of the frequency with which bacterial antigen can be detected in a large group of subjects with chronic interstitial nephritis of unknown etiology will require more extensive studies. These results do provide confirmation of the clinical observation that the absence of clinical symptoms is not sufficient to exclude renal infection.

Discussion

Although this presentation has raised more questions concerning pyelonephritis than it has answered, this does not seem inappropriate in the light of many of our quandries about this disease. Identification of problem areas is necessary before solutions can be sought for some of the enigmas of chronic pyelonephritis. The questions which appear to be of greatest importance are:

1. Determination of long-term effects of bacteriuria on renal function (other than concentrating ability).

a) What proportion of the large number of patients with bladder and renal bacteriuria ultimately develop renal insufficiency and how long must uncomplicated renal infection be present before significant renal damage results?

b) What factors are responsible for limiting extension of bladder infection and preventing renal infection?

c) What are the factors which are responsible for preventing clinically apparent renal injury in the large numbers of the population with renal bacteriuria? How can those patients at greatest risk of development of renal injury be identified?

d) More exact definition of other hazards of bacteriuria (prematurity, etc.).

2. Does bacterial infection of the kidney produce hypertension and how often? Or does hypertension predispose to bacteriuria?

3. Do immunologic mechanisms play an important role in the progression of the renal lesion in pyelonephritis?

4. What criteria or techniques can be developed to allow differentiation of the effects of bacterial infection on the kidney from the lesions of interstitial nephritis due to other causes?

Despite the number of questions posed, similar but equally puzzling questions concerning pyelonephritis have been answered during the past two decades. Continued careful investigations during the next few years can clarify our present questions.

References

1. Kass, E.H.: Trans. Assoc. Amer. Phys. 69: 56 (1956).
2. U.S. Public Health Service. Vital Statistics of the United States. Monthly Vital Statistics Report. (September 1971).
3. Angell, M.E., Relman, A.S., Robbins, S.L.: New Engl. J. Med. 278: 1303 (1968).
4. Barrington, F.J.F., Wright, H D.: J. Path. Bact. 38: 871 (1930).
5. MacDonald, R.A., Levitin, H., Mallory, G.K., Kass, E.H.: New Engl. J. Med. 256: 915 (1957).
6. Kass, E.H.: The role of asymptomatic bacteriuria in the pathogenesis of pyelonephritis. In: Quinn, E.L., Kass, E.H.: Biology of Pyelonephritis. (Little, Brown and Co., Boston, Mass. 1960).
7. Aascher, A.W., Sussman, M., Waters, W.E., Evans, J.A.S., Campbell, H., Evans, K.T., Williams, J.E.: J. infect. Dis. 120: 17 (1969).
8. Freedman, L.R.: Prolonged observations on a group of patients with acute urinary tract infection. In: Quinn, E.L., Kass, E.H.: Biology of Pyelonephritis (Little, Brown and Co., Boston, Mass.1960).
9. Kunin, C.M., Deutscher, R., Paquin, A., Jr.: Medicine 43: 91 (1964).
10. Stamey, T.A., Pfau, A.: Calif. Med. 113: 16 (1970).
11. Stamey, T.A., Govan, D.E., Palmer, J.M.: Medicine 44: 1 (1965).
12. Fairley, K.F., Carson, N.E., Gutch, R.C., Leighton, P., Grounds, A.D., O'Keefe, C.M.: Lancet II: 616 (1971).
13. Fairley, K.F.: The routine determination of the site of infection in the investigation of patients with urinary tract infection. In: Kincaid-Smith, P., Fairley, K.F.: Renal Infection and Renal Scarring (Mercedes Publishing Service, Melbourne 1970).
14. Kaitz, A.L.: J. clin. Invest. 40: 1331 (1961).
15. Clark, H., Ronald, A.R., Cutler, R.E., Turck, M.: J. invest. Dis. 120: 47 (1969).
16. Vosti, K.L., Goldberg, L.M., Monto, A.S., Rantz, L.A.: J. clin. Invest. 43: 2377 (1964).
17. Turck, M., Petersdorf, R.G.: J. clin. Invest. 41: 1760 (1972).
18. Stamey, T.A., Timothy, M.T., Millar, M., Mibara, G.: Calif. Med. 115: 1 (1971).

19. Bran, J.L., Levinson, M.E., Kaye, D.: New Engl. J. Med. 286: 626 (1972).
20. Cox, C.E., Hinman, F., Jr.: Factors in resistance to infection in the bladder. I. The eradication of bacteria by vesical emptying and intrinsic defense mechanisms. In: Kass, E.H.: Progress in Pyelonephritis (F.A. Davis Co., Philadelphia 1965).
21. Vivaldi, E., Munoz, J., Cotran, R., Kass, E.H.: Factors affecting the clearance of bacteria within the urinary tract. In: Kass, E.H.: Progress in Pyelonephritis (F.A. Davis Co., Philadelphia 1965).
22. Kaye, D.: J. clin. Invest. 47: 2374 (1968).
23. McCabe, W.R., Jackson, G.G.: New Engl. J. Med. 272: 1037 (1965).
24. Teague, N., Boyarsky, S.: Invest. Urol. 5: 423 (1968).
25. Govan, D.E., Palmer, J.M.: Pediatrics 44: 677 (1969).
26. Weyrauch, H.M., Bassett, J.B.: Standford Med. Bull. 9: 25 (1951).
27. Glynn, A.A., Brumfitt, W., Howard, C.J.: Lancet I: 514 (1971).
28. Kaijser, B.: J. infect. Dis. 127: 670 (1973).
29. Beeson, P.B., Rowley, D.: J. exp. Med. 110: 685 (1959).
30. Rocha, H., Fekety, F.R., Jr.: Delayed granulocyte mobilization in the renal medulla. In: Kass, E.H.: Progress in Pyelonephritis (F.A. Davis Co., Philadelphia 1965).
31. Chernew, I., Braude, A.I.: J. clin. Invest. 41: 1945 (1962).
32. Sanford, J., Jr., Hunter, B.W., Sonda, I.K.: J. exp. Med. 115: 383 (1962).
33. Williamson, J., Brainerd, H., Scaparone, M., Chuck, S.P.: Arch. intern. Med. 114: 222 (1964).
34. Vosti, K.L., Monto, A.S., Rantz, L.A.: J. Lab. clin. Med. 66: 613 (1965).
35. Sanford, J.P., Hunter, B.W., Donaldson, P.: J. exp. Med. 116: 285 (1962).
36. Cotran, R.S.: J. exp. Med. 117: 813 (1963).
37. Aoki, S., Merkel, M., Aoki, M., McCabe, W.R.: J. Lab. clin. Med. 70:204 (1967).
38. Aoki, S., Imamura, S., Aoki, M., McCabe, W.R.: New Engl. J. Med. 281: 1375 (1969).
39. Kalmanson, G., Glasscock, R., Montgomerie, J., Guze, L.: Is chronic pyelonephritis an immunologic disease? Abstracts V International Congress of Nephrology, p. 107 (1972).
40. Gloor, F.J.: Some morphologic features of chronic interstitial nephritis (Chronic Pyelonephritis) in patients with analgesic abuse. In: Kass, E.H.: Progress in Pyelonephritis (F.A. Davis Co., Philadelphia 1965).
41. Hamburger, J., Richet, G., Crosnier, J., Funck-Bretano, J.L., Antoin, B., Ducrot, H., Mery, J.P., de Montera, H.: Nephrology, 2 vols. (Saunders, Philadelphia 1968).
42. Freedman, L.R.: Ann. intern. Med. 66: 697 (1967).
43. Jackson, G.G., Dallenback, F.D., Kipnis, G.P.: Med. Clin. N. Amer. 39: 297 (1955).
44. Saphir, O., Taylor, B.: Pyelonephritis lenta. Ann intern. Med. 36: 1017 (1952).
45. Freedman, L.R.: Urinary tract infection, pyelonephritis, and other forms of chronic interstitial nephritis. In: Strauss, M.R., Welt, L.G.: Diseases of the Kidney (Little, Brown and Co., Boston 1971).

Table I.

Estimates of the Number of Cases of Asymptomatic Persistent Bacteriuria in the U.S.

Age Group	Number of Cases		Prevalence Rates %	
	Male	Female	Male	Female
0 - 14	12, 000	300, 000	< 0. 1	1
15 - 39	150, 000	600, 000 +	0. 5	2 - 3
40 +	150, 000	2, 000, 000	0. 5	6 - 8
Total	312, 000	2, 900, 000	<0. 5	3 - 4

Fatalities from Chronic Renal Disease (1970)

Nephritis and Nephrosis	7, 880
Infections of the Kidney	7, 920

Table II.

Reported Causes of Interstitial Renal Inflammation

Obstruction

Hypertension and Renal Vascular Disease

Nephrocalcinosis

Congenital and Hereditary Disorders

Potassium Depletion

Analgesics

Hyperuricemia

Diabetes Mellitus

Acute Tubular Necrosis

Sulfonamides

Irradiation

Choline Deficiency

Hypersensitivity or Immunologic Reactions

Hypophosphatemia

Medullary Cystic Disease

Sickle Cell Anemia

"Balkan Nephritis"

Lead and Cadmium Poisoning

Bacterial Infection

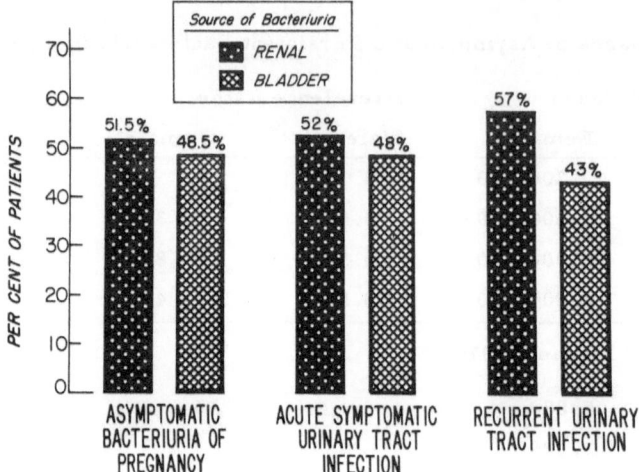

Fig. 1. The frequency of bacteriuria originating from either the bladder or the kidney in three groups of patients with urinary tract infection (modified from Fairley (13)).

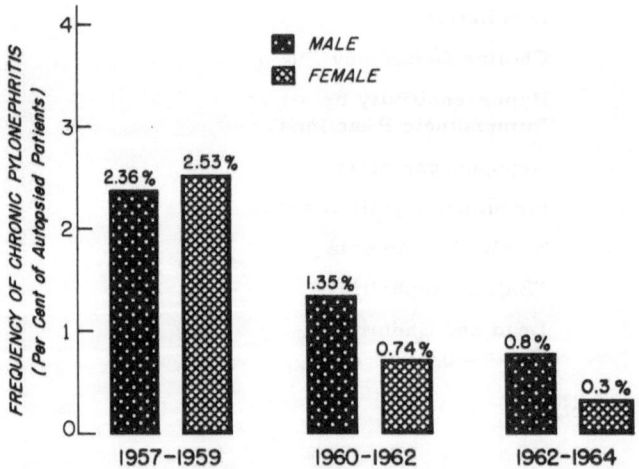

Fig. 2. The frequency of chronic pyelonephritis among autopsied male and female patients at Yale University School of Medicine during three periods from 1957 to 1964 (modified from Freedman (42)). During this time, a pathologic diagnosis of chronic pyelonephritis was made progressively less frequently.

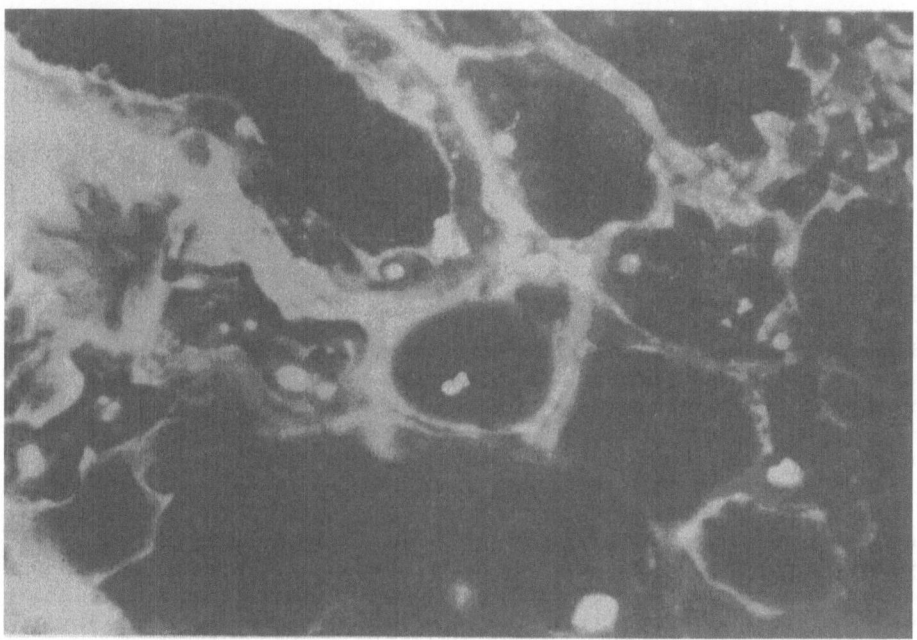

Fig. 3. Immunofluorescent staining of a section of kidney from a patient with acute pyelonephritis. Antiserum to an antigen (Common Enterobacterial Antigen) shared by most enterobacteria was utilized for immunofluorescent detection of residual bacterial antigen which appears white. Diffuse deposition of antigen is present in the interstitial tissue, tubular epithelium, and in macrophages. Whole bacteria may be seen in tubular cells and the tubular lumen.

Endotoxin Shock in Obstetrics

H. Graeff, W. Kuhn, and J. Zander

Septicemia and/or endotoxinemia by gram-negative bacteria exhibit a severe
course in pregnancy and, in addition, are frequently complicated by pronounced intravascular coagulation. They are mainly observed following intrauterine infections
in early and late pregnancy.

The intrauterine infection in early pregnancy is very often caused by the induction of abortion by inadequate means. After an interval of 24 and more hours a bacterial infection of the uterine content as well as of the endo- and myometrium follows; in severe cases a necrotising endomyometritis with additional pelvic inflammation may occur. This intrauterine infection in early pregnancy is defined as "infected abortion" and may be labelled "septic abortion" in cases with temperatures of
more than 39°C and/or chills.

The escape of bacteria or endotoxins from the uterus into the maternal circulation may easily occur in these cases either intermittently or continuously. Occasionally, the endotoxinemia may even be timely correlated to instrumental interventions
or to the delivery of fetus and placenta. The relationship between shock and intrauterine infection was described in the classical paper of Studdiford and Douglas (1)
by showing bacteria in the uterine vessels of a patient with endotoxin shock.

Premature rupture of the membranes is the main cause of intrauterine infection
in late pregnancy. Chorioamnionitis (or amnion infection syndrome) occurs in approximately 10 % of the cases if the time interval between rupture of the membranes
and delivery is less than 24 hours and in up to 30 % if the time interval exceeds 24
hours. Though the fetal mortality in chorioamnionitis is high (16 - 20 %), the maternal mortality is fortunately rather low (1 in 2 - 500 cases of chorioamnionitis)
and may be seen only once in 20,000 deliveries.

Endotoxin shock in pregnancy may also, though infrequently, be seen in cases of
severe pyelonephritis.

In addition to the anatomical conditions which facilitate the spreading of an intrauterine infection the pregnant organism has been known to be already prepared for
the generalized Shwartzman-reaction (GSR). Apitz (2) was able to demonstrate that
GSR can be elicited in pregnant animals by a single injection of endotoxin, whereas
two spaced injections are required in non pregnant rabbits. Recently it was shown
by exchange infusion experiments that the increased tendency of pregnant animals
to undergo disseminated intravascular coagulation (DIC) after the administration of
endotoxin is due to a transferable factor in the blood (3). Whether this factor is related to certain changes in pregnancy like the increased inhibition of fibrinolysis,
hypercirculation, hyperlipemia or increased cortisol binding capacity in serum is
still open for discussion. The so-called state of hypercoagulability in pregnancy
which might be related to one or more of the above-mentioned changes may be identified by the demonstration of soluble fibrin monomer complexes (SFMC) in plasma
(4, 5). In man as well as in the experimental animal (see von Hugo et al., this volume) maternal plasma samples contain SFMC as evidenced by the elution of FR-an-
tigen material in a shoulder prior to the fibrinogen peak after agarose gel filtration.
According to electrophoretic studies of eluted fractions, this material consisted
mainly of a component with the relative electrophoretic mobility of 0.28×10^{-5}
cm^2/V x sec. It might be assumed that this 0.28-derivative is a dimeric structure
which may occur either as a fibrin monomer complex or as a covalently linked dimer.
In normal pregnancy it amounts to up to 5 % of the total fibrinogen content (4). A

similar derivative was also observed after in vitro addition of small amounts of thrombin to plasma (5). It might, therefore, be assumed that the in vivo occurrence of the 0.28-derivative results also from the action of thrombin on fibrinogen. Hypercoagulability and enhanced susceptibility to DIC during pregnancy may very well be related to the increased amount of the 0.28-derivative.

DIC may, therefore, occur more frequently in an organism prepared by hypercoagulability if additional factors like endotoxin-induced disturbances of microcirculation with relative hypovolemia and acidosis come into effect. In order to evaluate the frequency of shock and mortality in patients with infected abortion a retrospective study was performed. In addition, it was the aim of this study to find criteria for the recognition of increased-risk patients with infected abortions. The evaluation of the data from 6255 patients with abortion showed that shock and mortality were related to fever and/or to the occurrence of chills. Patients with temperatures of more than 39.9°C and/or chills had a frequency of shock of 23 % and a mortality rate of 12 % (7).

It was our feeling that this group of patients would benefit from the prophylactic application of heparin to prevent DIC. In order to gain a certain safety margin, patients with a body temperature of 39.0°C (and/or chills) were included in the patient group with "septic abortion" and prophylactically treated with heparin (see Table I). It should be pointed out that all patients were treated in the same way and that an alternating series with an untreated control group still has to be performed. Nevertheless, the data of 371 patients with septic abortion who were treated in four years in Heidelberg (1966-1969) as well as in the following four years in Munich (1970-1973) show that in this series no maternal death occurred. The patients received 30,000 I.U. of heparin by a continuous infusion over 24 hours. Under this treatment 59 of the patients developed a shock-like syndrome, some of these received prednisone (5 mg/kg body weight) additionally. This syndrome exhibited some of the criteria of endotoxin shock, though it was characterized by a complete reversibility and by absent disturbances of organ function.

Fig. 1 demonstrates the course of illness in a patient with chorioamnionitis following premature rupture of the membranes who developed a shock-like syndrome during the treatment with heparin. Timely correlated to delivery and to the manual removal of the placenta, tachycardia, hypotension, hyperventilation with respiratory alkalosis and increase of central venous pressure occurred. These changes with inclusion of the pulmonal hypertension do probably correspond to the early reaction observed by Neuhof after a single injection of endotoxin (Neuhof, this volume). None of the 59 patients with a shock-like syndrome under the heparin treatment developed acidosis, acute renal failure or pulmonal insufficiency and the plasma fibrinogen content remained unchanged in the investigated cases. The decline of platelets was not influenced by the application of heparin (8) and is, therefore, also during heparin prophylaxis a reliable indicator of endotoxinemia.

Summary

1. Intrauterine infection in early or late pregnancy is often caused by gram-negative bacteria.
2. Pregnancy favours the occurrence of DIC during endotoxin shock.
3. 371 patients with septic abortion (temperatures of more than 39°C and/or chills) were treated by prophylactic application of heparin (30,000 I.U./24 hrs). 59 of these patients developed a shock-like syndrome. None of the patients died.

References

1. Studdiford, W.E., Douglas, G.W.: Amer. J. Obstet. Gynec. 71: 842 (1956).
2. Apitz, K.: J. Immunol. 29: 255 (1935)

3. Theiss, W., Beller, F.K.: Surg. Gynec. Obstet. 135: 713 (1972).
4. Graeff, H., von Hugo, R.: Thromb. Diath. haemorrh. 27: 610 (1972).
5. von Hugo, R., Graeff, H.: Thromb. Res. 3: 183 (1973).
6. Graeff, H., von Hugo, R., Hafter, R.: Thromb. Res. 3: 465 (1973).
7. Kuhn, W., Graeff, H.: In: Zander, J.: Septischer Abort und bakterieller Schock, p. 74 (Springer-Verlag, Berlin/Heidelberg/New York 1968).
8. Kuhn, W., Neuhof, H.: Geburtsh. Frauenheilk. In press.
9. Kuhn, W., Graeff, H.: Med. Welt 22: 1199 (1971).

Table I.

Patients with septic abortion, 1966-1973 (Jan. 1966-Dec. 1969, Universitäts-. Frauenklinik, Heidelberg; Jan. 1970-Oct. 1973, I. Universitäts-Frauenklinik, München).

	Number of Patients	Per Cent
Patients with septic abortion (temperatures 39°C and/or chills) treated by prophylactic application of heparin	371	100 %
Treated patients with a shock-like syndrome	59	15. 9 %
Deaths	0	-

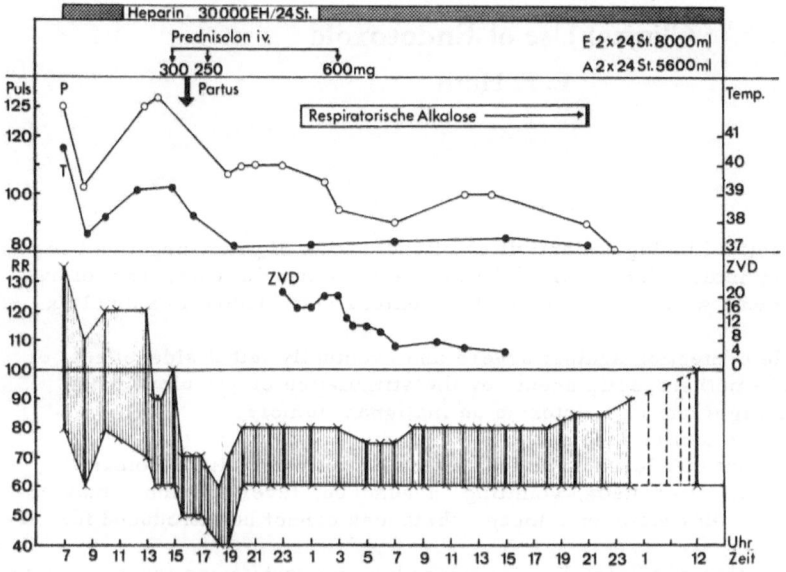

Fig. 1. Course of illness in a patient with chorioamnionitis following premature rupture of the membranes. The patient developed a shock-like syndrome during the treatment with heparin. Timely correlated to delivery and to the manual removal of the placenta tachycardia, hypotension, hyperventilation with respiratory alkalosis and increase of central venous pressure occurred. Within approximately 48 hours the patient recovered completely. (P: pulse rate; T: body temperature; E: infusion volume; A: excretion volume; ZVD: central venous pressure; RR: arterial blood pressure; Uhr Zeit: time in hours) (from 8).

Clinical Use of Endotoxoid
E. F. Huth

Some of the manifold biological activities of endotoxins of gram-negative bacteria would be very interesting for therapeutic use in human patients. The following are the phenomena which have already been extensively studied in animal experiments:

1. Considerable protection against severe and eventually lethal side-effects of radiotherapy or chemotherapeutic agents by the stimulation of granulopoiesis.
2. The necrotizing effect of endotoxins on malignant tumors.
3. The pyrogenic effect of endotoxins.
4. The development of tolerance during repeated treatment with endotoxins.

Since side-effects, like nausea, vomiting, headaches, fever and shock may occur after treatment with endotoxins, these substances cannot be introduced for clinical use.

After the description of a detoxification procedure for endotoxins (conversion to endotoxoid) by Nowotny (5), clinico-therapeutic research appeared to be justified. In animal experiments with this endotoxoid the above-mentioned side-effects were considerably diminished.

The Enhancement of Resistance against X-Rays and Chemotherapeutic Agents by Endotoxoid

During therapy of children with leukemia or solid malignant tumors with radiotherapy and chemotherapeutic agents, leukopenia or agranulocytosis often results. This forces the physician to discontinue this mode of treatment. In a great number of cases with acute lymphatic leukemia, bone marrow aplasia develops during the initial phase of treatment. After some days of observation a new population of normal cells starts to appear in the marrow. In the treatment of leukemia using various chemotherapeutic regimes one often encounters in remission a severe leukopenia. Thus, it is very important to look for a treatment which may be used prior to and during an X-radiation or chemotherapy of malignant tumors or leukemia. This treatment should effectively stimulate granulopoiesis. At present and according to many observations in animals detoxified endotoxins seem to be suitable substances for treatment in these situations. The underlying principle of this therapy was first demonstrated by experiments of Urbaschek (12), Urbaschek and Nowotny (10) (Fig. 1).

This effect, namely, stimulating granulopoiesis after administration of endotoxin in patients was described in 1965 by Wolff, Rubenstein, Mulholland and Alling (14). Using endotoxoid in mice the same effect was demonstrated in 1971 by Ringert (9), Urbaschek and Ringert (13). In kinetic studies of granulocytes Ostlund et al. (7) observed higher cell counts following injection of non-detoxified endotoxin than after injection of detoxified endotoxins.

Fritsch et al. (3) used endotoxoid in human volunteers and also observed the development of leukocytosis and granulocytosis.

Further animal experiments with monkeys were carried out by Urbaschek et al. (11) using endotoxoid in long-term toxicity studies. These experiments extended for a period of over 12 weeks. During and after injection of the endotoxoid a great number of hematologic and metabolic parameters were examined revealing no pathologic changes.

Thus the use of endotoxoid in human patients appeared to be justified. As had been anticipated from previous observations using endotoxin, the side-effects of fever were again observed using endotoxoid. As reported in the literature, it seemed to be possible using endotoxins to elicit enhanced toxic reactions in patients with malignant tumors (4, 6). In combinations with other chemotherapeutic agents, for example actinomycin and endotoxin, similar results were observed.

Methods

Ten patients (children with solid malignant tumors or leukemia) received i. v. injections of endotoxoid (charge endotoxoid U89) which had been prepared from Escherichia coli by Urbaschek* according to the method of Nowotny (5). The following doses of endotoxoid were used in a final volume of 1. 0 ml each of sterile pyrogen-free physiological saline: 50 /ug, 5 /ug, 0.5 /ug, 0. 05 /ug and 0.005 /ug. In all children the first dose of endotoxoid was 0. 005 /ug i. v.. Rectal temperatures were recorded prior to each injection, at 30 minutes, and at hourly interval for three hours thereafter. The injections were given daily. The dose was increased cautiously until a febrile reaction of 38. 0°C to 39. 0°C was observed. In spite of the fever which lasted 1 to 2 hours, the children were free of the serious side-effects, such as nausea, vomiting, headache, or shock. There were virtually no symptoms, aside from slight diaphoresis as the temperature returned to normal.

Results

In children without metastases, as for example from an extirpated solid malignant tumor or who were in leukemic remission, it was necessary to inject on the average 0.5 - 1. 0 /ug endotoxoid per kg body-weight in order to obtain a febrile reaction of 38. 0°C to 39. 0°C.

As may be seen from Table I the number of granulocytes in the peripheral blood markedly increased as daily injections of endotoxoid were given.

In bone marrow a moderately pronounced increase of neutrophils was observed. In treating children with the above-mentioned diseases, it appeared with regard to bone marrow depression that during and following treatment with endotoxoid the application of X-rays and chemotherapy was better tolerated. Of course this observation needs to be confirmed in a larger number of patients.

The Endotoxoid-Induced Necrotizing Effect on Malignant Tumors

If a certain amount of endotoxin - prepared according to the method of Shear (8) - is injected into mice having a transplanted Sarcoma 37 or a spontaneous mammary-carcinoma, the tumor becomes necrotic. After 24 hours the tumor tissue is hemorrhagic and the consistency is softer. Nowotny et al. (6) described the same phenomenon following endotoxoid administration.

Using endotoxoid in two cases of solid malignant tumors (one child with a sarcoma of the soft tissues and another with an osteogenic sarcoma), we did not observe tumor necroses; most probably the dose of endotoxoid was insufficient.

Febrile Reactions after Endotoxoid Administration

Recording febrile states after i. v. injections of endotoxoid is the best way to clinically assess the resulting reactions in patients. Strictly speaking, endotoxoid by itself is not fever-producing. According to the investigations of Beeson (2), Atkins (1) and other authors endotoxin stimulates granulocytes and monocytes to produce a special protein, the endogenous pyrogen. This endogenous pyrogen produces a febrile state by influencing the temperature-center in the brain.

---

*I am grateful to Dr. B. Urbaschek for supplying the endotoxoid.

Multiple injections of endotoxin induce tolerance to pyrogenicity, even higher doses of endotoxin producing less fever. It was possible to produce fever in patients with endotoxoid as well as with the non-detoxified endotoxin. However, according to experiments in healthy volunteers it was necessary to inject a dose of endotoxoid 140 times higher than that of endotoxin to achieve the same febrile reaction (3). In our observations the degree of fever showed a good correlation to the dose of injected endotoxoid (Fig. 2).

The ability to produce fever after injection of endotoxoid seems to be influenced by the size of the tumor mass in the body. This response of the body, namely the febrile reaction, decreases gradually as the growth of a malignant tumor increases and vanishes as cachexia develops.

This observation may be important for judging the degree of tolerance developing against endotoxoid in patients with malignant tumor.

The Development of Tolerance to Endotoxin and Endotoxoid

As has been often observed, daily injections of endotoxin rapidly produce tolerance, for example the decreased febrile response. This is due to antibody formation in response to the injected endotoxin. As a consequence, the effect of endotoxin on the granulocytes and the formation of endogenous pyrogen is blocked. During the clinical use of endotoxoid it was, however, observed that tolerance developed gradually.

As may be seen from Table II a child reacted with 39. 0°C fever after injection of 1 /ug per kg body-weight of endotoxoid. His reactive capacity was unchanged or even enhanced after 4 months of daily treatment with i. v. injections of endotoxoid.

During and after prolonged use of endotoxoid no side-effects with respect to functions of the liver and other organs could be detected.

Summary

1. It was possible to stimulate marrow granulopoiesis and thus increase the number of circulating granulocytes in the peripheral blood by i. v. injections of endotoxoid from E. coli in patients with leukemia and solid malignant tumors before and during X-radiation and chemotherapy.

2. A necrotizing effect of endotoxoid on solid tumors could not be detected with the doses of endotoxoid used.

3. Endotoxoid produced fever with great consistency and resulted in only minimal development of tolerance with respect to febrile response. No side-effects were seen for several months after the treatment with endotoxoid.

References

1. Atkins, E.: Pathogenesis of fever. Physiol. Rev. 40: 580 (1960).
2. Beeson, P. B.: J. clin. Invest. 27: 524 (1948).
3. Fritsch, H., Krecke, H.-J., Becker, B., Urbaschek, B., Nowotny, A.: Verh. dtsch. Ges. inn. Med. 74: 1151 (1968).
4. Havas, H. F., Donelly, A. J., Levim, S. I.: Cancer Res. 20: 393 (1960).
5. Nowotny, A.: Nature (London) 197: 721 (1963).
6. Nowotny, A., Golub, S., Key, B.: Proc. Soc. exp. Biol. Med. 136: 66 (1971).
7. Ostlund, R. E., Biskop, C. R., Athens, J. W.: Proc. Soc. exp. Biol. Med. 137: 763 (1971).
8. Perrault, A., Shear, M. J.: Cancer Res. 9: 626 (1949).
9. Ringert, R. H.: Untersuchungen zum Wirkungsmechanismus des Strahlenschutzes eines detoxifizierten Endotoxins. Inaug. Diss. Universität Heidelberg (1971).
10. Urbaschek, B., Nowotny, A.: Endotoxin tolerance induced by endotoxoid. In:

Chedid, M. L. : La structure et les effets biologiques des produits bactériens provenant de germes gram-négatifs, p. 357 (Editions du Centre National de la Recherche Scientifique 1969).

11. Urbaschek, B. , Urbaschek, R. , Mauff, G. , Gerrlach, Ch. , Huth, K. , Jung, F. , Ringert, R. H. , Nowotny, A. , Fritsch, H. : Experientia 27: 803 (1971).
12. Urbaschek, B. : Zur Frage des Wirkungsmechanismus bakterieller Endotoxine und seiner Beeinflussung. Habilitationsschrift Universität Heidelberg (1967).
13. Urbaschek, B. , Ringert, R. H. : Influence of a detoxified endotoxin on the bone marrow. In this volume.
14. Wolff, Sh. M. , Rubenstein, M. , Mulholland, J. H. , Alling, D. W. : Blood 26: 190 (1965).

Table I.

Stimulation of Granulopoiesis by Endotoxoid.

Nine-year-old boy with acute lymphoblastic leukemia in remission. Body-weight 33. 6 kg. Treatment with 15 µg endotoxoid daily.

Date:	5/21/1973	7/3/1973	9/7/1973
	(one day before endotoxoid)	(during endotoxoid treatment)	
Peripheral blood			
Leukocytes	2300 / mm^3	6500 / mm^3	5800 / mm^3
Eosinophils	--	1	2
Granulocytes Band forms	4	--	1
Segmented forms	39	70	60
Lymphocytes	49	27	34
Monocytes	8	2	3

Table II.

Tolerance and Endotoxoid

Thirteen-year-old boy, after extirpation of an osteosarcoma with metastases.

1973	Weight	Endotoxoid	Fever
May	44. 6 kg	40 µg	39. 0° C
August	46. 0 kg	25 µg	39. 0° C

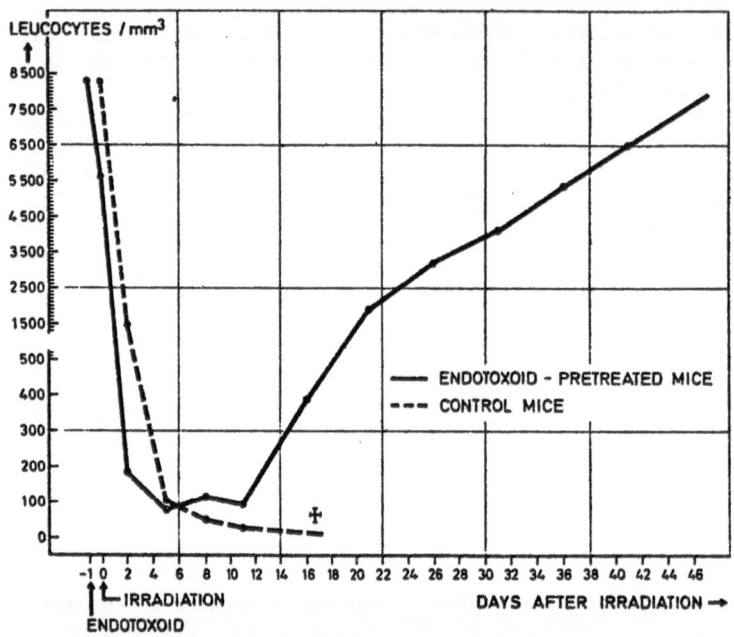

Fig. 1. Granulocyte count of endotoxoid-pretreated and control mice after irradiaction with 810 r. (Urbaschek, B., 12).

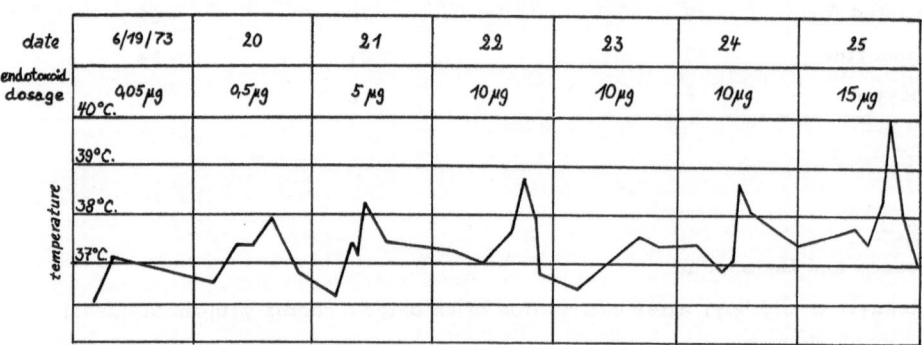

Fig. 2. Eight-year-old boy with acute lymphatic leukemia in complete remission. Body-weight 27.4 kg. Relationship between endotoxoid-dosage and subsequent febrile response.

Discussion

SHUBIN

Hennemann: I agree with your opinion concerning etiology and therapy, and also with what you said about the course of sepsis, for which as you said the focus is decisive. Where the bacteria come from may explain why our results are different from yours. In a medical clinic chronic sepsis is more frequent than acute sepsis. We had 8 cases of E. coli sepsis of these 5 were in shock and 2 died. I believe it is very important for lethality and outcome to know the source of the organisms.

Lundsgaard-Hansen: I am interested in the fact that you have reservations in the use of isoproterenol, which not only stimulates the myocardium, but also opens up the peripheral circulation. Recently McLean of Montreal stated that, if isoproterenol fails, nothing else helps. I would appreciate if you would comment, why you take a stand against isoproterenol in septic shock. I think in Europe, it is still the general impression that this is the best drug. A couple of years ago your group formulated the excellent VIP concept which we use in teaching our students and young army medical officers. Do you still stick to the order of priority which you established there?

Shubin: We have used isoproterenol in some patients with bacterial shock. In many cases we can increase the cardiac output. However, lactate may not be reduced but may actually increase. This would imply an even more severe degree of perfusion failure. There also is evidence that myocardial oxygen consumption may increase disproportionately and in patients with limited coronary circulation this may be detrimental. In reference to the VIP approach, this refers to ventilation, infusion, and pumping. When confronted with a patient who is in shock, one often does not have time to make a detailed assessment of what is going on. In looking at the critically ill patients, the most common immediate life-threatening condition often is the impaired respiratory status. We would want to immediately support ventilation. Only when this is secure do we turn our attention to the infusion of fluids and the administration of the specific drugs, such as steroids or antibiotics. Among patients in shock of diverse etiology, at least 80 % require substantial ventilatory support.

Irwin: There is an old clinical adage if you do not understand how a compound or a procedure works, you do not use it. You have stated that the steroids are quite effective here, but you did not tell us how they work. The only reason I can think of as to how they might work here is that they are antiinflammatory. However, they might also encourage infection. What do you think steroids are doing in bacterial shock?

Shubin: I am sorry you stopped in your explanation of how steroids work. I was hoping you would go further because I can only conjecture as to their action. Dr. Chedid and I were discussing this earlier and perhaps he will comment on it as well. It has been suggested that the steroids stabilize the cell membranes and reduce the lysosomal damage. They may increase cardiac output. There is also some evidence that they might influence the oxygen-hemoglobin equilibrium curve. I would like to invite members of the group to comment further on this.

Berry: I feel obliged to comment on the possible metabolic effects of the steroids. They are very potent inducers of regulatory enzymes. I do not think these can be overlooked in situations of this type.

Koslowski: I think the effect of the corticoids may be to decrease the permeability of the membranes. If I understood you correctly, you give large amounts of dexamethasone, 40 mg initially and then 20 mg every four hours.

Shubin: Yes. This is a substantial dose. I might say, however, that this dose is much less than that which is recommended by Dr. Richard Lillehei.

Koslowski: I think that in Germany, especially in our surgical clinics, we do not use such large amounts of corticoids. With regard to the catecholamines, Dr. Lundsgaard-Hansen has asked you whether you doubt their beneficial effect, and I think you are correct. We do not use catecholamines in septic shock because we are afraid of the metabolic consequences. In our experience, the acidosis increases despite the beneficial effect on the myocardium. With regard to the antibiotics, we also have the opinion at the present time that gentamycin is the most effective antibiotic and we combine it with cephalothin. This combination of cephalothin and gentamycin in a dosage of 80 mg intramuscularly three times daily has been the most effective therapy against Klebsiella and Pseudomonas infection.

Shubin: In considering the effect of the acidemia on the catecholamines, perhaps a comment is in order. It has been stated in the literature that in the presence of acidemia there is a reduction in the effectiveness of vasopressor drugs. Perhaps this ought to be clarified. In the presence of acidemia, there is a reduction in the effectiveness of the alpha-component of the vasoactive drug. In observations in animals, Dr. Weil noted that with mild acidemia, there seemed to be a potentiation of the beta-effect of the catecholamines. The presence of mild acidemia, therefore, is not necessarily detrimental. We must also consider the Bohr effect on the oxygen-hemoglobin curve, whereby in the presence of acidemia, we have a shift of the curve to the right and an increase in the release of oxygen.

Chedid: Three types of questions are raised by these problems. The first one is the biphasic effect of corticoids which can be either favorable in experimental animal models or unfavorable. The second one is the mechanism and why we use such large dosages. As far as the mechanisms, I think they have been covered by Dr. Berry's comments on metabolic reducable enzymes, capillary permeability and permeability at the cell and at the lysosome level. There are basic experimental data which give comfortable answers. They can be antiinflammatory or unfavorable according to the pathogen. The pathogen, as Dr. Shubin said, is very important. That means enteric diseases are quite different from gram-positive diseases. One must also keep in mind the possibility of intercurrent viral infections in which corticoids would be especially obnoxious. Concerning the large dosages, we may be able to afford some explanation by referring to Engel's experiments in which he showed very nicely that the animal became refractory to corticoids under stress, which is the usual case in infection. In hypercortical states, endogenous or exogenous hypercortical states, you get the same effect with, sometimes more than a log higher levels of corticoid administration.

Spaulding: How frequently in bacterial shock do you recover more than one species of bacteria from blood cultures and how often do you isolate gram-negative anaerobes? Are special anaerobic methods being used?

Shubin: In about 20 % of patients we have more than one organism present. The gram-negative anaerobes constitute a small number of the shock cases which we encounter. This is despite appropriate culture for the anaerobic organisms. Special anaerobic methods are being used routinely.

Braude: I would like to comment on this question of Bacteroides as a cause of septic shock. I have been impressed with the fact that it does not cause it. In addition, I have been impressed with the fact that it is very hard to demonstrate an endotoxin in Bacteroides fragilis.

CLOWES

Steinbereithner, Wien/Austria: I was greatly impressed with the presentation which was given by Dr. Clowes, and I would like to congratulate him on his opinions. With Dr. Shubin's paper, there is a very important question. From Dr. Clowes' presentation, we have the impression that we have to put every severe enteric septic case at the onset on the respirator, which implies that every surgical clinic dealing with major abdominal operations should have large respiratory units. In regard to Dr. Shubin's presentation, I think one has to distinguish whether you have enteric shock which can be corrected by surgical intervention. In our own material, the mortality is similar to Dr. Shubin's material almost 90 to 100 %. On the other hand, if you have a lesion which can be handled by a bold surgeon then our survival rate was about 80 %. I think it depends on the skill and boldness of the surgeon.

Clowes: Well, one can not disagree with your view. It seems to me that surgical drainage, removal or exteriorization of a gangrenous or septic lesion is fundamental to removing the stimulus which is producing the lung lesion. Our indication for the use of the respirator is when the patient's arterial blood-oxygen tension has fallen below 65 mm Hg while breathing air. Has it reached 60 mm Hg definitely he goes onto the respirator. Since adopting this viewpoint, I think we have seen far less of the late phase bronchopneumonia develop.

Shubin: Critically ill patients in general have a high incidence of respiratory failure. It seems almost an anachronism in 1973 to think of any intensive care unit, be it medical or respiratory, which does not have ventilatory support.

Schmahl: I have one question regarding the clinical studies of the treatment of septic shock. Is it related to the interaction between antibiotics and gram-negative bacteria? If you begin with massive antibiotic treatment, did you see in your cases any initial impairment of the situation? As you all know, there has been much speculation about the possibility of the setting free of toxins from the bacterial wall in the interaction with antibiotics. I refer to the name of Herxheimer. First, what is your experience with this massive onset of antibiotic treatment regarding the relations of bacteria and endotoxin? Second, what is your theoretical concept?

Braude: I was most impressed with this in Mexico in 1948 when we were trying aureomycin in the treatment of Brucella melitensis infections. We gave aureomycin to several hundred people and a fair percentage of them developed hypotension and all of them developed hyperpyrexia. That is, the fever went on up after the tetracycline was started. We thought this was due to the lysis of the organisms and release of toxins, probably endotoxin. The same thing has been seen with typhoid fever. Have you seen this?

Shubin: We have not been impressed with a hyperpyrexic response after administration of the antibiotic.

Braude: Outside of this experience with Brucella, I have not been impressed with this phenomenon in the ordinary case of gram-negative bacteremia due to coliform bacteria. I would say that it is probably not an important problem mainly because we do not have antibiotics that are as effective against E. coli, Klebsiella and Pseudomonas as the tetracyclines were against Brucella.

Gans: It should be mentioned that in surgery we see this respiratory distress syndrome or respiratory insufficiency syndrome in other conditions besides shock. As Dr. Clowes has already indicated we see it after trauma. We see it also after certain surgical operations. It has become very prominent in transplantation rejection and we have recently found that it plays a role in some of the difficulties that we see after hepatic resections. Particularly, with respect to the way the liver is divided. It is a problem which is becoming more and more prevalent in surgery and is not only seen in patients in shock.

Greisman: I would like to raise a question here. We keep talking about the fact that endotoxin must be released by lysis of the organisms, I do not know of any evidence that the organism has to lyse to release endotoxin that is going to be effective. There are data which show that the intact organism having the endotoxin on their surface can be just as active, at least in terms of pyrogenicity, as the purified endotoxin preparation. I am raising this because of the question that the organisms have to be lysed before the endotoxin is released to be effective. I do not know of any definitive data that support this clinical impression that lysis has to occur before endotoxemia or endotoxin produces its toxic effects. I would like to again second Dr. Braude's statement that as far as Herxheimer type reaction is concerned, I only know of brucellosis in which this has occurred and this may be a hypersensitivity phenomenon and not an endotoxemic type of phenomena.

GANS

Neter: Dr. Gans, I would like to ask you regarding your introductory statement, whether the elevated antibody titers against E. coli in these patients are directed against their own organisms or due to antibodies from different stimuli.

Gans: I really cannot comment on that. These are reports in the literature, I would have to go over this in a more detailed fashion, but the article by Bjørneboe and the other one by Triger appeared in the same issue of Lancet and I believe it was in the beginning of last year.

Brobmann: Dr. Gans, I would like to ask you if you think that the increase in absorption in your animals may be probably due to the bile-deficient status you had in your experiment. You had only an isolated loop of the small bowel and if you remember the literature, Koscár published in 1969 a paper in which he proved that in rats the absorption of endotoxin from the small intestine was increased in bile-deficient rats.

Gans: I am glad you brought this up. We tried to repeat some of these experiments by doing bile fistulas just the same as Dr. Koscár has indicated but we have had some difficulty in repeating these experiments, quite a number of the animals survive. So really I do not know what the role of bile is. Whether lipase in the intestine is one of the mechanisms by which endotoxin is detoxified or broken down, we do not know at present, but we intend to pursue this.

Greisman: I want to ask again Dr. Gans, how long after the operative procedure did you do these studies on the absorption on the endotoxin from the gut? I am wondering about the problem of whether the gut is really now in a normal physiologic state at this time. What evidence do you have that if you waited longer this would show changes in the absorption rate?

Gans: Well, this model, of course, is not a new model. I indicated that an Austrian physiologist was the first to use this more than 100 years ago. A great deal of the basic GI physiology that has been obtained in this field has been obtained with these loops. Glissen published a series of articles in the British literature on this in which he indicates that until twelve weeks, there is no essential change in the enzymic activity and in the absorptive activity for the substance of the studies and there was a wide range of molecules. I cannot, of course, state specifically for endotoxin. It is very difficult, but it seems that the absorptive function of this loop does not change whether one studies this on day one, or twelve weeks later. There is very little change. What does change is the remaining intestine, we see this in patients too where we do this operation, for instance for obesity, the remaining intestine that stays behind hypertrophies, especially the villi. By taking away a part of the resorptive surface, the remaining part tends to hypertrophy and to compensate for the loss of function that we have established with this particular loop.

McCABE

Sietzen, Frankfurt/Germany: I would like to make the following remarks concerning the K-antigen of E. coli in pyelonephritis. I believe that the studies referring to the influence of the quantity of the antigen are not so far advanced that one can consider this as finally proven. Because the K-antigen of E. coli in pyelonephritis is for the most part a K-antigen of the L-type. This K-antigen of the L-type dissociates from the cell-wall very easily at normal temperatures. I would like to add that we recently were able to detect a specific immunofluorescence of the K-antigen by means of homologous, absorbed sera and this shows in addition to your slides, that the K-antigen is remarkedly earlier detectable intracellulary than the 0-antigen, and that it behaves in the kidney more like a substance, which diffuses through the kidney tissue. On the other hand one can detect the persistence of the K-antigen over a period of about 4 weeks, though the degree of immunofluorescence is then less intense than that of the corresponding 0-antigen.

McCabe: What you mention, reflects one of the problems. I think there is no question at all that it has been clearly demonstrated that strains of E. coli which contain large quantities of K-antigen are more apt to produce upper tract infection than those which produce lower tract infection. This was first demonstrated by Glynn and Brumfitt and was secondly demonstrated by Kayser et al. but by a much more specific technique. Using the Glynn and Brumfitt technique we have had similar observations that strains of E. coli isolated from patients with bacteremia have a higher content of K-antigen than strains isolated from the feces or strains isolated from the urine. What do these demonstrate? They demonstrate only one thing, that there is a correlation between this marker and the phenomenon that you are looking at. They do not necessarily prove causal relationship.

Bitter-Suermann: One comment to the effect of elevated ammonia concentration of C4. Ammonia in an elevated concentration has an even more pronounced effect on C3 and I think the inactivation of C3 with the biological consequences of the effect of C3 is more important, if at all.

McCabe: I do not disagree, however, only measurements were made of C4. And that is all I can report.

GRAEFF

McCabe: Did I understand you correctly, were you saying that you recommended routine heparinization of patients with suspected gram-negative bacteremia? We have looked at a large series of patients in terms of the frequency of the occurrence of consumptive coagulopathy as mentioned earlier. While we find thrombocytopenia in perhaps 60 % or 70 % of the patients, we find less than 10 % with the syndrome of consumption coagulopathy. This is only one out of every four patients with shock.

Of those who have the changes in clotting parameters, only about half have significant bleeding. This means we would be administering heparin to 95 patients to prevent complications in five. I am not at all certain that the prophylactic administration of heparin might not result in more fatalities from the procedure than from the disease.

Graeff: I would recommend heparin prophylactically in patients with septic abortion and I would not like to comment any further or to give any further recommendations when to use heparin in other patients.

McCabe: It seems that consumptive coagulopathies are also much more frequent in meningococcemia than among patients with common enteric bacteremia.

Graeff: Yes, you have the Waterhouse-Friderichsen syndrome in mind.

HUTH, E.

Bloch: Have you observed the formation of antibodies during this longlasting administration of endotoxoid?

Huth, E.: We have not measured them.

Urbaschek: In chronic studies which were done in this connection in 10 monkeys, Macacca rhesus, we did not observe an increase in titer after finishing the administration of endotoxoid.

Chapter IX

Burn Disease

Bacterial Infection in Burn Disease

E. J. L. Lowbury

Introduction

The role of infection in burn disease is a subject on which there have been wide differences and fluctuations of opinion. This is not surprising, because infection is one of the components of a complex pathology in burns, with bacteria colonising tissues already damaged and inflamed as a result of thermal injury; sepsis is superimposed on the local and general effects of this damage. In the early days of bacteriology some workers - e. g. Lustgarten (43) and Stockis (52) - recognised infection to be an important and potentially fatal complication in severely burned patients; but the subject was widely ignored for many years while clinicians, in this post-Listerian era, were pre-occupied with the problems of shock due to fluid loss and the more conjectural toxaemia attributed to products of tissue destruction. Even the bacteriological studies on burns reported by Pack (48), Aldrich (4), Cruickshank (15) and others did little to change the situation; but in the 1940's and 1950's, when hypovolaemic shock was brought under effective control, sepsis - in particular sepsis caused by gram-negative bacilli - became a new focus of interest in the study of burn disease; some reports placed it first among the causes of death in severely burned patients (5, 17, 35, 51).

It is clear, however, that the hazards of infection have varied widely in respect both of the environments where patients were treated and of different severities and conditions of burn injury treated in the same environment. An additional source of confusion about the role of bacteria is the ambiguity of the word "infection" by which some people mean "sepsis" and others mean merely "bacterial colonisation".

Types and Sources of Bacteria in Burns

Bacteria on the skin are destroyed by the heat which causes the burn, but contaminants are quickly acquired by the fresh burn from adjacent skin and from the environment, human and inanimate, in which the patient is treated. These contaminants include a variety of organisms commonly classed as pathogens (Staphylococcus aureus, Pseudomonas aeruginosa and other gram-negative bacilli, Clostridium welchii, and - sporadically - Streptococcus pyogenes), as well as others (micrococci, aerobic sporing bacilli, corynebacteria, non-haemolytic streptococci, etc.) which are commonly classed as saprophytes or commensals. The incidence of colonisation by the main groups of burn flora over a period of 20 years in the Birmingham Burns Unit is shown in Fig. 1.

Gram-negative bacilli, including Ps. aeruginosa, Proteus spp, Klebsiella spp and others (see Table I) flourish in moist slough, and are therefore most abundant during the first two or three weeks (i. e., before the separation of slough) in burns treated with moist dressings which do not contain antimicrobial agents active against these organisms; they may also be abundant under the dry surface of burns treated by exposure (39). Staph. aureus and Strep. pyogenes can also grow rapidly in the slough. In the granulation tissue which is exposed after the separation of slough Staph. aureus (also Strep. pyogenes, should it appear) can grow more readily than gram-negative bacilli, which are usually sparse at this stage. The relative proportions of different genera, species and strains in burns vary greatly, depending on the amount of cross infection and on the selective action of antibiotics; multi-resistant variants of Staph. aureus and gram-negative bacilli tend to be the predominant flora in a burns unit.

The predominance and rapid acquisition by patients of antibiotic-resistant bacteria provides evidence that most of the bacterial flora of burns are acquired by cross infection from other patients. This is shown also by phage typing and serological typing of strains. For example, in the Brimingham Burns Unit, which has two wards, most of the patients whose burns became infected with Ps. aeruginosa acquired strains of types which had previously been isolated from other patients in the same ward (16) (Table II). Sometimes gram-negative bacilli are acquired by burns from the patient's faecal flora, the patient having previously acquired the organisms in hospital; this mechanism is probably commoner with Enterobacteria than with Ps. aeruginosa.

Pathogenenesis and Pathogenicity

Two factors, the pathogenicity of the organism and the susceptibility of the host, are relevant. In most patients bacteria colonising burns cause no obvious pathological effects. In earlier studies (e. g. 13, 15) particular attention was paid to Strep. pyogenes, which was recognised to have exceptional powers of invasion and of causing the complete failure of skin grafts. Oral chemotherapy, especiall with macrolides and with penicillinase-stable penicillins, made it possible to eliminate Strep. pyogenes from burns, even in the presence of penicillinase-producing bacteria, within about 3 days (41). Though still an important potential cause of skin graft failure, Strep. pyogenes infection became relatively rare and much less likely to cause invasive effects than formerly. Staph. aureus was, and still is, a very common organism in burns, but though it can cause important clinical infection (24), this is not common even in severely burned patients. Much of the severe infection to-day is due to relatively poor pathogens, such as Ps. aeruginosa, Klebsiella spp and occasionally fungi, which cause 'opportunistic' infections in patients with very extensive burns and an enhanced susceptibility to microbial infection; these opportunist organisms are usually very hard or impossible to control by antibiotic therapy.

Of the gram-negative bacilli Ps. aeruginosa has been most widely associated with severe infection in extensively burned patients, causing characteristic 'cuffing' and invasion of the walls of small blood vessels, necrosis of unburned tissue adjacent to the burn slough, and sometimes focal haemorrhagic, necrotic lesions which may cause ecthymatous ulceration through unburned skin (23, 54). In patients with severe Pseudomonas infection, whether or not septicaemia is present, there may be characteristic general signs and symptoms, including leucopenia and hypothermia (often supervening on leucocytosis and pyrexia), paralytic ileus and a sudden onset of hypotension (50). Stone (53) has described verdoglobinuria, the appearance in the urine of green, fluorescent pigmentation due to aberrant degradation of haemoglobin, as a characteristic finding in patients with invasive Ps. aeruginosa infection. Tumbusch et al. (55) have pointed out that some of the features of Pseudomonas sepsis resemble and might be due to local and generalised Shwartzman reactions caused by circulating endotoxin; this presents the clinician with a dilemma, because successful chemotherapy, by destroying bacteria, would release more endotoxin into the circulation. Liu and his colleagues (6, 36), however, regard endotoxins as unimportant in Pseudomonas pathogenesis and consider the pathogenic activity of these organisms to be mainly caused by diffusible 'exotoxins' present in the bacterial slime. Studies by my colleagues in Birmingham (11, 32) confirm the presence of a toxin in the non-dialysable portion of the culture filtrates of Ps. aeruginosa, which also has strong immunogenic properties. The relative importance of endo- and exotoxin in Pseudomonas sepsis of burns needs further study. Using the sensitive Limulus test, Caridis and others (10) have shown endotoxaemia in a patient and in laboratory animals with burns, associated (in the patient) with E. coli and Proteus infection. Markley and others (45) have shown that endotoxin from gram-negative bacilli is up to 1,000 times more lethal in burned mice than in unburned mice. It remains to be shown, however, whether the small amounts of endotoxin detectable by the Limulus test in the serum of burned patients are capable of causing pathological effects in burned animals.

Strains of Ps. aeruginosa vary in their virulence (e. g. 31) as do patients in their susceptibility to these essentially opportunistic infections. In addition to its properties of systemic invasion, Ps. aeruginosa has also been shown, from the results of controlled chemoprophylactic trials, to interfere with the success of skin grafting in some patients (25). Local sepsis can cause suppuration, marginal cellulitis and the presence of poor granulation tissue under separating slough, but these signs may be hard to interpret. Their absence cannot be assumed to exclude the pathological effects of bacterial colonisation; nor can the presence of pus be assumed to have entirely deleterious effects, for digestion of slough by the proteinase of bacteria undoubtedly accelerates the separation of slough and allows earlier skin grafting (42). A test for bacterial involvement in the pathogenesis of burn disease would be valuable; the nitro-blue-tetrazolium test (49) deserves study in burned patients.

Though Ps. aeruginosa has occupied the centre of the field in the study of burn sepsis, recent success in controlling Pseudomonas infections by topical chemoprophylaxis has brought other gram-negative bacilli (notably Klebsiella spp, Providencia and Serratia) into the focus of attention as opportunists which are more difficult to control by chemoprophylaxis or other methods (e.g. 12, 14).

Control of Infection

Methods by which one may prevent or try to prevent infection of burned patients can be classified under two headings: 1. defences against contamination ('first line of defence'), and 2. defences against invasion of tissues and blood stream by bacteria already colonising burns ('second line of defence') (see Fig. 2). Each of the three components of the first line of defence shown in the Figure has both virtues and shortcomings. For example, excision of dead tissue and skin grafting soon after the burning injury give excellent protection, but it is a method possible only for relatively small burns which are also less liable to septic complications. Certain forms of topical chemoprophylaxis have been found highly effective against Ps. aeruginosa and other bacteria, but the selection of resistant variants and species and the possibility of toxic effects limit the application of this method. The use of isolators has been found effective as an 'aseptic' method against Ps. aeruginosa, but not against some other bacteria (especially intestinal gram-negative bacilli) (40); with supplies of sterile food and other sterile materials this deficiency might be met, but even without such a complete application of the method, the use of isolators for burned patients has proved excessively cumbersome. Nevertheless, improvements in the first line of defence (in particular topical chemoprophylaxis) led to a reduced incidence of infection in the Birmingham Burns Unit, with no Ps. aeruginosa septicaemia for some years and a reduced mortality from burns (9, 12).

In the second line of defence systemic chemoprophylaxis with narrow-range antibiotics (penicillin, cloxacillin or macrolides) to prevent multiplication of Clostridium tetani and possible invasion by Strep. pyogenes is rational. Prophylaxis with broad-spectrum antibiotics, however, is a potentially dangerous procedure, encouraging the emergence of mutants and opportunists resistant to a wide range of antibiotics. For this reason attention has, in recent years, been directed towards immunological protection of severely burned patients as a possible method in the second line of defence against gram-negative bacilli.

Immunological Protection against Gram-Negative Bacilli

Since a patient with burns is exposed to the risk of infection from the time of injury, passive protection with a hyperimmune serum would seem to offer the best prospects; indeed, the patient is especially prone to infection during the first week after injury when the blood is likely to have reduced immunoglobulins (7) and a raised corticosteroid level (18). Experimental Ps. aeruginosa infection of burned mice, by intraperitoneal injection or by inoculation of burns, causes fatal invasive infection which can be prevented by intraperitoneal injection of a specific antiserum (31, 47). Some humans possess enough natural antibody to protect mice in this way (30), and may therefore be presumed to have considerable self-protective power against Ps. aeruginosa. Kefalides et al. (34) reported an inverse relationship between deaths from septicaemia and titre of antibodies to Ps. aeruginosa. Though mice usually survive for 6 or 7 days after inoculation of Ps. aeruginosa on burns, antiserum gives good protection only if it is injected on the same day as the infection is initiated; if antiserum is injected later, the mice will usually die, even though their serum has a high level of antibodies. This observation is consistent with the previously reported death of severely burned patients with Pseudomonas septicaemia in spite of the presence of a high titre of antibodies developing in response to the infection (21).

Feller and his colleagues (19, 20) have reported a decrease in the incidence, morbidity and mortality of Pseudomonas septicaemia in patients with burns given combined active and passive immunisation with a univalent hyperimmune serum and a vaccine prepared from the same strain; a smaller but appreciable effect was noted also in patients who had the vaccine only. This was surprising, in view of the two or three weeks which might be expected before active immunity could be established. However, early protection against intraperitoneal challenge, appearing two or three days after the initial dose of a Pseudomonas vaccine, has been reported by Markley and Smallman (46), and Jones (26) showed a similar specific early protection by vaccine of Ps. aeruginosa, Proteus mirabilis and Klebsiella aerogenes prepared from fractions extracted from cultures in a defined medium. Serum collected from mice 3 or 4 days after a single injection of vaccine passively protected unvaccinated mice against Pseudomonas infection (28).

The serum from mice showing early active immunity did not agglutinate the immunizing strain; it was shown that early protection was given mainly by IgM, while the protection which occurred after the first week was given mainly by IgG and associated with the appearance of agglutinins (27). Jones and Lowbury (33) have shown that freshly burned mice could be successfully protected by a Ps. aeruginosa vaccine against challenge two or three days later with the strain applied to a burn, provided that the burned area was kept moist with an inert cream and dressings (see Table III).

These findings are relevant to Feller's apparent success in the use of active immunisation, though it is surprising that a univalent vaccine should have been effective in his study. In another burns unit a number of different phage types of Ps. aeruginosa were found to be present at the same time, sometimes more than one type being present in the same patient's burns (16, 37). Alexander and his colleagues (1, 2) in Cincinnati have reported the successful use of a heptavalent Pseudomonas vaccine which covered the range of immunotypes (i. e. types distinguished on the basis of mouse-protective action) likely to cause infection. A new multivalent vaccine is also being developed in Britain.

Comments

The pathogenesis of gram-negative bacillary infection of burns is complex and varies with the bacterial species. In studies on mice with small full-skin-thickness burns, Ps. aeruginosa and Proteus mirabilis applied to the fresh burns caused local inflammation and a high incidence of fatal septicaemic invasion, while E. coli and Klebsiella aerogenes caused local inflammation but rarely invaded. Proteolytic enzymes, which are produced by the Pseudomonas and the Proteus but not by the other two species, may have been the factor which determined the invasiveness of the former; but Ent. cloacae, which produces a proteinase, did not cause invasive infection from the burn; the same was true of an avirulent strain of Ps. aeruginosa, though this was found to produce a smaller amount of proteinase than a virulent strain (11). A number of independent factors seem to be required for successful invasion of the blood stream by gram-negative bacilli colonising a burn.

The antibodies which prevent invasive infection with Ps. aeruginosa apparently do not kill the bacteria by direct bactericidal action, but do so by opsonic action (22, 28, 29). In severe burns the bactericidal activity of neutrophils on phagocytosed Ps. aeruginosa is reduced (3), and in severe Pseudomonas infection there is a leucopenia; both factors combine to render the tissues highly susceptible to invasive infection. Immunoprophylaxis may protect the host both by stimulating the neutrophilic leucocytes to phagocytose the Pseudomonas and by protecting the leucopoietic cells against toxic effects of the bacterial infection.

Patients with severe burns who are successfully protected against shock - as most are to-day - meet their next major challenge from gram-negative infections. Those who are protected by silver nitrate or other local applications against Ps. aeruginosa may still fall prey to Klebsiella spp or possible to Candida, which may cause fatal septicaemia. But the use of improved methods of controlling infection has led to a reduced incidence of septicaemia and a reduced mortality (Fig. 3 and 4); from which it appears that opportunists still beyond the range of our prophylactic methods are less pathogenic than those which have been effectively controlled.

The incidence of severe gram-negative bacillary infection is clearly related to treatment - e.g. covering the burns with effective antimicrobial agents such as silver nitrate or silver sulphadiazine reduces the incidence of infection (e.g. 40), but covering the burns with moist dressings which contain no topical chemoprophylactic agent may actually increase the amount of infection. Topical chemoprophylaxis is effective only if it can be started early - preferably in the first few hours after injury and before bacterial colonisation has developed. In a burns unit with good isolation and nursing facilities admitting only fresh burns, good clinical results have been obtained without topical chemoprophylaxis or immunoprophylaxis (8), but both of these methods used together had a limited value against Ps. aeruginosa, in a burns unit to which patients were usually admitted after initial treatment had been started elsewhere (2, 44). Another important factor is the number of patients with severe burns present in the ward; if an extensively burned patient is treated in an environment where there are few severe burns, the chances of acquiring infection will be smaller than in a similar burned patient treated in the same way but in a ward with many badly burned patients.

Gram-negative sepsis, especially that due to Ps. aeruginosa, is notoriously resistant to chemotherapy; even with antibiotics which are active against the organisms in vitro, treatment is often likely to fail because irreparable damage has already been caused by the time the diagnosis is made. For this reason special emphasis is placed on the need for prophylaxis, to achieve which patients must be admitted as soon as possible after injury to a hospital which can provide effective prophylactic treatment. Where measures effective against contamination (the first line of defence) are not practicable - e.g. in developing countries - there may be a special indication for immunoprophylaxis. Further studies and, in particular, controlled clinical trials are needed to determine their value and practicability.

Acknowledgments

I wish to thank Dr. J. P. Bull and the Editor of Proceedings of the Royal Society of Medicine for permission to reproduce Fig. 4; the Editors of the same journal and of The Lancet for permission to reproduce Fig. 1 and 3 respectively. I am grateful to Dr. Roderick Jones and other colleagues who collaborated with me in studies discussed in this paper.

References

1. Alexander, J. W., Fisher, M. W : In: Matter, P., Barclay, T. L., Konickova, Z.: Research in Burns, p. 226 (Huber, Bern 1971).
2. Alexander, J. W., Fisher, M. W., MacMillan, B. G., Altemeier, W. A.: Arch. Surg. 99: 249 (1969).
3. Alexander, J. W., Wixson, D. : Surg. Gynec. Obstet. 130: 431 (1970).

4. Aldrich, R. H.: New Engl. J. Med. 208: 299 (1933).
5. Altemeier, W. A., MacMillan, B. G.: In: Artz, C. P.: Research in Burns, p. 206 (Livingstone, Philadelphia 1962).
6. Atik, M., Liu, P. V., Hanson, B. A., Amini, S., Rosenberg, C. F.: J. amer. med. Ass. 205: 134 (1968).
7. Arturson, G., Hogeman, C. F., Johanson, S. G. O., Killander, J.: Lancet I: 546 (1969).
8. Birke, G., Liljedahl, S. O., Backdahl, M., Nylen, B.: Acta chir. scand. Suppl. 337 (1964).
9. Bull, J. P.: Lancet II: 1133 (1971).
10. Caridis, D. T., Reinhold, R. B., Woodruff, P. W. H., Fine, J.: Lancet I: 1381 (1972).
11. Carney, S. A., Jones, R. J.: Brit. J. exp. Path. 49: 395 (1968).
12. Cason, J. S., Lowbury, E. J. L.: Lancet I: 65 (1968).
13. Colebrook, L., Gibson, T., Todd, J. P., Clark, A. M., Brown, A., Anderson, A. B.: Spec. Rep. Ser. Med. Res. Coun. (Lond) 249 (1944).
14. Curreri, P. W, Bruck, H. M., Lindberg, R. B., Mason, A. D., Pruitt, B. A.: Ann. Surg. 177: 133 (1973).
15. Cruickshank, R.: J. Path. Bact. 41: 367 (1935).
16. Davis, B., Lilly, H. A., Lowbury, E. J. L.: J. clin. Path. 22: 634 (1969).
17. Decoulx, P., Razemon, J. P., Amourdu, C., Clayes, A. C., Demailles, A., Houcke, M., Hamon, G.: Soc. Med. Milit. Francaise 6: 196 (1963).
18. Evans, E. I., Butterfield, W. J. H.: Ann. Surg. 134: 586 (1951).
19. Feller, I. In: Wallace, A. B., Wilkinson, A. W.: Research in Burns, p. 470 (Livingstone, Edinburgh 1966).
20. Feller, I., Vial, A. B., Callahan, W., Waldyke, J.: J. Trauma 4: 451 (1964).
21. Fox, J. E., Lowbury, E. J. L.: J. Path. Bact. 65: 519 (1953).
22. Fraenkel, E.: Z. Hyg. 72: 486 (1912).
23. Fox, J. E., Lowbury, E. J. L.: J. Path. Bact. 65: 533 (1953).
24. Heggie, R. M., Heggie, J. F.: Lancet II: 664 (1942).
25. Jackson, D. M., Lowbury, E. J. L., Topley, E.: Lancet II: 137 (1951).
26. Jones, R. J.: Brit. J. exp. Path. 52: 100 (1971).
27. Jones, R. J., Hall, M., Ricketts, C. R.: Immunology 23: 889 (1972).
28. Jones, R. J., Lilly, H. A., Lowbury, E. J. L.: Brit. J. exp. Path. 52: 264 (1971).
29. Jones, R. J., Dyster, R. E.: Brit. J. exp. Path. (1973 in press).
30. Jones, R. J., Lowbury, E. J. L.: Lancet II: 623 (1965).
31. Jones, R. J., Lowbury, E. J. L. In: Wallace, A. B., Wilkinson, A. W.: Research in Burns, p. 474 (Livingstone, Edinburgh 1966).
32. Jones, R. J.: Brit. J. exp. Path. 49: 411 (1968).
33. Jones, R. J., Lowbury, E. J. L.: Brit. J. exp. Path. 53: 659 (1972).
34. Kefalides, N. A., Arana, J. A., Bazan, A., Velarde, N., Rosenthal, S. M.: Ann. Surg. 159: 496 (1964).
35. Liedberg, N. C. F., Reiss, E., Artz, C. P.: Surg. Gynec. Obstet. 99: 159 (1964).
36. Liu, P. V.: J. Bact. 88: 1421 (1964).
37. Lowbury, E. J. L., Fox, J. E.: J. Hyg. Lond. 52: 403 (1954).
38. Lowbury, E. J. L., Babb, J. R., Ford, P. M.: J. Hyg. Camb. 69: 529 (1971).
39. Lowbury, E. J. L., Crockett, D. J., Jackson, D. M.: Lancet II: 1151 (1954).
40. Lowbury, E. J. L., Jackson, D. M., Lilly, H. A., Bull, J. P., Cason, J. S., Davies, J. W. L., Ford, P. M.: Lancet II: 1105 (1971).
41. Lowbury, E. J. L., Miller, R. W. S.: Lancet II: 640 (1962).
42. Lowbury, E. J. L., Miller, R. W. S., Cason, J. S., Jackson, D. M.: Lancet II: 958 (1962).
43. Lustgarten, S.: Wien Klin. Wschr. 4: 528 (1891).
44. MacMillan, B. G., Hummel, R. P., Altemeier, W. A. In: Matter, T. L., Barclay, Konickova, Z.: Research in Burns, p. 248 (Huber, Bern 1971).

45. Markley, J., Smallman, E., Evans, G. In: Wallace, A.B., Wilkinson, A.W.: Research in Burns, p. 461 (Livingstone, Edinburgh 1966).
46. Markley, K., Smallman, E.: J. Bact. 96: 867 (1968).
47. Millican, R.C., Rust, J.D.: J. infect. Dis. 107: 389 (1960).
48. Pack, G.T.: Arch. Path. Lab. Med. I: 767 (1926).
49. Park, B.H., Fikrig, S.M., Smithwick, E.M.: Lancet II: 532 (1968).
50. Polk, H.C., Stone, H.H.: Contemporary Burn Management, p. 303 (Little, Brown & Co. Boston 1971).
51. Sevitt, S.: In: Medicine, Science and the Law (Sweet and Maxwell, London 1966).
52. Stockis, E.: Arch. Int. Pharm. Therap. 11: 201 (1903).
53. Stone, H.H.: Ann. Surg. 163: 297 (1966).
54. Teplitz, C.: Arch. Path. 80: 297 (1965).
55. Tumbusch, W., Vogel, E.H., Butkievicz, J.V., Graber, C.D., Larson, D.L., Mitchell, E.T. In: Artz, C.P.: Research in Burns, p. 235 (Blackwell Scientific Publications, Oxford 1962).

Table I.

Identification of 865 strains of gram-negative bacilli (excluding Ps. aeruginosa) isolated from burns   Sept. 1965 to Sept. 1967

Organism	No. of strains	Percentage of total strains
Proteus spp		
P. mirabilis	280	32.4
P. morganii	17	2.0
P. vulgaris	10	1.2
Providencia	38	4.4
Unclassified	15	1.7
Other gram-negative bacilli		
Escherichia coli	153	17.7
Enterobacter cloacae	119	13.8
Enterobacter aerogenes	54	6.2
Enterobacter spp.	6	0.7
Klebsiella aerogenes	58	6.7
Acinetobacter anitratum	37	4.3
Serratia marcescens	14	1.6
Alkalescens dispar	14	1.6
Citrobacter spp.	9	1.0
Alcaligenes spp.	5	0.6
Unclassified coliform bacilli	36	4.2

Table II.

Pre-existing infection with Ps. aeruginosa in Burns Wards as probable source of new infection, during four months

| Ward | Patients in whom infecting types of Ps. aeruginosa were[1] | | |
| | previously isolated from other patients[2] | | not previously isolated from other patients in Wards |
	in Ward E	in Ward F	E or F
Ward E	21	5	6
Ward F	3	17	7

[1] Patients acquiring more than one type of Ps. aeruginosa are represented separately for each type acquired.

[2] within previous 2 months.

Table III.

Early protection of burned mice by vaccine of Ps. aeruginosa (P14)

| Time from vaccination to challenge | Proportion of mice which died after challenge with Ps. aeruginosa (P14) | | |
| | by intraperitoneal injection | | by spreading on burns of mice vaccinated on day of burning |
	in vaccinated mice	in unvaccinated controls	
2 hours	-	-	5/5, 5/5
24 hours	0/5	5/5	3/5, 4/5
2 days	0/5	5/5	0/5, 0/5, 0/5
3 days	0/5	5/5	0/15
7 days	0/5	5/5	-

See Jones, R.J.: Brit. J. exp. Path. 52: 100 (1971).
    Jones, R.J., Lowbury, E.J.L.: ibid, 53: 659 (1971).

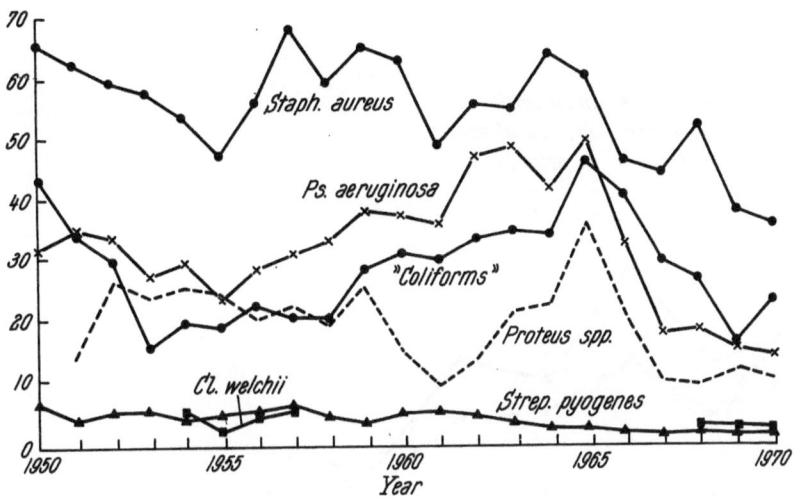

Fig. 1. Bacteria isolated from burns at Birmingham Accident Hospital, 1950-1970 (Lowbury, E.J.L.: Proc. Roy. Soc. Med. 65: 25 (1972).

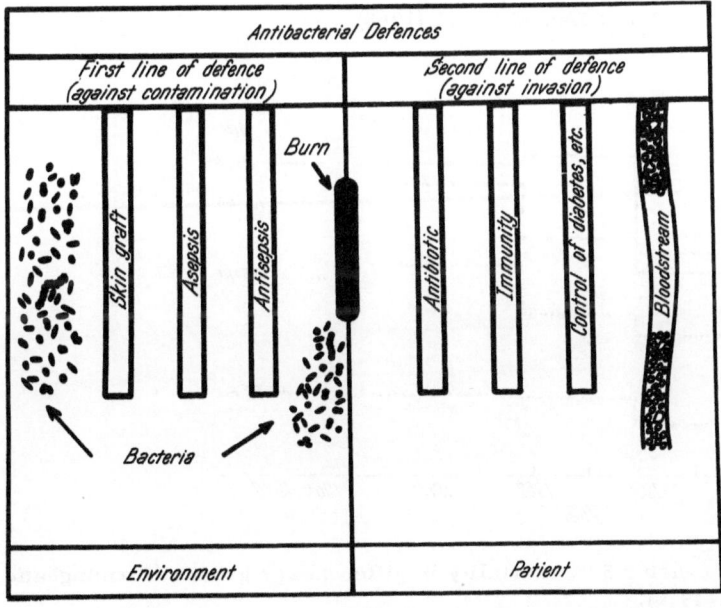

Fig. 2. First and second lines of defence against infection for patients with burns.

472

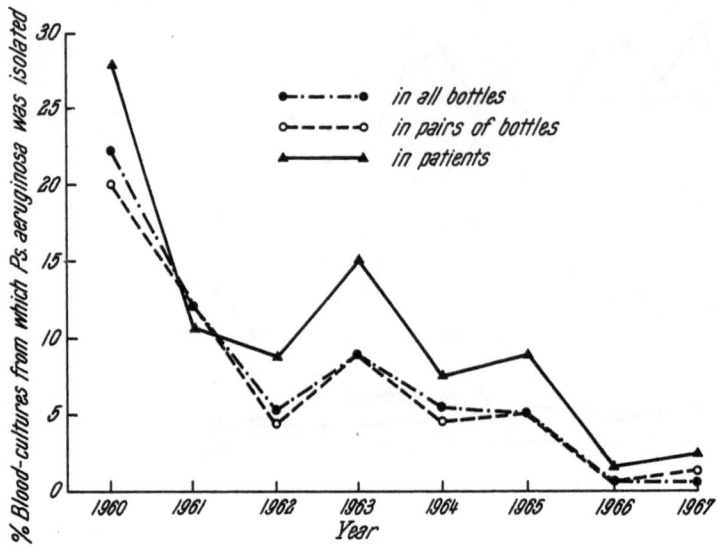

Fig. 3. Percentage of blood cultures from which Ps. aeruginosa was isolated in
Birmingham Burns Unit, 1960-1967.
(Cason, J.S., Lowbury, E.J.L.: Lancet I: 651 (1968).

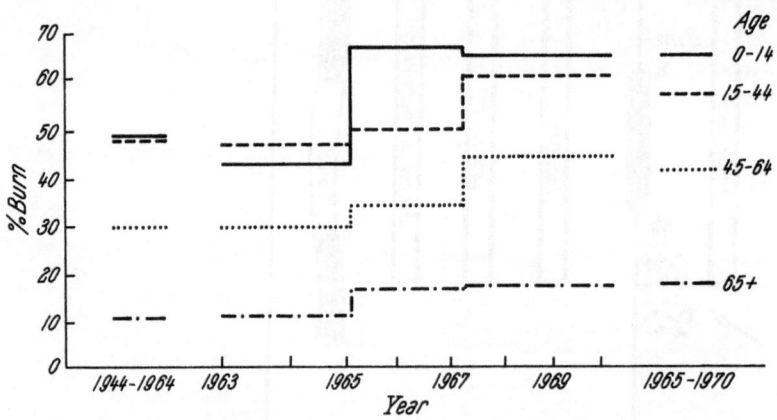

Fig. 4. Areas of burn causing 50% mortality in different age groups (Birmingham
Accident Hospital 1944-1970).
(Bull, J.P.: Proc. Roy. Soc. Med. 65: 27 (1972).

Bacteremia in the Course of Burns with Special Regard to the Initial Release of Mediators by Burn Trauma

L. Koslowski and G. Riethmüller

During World War 2 research in burns turned to the causes of increased vascular permeability, which dominates the initial phase of burns. Exudation, local edema and general increase of vascular permeability are characteristic events. These features distinguish burns from mechanical lesions of tissue.

It was rather obvious that vasoactive mediators are responsible for this increase in permeability. In 1945 Dekanski (4) reported that increased histamine concentrations occurred after burns in mice. Since that time numerous investigators have worked on this problem.

The increase of histamine has been ascertained in the circulating blood during the first hour after burn. We were able to demonstrate this increase in rats (Fig.1). However, reports concerning climax and duration of increased histaminemia differ widely. While we found that the increased histamine concentrations turned to normal values within 3 hours, Birke (3) observed a decreased secretion of histamine in the urine within 24 hours. Other authors, however, reported a histamine-peak after 48 to 96 hours. Although the content of blisters shows an extremely high concentration of histamine, in the injured tissue itself histamine is reduced. Dekanski (4) speculated that histamine is not only released during burn from its inactive or bound form, but that it is also synthesized de novo, as long as the temperature in the tissue does not exceed 60° C.

The effects on the circulation, observed after a thermal skin injury originate in the skin, since excision of the injured skin prevents a shock sydrome as well as the rise of histamine levels in the plasma, as long as the excision is performed immediately after burning (8). We postulate therefore with regard to the development of a burn sydrome that histamine itself is hardly responsible for protracted disturbances occurring days or weeks after a severe burn. Histamine may well be the initial promotor for the locally increased permeability of the injured tissue and it may be responsible also for the exudation, but should not be regarded as a toxic factors itself.

This concept is supported by our experiments with the histamine liberator compound 48/80 in rats. In these experiments rats were subjected to standardized burns. Blood pressure was measured by means of a manometer at their tails. Within 15 minutes after burn we observed a decline of the blood pressure to nearly half of the initial systolic values. To find out, whether this decline of blood pressure was due to increased histamine release, the rats were pretreated with compound 48/80.

This dehistamination, originally introduced by Feldberg and Talesnik (5), leads to a nearly complete disappearance of histamine in the skin (Fig. 2). In the plasma, however, we found after burn similarly increased histamine values in both pretreated and non-pretreated animals.

Under these conditions the question arises whether the fall of blood pressure, occurring immediately after burn and therefore not being due to a loss of blood at that time, is caused by histamine. It may well be that histamine plays a mayor role in the local ischemia and in local changes of permeability.

Stern and Valjevac (14) reported experiments, in which they compared burns of skin areas of high and low histamine content. In burns obtained by corresponding thermal energy, the injuries found in skin areas with high histamine concentrations were more severe and therefore these authors concluded that histamine takes part in the early tissue defect.

According to my knowledge no other tissue hormones have been investigated in a burn syndrome. Little is known about acetylcholine in connection to burn trauma. The release of proteases in the burned skin, however, played an important role in burn research during the last 30 years, particularly with regard to the biochemical lesion induced by thermal energy.

In 1945 Peters (12) reported on the disappearance of proteases of the skin after burn. All experiments, however, to demonstrate skin proteases in the circulating blood have failed, probably due to natural protease inhibitors of the serum, which inactivate the released proteases.

Although the release of proteases in the burned skin could not be proved, it was ascertained that the effect of proteases on burned skin, i.e. heat-coagulated, denatured skin, leads to a high increase in toxic substances from the burned skin tissue.

Extracts of healthy, non-injured skin - apart from bacterial contamination - are not toxic. Tryptic extracts of burned and subsequently excised skin are lethal for mice after intraperitoneal injection.

The beneficial effect of a protease inhibitor has been considered as indirect indication of the involvement of proteases after burns. We observed, indeed, using standardized burns in animals, a lower mortality after intraperitoneal injection of a protease inhibitor. However, we could not detect - even not with high doses - a definite effect of inhibitor therapy in burned patients.

Last not least we should mention the kinins, formed by bradykinin and kallikrein from precursors, the kininogens. We must have some reservations concerning substantial participation of kinins in circulatory disorders after trauma, as well as after burns, because the reserves of kininogens of the body are not high enough to produce such a large quantity of kinins, which are necessary to obtain a manifest circulatory effect (7).

Ten years ago sensational reports had been published about the therapeutic effect of convalescent serum on severe burns. These reports originated especially from the Soviet Union and from Czechoslowakia, and they were based on experimental and clinical investigations. Blood of patients or animals, which had recovered from severe burns, had been used in patients with acute burns.

The reported results were very manifold, ranging from an improvement of the general and psychical condition to significantly higher rates of survival. No exact data, however, were presented.

Our own investigations with Urbaschek and Versteyl showed (9) that convalescent serum had no significant effect on water content of organs after heat traumas.

The question of autotoxins, occurring after burns in the skin, remains unsettled. Whereas skin of burned conventional rats was lethal to mice, skin of burned germ free rats was not lethal (16). Dr. Schoenenberger will report on the latest results in this field. Bacteria and their toxins are obviously of particular interest in burn. The general infection, derived from the burned skin, is the most frequent cause of death after severe burns. As Fig. 3 by Stone and Humphrey (15) illustrates in almost 100 % of all burned patients gram-negative bacteria can be cultivated from the venous blood already during the first week. Most of the positive bacterial results are to be found at the 4th day after burns. Within the 2nd week the positive bacterial counts decrease and are rising again during the 5th week and thereafter. In this late phase many burn patients still die of sepsis.

It is easy to understand that the risk of a septic infection increases with the extent of the burn.

The data in Fig. 4 demonstrate that every 2nd patient with burns of 50 % and more of the body surface will develop sepsis. The insufficient effect of antibiotics on such general infections leads to therapeutic trials of vaccines and antisera against typical hospital strains, particularly against Pseudomonas. The results are very encouraging, especially those reported by Feller (6) in Ann Arbor. This immunotherapy has not yet found widespread application (1).

With regard to the body's own defence mechanisms the decreased concentration of immunoglobulins in serum during the first week after burns is well-known. The functions of the RES, such as phagocytosis, antibody formation and detoxification can be altered by antibiotics, by dye-injection for diagnostic purposes and other therapeutical intervention. Lemperle (10) recommended the stimulation of RES after burns by Glucan and Restim. There is no doubt that a secondary immune deficiency syndrome exists in the first phase of a severe burn. This secondary immune deficiency syndrome develops not only due to increased catabolism and loss of immunoglobulins, but also to an important defect of immunoglobulin synthesis.

More recently the importance of cellular immunity for the defence against bacterial infections has been emphasized. The humoral antibody deficiency syndrome can be compensated partially by administration of gamma-globulin preparations, though commercial gamma-globulin preparations usually lack the important secretory immunoglobulins like IgA.

It is not possible to compensate the suppression of the cellular immune response after severe burns by substitution of plasma or of gamma-globulins. We have to conceive according to a series of new findings that there are various factors in the serum after burns, which are able to influence the reactions of immune-competent cells directly. As pointed out by other speakers of this symposium, certain bacterial endotoxins are important stimulators of B-lymphocytes.

Further experiments are required to settle the question whether nonspecific B-lymphocyte stimulation by endotoxins occurs in vivo after burns.

The observation of Arturson (2), that serum IgM levels are abnormally increased during the 2nd week after burns, may indicate that this increase of IgM concentration is due to nonspecific stimulation of B-lymphocytes by endotoxins. To elucidate this question it would be helpful to analyze the antigenic specificity and affinity of these IgM antibodies synthesized during the 2nd week after burns. Beside lymphocytes, other serum factors have to be considered such as histamine or prostaglandin E_1 and E_2, which may affect the immune-competent lymphocytes. Melmon et al. (11) showed, for example, that lymphocytes bear histamine receptors and that the enzyme adenylcyclase in lymphocyte cell membranes is stimulated by histamine.

Recent experiments of Plat, Lichtenstein and Henney (13) show that histamine at a rather low concentration is able to inhibit the cytotoxic activity of T-lymphocytes in vitro.

According to our experiments particularly B-lymphocytes possess a high affinity for histamine coupled to Sepharose beads (Fig. 5 and 6). Also prostaglandin E_1 and E_2 inhibit T-cellbound cytotoxicity in vitro.

Therefore the unphysiologically high concentration of histamine, prostaglandin and other vasoactive amines may inhibit certain lymphocyte populations after burn trauma. The decreased mitogenic response of lymphocytes in vitro to phytohemagglutinin, for example, could be explained by such inhibitors.

The complexity of factors, which lead to an immunodeficiency after severe burns, may be so great that at present an evaluation of single factors is impossible. In general, the suppression of immunological reactions by mediators seems to be reasonable. Also under normal conditions the immune response is considered to be regulated by soluble mediators in order to prevent an excess production of effector molecules. It should be possible to test this hypothesis by in vitro experiments directly.

In addition, these experiments may show us a new way for a therapeutic approach of this secondary immunodeficiency in patients with severe burns.

476

References

1. Alexander, J. W., Fisher, M. W. In : Matter, P., Barcley, T. L., Konickova, Z. : Research in burns, p. 226 (Huber, Bern 1971).
2. Arturson, G., Hogeman, C. F., Johanson, S. G. O., Killander, J. : Lancet I: 546 (1969).
3. Birke, G., Dunér, H., Liljedahl, S. O., Pernow, B., Planin, L. -O., Troell, L. : Acta chir. Scand. 114 : 87 (1957).
4. Dekanski, J. : J. Physiol: (Lond.) 104 : 151 (1945).
5. Feldberg, W., Talesnik, J. : J. Physiol. (Lond.) 120 : 550 (1953).
6. Feller, I. In : Wallace, A. B., Wilkinson, A. W. : Research in burns, p. 470 (Livingstone, Edingburgh 1966).
7. Habermann, H. In : Haberland, G. L., Rohen, J. W. : Probleme der Pathophysiologie des Kininsystems, p. 37 (Schattauer, Stuttgart 1969).
8. Kisima, H. : Fukuoka Acta med. 31 : 5 (1938).
9. Koslowski, L., Urbaschek, B., Versteyl, R. : Nature 200 : 273 (1963).
10. Lemperle, G. In : Matter, P., Barcley, T. L., Konickova, Z. : Research in burns, p. 208 (Huber, Bern 1971).
11. Melmon, K. L., Bourne, H. R., Weinstein, J., Selc, M. : Science 177 : 707 (1972).
12. Peters, R. A. : Brit. med. Bull. 3 : 81 (1945).
13. Plaut, M., Lichtenstein, L. M., Henney, C. S. : Nature 244 : 284 (1973).
14. Stern, P., Valjevac, K. : Pharmacological treatment in burns. Excerpta Med. Amsterdam, p. 33 (1969).
15. Stone, N. H., Humphrey, C. R. In : Matter, P., Barcley, T. L., Konickova, Z. : Research in burns, p. 201 (Huber, Bern 1971).
16. Urbaschek, B., Koslowski, L., Versteyl, R., Haussmann, P., Sacquet, E., Charlier, H. : Klin. Wschr. 43 : 748 (1965).

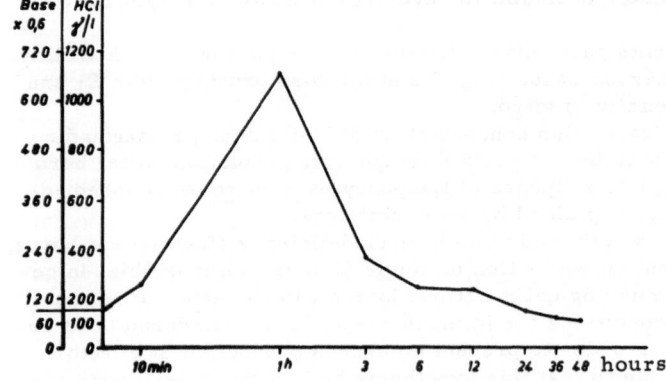

Fig. 1.

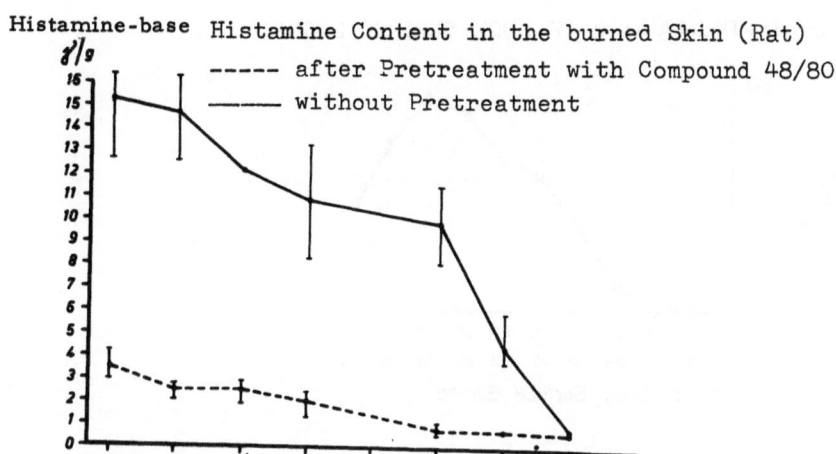

Fig. 2.

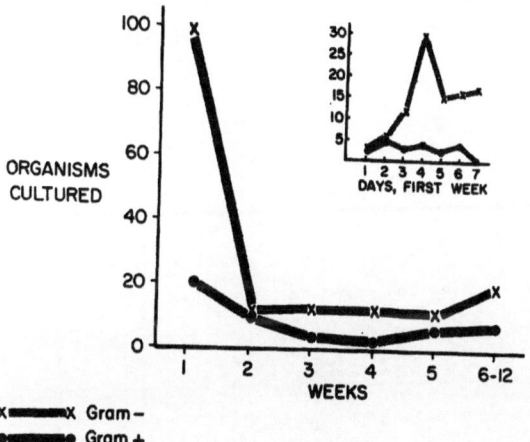

Fig. 3.

478

RISK OF SEPTICEMIA vs. % OF BODY SURFACE BURNED

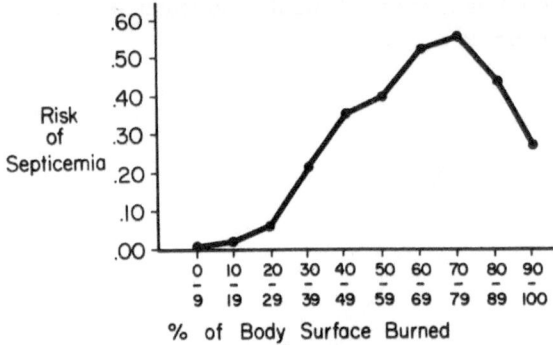

Fig. 4.

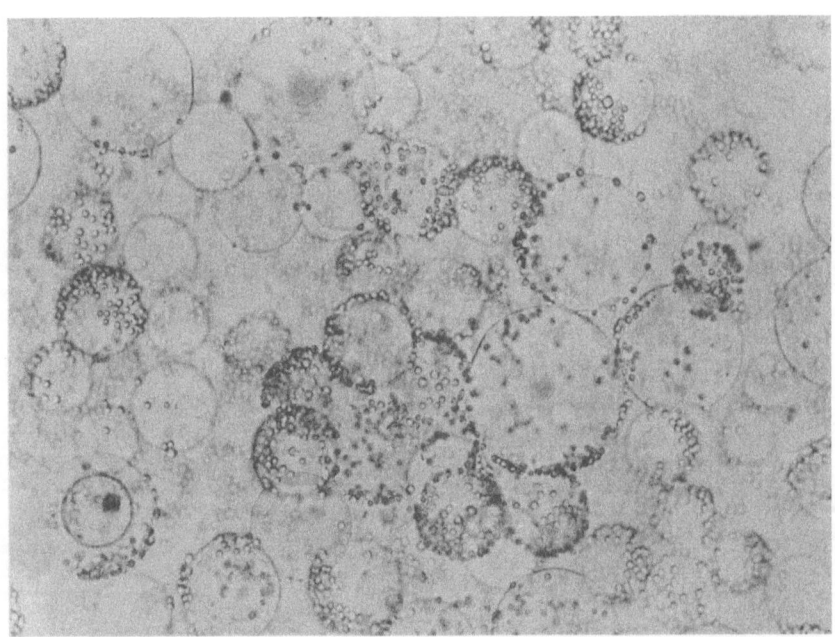

Fig. 5. 8 x 10^6 lymphocytes from mouse spleen (in 0.1 ml of culture medium
and 0.2 ml of a 25 % histamine-Sepharose/PBS suspension) were incubated at
37°C for 15 min. The counting resulted in 35-45 % of positive Sepharose-beads
(x 100).

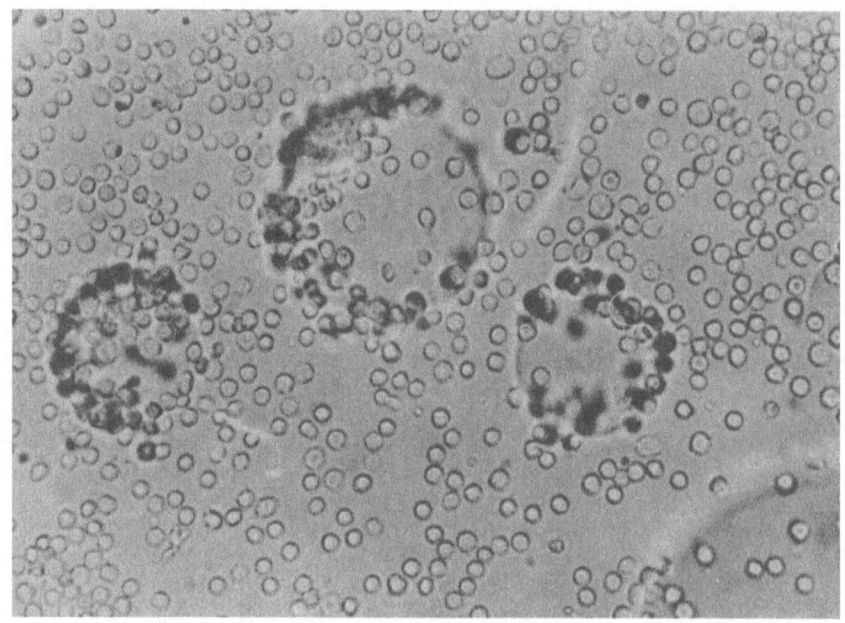

Fig. 6. Magnification of three positive Sepharose-beads from Fig. 5 (x 200).

Pathogenetic and Immunotherapeutic Significance of a Specific Cutaneous Human Burn Toxin

G. A. Schoenenberger, M. Allgöwer, F. Burkhart, W. Müller, and K. Städtler

Therapeutic progress has virtually eliminated death from primary shock and early infection following severe human burns (1). In contrast, recent statistics reveal the depressing fact that the mortality rate, after full thickness burns involving 50 % or more of the body surface, has not improved substantially during the last decade (2, 3). Thus, the interest in burn research has shifted to the problem of the "late mortality", sepsis appearing as the most obvious cause of this phenomenon (4). However, the lack of resistance of the burned patient seems to indicate a "host problem" rather than one of an increased virulence of the invading micro-organisms (5).

In severe burn patients who survive the initial fluid shift, Pseudomonas aeruginosa infections are especially refractory to treatment and are lethal either because of resistance to antibiotic therapy or increased susceptibility to infections together with the presence of these gram-negative bacteria in the burn eschar (6). Thus, experimental and therapeutic efforts were inaugurated to reduce Pseudomonas septicemia and overall mortality. A major effort has been directed to vaccination and passive immunotherapy, using a polyvalent heat-killed phenolized Pseudomonas vaccine and/or the corresponding hyperimmune plasma (6, 7, 8, 9). While "p o l y-v a l e n t" immunotherapy significantly improved overall survival and dramatically reduced the mortality from Pseudomonas sepsis, the incidence of other serious infections has not diminished. Indeed, the control of the Pseudomonas infections tended to result in a slightly higher incidence of the serious gram-negative and fungal infections, especially in the late postburned state (6). Moreover, while m o n o-v a l e n t immunotherapy in combination with optimal care for the burn patient resulted in a 43 % reduction in Pseudomonas septicemia incidence and a 27 % reduction in deaths due to Pseudomonas infections, total mortality during the years 1967-1970 in which this therapeutic regime was followed was not reduced as compared to the "control" years 1960-1962 without specific anti-Pseudomonas therapy (10). During the period 1960-1970 covered by this report (10), the overall mortality seemed rather to be proportional to the percentage of body surface burned as one would expect when assuming a causal relationship between mortality rate and a specific toxic compound produced in the skin by thermal energy.

We have previously reported the isolation and purification of a novel lipid-protein complex from mouse skin which is produced by the controlled application of thermal energy "in vitro" (11). For isolation purposes, a bioassay was used in which thorotrast was injected i. v. to recipient animals 4 hours prior to the fractions to be tested. This modification permitted to take a 100 % mortality of the recipient animals within 48 hours as means to monitor the isolation procedure (12). The final isolation product has lethal effects when injected into genetically different or litter-match recipient animals.

By the identical isolation technique, a naturally occurring, biologically inactive lipoprotein from thermally not altered or "native" skin was obtained and hence it was possible to compare the physical and chemical properties as well as the biolog-

This project was supported by the SNSF, grant No. 3.272.72

ical activity of the isolation products from both sources during the isolation procedure. This approach enabled the site of production of toxin to be restricted to a few skin components (13). Methods which have a tendency to dissociate lipid-lipid and lipid-protein bonds abolished the toxic activity of the derivative from thermally altered skin during isolation but they revealed, when applied to the chemically purified products, that both apparently uniform compounds were made up of six different oligomeric peptides and six different lipid classes, identical in composition, in both derivatives (14). However, the toxic compound differed from the inactive or native isolation product by possessing a larger size (MW) and higher density (15). These results suggested the hypothesis that thermal energy transformed a structural "pseudo-unit" of membranes, present in normal skin, into a polymeric form which has toxic properties (16).

Fig. 1 shows an electron micrograph of an aggregate of toxin molecules, isolated from g e r m f r e e mouse skin which was exposed to controlled thermal energy in vitro. By using germfree raised "donor animals" of different inbred strains and processing the skins under sterile conditions, bacteria were excluded as a possible source for toxin formation and as the cause of toxic activity. Immunological incompatibilities between donors and recipients did not occur in our test model.

Delipidation and elimination of "soluble" cell constituents by milling the skins prior to heat application reduced the starting material to a few skin components, i.e. cell wall membranes of the basal cell layer of the epidermis. From the electron microscopic pictures (magnification = 1:120,000; 3 x 40,000) a diameter of ~330 Å for each toxin molecule was calculated, a value which is in reasonable agreement with that found for cell walls (monolayer; lipid-protein =~100 Å thick) (17), in view of the fact that the toxin was found to be a polymeric form of a naturally occurring precursor. The photograph also proves the purity and uniformity of the molecular population which was later on used for active and passive immunotherapy of lethally burned mice.

Human split thickness skin (Thiersch) of 1 mm was subjected to controlled thermal energy and to the identical chemical isolation procedures as described for mouse skin (11, 16). Normal unburned skin was processed following the same procedure.

The separate products isolated from thermally exposed and unexposed skin by chemical isolation were further purified by ultracentrifugation in a density gradient containing 33-62 % sucrose for the "native" and 57-67 % (w/w) for the thermally altered skin extracts. The compounds were suspended in Tyrode I plus II (30 mg/ml) according to the method of Potter. The pH of the suspension fell immediately from 7.35 (mixture Tyrode I plus II) to 4.5. The centrifuge tubes were punctured at the bottom and fractions of 0.3 ml were collected into an LKB fraction collector. The 280 nm absorbance and ninhydrin reaction of each fraction were carried out. The fractions containing protein and corresponding to visible bands were pooled and the sucrose removed by dialysis.

The distribution pattern after purification by ultracentrifugation is shown in Fig. 2a and b. Common to both derivatives is a floating band of low density on the top of the tube and a second band with a density of 1.25. This is predominantly present in the isolation product from native skin (Fig. 2a). A third new band, with the highest density (1.31) and molecular weight is seen in the toxic material only (Fig. 2b). Thus, controlled thermal energy transforms between 50-80 % of a naturally occurring precursor into a polymeric burn toxin. Only the second band (d = 1.31) was toxic for recipient animals across the species barrier (man - mouse). These results additionally suggest that only a partial transformation of the "precursor" into toxin occurs, as would be expected in a "biological system", i.e. 20-50 % of the non-toxic lipid-protein complex, a different antigen, will be absorbed simultaneously with the toxin after burn injuries. We proved this assumption by isolating toxin and precursor from pooled sera from severely burned patients who died between the 5th and 15th postburned day.

Sera were desalted and freed of other low molecular weight substances (drugs etc.) by dialysis against H_2O (4°C). The material was lyophilized and redissolved in a deproteinized, serum-like salt solution (0.1 % w/v). The fraction containing the very high molecular weight compounds was separated by Diaflo (XM 300 Membrane / Amicon Corp., Lexington/Mass., USA) and subjected to an identical isolation procedure used for skin homogenates. The material obtained after chemical isolation was finally separated by ultrazentrifugation in a sucrose gradient, yielding the non-toxic precursor and the toxin (Fig. 2c, d = 1.14 and 1.18). These densities were apparently n o t identical with those of the corresponding compounds, isolated from human split thickness skin b u t they were identical with the densities of the mouse toxin and precursor respectively (11). This discrepancy can easily be explained by the fact that in mouse as well as human skin this tissue was removed prior to thermal energy application. Indeed, the lipid moieties of the compounds isolated from human skin contain only 10-15 % apolar lipids as compared to 40 % of the mouse toxin and precursor and to the corresponding compounds obtained from human sera. We have previously shown that the toxicity resides in the apoprotein and that the lipid portion contributes only a nonspecific and non-quantitative way to the toxic activity (12). Thus, the serum-toxin and precursor are essentially identical to the corresponding compounds isolated from human and mouse skin as was additionally shown by identical lipid- and aminoacid analyses.

The minimum molecular weight identical for both apoproteins and, computed from the weights of six oligomeric subunits, was 550,000 ± 8 %. Some of the oligomers were thought to be linked by disulfide bridges. Thus, the identity of the "substructure" of both protein cores was confirmed (13, 14, 16).

The biological and pathogenic properties of the isolated compound have now been investigated in vivo. For this purpose, a new experimental "model" was designed, where the input of thermal energy could be restricted exclusively to the skin by protecting the animal's body, i.e. central nervous system and organs from direct heat. Two identical transverse incisions of 1 or 2 cm were made on the animal's back near tail and neck (narcosis: Pentobarbital 60 mg/kg). The skin was mobilized and a Pyrex glass spatula of a given size inserted subcutaneously and pushed through the second incision to the surface. The spatula was fixed between two metal clamps and the animals were hung on the in situ mobilized skin. A cold airstream was blown through a channel between mobilized skin and the animal's body. Thermal energy was then applied under identical temperature, pressure and time conditions which guaranteed an optimum toxin formation in vitro. The incisions were closed with clamps and the animals were subsequently kept in separate cages. All animals were treated for shock with i.p. injections of balanced salt solution for the first three days in amounts of 20 % of their weight . Penicillin-streptomycin (400,000 units/400 mg/kg) was injected intramuscularly on the first and third day and sulfonamides were administered in the tap water.

All in vivo experiments were carried out with series of 3 x 20 Swiss albino mice of either sex, weighing 24-26 g. The body surface of the animals was calculated with Rubner's (19) formula: $s = k \times w^{2/3}$ where s = total body surface of the animal, k = a constant of 6.5, and w = the body weight. The experimentally injured skin was then expressed as per cent of the total body surface and/or the weight of the animals.

When 18 % of the body surface or 0.4 cm^2/g body weight were injured, a mortality rate of 90-100 % within 5 days was achieved, in spite of shock treatment. This size of injury was used as "lethal" burn for the subsequent experiments. Antibiotic therapy did not influence the mortality rate. All animals showed a substantial weight loss after the "dry injury", which could not be completely replaced. For lethal injuries, the weight loss within 6 hours was 10 % and within 24 hours 30 % for the animal's body weight. We then attempted to show the direct relationship between the human burn toxin isolated from skin subjected to controlled thermal injury in vitro and the mortality rate by using the in vivo model in two different experimental set-ups.

In the first experiment, animals of either sex weighing 14-16 g were injected twice i.p. with 10 mg of toxin (sublethal dose) each 8 days apart. Twelve days after the second injection, a lethal burn injury was set. The controls were injected with the non-toxic product according to the same scheme. This indirect proof for the toxic activity by protection through active immunization was then confirmed by showing a beneficial effect of a passive serum therapy with a heterologous (rabbit) immune serum or a corresponding IgG fraction obtained by Rivanol precipitation which was shown to contain 92-96 % IgG by cellulose acetate-, immuno-, and polyacrylamide electrophoresis. The animals were injected i.p. with 10 mg IgG/animal immediately after the injury and 5 mg five days thereafter. A corresponding heterologous "antihuman native" IgG was injected to the control series. The results of preimmunization are shown in Fig. 3. While all nonimmunized animals died within 6 days, 70 % of the animals immunized with the toxin were still alive at this time. The protected animals survived for weeks. Preimmunization with the "native" control substance or Freund's adjuvant had no effect.

The effect of passive immunotherapy is shown in Fig. 4. Only 20 % of the treated animals died as compared to 100 % of the controls which received normal rabbit or "antinative" serum. These animals survived for weeks. Importantly, none of the survivors of either active immunization or passive immunotherapy had elevated serum urea or creatinine values and no changes of the tubular or glomerular apparatus of the kidney could be observed up to 16 days after the lethal burn injury.

By showing a clear quantitative correlation between the mortality rate and the extent of skin injured in vivo by the same thermal exposure as applied to the skins in vitro, the production of a similar compound after burn injuries in living animals was suggested, i.e. the burn toxin must be assumed to have been formed. Moreover, the pronounced protective benefit of preimmunization with the human toxin in animals expected to have a mortality rate close to 100 % must be interpreted as strong evidence of a crucial pathogenic contribution of this compound to the fatal outcome in burn injuries. No significant protection was attained in the injured control animals immunized with the inactive precursor or another macromolecular gluco-lipoprotein as an antigen. Moreover, the significant therapeutic effect of a heterologous "antitoxic" IgG (human) confirmed this conclusion.

Immunoelectrophoresis on agargel is not suitable for detection of the corresponding antibodies in serum due to the "size" of the antigen which does not migrate in this medium.

From these experiments, we must conclude that the toxic lipid-protein complex isolated in our laboratories from human skin exposed to controlled thermal energy in vitro has a substantial and crucial effect upon the mortality rate after large high temperature burns. The mechanism by which it acts, most likely a generalized submicroscopic damage, is not yet clear and under investigation.

In order to investigate the influence of the burn toxin and/or Pseudomonas infections upon mortality, we have mimicked a burn injury "in vitro" by preinjecting sublethal, harmless doses of human burn toxin to mice which were subsequently infected with a standardized dose of Pseudomonas. All experiments were carried out on 5-weeks old Swiss albino mice of both sexes with an average weight of 25 g. These mice were bred and kept in our laboratories. In these experiments, a reticuloendothelial system (RES) block was never applied to any of the recipient mice. A sublethal dose i.p. of 15 mg of the isolated toxins in 1 ml of a deproteinized, serum-like balanced salt solution was found to be harmless in a control group of 20 animals.

To produce a wound for the infection, a 5 cm incision was made longitudinal on the back of each animal (pentobarbital 60 mg/kg of body weight). Skin flaps measuring 1 cm of both sides were mobilized from the fascia, thus, a wound of 10 cm^2 re-

sulted. Pseudomonas organisms were applied uniformly in the wound with a micro-
syringe. The skin flaps were then closed with two wound clamps.

I. In a first set of experiments, (n = 10/group) we established the correlation
between mortality rate and size of infection. To obtain a standardized infection, we
used a green strain of Pseudomonas that was cultured for 24 hours on blood agar
(20). The strain was found to be resistant primarily against tetracycline, chloramphenicol, gentamycin and nalidixic acid, a gram-negative antibacterial. The organisms were suspended in physiological saline solution and counted by nephelometry
and in a Petroff-Hauser chamber (21). The suspensions used in these control groups
contained 3×10^6, 1×10^7, 3×10^7, 1×10^8, and, according to Jones and Lowbury
(9), 7×10^8 organisms in 0.1 ml. In all consecutive experiments, 1×10^7 organisms
were applied. This infection was found to yield a mortality rate of 0-20 %, within
72 hours, i.e., a moderate mortality rate that was thought to be optimal for experimental purposes. The mortality rates were estimated as percentages at intervals of
24 hours, ending at 72 hours. The animals surviving after 72 hours were observed
for 10 days.

II. In a second set of experiments, three groups of 10 animals each were pretreated as follows:

Group 1 was injected i.p. with 15 mg of burn toxin derived either from mouse
or human skin and suspended in 1 ml of deproteinized, serum-like balanced salt solution.

Group 2 received an indentical amount of the non-toxic precursor derived from
not burned or normal skin by the same isolation procedure as described for the toxin isolation under otherwise identical conditions.

Group 3 was injected i.p. with 1 ml of saline solution alone and served as control.

72 hours after this pretreatment, all three groups were infected with Pseudomonas aeruginosa under standardized conditions (1×10^7 organisms) as described
and the mortality rates are shown in Table I. It is evident from the data shown in
Table I that the toxin increased the mortality rates.

III. In a third set of experiments, the effects of a specific antitoxic immunotherapy upon mortality due to Pseudomonas infection were investigated (n = 10/group).

Group 1 was again injected i.p. with 15 mg of either burn toxin.

Group 2 received also 15 mg of either burn toxin.

Group 3 (controls) received 1 ml of deproteinized, serum-like saline solution,
i.p.

72 hours after this pretreatment, all three groups were infected with 1×10^7
gram-negative organisms. Additionally, group 2 was treated with 7 mg of antitoxic
IgG i.p. immediately after infection, followed by 5 mg at 24 and 48 hours.

Table II summarizes the mortality rate obtained with preinjection of burn toxin
followed by a standardized Pseudomonas injection with or without antitoxic IgG treatment. In these series, injection of 15 mg of toxin i.p. three days prior to infection
produced a mortality rate of 80 % (group 1), compared with 20 % for the controls
(group 3). However, animals (group 2) which also were preinjected with 15 mg of
burn toxin, but then in addition were treated with antitoxic IgG as described earlier,
showed a mortality rate of 20 % instead of the expected 80 %, i.e., the mortality
rate following infection was reduced to that of the controls (group 3).

Our results suggest strongly the role of burn toxin in susceptibility to infection.
In animals pretreated with a single i.p. sublethal dose of a specific, purified, and
characterized burn toxin a superimposed standardized Pseudomonas infection caused
the mortality rate to rise from 0 to 50 %, confirming the hypothesis that a specific
cell damage is the triggering mechanism for generalization of the wound infection.
Furthermore, the experiments with the toxin isolated from burned human skin demonstrated a biological activity across the species barrier and suggested that thermal energy produces a generalized pathogenic principle that increases mortality and

impairs the host defense against infections. A fourfold increase in the mortality rate of mice after preinjection with the human toxin with superimposed infection strongly supports this assumption.

Finally, the therapeutic effect of a specific purified heterologous antitoxic IgG, i. e., an IgG produced by an antigen quite different from the antigen obtained from Pseudomonas aeruginosa completely neutralized the pathogenic effect of the toxin on a superimposed Pseudomonas sepsis. This led us to speculate that a possible additional antitoxic IgG treatment of severely burned patients might add a new aspect to the specific anti-Pseudomonas treatment in severe burns. Therapeutic measures established during the past few years might become even more effective in a combined therapy with antitoxic IgG.

References

1. Moyer, C. A., Butcher, H. R., Jr.: Burns, shock and plasma volume regulation. (C. V. Mosby Co., St. Louis, USA 1967).
2. Kernahan, D. D.: Canad. med. Ass. J. 97: 433 (1967).
3. Ramirez, A. T., Tamondong, C. T., Del Castillo, A. N. L., Dino, B. R.: Surgery 68: 813 (1970).
4. Pruitt, B. A., Jr., Moncrief, J. A.: J. surg. Res. 7: 332 (1967).
5. Sevitt, S.: Modern trends in plastic surgery 2: 126 (1966).
6. Alexander, J. W., Fisher, M. W., McMillan, B. G.: Arch. Surg. 102: 31 (1971).
7. Feller, I.: The use of Pseudomonas vaccine and hyperimmune plasma in the treatment of severely burned patients. In: Wallace, A. B., Wilkinson, A. W.: Research in burns, p.470 (Livingstone, Edinburgh 1966).
8. Feller, I., Pierson, C.: Arch. Surg. 97: 225 (1968).
9. Jones, R. J., Lowbury, E. J.: Antiserum and antibiotic in the prophylaxis against Pseudomonas aeruginosa. In: Wallace, A. B., Wilkinson, A. W.: Research in burns, p.474 (Livingstone, Edinburgh 1966).
10. Pierson, C., Feller, I.: Surg. clin. N. Amer. 50: 1377 (1970).
11. Schoenenberger, G. A., Cueni, L. B., Bauer, U., Eppenberger, U., Allgöwer, M.: Biochim. biophys. Acta 263: 149 (1972).
12. Cueni, L. B., Allgöwer, M., Schoenenberger, G. A.: Z. ges. exp. Med. 156: 110 (1971).
13. Schoenenberger, G. A., Cueni, L. B., Bauer, U., Eppenberger, U., Allgöwer, M.: Surg. Forum XXI: 515 (1970).
14. Schoenenberger, G. A., Bauer, U. R., Cueni, L. B., Eppenberger, U., Allgöwer, M.: Biochem. biophys. Res. Commun. 42: 975 (1971).
15. Cueni, L. B., Allgöwer, M., Eppenberger, U., Städtler, K. E., Schoenenberger, G. A.: Physico-chemical characterization of a toxic lipoprotein produced by heat in mouse skin. In: Matter, P., Barclay, T. L., Konickova, Z.: Research in burns, p.471 (Huber, Bern 1971).
16. Schoenenberger, G. A., Cueni, L. B., Bauer, U., Eppenberger, U., Städtler, K., Allgöwer, M.: Biochim. biophys. Acta 263: 164 (1972).
17. Vanderkooi, G.: Ann. N. Y. Acad. Sci. 195: 6 (1972).
18. Slater, G. G.: Analyt. Biochem. 24: 215 (1968).
19. Rubner, M.: Z. Biol. 19:535 (1883).
20. Müller, R.: Medizinische Mikrobiologie, p.445 (Urban und Schwarzenberg, München 1950).
21. Simmons and Gentzkow: Laboratory methods of the U. S. Army, p.404, Lee and Febiger, Eds., 1944.

Table I.

Mortality Rate in Percent after a Standardized <u>Pseudomonas</u> Infection

Conditions:
Pretreatment with 1. burn toxin 15 mg/ml i.p.
        2. native (nontoxic) compound 15 mg/ml i.p.
        3. saline solution 1 ml i.p.

Mortality in % hrs after infection	1.	2.	3.
24	20	0	0
48	30	0	0
72	50	10	0

Table II.

Mortality Rate in Percent after a Standardized <u>Pseudomonas</u> Infection

Conditions:
Pretreatment with 1. burn toxin 15 mg/1 ml i.p.
        2. burn toxin 15 mg/1 ml i.p. + specific IgG-therapy 17 mg i.p.
        3. saline solution 1 ml i.p.

Mortality in % hrs after infection	1.	2.	3.
24	20	0	10
48	50	10	20
72	80	20	20

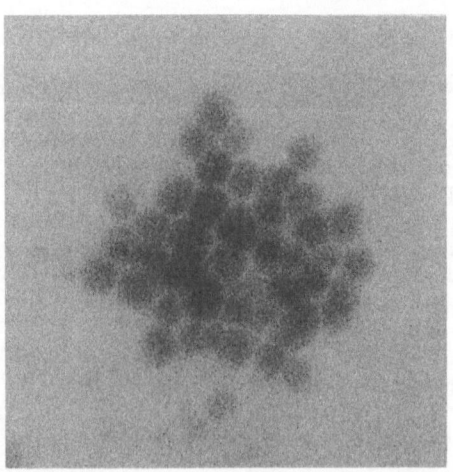

Fig. 1. Electron microscopic photograph of an aggregate of toxin molecules isolated from mouse skin (magnif.: 1:120,000; osmium tetroxyde staining).

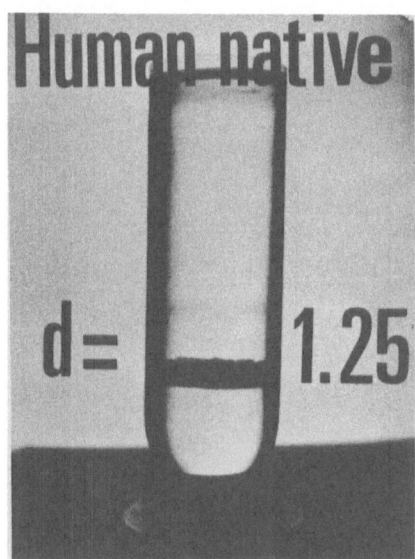 

Fig. 2a.   Fig. 2b.

Fig. 2a.   Ultracentrifugation of the isolation product derived from normal human skin (precursor) in different sucrose concentrations. From top to bottom: = 33, 41, 52, 62 % sucrose (w/w).

Fig. 2b.   Ultracentrifugation of human toxin derived from thermally altered skin in different sucrose concentrations. From top to bottom: = 57, 61, 64, 67 % sucrose (w/w).

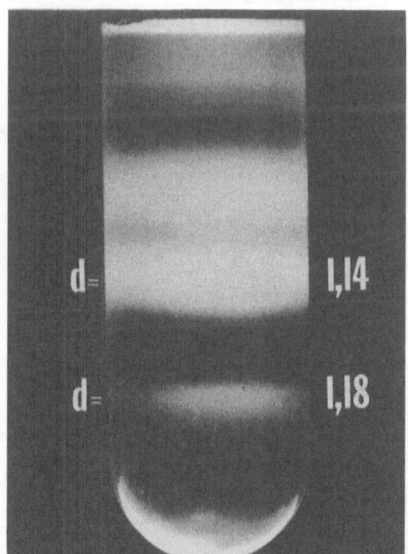

Fig. 2c.   Ultracentrifugation in a sucrose gradient of the very high molecular weight fraction from sera of severely burned patients after isolation by molecular sieve membranes and chemical fractionation (4°C; 3 hrs; 70'000 x g). d = 1.14 = nontoxic precursor; d = 1.18 = burn toxin.

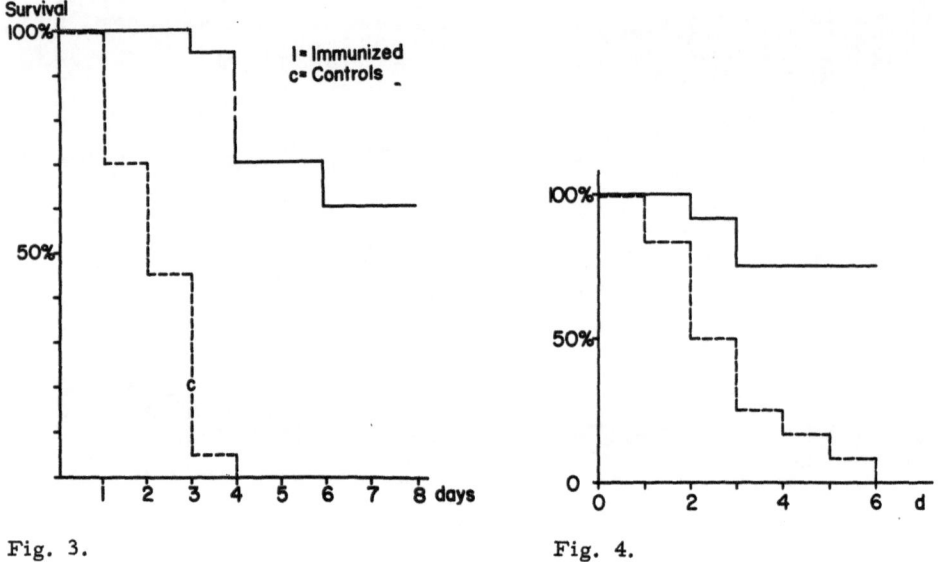

Fig. 3.   Fig. 4.

Fig. 3. Effect of an active immunization with the human toxin upon the survival
of lethally burned mice (———) as compared to the nontreated animals and mice
pre-immunized with Freund's adjuvant or the nontoxic compound isolated from
native skin (–––––) (n = 3 x 10 for each group).

Fig. 4. Effect of a specific immunotherapy with an "antitoxic" (human) rabbit
IgG upon the survival of lethally burned mice (———) as compared to the controls
(–––––) treated with the same dose of normal rabbit or "antinative" IgG. (n = 3
x 10 for each group).

Immunization against Pseudomonas in Burn Disease

J. W. Alexander and M. W. Fisher

Pseudomonas aeruginosa is a ubiquitous microbe which rarely causes diseases in healthy individuals. However, in patients with depressed immunity, this organism frequently becomes pathogenic, and such infections often result in septic death (1). Patients with major thermal injury have been particularly prone to develop Pseudomonas infections because many of their host defense mechanisms are compromised, and there is an ever present portal of entry via the burn eschar which can rarely be sterilized by topical antimicrobial therapy.

The prevalence of serious Pseudomonas infections in burn patients and the associated high mortality stimulated a search by several investigators for a better means to control this pathogen (5, 8, 10-12). Our own studies began in 1964, and ultimately resulted in a clinical and laboratory investigation of a heptavalent Pseudomonas vaccine (3) (Pseudogen[R], Parke-Davis & Co.) and a hyperimmune anti-Pseudomonas gammaglobulin (9) for immunotherapy against Pseudomonas in burn injury. Patients having major thermal injury at the University of Cincinnati Burn and Trauma Unit and the Cincinnati Shriners Burns Institute have now been immunized with Pseudogen[R] since March, 1968. The hyperimmune anti-Pseudomonas gammaglobulin has been available for use in selected patients with Pseudomonas bacteremia or Pseudomonas burn wound sepsis since July, 1970. This report summarizes the clinical and laboratory results of the initial five year experience in burn patients.

Materials and Methods

All consenting patients who entered the University of Cincinnati Burn and Trauma Unit and the Cincinnati Shriners Burns Institute after March, 1968, with burn injuries involving 20 % total body surface area or greater were given the heptavalent Pseudomonas vaccine with the exception of 20 patients in 1968 who were used as non-vaccinated controls. Since these were the initial studies of this vaccine in humans, a variety of routes and doses for the administration of the immunogen were evaluated which resulted in six basic phases of study (Table I). Phase I patients were randomized with non-vaccinated controls and were used to determine basic antibody response to relatively low doses of immunogen. In Phase II, the number of injections was increased, and the dose was slightly increased in some patients. In addition, an initial comparison was made of intramuscular, intradermal, and subcutaneous routes. In Phase III, the dosage for each injection was increased to what was regarded to be a maximal tolerable level for a diversified burn population. This dose of vaccine caused some detectable local or systemic response approximately 50 % of the time. A combination of intramuscular and intradermal injections was given empirically to achieve the highest possible antibody titer. In Phases IV a and IV b, only intramuscular injections were given, and an attempt was made to determine whether administration of larger amounts of the antigen during the initial days of immunization would increase the resistance to infection and the corresponding levels of antibody. In Phase V, all patients were given the same total dose of antigen intradermally using a more concentrated vaccine but adhering to the same schedule for injections used for Phase IV b. To avoid possibility of an overdose with the concentrated vaccine, seven patients begun on the IV b protocol received their remaining immunizations by the intradermal route. This also allowed evaluation of the ability of intradermal injections to cause an increase in the antibody titer at a later time. In the last group of patients (Phase VI) all injections were given intramuscularly according to the original schedule in Phase III. In all patients, the first injection was given as soon as informed consent was obtained from the patient or his family, but the time after burn injury varied considerably because of referral of many patients to our center several days postburn.

Observations were made for local and systemic reactions for 48 hours following each injection, and serum samples were taken for the determination of circulating antibody on days 1, 5, 8, and weekly thereafter as long as the injections were continued. Survival, cause of death, clinical evidence of invasive infection from any organism, positive blood cultures for any organism, and positive cultures for <u>Pseudomonas aeruginosa</u> in the blood, wound, respiratory tract, and urinary tract were documented for each patient.

To provide a basis for comparison, the records of 75 non-vaccinated patients admitted between January, 1967 and April, 1968, with burn injuries of 20 % or more were carefully evaluated to obtain similar information regarding colonization and infection.

All patients who developed <u>Pseudomonas</u> septicemia within the first five days after immunization began, or who died during the first five days of hospitalization from any cause were excluded from analysis in both the control and vaccinated groups because immunization could not be expected to be capable of preventing infection before the appearance of antibody in the circulation.

Eight of the ten patients in Phases IV - VI who developed <u>Pseudomonas</u> bacteremia were treated with hyperimmune anti-Pseudomonas gammaglobulin. On the initial day of treatment, these patients were usually given 1 ml globulin/kg body weight intramuscularly. An additional injection of 0.5 ml/kg was given on the second and third days if there was no evidence of improvement of the sepsis within 24 hrs.

Measurements of Circulating Antibody

Serial measurements of circulating antibody were made in all patients in Phases I - V by a passive hemagglutination (HA) technique. In earlier patients, the HA titers were determined both before and after 2-mercaptoethanol inactivation, but after it was appreciated that IgG was far more important than IgM for protection against <u>Pseudomonas</u> infection (2), samples were measured for hemagglutinating activity only after 2-mercaptoethanol inactivation. Additional manifestations of antibody activity in selected patients were evaluated by studying opsonic support of human neutrophil bactericidal function <u>in vitro</u> and by mouse protection assays.

Results

Comparison of the Patient Population during Various Phases of the Study

Since the various phases of the study were not done concurrently and antecedent controls were used as a basis for comparison, it is important to evaluate the comparability of the patient populations. These data are shown in Table II. The basis for patient comparison could be divided into four basic groups: antecedent controls, patients in Phases I and II who received submaximal and inadequate vaccination, patients in Phase III who received maximal vaccination, and patients in Phases IV-VI who received maximal vaccination and for whom hyperimmune anti-Pseudomonas globulin was used to treat vaccine failures associated with <u>Pseudomonas</u> bacteremia. During the period of observation, there has been an increase in the average age from 12.3 to 19.5 years, reflecting an increase in the percentage of elderly burn patients. In addition, the size of burn injury in vaccinated patients was slightly greater than in the antecedent controls. Despite these adverse factors, there was an overall improvement in mortality of almost 5 % associated with vaccination. A more critical measure of improvement in burn survival is the percentage of total body surface area burn associated with a 50 % mortality (LA_{50}). This has risen from 55 % to 72 % with a stepwise improvement during development of the immunotherapy program. However, it is unlikely that all of the improvement in LA_{50} resulted from immunization against Pseudomonas aeruginosa since the incidence of positive blood cultures to organisms other than Pseudomonas has decreased from 42.7 % to 33.7 % during the same period of observation.

The primary agents used for topical therapy are shown in Table III. The spectrum of topical therapy varied considerably during the period of observation, but these changes were not felt to contribute to differences in infection from Pseudomonas aeruginosa.

All things considered, the patient populations, except in Phase VI, were felt to be basically similar in their susceptibility to serious Pseudomonas infections despite lack of concurrency. Patients in Phase VI were suspected to be less susceptible to infection because of a better dietary management.

Incidence of Pseudomonas Infections

The effect of immunotherapy on colonization and infection by Pseudomonas aeruginosa is shown in Table IV. Pseudomonas colonization was considered to be present if any culture from blood, wound, respiratory tract, or urinary tract was positive for Pseudomonas aeruginosa during the course of hospitalization. Colonization was basically similar during each of the control and treatment phases except for slight decreases in Phase V and VI patients. On the other hand, the recovery of positive blood cultures for Pseudomonas was significantly decreased in patients receiving optimal vaccination. Even more striking was the reduction in death from Pseudomonas infection from 14.7 % in non-vaccinated controls to 3.1 % in patients receiving maximal vaccination and 0.0 % in patients receiving maximal vaccination plus hyperimmune globulin therapy for episodes of bacteremia. These results strongly indicate that Pseudomonas vaccination is effective in reducing serious infections and that hyperimmune globulin is a valuable adjunct in the therapy of established infections. A further indication of the effectiveness of therapy is the observation that during the last three years, there have been six additional patients who had Pseudomonas bacteremia at the time of admission or within the first five days. Three of these six patients died, two as a direct result of Pseudomonas sepsis. Of the two patients who died, one did not receive hyperimmune globulin, and the other patient received it while in a moribund condition only a few hours before death.

One of the most dramatic aspects of the study has been a prompt recovery of patients with Pseudomonas septicemia who have been treated with immune antibody. Eight of the ten vaccinated patients in groups IV - VI who developed Pseudomonas bacteremia were treated with hyperimmune globulin, and all survived. An eleventh patient who died suddenly from a pulmonary embolus had a positive blood culture. This patient was included as having a positive blood culture for Pseudomonas, but the death was clearly not caused by Pseudomonas. The other two patients not treated with Pseudomonas hyperimmune globulin were felt to have had either transient bacteremias without sepsis or contaminated cultures. One of these patients later died as a result of sepsis from Candida albicans.

Effect of Immunization on Serum Antibody Levels

A careful analysis was made of the influence of age, sex, dose, route of immunization, number of immunizations, frequency of immunization, size of burn, and postburn day of immunization on the antibody responses to the antigen. As might be expected, there was considerable variation among patients in their ability to synthesize anti-Pseudomonas antibody. Taking into account this variability, there was no significant relationship between the average antibody response elicited by vaccination and age, sex, route, or frequency of early injections. However, the subcutaneous route was not sufficiently evaluated because five patients so vaccinated in Phase II died from Pseudomonas infection, suggesting that this route was inferior.

In patients not given sequential injections, the rise in antibody titer was relatively brief and insufficient for continuing protection. Because of this, it was felt to be clear that immunization should continue at periodic intervals until the risk of Pseudomonas infection was over. Weekly intervals were chosen arbitrarily for subsequent studies. A rapid increase in antibody titer was usually achieved within 8 days of the first immunization, and the first detectable rises in titer were usually found 4 - 5 days after the initial injection. Surprisingly, the rise in antibody titer after early vaccination was proportionately similar between IgG and IgM types based on 2-ME inactivation. There were no differences in the rate of rise or ultimate antibody levels in patients with burn indexes (% 3° + 1/2 % 2^O) less than 40 regardless of when vaccination was begun. However, in those patients with burn indexes greater than 40, the response to immunization was significantly poorer when immunization was initiated after the 6th postburn day. Patients with burn indexes greater than 40 whose immunization was initiated before the sixth postburn day usually had a quicker and higher initial rise in titer than the other patients, but these titers later fell significantly, reaching lowest levels between the 22nd and 36th post-immunization day. With coverage of the burn wound, there was again a gradual rise in average antibody titer.

Correlations were made between the apparent clinical resistance to Pseudomonas sepsis, the titers of IgG and IgM antibody measured by the HA test and passive protection assays in mice. From an analysis of several hundred thousand measurements, it was apparent that the IgG type of antibody provided significantly more protection than equivalent hemagglutinin units of the IgM type. This analysis also allowed postulation that an IgG HA antibody titer of 1:10 usually provided protection against invasion of Pseudomonas infection by a homologous immunotype. However, three of the patients who were studied developed Pseudomonas infection with IgG antibody titers greater than 10 for the homologous immunotype.

Approximately 90 % of the isolates from the burn patients showed serological activity with one of the seven immunotypes represented in the vaccine. Of the patients who died from Pseudomonas infection, only one was caused by an untypable strain. Another was caused by a typable strains, but experimental infections with this strain did not show protection with the homologous IgG antibody.

Safety and Reactivity

The immunogen in the heptavalent Pseudomonas vaccine is a purified, high molecular weight polysaccharide which resembles endotoxin in some, but not all aspects. It is not surprising, therefore, that non-specific reactions occurred which consisted primarily of malaise, fever, and local discomfort or induration. Approximately 1/2 of the injections consisting of 25 mcg/kg caused a local or systemic response, and nearly ninety percent of the patients had a detectable local or febrile reaction at some time. Febrile reactions occurred sporadically, but in 3.2 % of the patients, they were impressive enough to discontinue the vaccine or decrease its dose. The appearance of febrile reactions had no relationship to age of the patient or the presence of pre-existing antibody, but they tended to occur more frequently in patients with smaller burns. Local reactions occurred predominantly when injections were given by the ID route. Despite these local and febrile responses which were occasionally impressive, approximately 3000 injections of the vaccine have been given to our burn patients without a serious complication.

Except for the volume of the injection, there have been no untoward effects as a result of the intramuscular administration of the hyperimmune globulin.

Discussion

The administration of Pseudogen^R stimulates a rapid antibody response which is dose related and capable of being sustained by multiple injections. In burn patients, this antibody response appears to be associated with significant protection against invasive infection. During a five year experience with vaccination of 322 seriously burned patients, the overall survival has improved, and there has been a dramatic reduction in the mortality from Pseudomonas infection. Based upon this experience, we feel that Pseudogen^R is effective and can be given safely to burn patients, but it is unlike other vaccines in that doses nearing the maximum tolerable level must be administered to achieve demonstrated effectiveness.

Criticism can be made of this study because various phases were not run concurrently with non-vaccinated controls. However, it must be emphasized that the primary purpose of this initial series of investigations was to prove safety and to study various protocols for immunization rather than to prove efficacy. Nevertheless, the impressive results leave little doubt of efficacy since death from Pseudomonas infection continues to occur at the rate of 5 % to 10 % in patients with burns involving greater than 20 % total body surface area in other major burn centers in the United States. Further evidence of efficacy of the vaccine have been provided by Young, Meyers, and Armstrong (14) in a controlled prospective trial of this same vaccine in cancer patients, and by Polk et al. (13) in a double blind study performed on intensive care patients. On the other hand, a double blind cooperative study in burn patients, using a lower dose of vaccine per injection than in our study and limited to five injections, showed indicative but not statistically significant evidence of protection against Pseudomonas sepsis (12).

Evidence for the effectiveness of the immunological approach to controlling Pseudomonas infections was also provided by the prompt and effective response to therapy with hyperimmune anti-Pseudomonas gammaglobulin in most burn patients who developed Pseudomonas bacteremia. At first, it was a perplexing observation that an occasional patient with substantial levels of circulating IgG antibody against an infecting immunotype would not be protected against the development of septicemia. However, this may be explained by the observations of Bjornson and Michael (7) that natural and immune antibody cannot be differentiated by passive hemagglutination and that natural IgG anti-Pseudomonas antibody requires participation of the alternate pathway of complement to promote bacterial phagocytosis by leukocytes whereas immune IgG anti-Pseudomonas antibody does not. Recently, Bjornson and Alexander (6) have demonstrated that patients with severe thermal injury have a marked deficiency in the activity of C3 proactivator and other components of the alternate pathway. Furthermore, we have demonstrated that septic episodes in these patients have occurred in association with alterations of these non-specific opsonins. It may well be that natural IgG antibody was being measured in passive hemagglutination tests rather than immune IgG antibody, and patients became infected because of an abnormality in the activity of their alternate pathway of complement. This working hypothesis could also explain the prompt response of the majority of these patients to the passive administration of immune IgG antibody. This mechanism might also help to explain the favorable results reported by Feller et al. (8) who used a monovalent anti-Pseudomonas vaccine to protect burn patients but also routinely administered fresh frozen plasma collected from immunized donors. In addition to the abnormalities of the alternate pathway of complement following burn injury, marked abnormalities of neutrophil function have been documented in our laboratory (4). These abnormalities have preceeded the development of bacterial sepsis and have been shown to account for the development of invasive infection by Pseudomonas in patients with heavy colonization of the wound despite the presence of circulating antibody. In a few patients, consumption of the circulating antibody has been demonstrated (9).

The complex mechanisms of resistance to Pseudomonas infection make it apparent that no single procedure will suffice to completely protect all patients. Nevertheless, the results of this clinical trial suggest that immunization against Pseudomonas aeruginosa should become a routine component of burn therapy.

Summary

Three hundred twenty-two patients with burn injuries involving 20 % or more total body surface area have been actively vaccinated with a polyvalent Pseudomonas vaccine at our center during the five year period preceding April, 1973. In a similar group of 75 consecutive non-vaccinated controls treated between March 1967, and April 1968, deaths from Pseudomonas occurred in 14.7 %, and Pseudomonas bacteremia occurred in 18.7 %. With increasing effectiveness of vaccination, the incidence of Pseudomonas bacteremia has declined to 6 %. Deaths from Pseudomonas occurred in 3.1 % of 96 patients immunized with maximal dose vaccination, but mortality from this cause has not occurred in the last 186 vaccinated patients where hyperimmune anti-Pseudomonas.gammaglobulin was available for the adjunctive treatment of bacteremic patients. These clinical observations have been accompanied by a variety of immunological studies which substantiate that immunization with the polyvalent vaccine provides protection by an immune IgG antibody response. Invasive infection with Pseudomonas aeruginosa is accompanied by or preceded by a variety of immunological deficiencies which can be compensated by the administration of immune IgG antibody. Predominate among these are abnormalities of the alternate pathway of complement and of the antibacterial function of neutrophils. The results of this study strongly suggest that all patients with major thermal injuries should be actively vaccinated against Pseudomonas. They further suggest that anti-Pseudomonas hyperimmune globulin is an effective adjunctive measure to indicated therapy for the treatment of life-threatening infections caused by Pseudomonas aeruginosa.

References

1. Alexander, J.W.: Pseudomonas infections in man, p. 103 (Waverley Press Inc., Baltimore 1970).
2. Alexander, J.W., Fisher, M.W.: J. Trauma 10:565 (1970).
3. Alexander, J.W., Fisher, M.W., MacMillan, B.G.: Arch. Surg. 102:31 (1971).
4. Alexander, J.W., Meakings, J.L.: Ann. Surg. 176:273 (1972).
5. Alms, T.H., Bass, J.A.: J. infect. Dis. 117:249 (1967).
6. Bjornson, A.B., Alexander, J.W.: J. Lab. clin. Med. To be published.
7. Bjornson, A.B., Michael, J.G.: J. infect. Dis. 128:S182 (1973).
8. Feller, I., Pierson, C.: Arch. Surg. 97:275 (1968).
9. Jones, C.E., Alexander, J.W., Fisher, M.W.: J. Surg. Res. 14:87 (1973).
10. Jones, R.L., Jackson, D.M., Lowbury, E.J.L.: Brit. J. plast. Surg. 19:43 (1966).
11. Millican, R.C., Evans, G., Markley, K.: Ann. Surg. 163:603 (1966).
12. Parke-Davis & Co. --- Information on file.
13. Polk, H.C., Jr., Borden, B.S., Aldrete, J.A.: Ann. Surg. 177:607 (1973).
14. Young, L.S., Meyers, R.D., Armstrong, D.: Ann. int. Med. To be published.

Table I.

Phases of Study

	Phase	Number Patients	Date of Study Mo/yr	Dosage and Route of Each Injection	Schedule for Immunization
Not Vacci-nated	Antece-dent control	75	1/67 - 3/68	None	None
Low Dose, Sub-Maxi-mal Vacci-nation	I	17	3/68 - 10/68	4.3-11.1 mcg/kg I.M.	Days 1, 5, 9
	II	23	10/68 - 2/69	7.2-20.3 mcg/kg I.M., I.D. or S.C.	Days 1, 5, 9 then weekly X 0-6
Maximal Vaccination	III	96	2/69 - 7/70	25 mcg/kg I.D. or I.D. and I.M.	Days 1, 5, 9 then weekly X 7
Vaccination Considered Maximal. Hy-perimmune Globulin Gi-ven to Pa-tients with Pseudomo-nas Bactere-mia	IV a	21	7/70 - 12/70	25 mcg/kg I.M.	Days 1, 3, 5, 7 then weekly X 7
	IV b	26	12/70 - 5/71	25 mcg/kg I.M.	Days 1, 2, 3, 4, 7 then weekly X 7
	IV b - V	7	5/71 - 7/71	25 mcg/kg I.M. switched to I.D.	Started IV b schedule, switched to V
	V	49	7/71 - 4/72	25 mcg/kg I.D.	Days 1, 2, 3, 4, 7 then weekly X 7
	VI	83	4/72 - 4/73	25 mcg/kg I.M.	Days 1, 5, 9 then weekly X 7

Table II.

Comparison of Patient Data

Phase	Number Patients	Average Age	Average Size Burn T/3°	BI	Positive Blood Cultures Other Than Pseudomonas	Deaths	LA50
Antecedent Control	75	12.3	39.2/27.7	33.5	42.7 %	26.7 %	55 %
I	17	11.0	42.2/29.3	35.8	52.9 %	0.0 %	60 %
	} 40	} 14.1	} 42.8/26.9	} 34.8	} 42.5 %	} 20.0 %	} 60 %
II	23	16.4	43.3/25.1	34.3	34.8 %	34.8 %	60 %
III	96	14.6	42.8/28.3	35.5	39.2 %	21.7 %	63 %
IV a	21	15.8	44.9/21.4	33.1	47.6 %	19.0 %	69 %
IV b	26	15.3	39.5/21.8	30.6	19.2 %	11.6 %	70 %
IV b - V	7	31.0	42.8/16.4	29.6	42.8 %	28.5 %	---
	} 186	} 19.5	} 42.5/25.9	} 34.2	} 36.0 %	} 22.6 %	} 72 %
V	49	23.1	43.0/26.1	34.6	43.0 %	24.5 %	75 %
VI	83	18.6	42.5/28.9	35.7	33.7 %	25.3 %	71 %

Table III.
Topical Therapy

Phase	Gentamycin	Mafenide	AgSD	AgNO$_3$	Betadine	Other
Control	29 (39 %)	8 (11 %)	0	37 (49 %)	0	1
I	12 ⎫ (73 %)	2 ⎫ (10 %)	3 ⎫ (17 %)	0	0	0
II	17 ⎭	2 ⎭	4 ⎭	0	0	0
III	51 (53 %)	18 (19 %)	27 (28 %)	0	0	0
IV a	6 ⎫	4	10 ⎫	0	1	0
IV b	4 ⎪	1	17 ⎪	0	4	0
IV – V	4 (35 %)	2 (15 %)	1 (37 %)	0	0 (9 %)	0 (4 %)
V	8 ⎪	5	28 ⎪	0	6	2
VI	43 ⎭	16	13 ⎭	0	6	5

Table IV.

Effect of Immunotherapy on Colonization and Pseudomonas Infection

	Phase	Number Patients	Ps. Colonization (Any Site)	Blood Cultures Positive for Ps. After 5th Day	Deaths from Ps. Infection After 5th Day
Not Vaccinated	Antecedent Control	75	72.2 %	18.7 %	14.7 %
Submaximal Immunization	I	17 ⎫ 40	64.4 % ⎫ 72.5 %	5.9 % ⎫ 15.0 %	0.0 % ⎫ 12.5 %
	II	23 ⎭	78.2 % ⎭	21.7 % ⎭	21.7 % ⎭
Maximal Immunization	III	96	73.2 %	8.25 % P < 0.05	3.1 % P < 0.01
Vaccination Considered Maximal.	IV a	21 ⎫ 186	71.4 % ⎫ 66.1 %	9.5 % ⎫ 5.9 %	0.0 % ⎫ 0.0 %
	IV b	26	69.2 %	7.7 %	0.0 %
	IV - V	7 ⎭	71.4 % ⎭	0.0 % ⎭ P < 0.005	0.0 % ⎭ P < 0.0005
Hyperimmune Globulin Given to Patients with Ps. Bacteremia	V	49	67.3 %	4.0 %	0.0 %
	VI	83	63.0 %	6.0 %	0.0 %

Active Immunization against Pseudomonas aeruginosa in Burns

P. R. Zellner, E. Metzger, and O. Zwisler

Although during the last few years, the incidence of septicemia as a complication in full-thickness burns has been reduced, it is not yet possible to avoid it completely. The percentage of patients who still die during the course of infection requires the study of this problem intensively from the clinical and bacteriological points of view. It has to be pointed out that all measures to keep the burn surface free of pathogenic bacteria are not successful. It is possible, however, to reduce the number of bacteria by the so-called topic treatment. In general, primarily gram-negative bacteria are involved. In contrast to the increasing number of cases of infection with Klebsiella and Candida reported in the United States, we have observed predominant infections with Pseudomonas aeruginosa following extensive third degree burns. Before submitting our patients to an active immunization program similar to that of Feller (1), we performed an extensive epidemiologic study (3, 4). The reason for this approach was to get information about the frequency and distribution of different strains of Pseudomonas. We used the method according to Lányi (2) by which 13 0-groups can be differentiated. This method was applied in more than 8000 cultures taken from the wounds, the environment, and the in- and outlets of the waterpipes. Tables I and II show the summary of these results obtained in 1972. The most frequent 0-groups isolated from the wounds were groups 5 and 13. A similar prevalence was found in the environment. In contrast, there was a high percentage of 0-group 4 in the water in- and outlets. We conclude that Pseudomonas isolated from the latter contributes to wound infection only to a very minor degree.

Due to the frequent disinfection of the cubicles, the pathogenic bacteria in the environment as shown in Table I are, from our point of view, transferred by contamination from the patients' wound to the walls and lockers, and not vice versa.

We can also conclude that there is a significant difference in virulence because most of the Pseudomonas 0-groups present on the skin and burn wound at the time of admission were overgrown by the more aggressive strains which can be described as the resource of hospitalism (Table II). To overcome this problem we changed the regimen of the local care of the burn wound. From the middle of 1973 on, no patient was treated in the bathroom preoperatively by a shower. This modification of the organisation inside the burn unit resulted in a complete change of the epidemiology. No strains of hospitalism were observed. 0-groups 5 and 13 disappeared completely and wound cultures now show many different Ps. aeruginosa strains.

These preliminary examinations were necessary before using a polyvalent vaccine. This was given to volunteers and patients with a burn over 20 % on the day the accident occurred (4). A dosage of 0.5 ml was used at weekly intervals.

Fig. 1 shows a marked difference in the development of the antibody response of both groups. In the volunteers the titers of agglutinating antibodies rise slowly but continuously (Table III). In patients, however, there is a decrease of antibody titers after a brief rise.

In a second trial we vaccinated another 136 patients with 1 ml of Pseudomonas vaccine. First, Table IV shows that in many patients antibodies to Pseudomonas were already present before the vaccine was applied. Second, in these patients a higher increase (1 : 80) in antibody titer was observed after the first vaccination, a difference which continued after the second, third and fourth vaccinations. Third, statistical examinations showed that there was no difference between patients who had a burn up to 33 % and those with a burn between 33 % and 60 %. In order to obtain more information, we determined the antibody level in 72 patients separately (Table V). It should be mentioned that in 53 patients an antibody decrease occurred in the third and fourth week. After that time the titers increased in only 26 patients. Out of 72 patients 8 had positive blood cultures (Ps. aeruginosa). All patients survived. During the period of decrease of antibody titers in the third and fourth weeks the treatment with gamma-globulin did not show any results.

The overall evaluation of our investigation confirms the study by Allgöwer and Schoenenberger in relation to the burn toxin. Their experiments have clearly shown that there is a significant reduced antibody response to Ps. aeruginosa infection in burned animals. The mortality rate of burned mice infected with Ps. aeruginosa was statistically higher compared to another group which was subjected to the gram-negative infection only. The anticipated immuno-suppressive effect of the burn toxin probably is the explanation of the difference in antibody titers in burned patients and in volunteers in our study. From the clinical point of view this is an important consideration towards further improvement of therapy. Even an immune-serum against Ps. aeruginosa will have only a transitory effect due to the well-known changes of the bacterial flora, notably other gram-negative bacteria. The ultimate answer lies in the preservation of antibody response against any sort of bacteria by a burn antitoxin.

References

1. Feller, I., Pierson, C.: Arch. Surg. 97: 225 (1968).
2. Lányi, B.: Acta Microbiol. Acad. Sci. hung. 13: 295 (1966/67).
3. Zellner, P.R., Metzger, I.: Med. Klin. 69: 346 (1974).
4. Pranter, W., Staerk, J., Zellner, P.R., Zwisler, O.: Progr. immunbiol. Standard. 5: 414 (1972).

501

Table I.

Results of Swabs in Water Pipes (in- and outlet) 1972

Total number of swabs: 587
Result of the cultures:
gram-positive bacteria: 89   gram-negative bacteria: 86   sterile cultures: 304   Ps. aeruginosa: 109

Frequency of Ps. aeruginosa 0-groups

0-groups	1	2	3	4	5ad	5abd	6	7	8	9	10	11	12	13	polyaggl.
Percentage	-	1.8	0.9	81	4.6	9.0	-	-	-	-	0.9	-	-	1.8	-

Results of impression plates in the environment

Total number of the impression plates: 1982
Result of the cultures:
gram-positive bacteria: 940   gram-negative bacteria: 111   sterile cultures: 872   Ps. aeruginosa: 151

Frequency of Ps. aeruginosa 0-groups

0-groups	1	2	3	4	5ad	5abd	6	7	8	9	10	11	12	13	polyaggl.
Percentage	4	0.5	-	19.5	1.5	32	-	-	-	-	-	0.5	-	42	-

Table II.

Wound Swabs in Burned Patients 1972

Total number of patients: 176
Ps. aeruginosa in 150 patients

Total number of swabs: 5 482
Total number of cultures with Ps. aeruginosa: 2 330

Frequency of Ps. aeruginosa 0-groups

0-groups	1	2	3	4	5ad	5adb	6	7	8	9	10	11	12	13	polyaggl.
Percentage	1.2	1.3	2.4	4.6	0.6	42.6	0.95	1.5	0.05	-	0.05	1.3	-	43.2	-

Patients with change of Ps. aeruginosa 0-groups: 98
Patients with only one 0-group of Ps. aeruginosa: 52

Day of onset of Ps. aeruginosa in 150 patients

1st day present when admitted	1st day anal	1st week	2nd week	3rd week	4th week	No Ps. aeruginosa
42	2	60	30	13	3	26

Table III.

Agglutination Titer in Volunteers Immunized with 0.5 ml Pseudomonas Vaccine at Weekly Intervals

	1: 10	1: 20	1: 40	1: 80	1: 160	1: 320	1: 640	1: 1280	1: 2560	1: 5120
no titer before immuni- zation	-	-	-	-	-	-	-	-	-	-
1 week after 1st immun.	23	9	13	6	2	-	-	-	-	-
1 week after 2nd immun.	4	6	10	11	11	5	2	-	-	-
1 week after 3rd immun.	2	2	8	9	14	10	4	4	-	-

Table IV.

Number of Patients with Burns and their Antibody Titer before, 2, 3, 4, and 5 Weeks after Immunization with Pseudomonas aeruginosa

Attained Titer \ Titer before Immunization	Weeks after immunization							
	2		3		4		5	
	<1:20	≧1:20	<1:20	≧1:20	<1:20	≧1:20	<1:20	≧1:20
≦1:80	27	14	25	16	23	14	13	13
>1:80	26	49	21	37	15	30	14	26
Total Number of Examinations	53	63	46	53	38	44	27	39
Significance	p < 0.01		p < 0.05		p < 0.01		cannot be proved (no figure)	

504

Table V.

Agglutination Titer in Burned Patients Immunized with 1.0 ml Pseudomonas Vaccine at Weekly Intervals

	1: 10	1: 20	1: 40	1: 80	1: 160	1: 320	1: 640	1: 1280	1: 2560	1: 5120
titer before immunization	39	11	9	9	6	-	-	-	-	-
1 week after 1st immun.	3	7	1	18	28	10	1	-	-	-
1 week after 2nd immun.	2	6	4	22	20	9	5	-	-	-
1 week after 3rd immun.	4	5	8	14	22	8	2	1	-	-
1 week after 4th immun.	3	5	9	11	21	11	8	3	1	-
1 week after 5th immun.	6	-	5	8	17	11	8	3	2	-

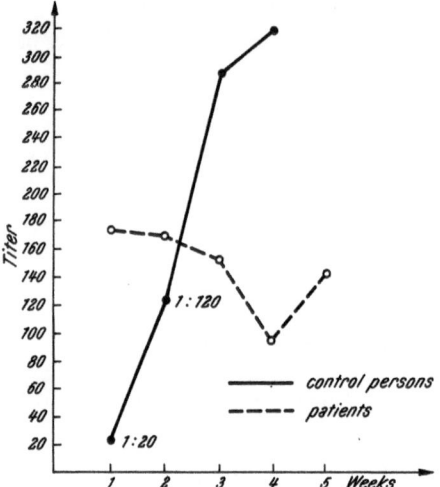

Fig. 1. Average anti-Pseudomonas titers in patients and control persons immunized with <u>Pseudomonas</u> vaccine (0.5 ml).

Discussion

IMMUNOPROPHYLAXIS

Commenting on Dr. Zellner's finding of reduced antibody titres in immunised burned patients, the Chairman referred to falls in antibody level which he had attributed to absorption of antibody by bacteria growing in burns. Dr. Zellner said that the degree to which antibody titres fell was unrelated to the area of burn (in the range of 20 % to 50 % body surface). In more extensively burned patients Dr. Alexander had found that a reduced antibody titre after vaccination was apparently related to a metabolic deficit with associated hypermetabolism. The type of vaccine used for immunisation against Pseudomonas aeruginosa was discussed. Dr. Lowbury said that vaccines prepared from culture filtrates and from extracts of killed bacterial cells (especially the latter) had been very effective.

Dr. Alexander said that patients with established Pseudomonas septicaemia were treated with all available methods, but in his experience such patients, and especially those who had developed ecthyma gangrenosum, did not survive unless treated with immunoglobulin. Dr. Lowbury referred to experiments in which burned mice received no protection from antiserum if this was given 48 hours or more after burn infection had been initiated. Dr. Alexander found that if globulin was given late in the course of infection, much larger doses were required; a patient could probably not be successfully treated with immunoglobulin unless he had been kept alive for at least 24 hours from the time of starting this treatment.

BURN TOXINS

In answer to questions from Dr. Urbaschek and Dr. Lundsgaard-Hansen as to whether the "burn toxin" could be produced by heating skin at temperatures below 250°C or by altering the time-temperature ratio, and as to the relationship between the amount of the "burn toxin" and the clinical course of burn disease, Dr. Schoenenberger said that exposure to lower temperatures for longer times could lead to toxin-formation; with mouse skin, exposure to 250°C for 15 seconds gave optimal conditions for experimental purposes. The essential requirement was a temperature high enough to cause complete removal of water from the skin but not so high as to cause charring. Scalds did not provide these conditions; nor did temperatures that destroyed the lipid-protein complexes. Dr. Urbaschek had found that extracts from burned skin were toxic when obtained from "conventional" burned rats but non-toxic when obtained from germfree burned rats. These animals, however, had been burned at 95°C, a temperature at which the toxin described by Dr. Schoenenberger was not produced; toxicity of extracts from the "conventional" burned rats was presumably due to bacterial products. Dr. Koslowski wondered whether the "burn toxin" could be an artifact and what practical importance there could be in a toxin which required such special conditions for its formation; most clinical burns, in his view, involved lower temperatures. Dr. Urbaschek reported his studies on an acquired antigenic determinant group in burned skin in different species. Since Pick it is known that temperatures above 60°C alter the sequence of aminoacids. Antibodies produced with this antigen, which is not species-specific, had no protective effect in experimental burns.

In answer to question from Dr. Wissler, Dr. Schoenenberger said that removal of the lipid component of the toxin reduced but did not eliminate its toxicity; detergents, organic solvents, and extremes of pH could destroy its toxic action. He did not think that toxicity was due to small molecules absorbed onto the surface of the lipo-protein complex, especially in view of the different biological and immunological properties of the toxin preparation and the precursor. Dr. Wissler doubted whether the insoluble high molecular weight particles could, in themselves, cause any systemic reaction; the reactions described were more consistent with the effects of soluble mediators. In reply, Dr. Schoenenberger said that if a mediator were involved, one would expect it to be highly specific, or else released by the non-toxic presursor. His studies, however, showed that antibody to toxin from human or mouse burns decreased the burned animals' susceptibility to Ps. aeruginosa; antibody to precursor cells, on the other hand, had no such effect. In answer to a question from Dr. Lowbury, Dr. Schoenenberger said that studies had not yet been made on protection of burned animals against bacteria other than Ps. aeruginosa by antibody to the "burn toxin".

OTHER PROBLEMS

There was some discussion of topical chemoprophylaxis of burns. In answer to a question, Dr. Lowbury said that application of silver nitrate and other silver preparations had been associated with relatively little absorption of silver or toxic effects of silver absorption; the main hazard, partly controllable by oral supplements, appeared to be electrolyte imbalance due to the application of a hypotonic solution when 0.5 % silver nitrate compresses were used. The types of agent used for prophylaxis and the avoidance of systemic chemoprophylaxis were briefly discussed; Dr. Schauer discussed the mechanisms of oedema and the role of mast cells and of proteolytic enzymes in the burn wound.

Index

M

Macrophages, 165
- activation by lipid A, 132
- cytostatic activity, 132
- cytotoxic activity, 165
- endotoxin effect on, 132, 148, 149
- microbicidal activity, 131, 132
- role in nonspecific resistance, 131-133

Mast cells
- degranulation, 315, 316, 324, 346
- role in endotoxin shock, 278

Mediators
- in burn disease, 473-479
- of anaphylactic shock, 311
- of endotoxin activity, 271-278, 280, 311-314, 315-322
- - on lung, 422-424
- of interaction of serum with foreign macromolecules, 91-105

Metabolic changes
- in bacterial shock, 237-245, 412
- in endotoxemia, 248-255, 271-277, 279, 325
- - in mitochondria, 288-295
- in endotoxin shock, 231-236
- - and respiratory changes, 231-236

Microcirculation
- endotoxin effect on, 323, 324
- - pathomorphology, 383, 384, 385

Microthrombi, 353
- formation by endotoxin, 324, 354
- in shock, 385, 386, 387, 396
- ultrastructure, 385, 386, 387

Migration inhibition, 95, 96
- effect of endotoxin on, 148, 149, 153, 154

Mitochondria
- energy metabolism
- - in endotoxemia, 278, 279, 288-295
- - serotonin effect on, 278, 279
- metabolic alterations in shock, 288-295, 307

Mucosa
- attachment of bacteria to, 8, 9
- defence mechanisms of, 8
- invasion of bacteria into, 13, 16-23, 61, 80, 81
- penetration of bacteria through, 8, 9, 16-19, 21, 61, 80-83

N

Nephritis
- interstitial
- - induction by Lipid A, 42

Neurotoxin, 60, 338, 350

Nonspecific resistance
- induction by bacteria, 131
- induction by endotoxin, 147-154
- - mechanisms, 131-133, 165
- induction by unrelated organisms, 132
- induction by various agents, 131
- passive transfer, 166, 167
- role of cell mediated immunity, 132, 147, 148, 149
- role of macrophages, 131-133
- role of RES, 149

Nosocomial disease, 30-37, 64, 112
- Aeromonas hydrophila, 30
- Bacteroides fragilis, 31, 32
- definition, 30
- Herbicola-Lathyri, 30, 31
- Proteus rettgeri, 33
- Providencia stuartii, 33, 34
- Pseudomonas cepacia, 34
- Pseudomonas maltophilia, 34
- Serratia marcescens, 32, 33

Nutritional status
- fasting, effect on
- - lipid metabolism, 302
- - liver enzymes, 272-274, 281
- hyperalimentation
- - Candida septicemia, 221
- malnutrition
- - neutrophil dysfunction in, 191-193, 221
- starvation, effect on
- - hemodynamic response, 251
- - metabolic changes, 249-251
- - surfactant synthesis, 422

O

Opportunistic infections, 112
- diagnosis of, 5
- factors involved, 3
- in burn disease, 464
- with anaerobic bacteria, 38